Ummu Raihanah Hashim · Ahmad Kamil Arshad ·
Nor Hayati Abdul Hamid · Rohana Hassan ·
Ekarizan Shaffie · Anizahyati Alisibramulisi ·
Norshariza Mohamad Bhkari ·
Muhd Norhasri Muhd Sidek
Editors

Green Infrastructure

Materials and Sustainable Management

Springer

Editors
Ummu Raihanah Hashim
Institute for Infrastructure Engineering and Sustainable Management (IIESM)
Universiti Teknologi MARA
Shah Alam, Selangor, Malaysia

Ahmad Kamil Arshad
Institute for Infrastructure Engineering and Sustainable Management (IIESM)
Universiti Teknologi MARA
Shah Alam, Selangor, Malaysia

Nor Hayati Abdul Hamid
Institute for Infrastructure Engineering and Sustainable Management (IIESM)
Universiti Teknologi MARA
Shah Alam, Selangor, Malaysia

Rohana Hassan
Institute for Infrastructure Engineering and Sustainable Management (IIESM)
Universiti Teknologi MARA
Shah Alam, Selangor, Malaysia

Ekarizan Shaffie
Institute for Infrastructure Engineering and Sustainable Management (IIESM)
Universiti Teknologi MARA
Shah Alam, Selangor, Malaysia

Anizahyati Alisibramulisi
Institute for Infrastructure Engineering and Sustainable Management (IIESM)
Universiti Teknologi MARA
Shah Alam, Selangor, Malaysia

Norshariza Mohamad Bhkari
Institute for Infrastructure Engineering and Sustainable Management (IIESM)
Universiti Teknologi MARA
Shah Alam, Selangor, Malaysia

Muhd Norhasri Muhd Sidek
Institute for Infrastructure Engineering and Sustainable Management (IIESM)
Universiti Teknologi MARA
Shah Alam, Selangor, Malaysia

ISBN 978-981-99-7002-5 ISBN 978-981-99-7003-2 (eBook)
https://doi.org/10.1007/978-981-99-7003-2

This Springer imprint is published by the registered company Springer Nature Singapore Pte Ltd.
The registered company address is: 152 Beach Road, #21-01/04 Gateway East, Singapore 189721, Singapore

Paper in this product is recyclable.

Preface

The book *Green Infrastructure: Materials and Sustainable Management* is part of a sub-series Green Infrastructure. Materials and Sustainable Management offers comprehensive research-based practices highlighting the advanced and developed materials related to green infrastructure as well the sustainable approaches specifically for construction management, transportation, pavement, environment, timber, and seismicity.

This book covers chapters including Chapter "Assessment of Entry Timing Decisions (AoETD) Towards Sustainable Operations of Malaysian Construction Firms in International Markets" identifies the key and associated determinants for international ETD for Malaysian construction firms; Chapter "Pre-construction Complexity Factors Affecting Cost Performance of Infrastructure Projects"—identifies the most significant complexity factors contributing to project performance and develops a complexity assessment model for infrastructure projects; Chapter "Performance Measurement Criteria: Conceptual Framework for Subcontracting Management in the Malaysian Construction Supply Chain"—focuses on the identification of performance measurement criteria that contribute to successful subcontracting management in Malaysian construction projects; Chapter "Building Information Modelling Implementation Framework (BIMIF) for Government Building Construction Among Civil and Structural Engineering Consultants in Malaysia"—presents the initial stage of establishing a BIM implementation framework (BIMIF) for government building construction projects based on the conventional contract approach and focuses on developing the BIM process from a chosen case study; Chapter "Integrating Value Management: Determine Project Management Knowledge—Addressing Theory–Practice Gap"—identifies the Project Management (PM) knowledge areas that can be adopted in value management methodology; Chapter "Strategies of Carbon Reduction Management in Construction Operations"—investigates the current carbon emissions management practices and key strategies in reducing emissions effectively; Chapter "Green Infrastructure Development in Malaysia: A Review"—explains Malaysia's progress towards constructing green infrastructure and becoming a climate-resilient nation which holistically assessed as a result of the detailed study of incorporating green techniques into current assessments; Chapter "Why Current Procurement Systems Require Modifications to Suit the Natures of Malaysian Pre-fabricated Construction"—highlights the necessity and natures of the prefabricated projects in Malaysia, and why current procurement system needs to be adjusted so that it can help the projects to reap maximised benefits from prefabricated concept; Chapter "A Review of Green Open Space Implementation Towards Green City Development in Developing Countries"—understands the advantages and implementation of green open space (GOS) in green city development (GCD) in Malaysia and Indonesia; Chapter "Environmental Impacts of a Forensic Unit Construction at a Teaching Hospital in Malaysia"—educates on the material waste generation, energy and water consumption, and total carbon emissions from constructing a forensic unit at a teaching hospital in Malaysia before the COVID-19 pandemic; Chapter "Mechanical Properties of Concrete Containing Palm Oil Fuel Ash (POFA) as Cement and Sand Replacement"—analyses the mechanical properties of POFA in concrete as a partial replacement for cement and sand; Chapter "A Review of Graphene Research and Its Outputs: Waste Carbon Source and Synthesis Technique"—describes and reviews the potential of natural and synthetic waste to be converted to high-quality graphene; Chapter "Influence of Waste Paper Sludge Ash (WPSA) on Mechanical and Durability Properties of Self-consolidating Lightweight Foamed Concrete (SCLFC)"—investigates the effect of WPSA addition on workability, strength, ultrasonic pulse velocity, porosity, and water absorption characteristics of SCLFC; Chapter "The Effect of Tendon Directions to The Analysis and Design of Transfer Slab—A Case Study"—to analyse and design post-tensioned transfer slab by using RAM Concept as finite element analysis tool; Chapter "Perception on Impact Land Reclamation from Pan Borneo Highway Project-Pilot Study"—presents some of the effects of land reclamation resulting from the Pan Borneo Highway Project based on survey; Chapter "Challenges, Characteristics and Success Factors in Implementing Green Highway Using Structural Equation Modelling-Partial Least Squares (SEM-PLS)"—a study using triangulation research to obtain the primary data using unstructured interviews and questionnaire surveys, in which the data were analysed using SEM-PLS; Chapter "Proposed Development of an Integrated Framework for Public-Private Partnership (PPP) and Value for Money (VFM) Evaluation System of Urban Rail Transit in China"—proposes a study towards developing a framework integrating the public–private partnership (PPP) model and the VFM evaluation system for urban rail transit projects in China; Chapter "Pavement Maintenance in Malaysia: The Key to Pavement Sustainability"—presents various pavement maintenance techniques used in Malaysia to preserve the condition of road pavements so that the road can be continuously operated without disruption due to major road rehabilitation activities; Chapter "Evaluation on Volumetric Properties of Stone Mastic Asphalt Mix Containing Steel Fibre Using Response Surface Method"—studies the effects of different amounts of steel fibre on the volumetric properties of stone mastic asphalt (SMA) mixtures; Chapter "Envisaging the Potential Use of Resistance Micro Drilling On Wood Density (WD) Assessment: A Review"—provides an insightful review of the research methodologies on Resistograph, and discusses the use of a micro-drilling device measuring WD in standing trees; Chapter "Phytochemical Research for the Sustainability of Moringa Species Using Different Extraction Methods"—employs both maceration and ultrasonic-assisted techniques, followed by the phytochemical screening of the extracts by using thin-layer chromatography (TLC) and High-Performance Liquid Chromatography (HPLC); Chapter "Performance of Kapok Fibres and Kapok Ash Wood as Oil Absorption Materials"—examines the absorption capacity of kapok fibre and kapok wood ash as well as a combination of both materials into waste cooking oil; Chapter "Physical and Chemical Characteristics of Podo Wood-Xylem Filtered Water"—focuses on filtering water using wood-xylem of tropical timber, i.e. Podo species; Chapter "Effect of Tunnel Form Building (TFB) Under 10 Past Earthquake Records Analysed Using Ruaumoko 2D"—used historical earthquake records to analyse and forecast how TFB will behave in Malaysia during unpredictable earthquake; Chapter "Experimental Analysis of Seismic Responses Interior Beam-Column Joint with and Without Fuse Bars Under In-Plane Lateral Cyclic Loading"—presents two full scale super assemblage of interior beam-column joints with and without fuse bars; designed, constructed, analysed, modelled, and compared their performances under in-plane lateral cyclic loading.

The editors would like to thank all authors who are experts in green infrastructure and sustainable management for contributing their ideas and providing their knowledge and valuable insights. We are also grateful to Springer Nature for their support, especially Loyola D'Silva and Rajesh Manohar for helping us to finalise this book.

Shah Alam, Malaysia

Ummu Raihanah Hashim
Ahmad Kamil Arshad
Nor Hayati Abdul Hamid
Rohana Hassan
Ekarizan Shaffie
Anizahyati Alisibramulisi
Norshariza Mohamad Bhkari
Muhd Norhasri Muhd Sidek

Summary

The book *Green Infrastructure: Materials and Sustainable Management* is part of a sub-series that provides research-based practices highlighting advanced and sustainable materials for green infrastructure, specifically for construction management, transportation, pavement, environment, timber, and seismicity. The book includes various chapters covering different topics, such as assessment of entry timing decisions for Malaysian construction firms, pre-construction complexity factors affecting the cost performance of infrastructure projects, performance measurement criteria for subcontracting management in Malaysian construction projects, green infrastructure development in Malaysia, mechanical properties of concrete containing palm oil fuel ash, and many others.

Contents

Assessment of Entry Timing Decisions (AoETD) Towards Sustainable Operations of Malaysian Construction Firms in International Markets

Norizzati Ibrahim, Che Maznah Mat Isa, Nur Kamaliah Mustaffa, and Nur Izzati Ab Rani

Abstract Malaysian construction firms were found to have an in-depth grasp of the international market experience and cross-border networks. In line with the Construction 4.0 Strategic Plan (2021–2025), previous research indicates insufficient studies focused on developing a systematic assessment to measure entry timing decisions (ETD). This flaw was discovered because of low educational readiness and a lack of data from local researchers. Accordingly, the number of successful local construction firms competing in the global market has decreased. Therefore, the current study provides construction firms with exposure to the importance of entering foreign markets through an Assessment of Entry Timing Decisions (AoETD). The ETD for this study were divided into pioneer (Pi), early follower (EF), and late follower (LF). The study identifies the key and associated determinants for international ETD for Malaysian construction firms. In addition, this paper establishes appropriate decisional level scales for each key determinant, ranging from poor to excellent decision-making. Analysis of data and discussion of study findings were obtained from Smart PLS analysis. Correspondingly, AoETD measurement found that Pi sustained in the international market followed by LF and EF. As such, the development of AoETD

N. Ibrahim
School of Civil Engineering, Universiti Teknologi MARA Johor Baharu Branch, Pasir Gudang Campus, Johor, Malaysia
e-mail: izzati6752@uitm.edu.my

C. M. Mat Isa (✉)
Center of Civil Engineering Studies, Universiti Teknologi MARA Pulau Pinang Branch, Permatang Pauh Campus, 13500 Pulau Pinang, Malaysia
e-mail: chema982@uitm.edu.my

N. K. Mustaffa · N. I. Ab Rani
School of Civil Engineering, College of Engineering, Universiti Teknologi MARA, Shah Alam, Selangor, Malaysia
e-mail: nurkamaliah@uitm.edu.my

N. I. Ab Rani
e-mail: izzati.rani@uitm.edu.my

U. R. Hashim et al. (eds.), *Green Infrastructure*,
https://doi.org/10.1007/978-981-99-7003-2_1

can help firms implement strategic entry decisions into the global market. Furthermore, AoETD is developed in line with the Malaysian government's aims to create collaborations between academicians, government, industry, and society.

Keywords Assessment · Entry timing · Entry decisions · Construction · Sustainable decisions

1 Research Background

A firm considering entering a new global market must decide when to enter that market (Al Sadi & Dulaimi, 2019; Lo & Kletsova, 2018; Utama et al., 2019). Decision-making entails the integration of information from various perspectives and is critical to making good decisions before entering a foreign market (Low et al., 2004; Preece et al., 2016; Kaffash et al., 2012). The target of internationalisation is to expand the company's ideas, products, funds, and investment opportunities. However, this study only focuses on entry timing decisions (ETD). Construction firms confront significant obstacles in selecting strategic business ETD for the international market. According to Lilien and Yoon (1990), identifying the best time for companies to enter emerging industries has long been a major concern. Researchers are still struggling to develop a theoretical foundation that can fully integrate empirical findings in this field (Lo & Kletsova, 2018; Suarez et al., 2013). According to Lo and Kletsova (2018), pioneer firms have significant advantages over late-entry firms, but they also face greater risks and disadvantages. As a result, entry timing has become a popular research topic. The following questions must be addressed: (1) When is the best time to enter? (2) Is it better to be a pioneer or to wait and enter later, avoiding risk but sharing a more congested foreign market? Late entrants may provide adequate engineering support and investment for designing a better product or developing an effective marketing programme, lowering the risk of failure. As a result, the decision to enter the market should be timed to balance the risks of being a pioneer (Pi) and the problems associated with missed opportunities (late follower). According to Kalyanaram and Gurumurthy (1998), market pioneers generate the most revenue in the international market, followed by early and late followers. However, the findings of that study are restricted to industrial and consumer goods businesses. Consequently, ETD for this study consists of Pi, early follower (EF), and late follower (LF) used in the development of AoETD. Final findings of this study produce empirical findings to show the relationship between determinants that include firm-specific (FS), firm-resource commitment (FRC), project-specific (PS), target country (TC), home country (HC), and market-specific (MS) with ETD that need to be considered by construction firms to enter the international market. These determinants are used in AoETD and measured using scales based on empirical data.

2 Entry Timing Decisions (ETD)

The term "pioneer" refers to the first mover to commercialise an innovation. Pioneer entered a market supported by significant investments in the product's production, marketing, and distribution, as well as the elapsed time between its entry. Meanwhile, the terms "early follower" or "early entrant" refer to multiple firms entering a market in quick succession with significant investments in product production and distribution and the ability to achieve advantageous resource positions (García-Villaverde et al., 2017). Firms that enter after several other players have arrived are referred to as late entrants (Cleff & Rennings, 2011; García-Villaverde et al., 2017).

Previous research has shown that being the first to market in most cases provides a significant and sustained market-share advantage over later entrants (Lo & Kletsova, 2018; Kalyanaram & Gurumurthy, 1998). Later entrants, on the other hand, can succeed by using distinct positioning and marketing strategies. Pioneers in most industries are powerful once they have attained the status of incumbent. Numerous studies have found that later market entrants (pioneers or early followers) achieve a lower market share than earlier entrants (pioneers or early followers), and that this holds true across a wide range of product categories and industries, including consumer packaged goods and industrial goods. Even after accounting for a company's tangible (e.g., financial) and intangible (e.g., brand equity) resources and business skills, early entrants maintain a market-share advantage (Kalyanaram & Gurumurthy, 1998).

The entry timing market decision is one of the key determinants of new product success or failure (Lo & Kletsova, 2018; Lilien & Yoon, 1990). However, despite the fact that defining the optimal timing for businesses to enter new industries has long been a priority in the plan, researchers are still struggling to develop a theoretical foundation that can fully integrate empirical findings in this field (Suarez et al., 2013). The research and design and marketing investments will alter the level of the new product's opportunities and risks. A late entry, for example, may provide appropriate engineering support and investments for designing a better product or developing an effective marketing programme, reducing the risk of failure. As a result, the decision to enter the market should be timed to balance the risks of early entry (entry too soon) and the problems associated with missed opportunities (entry too late).

Lilien and Yoon (1990) identified several determinants, including R&D competition, entry competition, product competition, demand potential, and market evolution. As a result of the research and design, as well as marketing investments, the new product's prospects and risks will be altered. A late entry, for example, may provide sufficient engineering assistance and investment to develop a superior product or a successful marketing programme, lowering the risk of failure (Lo & Kletsova, 2018). As a result, market entry decisions should be made at a time that balances the risks of premature entry (over-entry) with the issues of missed opportunities (entry too late). Lilien and Yoon (1990) identify several drivers, including R&D competition, entry competition, product rivalry, demand potential, and market evolution. Contractors must consider various aspects of each entry decision before entering the international market in terms of risk exposure, resource commitment and investment risk control, and flexibility (Al Sadi & Dulaimi, 2019; Utama et al., 2019).

2.1 The Determinants of Entry Timing Decisions

Construction companies face significant challenges in selecting strategic business ETDs for the international market. According to Zander (2015), firm-specific (FS) advantages towards internationalisation that aim to reveal something new, such as a firm establishing a design centre in a new location known as a new resource or introducing technologies, capabilities, and products that offer different and better growth. Meanwhile, the need for firms to manage innovation across organisational boundaries and within an interdependent network of suppliers, customers, and regulatory bodies is referred to as project-specific (PS). The development of a firm's commitment to internationalisation in order to ensure effective decisions and profitable actions in the international market is referred to as firm-resource commitment (Bianchi et al., 2018) or refers to the extent to which organisations and managerial resources are devoted to internationalisation (Lages et al., 2008). Abdul-Talib et al. (2011) cited that larger firms with greater resources and competencies will be able to compete more efficiently and effectively in foreign markets than smaller firms. In addition, the target country (TC) also influences construction firms' decisions to enter the international market. Before entering a country, the legal environment of the target country (including legal issues such as foreign exchange rates, jurisdiction, corruption, the existence of strict time limits, strict quality requirements, etc.) must be considered (Zeqiri & Angelova, 2011). Firms that rely solely on the domestic market do not have the right sensible strategy (Durmaz & Tasdemir, 2014). Therefore, a firm needs to go abroad and look for opportunities to move forward. As a consequence, expanding into foreign markets is one of the most effective ways for a company to grow (Greening et al., 1996). There is a plethora of research demonstrating that firms that implement effective strategies are able to reap the competitive and profitable benefits of internationalisation. This means that businesses can choose to collaborate with others in foreign markets as a means of expanding and becoming more successful. Furthermore, the presence of domestic competitors or barriers to market entry in the home country causes firms to enter the international market due to demand in the host country (Asgari & Ahmad, 2010). As a result, key independent determinants are divided into six categories: firm-specific, firm-resource commitment, project-specific, target country, home country, and market-specific. Next, a total of 41 significant independent determinants have been identified in the development of AoETD, as shown in Fig. 1.

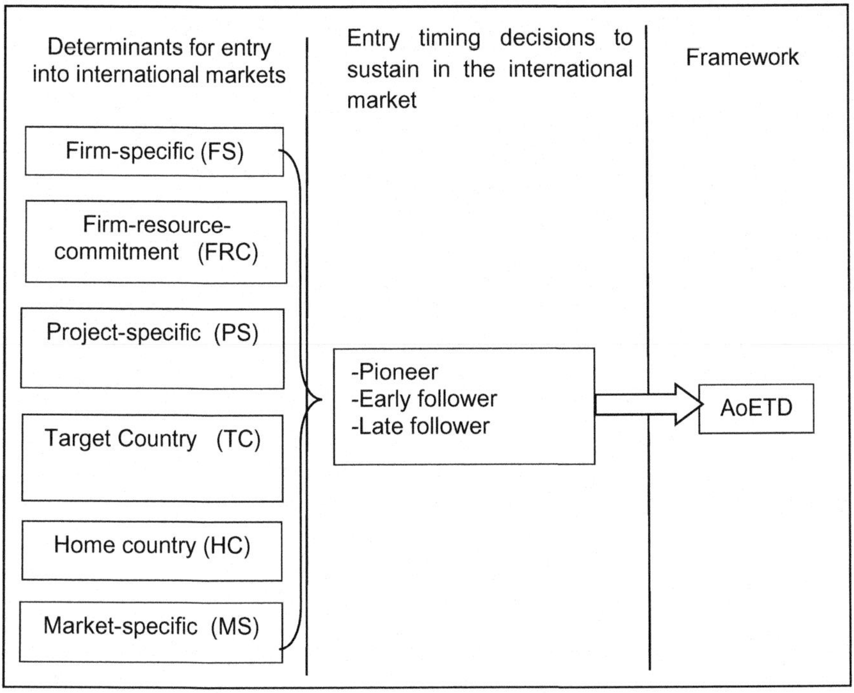

Fig. 1 Conceptual framework of classification of the entry timing decisions

3 Methodology

A quantitative approach was used as the selection was based on CIDB's (2020) record of 132 firms registered as global players operating internationally. The data for this study is focused on managers from construction firms to enquire about their opinions and perceptions of the international market ETDs adopted by their firms in international markets as recommended by Creswell (2009). Next, this study used the Smart PLS 2.0 M3 software to obtain the AoETD. PLS is a programme designed to strengthen the study's theory early, either through minor consolidation or limited data (Ringle et al., 2014). Furthermore, this model (Smart PLS) aids in the resolution of complex data (observed variables) and models with limited theoretical support (Ringle et al., 2014). Thus, this PLS model is helpful for this study. Accordingly, this model can relate a large number of linear equations at the same time. Notably, the model can connect the dependent and independent determinants, as well as measured items and structural models. As a result, this model is appropriate for this study because it can build complex models using more minor data (Ringle et al., 2014).

Table 1 The symbols used in the structural equation models

Symbol	Definition
	Construct or variable lantent (LV)
	Variable observed or measured or indicated (OV)
	Correlation between LV and OV (measuring model)
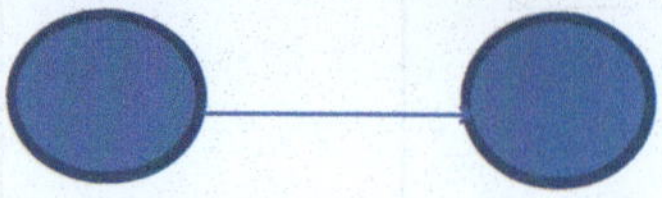	Causal relation-coefficient of the path between an independent LV to dependent variable (structural model)

3.1 Mounting the Measuring Models on the Smart PLS

Smart PLS software is used to obtain multiple simultaneous regressions and to construct linear regressions between models (Henseler et al., 2009). This software is also used to determine the relationship between constructs and items or determinants that can be measured or observed. Based on the research of Henseler et al. (2009), Hair et al. (2014), and Cohen, five (5) major steps were taken to structure AoEMD: convergent validity analysis, discriminant validity analysis, composite reliability analysis, T-test analysis, and finally the path coefficient (1988). To begin, the Smart PLS considers several symbols (see Table 1) to obtain the Structural Equation Modeling (SEM).

3.2 The Steps of Analysis by Using Smart PLS

Measurement of AoETD implicated six (6) main analyses, including convergent validity, discriminant validity, model reliability, T-test, Pearson closure coefficient (R2) evaluation, and path coefficient) highlighted during data analysis using Smart PLS, as shown in Table 2, referring to previous researchers' statements or studies.

The T-ratio tests for evaluating the significance of correlations and regressions are critical for this study because they ensure that the threshold values are +1.96.

Table 2 Summary of the steps of analysis by using smart PLS

No.	Indicator/analysis	Purpose	Referential values/criteria	References
1	AVE measurement	Convergent validity	AVE > 0.50	Henseler et al. (2009)
2	Criteria of Fornell and Larcker	Discriminant validity	Compare the R^2 of the AVE values of each item construct with the correlation between the construct (latent variables). Next, the R^2 of AVE values should be > correlation of the construct	Fornell and Larcker (1981)
3	Cronbach Alpha and composite reliability	Model reliability	CA > 0.70 CR > 0.70	Hair et. al. (2014)
4	T-Test or T-analysis	Evaluation of significance of the correlations and regressions	Beta Coefficient, $\beta > \pm 1.96$	Hair et. al. (2014), Kock (2016)
5	Evaluation of the coefficient of the Pearson's determination (R^2)	Evaluate the range of R^2	$R^2 = 2\%$ (small effect) $R^2 = 13\%$ (median effect) $R^2 = 26\%$ (large effect)	Cohen (1988)
6	Path coefficient/ P-value analysis	Evaluation of relation	Value interpretation. The P-value test is used to examine the hypothesis that 0. We compute the one-tailed P-value associated with the path coefficient at the 0.05 significance level (i.e., 1–95%). In general, this quantity can be interpreted as the likelihood of belonging to a distribution with a mean of zero. If P 0.05, the hypothesis is accepted; otherwise, it is rejected	Cohen (1988), Kock (2016)

3.3 *Measurement of AoETD*

AoETD models are designed to rank the most important determinants of success in the international market, increasing a company's ability to enter the international market. To begin, a questionnaire survey is used to test the ET used by firms to enter the international market. The Rasch Model was used to identify fit items for each determinant or item. Then, based on the beta coefficient value (ß), only fit items were used for Smart PLS analysis to identify the relationship between determinants and ET decisions. Each determinant is labelled on a scale of one to six to represent poor to excellent entry decisions made to the international market, ensuring the firms' long-term sustainability in the international market. The higher the ß value, the more significant the ET decision that allows firms to stay in the international market for an extended period of time. Equation (1) presents the AoETD calculation, where the X label represents each ET decision (Awang et al., 2018).

$$\text{AoETD} = \beta_1 X_1 + \beta_2 X_2 + \beta_3 X_3 + \ldots \beta_4 X_n \tag{1}$$

Following that, AoETD are labelled using the grade scores obtained by the firm, which are labelled from A to E as shown below:

A = can extremely sustain, B = high potential to sustain, C = moderately sustain, D = high risk to sustain, and E = very high risk to sustain

To determine the best ETD, 5 distinct scales are used. Scales are frequently used to evaluate behaviours, feelings, or actions using multiple variables or items (Boateng et al., 2018). As a result, item measurements were taken in order to obtain more accurate results. Data are organised in a matrix in the best ETD for entering the international market, from top to bottom or from A to E. In other words, the highest rank represents the best entry decision that the firm must consider when entering the international market, as shown in Table 3. The grid arrangement, as depicted in Table 3, should be carefully observed, where entry decisions to the international market are measured.

Although the firm's score is low, it has the potential to enter the international market because all these key determinants are important for entry. If the obtained

Table 3 A preference matrix of entry decision to international market

Level of the firm's score to sustain in the international market		Entry decision				
		Can extremely sustain	High potential to sustain	Moderately sustain	High risk to sustain	Very high risk to sustain
	Grade score	A	B	C	D	E
		100–81	61–80	41–60	21–40	0–20
Can extremely sustain	A					
(100–81)	✓					
High potential to sustain	B					
(61–80)		✓				
Moderately sustain	C					
(41–60)			✓			
High risk to sustain	D					
(21–40)				✓		
Very high risk to sustain	E					
(0–20)					✓	

score is low, the company must find a way to elevate it. Finally, based on a combination of critical determinants and ET decisions, the firm must determine the best ET decision.

4 Results and Analysis

AoETD was developed as a result of a combination of analysis using Smart PLS software. AoETD is divided into three phases, namely: the relationship between ETDs and the sustainability of Malaysian construction firms in the international market; significant determinants influencing ETDs; and measurement of ETDs.

4.1 Phase 1: The Relationship Between ET Decisions and the Sustainability of Malaysian Construction Firms in International Market

The following are three (3) hypotheses to be achieved in this study, which test the relationship between the three items. For example, such items are pioneers (Pi), early followers (EF), late followers (LF), and sustainability by Malaysian construction firms in the international market. The hypotheses are as follows:

H1: Pioneer (Pi) has the significance of firms' ability to sustain in the international market

H2: Early follower (EF) has the significance of firms' ability to sustain in the international market

H3: Late follower (LF) has the significance of firms' ability to sustain in the international market

First, convergent validity analysis was performed to test all three hypotheses (H1, H2, and H3) and identify all determinants as valid (AVE > 0.5). Thus, the results of the convergent validity analysis are as follows:

Surprisingly, according to Smart PLS analysis (shown in Fig. 2), all items are valid (>0.50). All items under all independent determinants have a significant impact on firms' ability to sustain themselves in international markets. Since the value average (AVE) for all items was greater than 0.5, the data are reliable (Fornell & Larcker, 1981). Next, all of the CR values are greater than 0.70, while the CA value is greater than 0.70 (Hair et al., 2014). Table 4 then displays the results of the beta coefficient (β) of Pearson's determination (R2) and the P-value analysis.

Based on Table 4, the P-values for all items are significant ($P \leq 0.05$). As a result, the hypothesis is accepted based on Kock (2016). The path (arrow) and its coefficients, which measure the correlation significance of each item construct, are also shown in the table. As a result, the findings show that all items have a significant impact

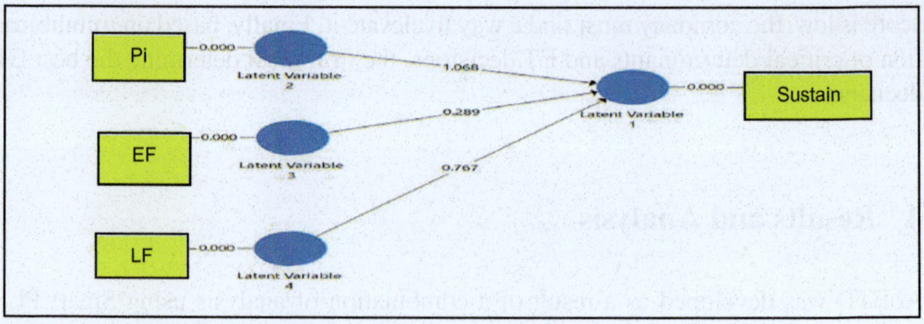

Fig. 2 The standardised regression weights. The relationship between ET decisions and the sustainability of Malaysian construction firms in the international market

Table 4 Testing the causes of sustainability of Malaysian construction firms on entry timing decisions

	Beta coefficient, β	P-value	Result
The relationship between Pi and the sustainability of firms in international markets	1.068	0.000	Significant
The relationship between EF and the sustainability of firms in international markets	0.289	0.024	Significant
The relationship between LF and the sustainability of firms in international markets	0.767	0.050	Significant

on firms' sustainability in international markets, with beta (β) equal to +1.96 (Hair et al., 2014; Kock, 2016). As shown in Table 5, all hypotheses are supported in this case.

Finally, the regression Eq. (2) constructed from the above findings is as follows:

$$\text{AoETD (ET decisions)} = 1.068\ \text{Pi} + 0.767\ \text{LF} + 0.289\ \text{EF} \tag{2}$$

Table 5 Results of hypotheses on the relationship between ET decisions and the sustainability of Malaysian construction firms in the international market

Hypotheses	Result of hypotheses
H1: Pioneer (Pi) has the significance of firms' ability to sustain in the international market	Supported
H2: Early follower (EF) has the significance of firms' ability to sustain in the international market	Supported
H3: Late follower (LF) has the significance of firms' ability to sustain in the international market	Supported

Overall, the Pis exhibited the most significant beta coefficient, 1.068, influencing firms' sustainability in international markets. The value also means that Pi is an effective primary ET, influencing the firm's success internationally compared to LF and EF.

4.2 Phase 2: Significant Determinants Influencing ET Decisions

The following are six (6) hypotheses to be achieved in this study to test the relationship between the six (6) determinants and ET decisions of Malaysian construction firms in international markets. These determinants include firm-specific (FSP), firm-resources commitments (FRC), project-specific (PS), home country (HC), target country (TC), and market-specific (MS) affecting ET decisions. The hypotheses are shown below:

H1: Firm-specific (FS) has significant effect on entry timing (ET) decision
H2: Firm-resources commitments (FRC) has significant effect on entry timing (ET) decision
H3: Project-specific (PS) has significant effect on entry timing (ET) decision
H4: Home country (HC) has significant effect on entry timing (ET) decision
H5: Target country (TC) has significant effect on entry timing (ET) decision
H6: Market-specific (MS) has significant effect on entry timing (ET) decision.

Notably, all items under the FRC are not valid (<0.50), and thus hypothesis 2 is rejected because the item does not influence the firm in its ET decision. Next, FRC items are eliminated in this analysis. Figure 3 indicates the fit or valid determinants (FS, PS, HC, TC, and MS) results from the Smart PLS analysis. Smart analysis shows all determinants are valid as they exceed 0.50.

The findings from Smart PLS in Fig. 3 show that all items (TC, FSP, PS, HF and MF) significantly influence the ET decisions (>0.50). Next, Table 6 shows the findings on the relationship between determinants and ETDs by Malaysian construction firms in the international market using analyses of convergent validity, T-test, Cronbach alpha, composite reliability, and P-value.

The P-value in Table 6 indicated that all variables had significant values of less than 0.05 (Hair et al., 2014). Meanwhile, according to Hair et al., all of the CR and CA values are valid because they exceed 0.70 (2014). Furthermore, bootstrapping modules are used to identify T-values in order to determine the significance of the correlation and P-value, as shown in Table 7.

The results in Table 7 show that all the P-values are significant (0.000) for all items and the t-values are greater than ± 1.96. Significantly, the results showed that all items significantly affected the ET decisions. In this case, the hypotheses (H1, H3, H4, H5, and H6) that significantly affect ET decisions are supported. Furthermore, the results of every hypothesis are presented in Table 8.

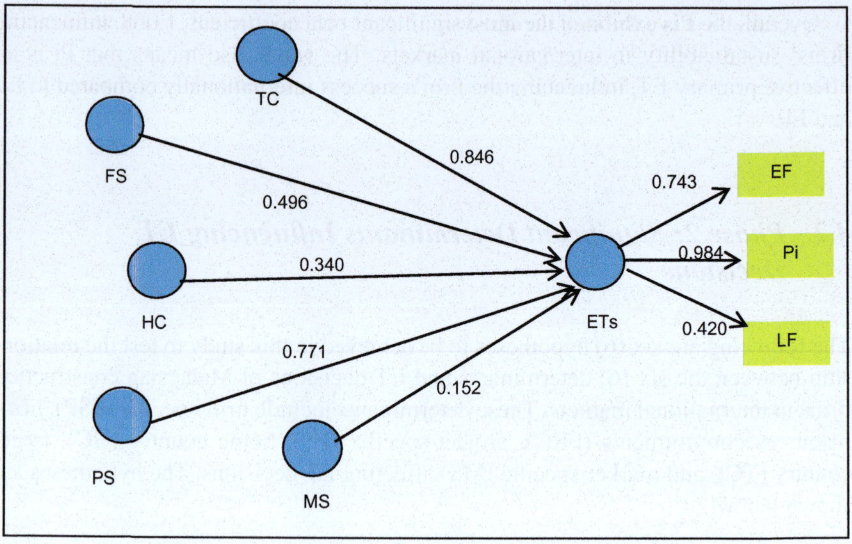

Fig. 3 The relationship between the determinants and ET decisions of Malaysian construction firms in international markets using standardised regression weights

Table 6 Findings on the relationship between determinants and entry timing decisions by Malaysian construction firms into the international market

Construct or variable lantent (LV)	Beta Coefficient, β	AVE	CR	CA	P-value	Result
FS	0.496	0.686	0.824	0.851	0.014	Significant
PS	0.771	0.687	0.820	0.779	0.018	Significant
HC	0.340	0.757	0.855	0.834	0.004	Significant
TC	0.846	0.676	0.811	0.764	0.025	Significant
MS	0.152	0.743	0.898	0.886	0.001	Significant

Finally, the regression Eq. (3) constructed from the above findings are:

$$\text{AoETD (determinants)} = 0.85\,\text{TC} + 0.77\,\text{PS} + 0.50\,\text{FS} + 0.34\,\text{HC} + 0.15\,\text{MS} \tag{3}$$

The TC significantly presented the most significant beta coefficient, 0.85, influencing ETDs. The value also means that TC is a critical determinant influencing the firm's ET decision to enter the international market. The second determinant influencing the ET decision was the PS (0.77), followed by FSP (0.50), HC (0.34), and MS (0.15).

Table 7 Results of hypotheses on the relationship between determinants and entry timing decisions by Malaysian construction firms into the international market

	Beta coefficient	T-value	P-value after bootstrapping	Result
FS → ET	0.496	23.150	0.000	Significant
PS → ET	0.771	16.642	0.000	Significant
HC → ET	0.340	18.518	0.000	Significant
TC → ET	0.846	12.640	0.000	Significant
MS → ET	0.152	32.196	0.000	Significant

Table 8 Results of hypotheses on the relationship between determinants and entry timing decisions by Malaysian construction firms into the international market after bootstrapping

Hypotheses	Results on hypotheses
H1: Firm-specific (FS) has significant effect on entry timing (ET) decision	Supported
H2: Firm-resources commitments (FRC) has significant effect on entry timing (ET) decision	Rejected
H3: Project-specific (PS) has significant effect on entry timing (ET) decision	Supported
H4: Home country (HC) has significant effect on entry timing (ET) decision	Supported
H5: Target country (TC) has significant effect on entry timing (ET) decision	Supported
H6: Market-specific (MS) has significant effect on entry timing (ET) decision	Supported

4.3 *Phase 3: Measurement of ET Decisions*

ET is estimated using the relationship of critical determinants (MS, HC, TC, FS, and PS). As ET decisions, other determinants include Pi, EF, and LF. The evaluation of ET decisions is broken down into three stages.

4.3.1 Stage 1: Measurement of the Key Determinants of ET Decisions to Analyse the Firm's Ability to Enter the International Market

In Stage 1, the firm can determine whether it has the potential or ability to sustain itself in the international market. Firm capabilities can be tested using the Eq. (4)

derived from Smart PLS analysis:

$$\text{AoETD (determinants)} = 0.85\,\text{TC} + 0.77\,\text{PS} + 0.50\,\text{FS} + 0.34\,\text{HC} + 0.15\,\text{MS} \tag{4}$$

Based on Eq. (4), the firm needs to choose key determinants that the firm considers to enter the international market. The firm must mark (√) if "yes" and (x) if "not applicable." An example of the assessment (AoETD) is shown in Table 9.

Based on Table 9, a value of 1 will be given if the firm chooses (√) and 0 if otherwise. The arrangement of independent determinants is arranged on a scale of 1 to 5, i.e., the larger the value of the scale, the greater the influence the determinant has on the firm's ET decision to enter the international market. In this case, scale 5 refers to TC, followed by PS, FS, HC, and MS.

Hence, AoETD measurements are based on the previous Eq. 4. Measurements for AoETD are as follows:

$$\text{AoETD (determinants)} = 0.85\,\text{TC} + 0.77\,\text{PS} + 0.50\,\text{FSP} + 0.34\,\text{HC} + 0.15\,\text{MS}$$

Hence,

$$\begin{aligned}\text{Final score, Y} &= 0.85(5) + 0.77(4) + 0.50(3) + 0.34(2) + 0.15(1)\\ &= 4.25 + 3.08 + 1.5 + 0.68 + 0.15\\ &= 9.66\end{aligned}$$

$$\begin{aligned}\text{Y in percentage} &= (9.66/9.66) * 100\\ &= 100\end{aligned}$$

An example of the measurement results in Table 9, the firm obtains a total score (100%). This score indicates that the firm will be extremely sustained in the international market if it successfully meets these critical determinants.

Table 9 Measurement of the key determinants of entry timing decision to analyse the firm's ability to enter the international market

Scale					Final score (Y)	Y (%)	Grade score	Potential of entry decisions
5	4	3	2	1				
TC	PS	FS	HC	MS				
0.85	0.77	0.5	0.34	0.15				
√	√	√	√	√				
1	1	1	1	1				
4.25	3.08	1.5	0.68	0.15	9.66	100	A	Can extremely sustain

4.3.2 Stage 2: Measurement of AoETD Based on the Relationship Between the Dependent Determinants and ET Decisions

Firstly, measurements for all ET are made based on Eq. (5):

$$\text{AoETD (determinants)} = 1.068\ \text{Pi} + 0.767\ \text{LF} + 0.289\ \text{EF} \quad (5)$$

Next, the measurements of AoETD were made, and an example of the calculation is shown in Table 10. Option 1 refers to all ET decisions, while Option 2 applies if the firm chooses one of the ET decisions. In this case, the firm chooses pioneer (Pi) as the ET decision to go to the international market.

Based on Table 10, a value of 1 will be given if the firm chooses (√) and 0 if otherwise. The arrangement of independent determinants is arranged on a scale of 1 to 3, i.e., the larger the value of the scale, the greater the influence the determinant has on the firm's ET decision to enter the international market. In this case, scale 3 refers to Pi, followed by LF, and EF.

Hence, AoETD measurements are based on the previous Eq. (5). Measurements for AoETD are as follows:

$$\text{AoETD (ET decisions)} = 1.068\,\text{Pi} + 0.767\,\text{LF} + 0.289\,\text{EF}$$

Hence, for Option 1:

Table 10 Measurement of the ET decisions (Option 1)

Opt	Scale			Final score, X	Score for each option	Total score, Z	Grade	Level of sustainability
Opt1	1	2	3					
	EF	LF	Pi					
	0.289	0.767	1.068					
	√	√	√					
	1	1	1					
	0.289	1.534	3.204	5.027	100	100.00	A	Can extremely sustain
	Scale							
	1	2	3					
	EF	LF	Pi					
	0.289	0.767	1.068					
	x	x	√					
	0	0	1					
	0	0	3.204	3.204	63.74	81.87	A	Can extremely sustain

$$\begin{aligned}\text{Final score, X} &= 1.068(3) + 0.767(2) + 0.289(1)\\ &= 3.204 + 1.534 + 0.289\\ &= 5.027\end{aligned}$$

$$\begin{aligned}\text{X in percentage} &= (5.027/5.027) * 100\\ &= 100\end{aligned}$$

Y refers to the score for the independent determinant (refer to Stage 1). For example, if score Y is 100%.

Hence,

$$\begin{aligned}\text{Total Score Z} &= ((\text{Y} + \text{X})/200) * 100\\ &= ((100 + 100)/200) * 100\\ &= 100\end{aligned}$$

For Option 2, the firm acts as a Pi to go to the international market. Hence,

$$\begin{aligned}\text{Final score, X} &= 1.068(3) + 0.767(0) + 0.289(0)\\ &= 3.204\\ \text{X in percentage} &= (3.204/5.027) * 100\\ &= 63.74\end{aligned}$$

Y refers to the score for the determinants (refer to Stage 1). For example, if score Y is 100%,

Hence,

$$\begin{aligned}\text{Total Score Z} &= ((\text{Y} + \text{X})/200) * 100\\ &= ((100 + 63.74)/200) * 100\\ &= 81.87\end{aligned}$$

Based on the assessment (AoETD), Pi is the best ET to enter the international market with an ET decision score of 81.87, or an A score, which ensures the firm's sustainability in the global market. It is followed by LF and EF. Pi also showed that the firm's extreme sustainability in foreign markets was evident.

4.3.3 Stage 3: Final Measurement of AoETD

The last stage is to combine the measurement of independent determinants with the ET decisions. Based on ranking, top managers must look at the level of ET decisions in helping firms sustain internationally before entering the global market. The same steps are taken for Option 3 and Option 4 (refer to the measurement examples in

Table 11 Overall outcome based on ETAC

	ET decisions	Score of ET decision	Grade score	Rank	Level of sustainability of firms in the international market
Option 1	All ET decisions	100.00	A	1	The firm can extremely sustain
Option 2	Pi	81.87	A	2	The firm can extremely sustain
Option 3	LF	65.26	B	3	High potential of firm to sustain
Option 4	EF	52.87	C	4	The firm moderately can sustain

Option 1 and Option 2), and the comparison for each option is illustrated based on the firm's score and level of sustainability in the international market. Finally, a total of four (4) decisions on ET were made as shown in Table 11.

Accordingly, it is relatively impossible for firms to apply all ET decisions (refer to Option 1). Therefore, the second option is to choose Option 2, which is Pi. These findings are made based on the most significant ET decision scale referring to AoETD. Thus, the result of the Pi shows that a firm can extremely sustain itself in the international market, followed by LF (Option 3) and EF (Option 4). This is similar to the study by Kalyanaram and Gurumurthy (1998). As cited by Kalyanaram and Gurumurthy (1998), late entrants can then be successful by adopting their own marketing positions and strategies, which proved findings of the AoETD.

5 Conclusions

MEDIF is developed to provide guidelines to local contractors in making strategic decisions in making the right ET (Pi, EL, and LF) decisions. Thus, AoETD highlighted six (6) key independent determinants to enter the international market arranged by scale; the key determinants chosen by firms to enter the global market were the TC, followed by PS, FS, HC, and MS. Hence, AoETD facilitates firms to decide on entry according to the firm's capabilities through the scale. AoETD measurements were made based on the scale with coefficient values from Smart PLS analysis. Based on the overall measurement of AoETD, firms can sustain in the international market and AoETD measurement could provide indicators or guidelines for local firms to operate globally from poor to excellent decision-making. According to AoETD measurements, Pi was found to be the most sustainable in the international market, followed by LF and EF.

Acknowledgements The authors gratefully acknowledge the Ministry of Higher Education Malaysia for providing the Fundamental Research Grant Scheme (600-IRMI/FRGS 5/3 (025/2019) and also to the School of Civil Engineering, College of Engineering, UiTM Shah Alam for their support in carrying out this research. We would like to thank all the managers who have participated in this study.

References

Abdul-Talib, A.-N., Salleh, M. F. M., Shamsuddin, F. M., & Ashari, H. (2011). The effects of firm size and international business experience on export attitudes. *American College of Radiology, 19*(1&2), 312–347.

Al Sadi, O., & Dulaimi, M. (2019). Choice of GCC construction market entry mode. In The international council for research and innovation in building and construction (CIB) world building congress 2019—constructing smart cities, 17 June 2019–21 June 2019, Hong Kong.

Asgari, M., & Ahmad, S. Z. (2010). Internationalization of Malaysian service firms: Business strategy and choice of foreign market entry mode. In *International conference on business, economics and tourism management (CBETM 2010)* (Vol. 1, pp. 1–6).

Awang, Z., Hui, L. S., & Zainudin, N. F. S. (2018). *Pendekatan Mudah structural equation modelling*. MPWS Rich Publication.

Bianchi, C., Carneiro, J., & Wickramasekera, R. (2018). Internationalisation commitment of emerging market firms. *Journal of Small Business and Enterprise Development*, *25*(2), 201–221. https://doi.org/10.1108/jsbed-07-2017-0221

Boateng, G. O., Neilands, T. B., Frongillo, E. A., Melgar-Quiñonez, H. R., & Young, S. L. (2018). Best practices for developing and validating scales for health, social, and behavioral research: A primer. *Frontiers in Public Health, 6*(149). https://doi.org/10.3389/fpubh.2018.00149

Cleff, T., & Rennings, K. (2011). *Theoretical and empirical evidence of timing-to-market and lead market strategies for successful environmental innovation.* Simon Fraser University.

Cohen, J. (1988). Statistical power analysis for the behavioral sciences (2nd ed.). Lawrence Erlbaum Associates, Publishers. Construction Industry Development Board Malaysia (2020). List of construction firm operating internationally.

Construction Industry Development Board Malaysia Report. (2020). Construction industry development board Malaysia record. http://international.cidb.gov.my/international

Creswell, J. W.(2009). Research design: Qualitative,quantitative,and mixed methods approaches (3rd.ed). Sage Publications, Inc.

Durmaz, Y., & Taşdemir, A. (2014). A theoretical approach to the methods introduction to international markets. *International Journal of Business and Social Science*, *5*(6(1)), 47–54.

Lo, F.-Y., & Kletsova, A. (2018). Entry timing into international markets: Evidence from the Taiwanese service industry. *Economic Research-Ekonomska Istraživanja, 31*(1), 1059–1077. https://doi.org/10.1080/1331677X.2018.1436456

Fornell, C., & Larcker, D.F. (1981). Evaluating SEM models with unobservable variables and measurement error. *Journal of Marketing Research, 18*(1), 39–50.

García-Villaverde, P. M., Parra-Requena, G., & Ruiz-Ortega, M. J. (2017). From pioneering orientation to new product performance through competitive tactics in Smes. *BRQ Business Research Quarterly, 20*(4), 275–290. https://doi.org/10.1016/j.brq.2017.05.002

Greening, D. W., Barringer, B. R., & Macy, G. (1996). A qualitative study of managerial challenges facing small business geographic expansion. *Journal of Business Venturing II, 11*(4), 233–256.

Hair, J. F., Hult, T. M., Ringle, C.M., & Sarstedt, M. A. (2014). *Primer on partial least squares structural equation modeling (PLS-SEM).* Sage

Henseler, J., Ringle, C. M., & Sinkovics, R. R. (2009). The use of partial least squares path modeling in international marketing. *Advances in International Marketing, 20*, 277–319.

Ibrahim, N., Mat Isa, C. M., Mohd Nusa, F. N., Preece, C. N., & Khairolden Ghani, M. (2021). Significant factors influencing entry decisions (SEFIED) into international market. In *2021 IEEE 12th control and system graduate research colloquium (ICSGRC)* (pp. 138–143). https://doi.org/10.1109/ICSGRC53186.2021.9515240

Kaffash, M. H., Haghighikhah, M., & Kordlouie, H. (2012). Identifying factors influencing entry mode selection in food industry of small and medium-sized enterprises (SMEs) in Iran. *International Journal of Marketing Studies, 4*(5). https://doi.org/10.5539/ijms.v4n5p47

Kalyanaram, G., & Gurumurthy, R. (1998). *Market entry strategies: Pioneers versus late arrivals.* Chapter in book of strategies. Third Quarter 1998/Issue 12.

Kock, N. (2016). Hypothesis testing with confidence intervals and P values in PLS-SEM. *International Journal of e-Collaboration, 12*(3), 1–6.

Lages, L. F., Jap, S. D., & Griffith, D. A. (2008). The role of past performance in export ventures: A short-term reactive approach. *Journal of International Business Studies, 39*(2), 304–325.

Lilien, G. L., & Yoon, E. (1990). The timing of competitive market entry: An exploratory study of new industrial products, *36*(5), 568–585.

Low, S. P., Hongbin, J., & Leong, C. H. Y. (2004). A comparative study of top British and Chinese international contractors in the global market. *Construction Management and Economics, 22*(7), 717–731. https://doi.org/10.1080/0144619042000202780

Preece, C. N., Mat Isa, C. M., Saman, H. M., & Che Ibrahim, C. K. (2016). Development of entry location, entry timing and entry mode decision model for construction firms in international markets. *Construction Management and Economics, 34*(4–5). https://doi.org/10.1080/01446193.2015.1084429

Ringle, C., Silva, D., & Bido, D. (2014). Structural eqaution modeling with the SmartPLS. *Brazilian Journal of Marketing, 13*(2), 56–73.

Suarez, F. F., Grodal, S., & Gotsopoulos, A. (2013). Perfect timing? Dominant category, dominant design, and the window of opportunity for firm entry. *Journal of Forthcoming, Strategic Management*, 1–26.

Utama, W., Chan, A., Sesmiwati, Zahoor, H., & Gao, R. (2019). Internationalization of construction enterprises: An overview of motivation. *International Journal of Technology, 10*(1), 36–46.

Zander, I. (2015). Internationalization: Patterns in business firms. In *International encyclopedia of the social & behavioral sciences* (pp. 580–586). https://doi.org/10.1016/b978-0-08-097086-8.73008-x

Zeqiri, J., & Angelova, B. (2011). Factors that influence entry mode choice in foreign markets. *European Journal of Social Sciences, 22*(4), 572–584.

Pre-construction Complexity Factors Affecting Cost Performance of Infrastructure Projects

Akhtarul Norfaiza Che Nen, Che Maznah Mat Isa, Che Khairil Izam Che Ibrahim, and Mohamad Shakri Mohmad Shariff

Abstract Cost overruns are typical in infrastructure projects all around the world. However, previous studies have shown a lack of understanding of the term complexity in the construction industry, specifically regarding infrastructure projects. This paper is part of the research project, which aims to identify the most significant complexity factors contributing to project performance and develop a complexity assessment model for infrastructure projects. This paper aims to categorize and rank the complexity factors affecting the cost performance of infrastructure projects during the pre-construction stage. A survey questionnaire was designed to identify complexity factors affecting infrastructure project cost performance during the pre-construction phases. One hundred and six (106) managers consisting of clients, consultants, contractors, sub-contractors, and others involved in infrastructure projects have responded to the survey. Using Rasch Model Analysis, the significant complexity factors with logit measure ranges show that these complexity factors are critical and vastly impact infrastructure project cost performance. The following six (6) most significant pre-construction complexity factors affecting cost performance (PRECO) have been identified: original design errors, low bid award that is qualified or non-compliant, redesign because over-budgeted, lack of optimization cost and time, lack of design coordination information between consultant and client, and unit prices that are not properly specified or evaluated. This study contributes to integrating complexity assessment for infrastructure projects during the pre-construction life cycle in line with the Twelfth Malaysia Plan for the construction industry.

A. N. Che Nen · C. K. I. Che Ibrahim
School of Civil Engineering, College of Engineering, Universiti Teknologi MARA, 40450 Shah Alam, Selangor, Malaysia

C. M. Mat Isa (✉)
Center of Civil Engineering Studies, Universiti Teknologi MARA Pulau Pinang Branch, Permatang Pauh Campus, 13500 Pulau Pinang, Malaysia
e-mail: chema982@uitm.edu.my

M. S. Mohmad Shariff
Centre for Geotechnics and Transportation, Faculty of Engineering and Quantity Surveying, INTI International University, Nilai, Negeri Sembilan, Malaysia

U. R. Hashim et al. (eds.), *Green Infrastructure*,
https://doi.org/10.1007/978-981-99-7003-2_2

Keywords Factors · Complexity assessment · Cost Overrun · Pre-construction stage

1 Introduction

Cost overruns are widespread and virtually always connected to construction projects, especially infrastructure projects. According to Flyvbjerg et al. (2005) and Flyvbjerg (2009), the average cost overrun for large-scale infrastructure projects could range from 20.4 to 44.7%, with overruns costing 90% of projects worldwide. They also said that cost overruns are observed in 20 countries and five continents. Bruzelius et al. (2002) stated that there had been no systemic change in the cost overrun of infrastructure projects over the past 70 years. Different causes of cost overrun were identified. According to Kamaruddeen et al. (2020), most construction projects in Malaysia have experienced cost overruns of 5–10% of the total contract sum. Therefore, many complexity factors can be identified related to cost overruns in project development, especially during the pre-construction stage.

There are numerous concepts of complexity, according to Saed et al. (2018), project complexity is a critical issue since it is intimately related to the project life cycle, particularly in the construction industry. The Collins English Dictionary (2006) defines complexity as "the state or quality of being intricate or complex", whereas complex is defined as "made up of many interconnecting parts" (Wood & Gidado, 2008). Complexity is a broad concept that can apply to any subject.

The key factors leading to the project's complexity can be considered, in effect, the difficulties that arise in construction projects before and during the construction. In its essence, the construction industry is complex as it includes vast numbers of parties as owners (customers), contractors, consultants, stakeholders, and regulators. In 2015, the civil engineering sub-sector contributed the most to the gross output of the construction sector with a share of 27.0% (RM48.1 billion) as compared to 27.2% (RM24.9 billion) in 2010 (Department of Statistics, 2022). However, many local construction projects report poor performance due to many evidential project-specific causes (Enshassi et al., 2009).

The focus of this paper is to clarify the complexity factors affecting infrastructure projects in order to determine the significant relationship between complexity factors and project performance in pre-construction infrastructure projects based on cost performance.

2 Literature Review

Cost performance is a critical factor in project success (Frimpong et al., 2003; Olawale & Sun, 2010). The construction industry has historically faced low-cost performance that defines the failure to complete a project within budget. On the other hand, according to Abdullah et al. (2009), in MARA's large-scale construction project report, it emerged that more than 90% of the significant MARA construction project had been delayed due to time and cost overruns since 1984. Complexity can be derived from complexity of the projects and processes. Cardoso (2006) defined process complexity as "the degree to which a process is challenging to analyze, understand or explain". Wood and Gidado (2008) stated that complexity could be hard to define as it has several different meanings. According to Wood and Gidado (2008), the application of complexity to the construction industry is not commonly studied, but some evidence has been found that the construction process can be considered a complex method. Construction processes have been considered the most complex undertaking in any industry, and recently the construction industry has had difficulty coping with the increasing complexity of major construction projects (Baccarini, 1996).

As stated by Mills (2001), the construction industry is one of the most dynamic, challenging, and riskiest industries. It has a deplorable reputation for managing risk, with many complex projects failing to meet the criteria for project performance. This issue is supported by Wood and Gidado (2008) stated that project success in terms of cost, time overruns, quality, performance, and safety are historically poor in the construction industry because of the exceptionally complex design and construction process. Rahman et al. (2013) and Masrom et al. (2015) stated that a successful project meets time, cost, quality, and safety requirements. The effectiveness of measuring efficiency is a matter of increasing interest for industrialists and academics alike. Cost, schedule, and efficiency in construction projects are the typical success areas.

Overall, construction projects and road infrastructure projects all consist of many interconnecting parts that fit well with the complexity concept. Previous research on the complexity of construction projects, on the other hand, has been limited, with most studies focusing primarily on the conceptual aspects that contribute to the construction industry's complexity (Baccarini, 1996; Brockmann & Girmscheid, 2007; Maylor et al., 2008; Sinha et al., 2006; Williams, 1999; Wood & Gidado, 2008).

Luo et al. (2017) summarized many types of project complexity. Most researchers agreed that organizational, technological and environmental complexity affected the project performance.

3 Methodology

3.1 Sampling Design and Target Respondent

A random sampling was used to target the respondents who were the clients, consultants, contractors, sub-contractors, and other parties involved in infrastructure projects based on a sampling frame provided by Construction Industry Development

Board (CIDB), Malaysia and Master Builders Association Malaysia (MBAM). The targeted number of respondents was around 384 based on Krejcie and Morgan (1970).

3.2 *Questionnaire Design*

The survey required respondents to answer it based on their working experience with complexity factors affecting cost performance during the pre-construction stage. The survey comprises 44 items constructed for the complexity factors affecting cost performance during the pre-construction stage, as shown in Table 1. All 44 items used the five-point Likert scale points was used: 1: Strongly Disagree; 2: Disagree; 3: Agree; 4: Strongly Agree; 5: Strongly Agree).

3.3 *Data Analysis*

The Rasch model was used to analyze the data collected from the questionnaire survey, which was run on WINSTEPS version 3.69.1.16 software. Rasch model analysis was utilized because the model shifts the concept of dependability from producing a line of fit to building a reliable measurement instrument (Said, 2016). According to Baghaei (2008), one of the benefits of the Rasch model is that it creates a hypothetical unidimensional line along which objects and people are located based on their difficulty and ability assessments, as seen in the Person Item Distribution Map (PIDM). As stated by Bond and Fox (2001), the Rasch model is a prescriptive model in which we study how the data fits the model rather than the more traditional statistical problem of how the model fits the data. Scholten (2011) went on to say that the data must fit the model, and if they do not, the elements that demonstrate misfit are removed until a good fit is reached.

The Rasch analysis is carried out by Bond and Fox (2007), who clarify that the logic value employed in the Rasch model is the unit of measurement at an interval level rather than the ordinary number. Thus, summary statistics, item characteristic curve scalogram, PIDM, and person and item measure order are used in Rasch model analysis. The analysis includes person and item measurements to show that the individual doing the survey understands the questions, while the item in the survey is understood and replied to by the respondent. For example, Othman et al. (2014) used Rasch analysis to investigate the construct validity and reliability of the competitiveness scale utilizing the Rasch model technique.

Table 1 Item construct for complexity factors affecting cost performance during pre-construction (PRECO)

Complexity factors	Code for (PRECO)
Lack of design coordination information between consultant and client	AC1
Insufficient collection and reviewing of all drawings	AC2
Insufficient review from a relevant expert about value engineering	AC3
Insufficient study on construction ability and feasibility	AC4
Lack of suggestion on sourcing and alternate material and agency	AC5
Lack of optimization of cost and time	AC6
Inadequate technical specification and BOQ with an estimate from the entire consultant	AC7
Lack of reviewing and making a consolidated budget and get it approved by the client	AC8
Lack of knowledge of tender preparation document and get it approved from the client	AC9
Insufficient of the original design	AC10
Change of original design from client	AC11
Errors in the original design	AC12
Redesign because of over-budgeted	AC13
Team members are not participating in technical discussions with owners	AC14
Tender document not complete	AC15
Insufficient time in preparing tender documents for sub-contractor	AC16
Lack of standardized systems during tender evaluation	AC17
Insufficient time to evaluate tenders from sub-contractor	AC18
Sub trade list is missing or inadequate	AC19
Bid Bond missing from bid	AC20
Surety's consent missing from the bid	AC21
Receipt of addenda not acknowledged	AC22
Bid not signed or sealed properly	AC23
Mathematical errors in the bid	AC24
Insufficient number of bids	AC25
Manipulated bid results using alternate or separate prices	AC26
Award to low bid that is qualified or non-compliant	AC27
Exercise privilege clause to award under non-disclosed criteria	AC28
Add an uninvited bidder after a prequalification process	AC29
Open a late bid either because too few bids were received, or the owner knows/prefers a bidder	AC30
Award contract for the same scope to a bidder who submitted outside of the tender process	AC31
Unit prices not correctly specified or evaluated	AC32

(continued)

Table 1 (continued)

Complexity factors	Code for (PRECO)
Criteria for determining compliance are not specified or clear	AC33
Consequences of non-compliance are not specified	AC34
Methods of remedying informalities in the bid are not specified	AC35
An addendum is issued late, e.g. too close to the submission deadline	AC36
Delay in preparing technical specifications, scope of work, or terms of reference	AC37
Failure to start the procurement process on time	AC38
Extension of bid or proposal submission date	AC39
Delay in opening bids or proposals received	AC40
Delay in starting or finishing the evaluation process	AC41
Delays during the approval process	AC42
Delay in contract negotiations	AC43
A contractor, supplier, or service provider challenges the procurement process	AC44

4 Results and Discussion

4.1 Respondent Demographic Analysis

In total, about 106 respondents returned the completed survey questionnaires giving a response rate of 27.6%. The response rate is reasonable since most of the survey done in Malaysia generated a rate that falls between 10 and 20% (Ainin et al., 2010; Ramayah et al., 2005).

Table 2 depicts the demographic characteristics of respondent which are stakeholder's types (client, consultant, contractor, sub-contractor, and others), current position, working experiences, types of infrastructure projects, sector (public or private), and average percentage of project cost performance.

One hundred and six (106) respondents participated in the survey representing 33% of public and 67% of private sectors in the construction industry. They held various positions which are project director (4.7%), project manager (20.8%), contract manager (3.8%), project engineer (30.2%), site supervisor (1.9%), quantity surveyor (4.7%), and others. The data collected reveals that 22.6% of the respondents have acquired below than 5 years of working experience, 27.4% of the respondents have experience between 6 and 10 years, 25.5% of the respondents have experience between 11 and 15, 5.7% of the respondents have experience between 16 and 20 years and 18.9% of the respondents have more than 20 years of working experience. Based on lengthy experience in years and from performing different leading positions within their firms, respondents were all knowledgeable about infrastructure projects. Thus, the respondents have gained the necessary knowledge and experiences to participate and give reliable opinions in the survey.

Table 2 Demographic of respondents

	Frequency	Percentage
Respondent background		
Client	28	26.4
Consultant	28	26.4
Contractor	35	33.0
Sub-contractor	4	3.8
Others	11	10.4
Respondent's position		
Project director	5	4.7
Project manager	22	20.8
Contract manager	4	3.8
Design manager	1	0.9
Project engineer	32	30.2
Construction manager	2	1.9
Risk manager	0	0
Site supervisor	2	1.9
Quantity surveyor	5	4.7
Others	33	31.1
Respondent working experience		
<5 years	24	22.6
6–10 years	29	27.4
11–15 years	27	25.5
16–20 years	6	5.7
>20 years	20	18.9
Type of sector		
Public sector	35	33.0
Private sector	71	67.0

4.2 Validity and Reliability Analysis

In Rasch analysis, both person (respondent) and item reliability were measured. Table 3 demonstrates that the reliability for a person is 0.95 with the Standard Error (SE) at 0.11. This indicated that the instrument had an excellent ability range with a sufficient rating length scale due to an adequate number of categories per item and exceptional target respondents, as shown in Table 4. In Table 5, the item reliability is 0.82 with SE = 0.05, suggesting that the instrument fits the model well (Fisher, 2007). Furthermore, the high item reliability indicated that the replicability of the items would occur if these items were tested by other respondents of the same size (Bond & Fox, 2007). Overall, this showed that the items in the questionnaire survey had fulfilled the criteria in identifying the complexity factors affecting cost performance during the pre-construction stage. Furthermore, it showed a good indication of the goodness of fit of the questionnaire survey, which measured what must be measured in the underpinning theory, contributing to its validity.

Table 3 Summary statistic of 103 measured (non-extreme) person to identify pre-construction complexity factor affecting cost performance in infrastructure projects

	Total score	Count	Measure	Model error	Input		Output	
					MNSQ	ZSTD	MNSQ	ZSTD
Mean	155.6	44.0	0.83	0.20	1.06	–0.6	1.14	−0.5
S.D	31.9	0.0	1.15	0.05	0.76	3.4	1.08	3.5
Max	215.0	44.0	3.92	0.45	3.94	8.9	6.14	8.8
Min	69.0	44.0	–2.17	0.17	0.07	–8.5	0.07	−8.5
Real SMSE	0.25	True S.D	1.12	Separation	4.58	Person reliability	0.95	
S.E of person mean = 0.11								

Table 4 Rating scale

Person and item measurement reliability	
Poor	<0.67
Fair	0.67–0.80
Good	0.81–0.90
Very good	0.91–0.94
Excellent	>0.94

Table 5 Summary statistic of 44 measured (non-extreme) items to identify pre-construction complexity factor affecting cost performance in infrastructure projects

	Total score	Count	Measure	Model error	Input		Output	
					MNSQ	ZSTD	MNSQ	ZSTD
Mean	379.3	106.0	0.00	0.12	1.00	−0.10	1.14	0.20
S.D	20.0	0.0	0.30	0.00	0.20	1.40	0.93	2.20
Max	435.0	106.0	0.71	0.13	1.81	1.81	6.98	9.90
Min	330.0	106.0	–0.89	0.12	0.65	−3.00	0.62	−3.10
Real SMSE	0.13	True S.D	0.27	Separation	2.15	Item reliability	0.82	
S.E of person mean = 0.05								

4.3 Variable Map: Person Item Distributions Analysis

Figure 1 shows the most significant complexity factors using Variable Map from Rasch analysis.

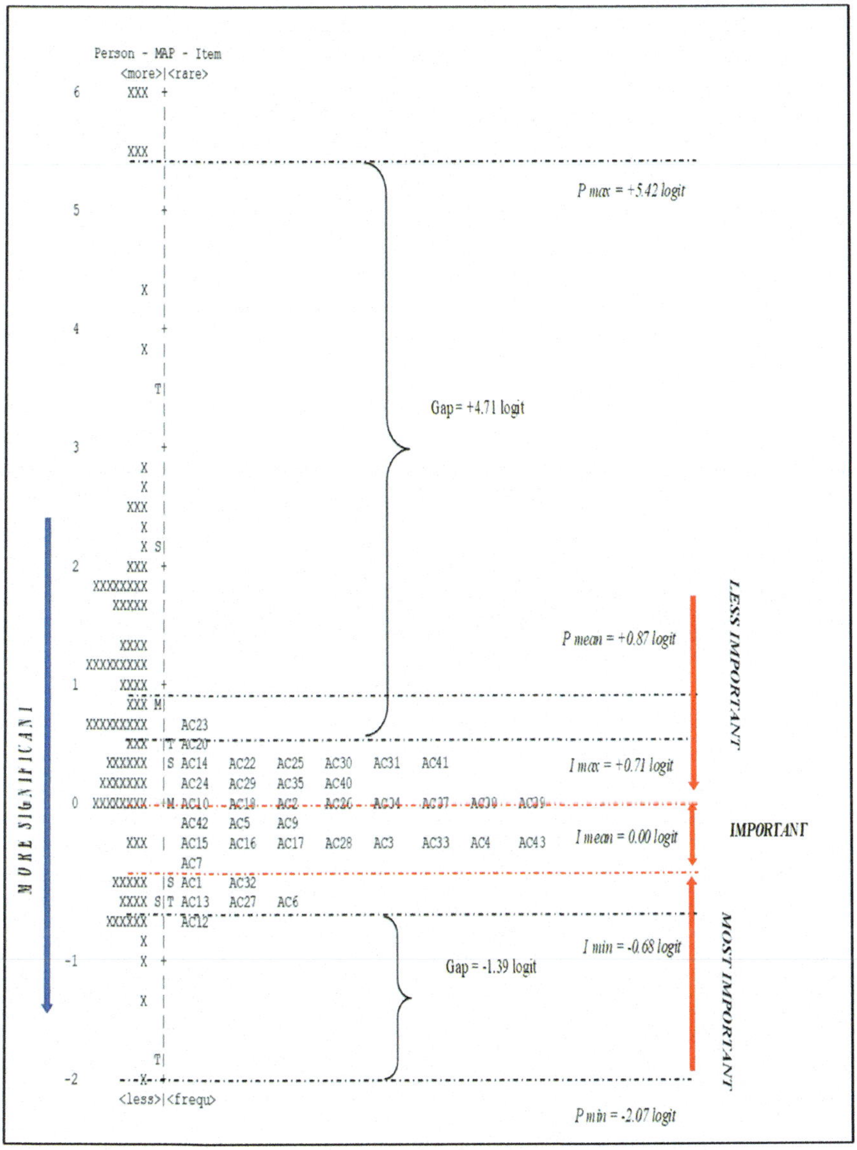

Fig. 1 Person item distribution map to identify the most significant complexity factors (PRECO construct)

In Rasch analysis, respondents/persons are arranged in measure according to their ability to endorse the items based on difficulties of the items in the logit scale. The maximum respondents/person ability was +5.42 logit, and the minimum measure was −2.07 logit. The separation statistic for the respondents/person was 4.58, indicating that 106 respondents were generally separated into three groups. The respondents/persons mean +0.87 logit indicated that the respondents could agree on the most significant complexity factors, as shown in Fig. 1.

The locations for 106 respondents are on the left side of the vertical line, while the right side represents the 38 items. Overall, six (6) most significant pre-construction complexity factors affecting cost performance (PRECO) with logit measure ranges from −0.26 to −0.68 have been identified as shown in Table 6.

Based on analysis by construct for PRECO, six (6) most significant complexity factors are (1) Errors of original design (AC12: −0.68 logit), (2) Award to low bid that is qualified or non-compliant (AC27: −0.54 logit), (3) Redesign because over-budgeted (AC13: −0.44 logit), (4) Lack of optimization cost and time (AC6: −0.43 logit), (5) Lack of design coordination information between consultant and client (AC1: −0.29 logit), and (6) Unit prices not properly specified or evaluated (AC32: −0.26 logit).

In this study the first ranked complexity factor during pre-construction stage is errors of original design. It is critical to minimize errors of original design during pre-construction phase. Poor project design and implementation will result in cost overruns (Cantarelli et al., 2010). Therefore, there is a need to discover design errors earlier to avoid cost overruns in the pre-construction phase (Li & Taylor, 2014). The design and specifications are incorrect or are not provided to the implementing party, which can result in poor project quality (Rauzana et al., 2018). It demonstrates that errors in the original design are the most significant complexity factor affecting both cost and quality performance during the pre-construction phase.

The second ranked complexity factor affecting cost performance during pre-construction stage is award to low bid that is qualified or non-compliant. Previous studies show that bidding decision is the most important factor that affected both cost and time performance that were influenced by the behaviors of contractors bidding. Nevertheless, accepting the lowest price bid does not guarantee maximum value (Jarkas et al., 2014; Bedford, 2009). However, Minchin et al. (2013) stated that the

Table 6 List of six (6) most significant complexity factors affecting cost performance during pre-construction phase (PRECO)

Code	Description	Logit measure
AC12	Errors of original design	−0.68
AC27	Award to low bid that is qualified or non-compliant	−0.54
AC13	Redesign because over-budgeted	−0.44
AC6	Lack of optimization cost and time	−0.43
AC1	Lack of design coordination information between consultant and client	−0.29
AC32	Unit prices not properly specified or evaluated	−0.26

low bid method affects quality performance of the final product to decline if the incentives are aimed only at time reduction.

In this study, lack of optimization of cost and time factor has significantly affected both time and cost project performance. Thus, efficient design optimization, cost estimation, and scheduling must be carried out by different stake-holders from the design to the construction phase to see the information and analysis results using BIM for tunnel infrastructure projects (Sharafat et al., 2021).

5 Conclusion

Cost overrun is a severe problem commonly faced by Malaysia's construction industry, especially in infrastructure projects during the pre-construction stage. It resulted from various complexity factors which had been identified in this study. The literature reviews have shown current or lack of understanding of the term complexity in the construction industry, especially regarding infrastructure projects. Complexity appears to provide a prolific new understanding of the construction process. This study has been undertaken as part of the overall research project, which aims to develop a complexity model to identify the most significant factors contributing to complexity and develop a theoretical model of complexity assessment for infrastructure projects. This paper aimed to identify the complexity factor that affected performance in infrastructure projects during the pre-construction stage based on cost. The findings show that various important complexity factors during the pre-construction stage have affected the cost performance of the infrastructure projects. Identifying and understanding these factors affecting cost performance are vital to encouraging construction firms, agencies, and other construction players to ensure project success. Thus, a deeper understanding of the factors influencing project performance in early project development, specifically during the pre-construction stage, is needed to provide successful projects. Knowledge of the complexity factors that impact cost performance during the pre-construction stage and an awareness of their potential implications is the first step to mitigating prospective values. Proactive measures to limit cost impacts may be taken to mitigate factors that occur during the design, tendering, and contract award stages of project development. This paper only presents the most crucial complexity factors that affected performance in infrastructure projects during the pre-construction stage concerning cost performance based on the respondents' feedback. The applications can be derived through effective cost performance could contribute toward the sustainability of the overall system or project. Hence, further analysis shall also include other project performance indicators such as time and quality. Further study can also be carried out by considering the construction stage in infrastructure projects by developing and utilizing the findings of this study.

Acknowledgements We would like to thank the Faculty of Civil Engineering, UiTM, for giving us support. We are also grateful to the professionals and managers from Malaysian construction firms, CIDB Malaysia, and other institutions for their participation in this research.

References

Abdullah, M. R., Abdul Azis, A. A., & Abdul Rahman, I. (2009). Potential effects on large mara construction projects due to construction delay. *International Journal of Integrated Engineering, 1*(Civil and Environmental Engineering), 53–62.

Ainin, S., Kamarulzaman, Y., Farinda, A. G., & Che Azmi, A. (2010). Business and entrepreneur characteristics influence on business performance of professional small medium enterprises. In *Proceedings of the 5th European Conference on Entrepreneurship and Innovation, Univ. of Athens, Greece* (pp. 31–41).

Baccarini, D. (1996). The concept of project complexity—a review. *International Journal of Project Management, 14*(4), 201–204. https://doi.org/10.1016/0263-7863(95)00093-3

Baghaei, P. (2008). The Rasch model as a construct validation tool. *Transactions of the Rasch Measurement SIG American Educational Research Association, 22*(1), 1145–1162.

Bond, T. G., & Fox, C. M. (2007). *Applying the Rasch model: Fundamental measurement in the human sciences* (2nd ed.). Lawrence Erlbaum Associates Inc.

Bond, T. G., & Fox, C. M. (2001). *Applying the Rasch model: Fundamental measurement in the human sciences* (1st ed.). Lawrence Erlbaum Associates Inc.

Bedford, T. (2009). *Analysis of the low-bid award system in by sector construction procurement.*

Brockmann, C., & Girmscheid, G. (2007). Complexity of megaprojects. In *CIB world building congress* (pp. 219–230).

Bruzelius, N., Flyvbjerg, B., & Rothengatter, W. (2002). Big decisions, big risks. Improving accountability in mega projects. *Transport Policy, 9*(2), 143–154. https://doi.org/10.1016/S0967-070X(02)00014-8

Cantarelli, C. C., Flybjerg, B., Molin, E. J. E., & van Wee, B. (2010). Cost overruns in large-scale transport infrastructure projects. *European Journal of Transport and Infrastructure Research, 10*(1), 5–18.

Cardoso, J. (2006). Approaches to compute workflow complexity. In *Dagstuhl Seminar Proceedings: The Role of Business Processes in Service Oriented Architectures, Internationales Begegnungs- und Forschungszentrum Fuer Informatik (IBFI), 06291*(July), 16–21.

Enshassi, A., Mohamed, S., & Abushaban, S. (2009). Factors affecting the performance of construction projects in the Gaza strip. *Journal of Civil Engineering and Management, 15*(3), 269–280. https://doi.org/10.3846/1392-3730.2009.15.269-280

Fisher, W. J. (2007). Rating scale instrument quality criteria. *Rasch Measurement Transaction, 21*(1), 1095.

Flyvbjerg, B. B. (2009). Survival of the unfittest why the worst infrastructure gets built—and what we can do about it. *Oxford Review of Economic Policy, 25*(3), 344–67*.pdf*. February, 1–31.

Flyvbjerg, B., Skamris Holm, M. K., & Buhl, S. L. (2005). How (In)accurate are demand forecasts in public works projects?: The case of transportation. *Journal of the American Planning Association, 71*(2), 131–146. https://doi.org/10.1080/01944360508976688

Frimpong, Y., Oluwoye, J., & Crawford, L. (2003). Causes of delay and cost overruns in construction of groundwater projects in a developing countries; Ghana as a case study. *International Journal of Project Management, 21*(5).

Jarkas, A. M., Mubarak, S. A., & Kadri, C. Y. (2014). Critical factors determining bid/no bid decisions of contractors in Qatar, *30*(4). https://doi.org/10.1061/(ASCE)ME.1943-5479.0000223

Kamaruddeen, A. M., Sung, C. F., & Wahi, W. (2020). A study on factors causing cost overrun of construction projects in Sarawak, Malaysia. *Civil Engineering and Architecture, 8*(3), 191–199. https://doi.org/10.13189/cea.2020.080301

Krejcie, R. V., & Morgan, D. W. (1970). Determining sample size for research activities. *Educational and Psychological Measurement, 30*, 607–610. https://doi.org/10.1891/9780826138446.0006

Li, Y., & Taylor, T. R. B. (2014). Modeling the impact of design rework on transportation infrastructure construction project performance. *Journal of Construction Engineering and Management, 140*(9), 04014044. https://doi.org/10.1061/(asce)co.1943-7862.0000878

Luo, L., He, Q., Jaselskis, E. J., & Xie, J. (2017). Construction project complexity: Research trends and implications. *Journal of Construction Engineering and Management, 143*(7), 04017019. https://doi.org/10.1061/(asce)co.1943-7862.0001306

Masrom, M. A. N., Rahim, M. H. I. A., Mohamed, S., Chen, G. K., & Yunus, R. (2015). Successful criteria for large infrastructure projects in Malaysia. *Procedia Engineering, 125*, 143–149. https://doi.org/10.1016/j.proeng.2015.11.021

Maylor, H., Vidgen, R., & Carver, S. (2008). Managerial complexity in project-based operations: A grounded model and its implications for practice. *Project Management Journal, 39*, 15–26. https://doi.org/10.1002/pmj.20057

Mills, A. (2001). A systematic approach to risk management for construction. *Structural survey, 19*(5), 245–252.

Minchin, R. E., Thurn, S. B., Ellis, R. D., & Lewis, D. W. (2013) Using contractor bid amounts to estimate the impact of night construction on cost for transportation construction. *Journal of Construction Engineering and Management, 139*(8), 1055–1062. https://doi.org/10.1061/(asce)co.1943-7862.0000688

Olawale, Y. A., & Sun, M. (2010). Cost and time control of construction projects: Inhibiting factors and mitigating measures in practice. *Construction Management and Economics, 28*(5), 509–526. https://doi.org/10.1080/01446191003674519

Othman, N., Salleh, S. M., Hussin, H., & Wahid, H. A. (2014). Assessing construct validity and reliability of competitiveness scale using Rasch model approach. *WEI International Academic Conference Proceedings,* 113–120.

Prime Minister's, M. D. Department of Statistic Malaysia. (2022). Quarterly Construction Statistics, Fourth Quarter 2021.

Rahman, I. A., Memon, A. H., & Karim, A. T. A. (2013). Significant factors causing cost overruns in large construction projects in Malaysia. *Journal of Applied Sciences, 13*(2), 286–293. https://doi.org/10.3923/jas.2013.286.293

Ramayah, T., Lim, Y. C., & Mohamed, S. (2005). SME e-readiness in Malaysia: Implications for planning and implementation. *Sasin Journal of Management, 11*(1), 103–120.

Rauzana, A., Husin, S., & Usni, D. A. (2018). Critical factors affecting quality performance in construction projects. *International Journal of Management and Administrative Sciences (IJMAS), 6*(3), 13–21.

Saed, M. A., Yong, K., & Othman, M. (2018). Categorization of project complexity factors in Malaysian construction industry. *International Journal of Property Sciences, 7*(1), 1–15. https://doi.org/10.22452/ijps.vol7no1.3

Said, R. F. M. (2016). Application of Rasch measurement model in evaluating student performance for Foundation of Computing II. In C. Fook, G. Sidhu, S. Narasuman, L. Fong, & S. Abdul Rahman (Eds.), *7th international conference on university learning and teaching (InCULT 2014) proceedings*. Springer.

Sharafat, A., Khan, M. S., Latif, K., & Seo, J. (2021). BIM-based tunnel information modeling framework for visualization, management, and simulation of drill-and-blast tunneling projects. *Journal of Computing in Civil Engineering, 35*(2), 04020068. https://doi.org/10.1061/(asce)cp.1943-5487.0000955

Sinha, S., Kumar, B., & Thomson, A. (2006). Measuring project complexity: A project manager's tool. *Architectural Engineering and Design Management, 2*(3), 187–202. https://doi.org/10.1080/17452007.2006.9684615

Williams, T. M. (1999). The need for new paradigms for complex projects. *International Journal of Project Management, 17*(5), 269–273. https://doi.org/10.1016/S0263-7863(98)00047-7

Wood, H., & Gidado, K. (2008). Project complexity in construction. *Cobra 2008, January 2008*, 1–13. https://doi.org/10.1136/bmj.d1875

Scholten, A. Z. (2011). Admissible statistics from a latent variable perspective, Thesis, fully internal, Universiteit van Amsterdam.

Performance Measurement Criteria: Conceptual Framework for Subcontracting Management in the Malaysian Construction Supply Chain

Daniel L, Siti Hamidah Abdull Rahman, Che Maznah Mat Isa, Musmuliadi Kamaruding, and Fatin Najwa Mohd Nusa

Abstract The construction industry employs workers of all levels and roles, from professionals to skilled and unskilled labourers. Subcontracting is increasingly being used in the construction industry to transfer risk from the main contractor to the subcontractor, and it will have a positive impact on specialisation in construction work. However, failure in subcontracting management may result in a negative relationship between them, uncoordinated on-site supervision, a lack of quality, and improper planning and scheduling. This chapter focuses on the identification of performance measurement criteria that contribute to successful subcontracting management in Malaysian construction projects, impacting the three dimensions of project performance, namely time, cost, and quality. This research used mixed methods, which included both quantitative and qualitative non-experimental components. The perception of subcontracting of main contractors was sought through a questionnaire survey. Next, semi-structured interviews with highly experienced contractors were conducted to validate the quantitative findings and the proposed subcontracting management assessment framework. The key determinants

D. L · S. H. Abdull Rahman · M. Kamaruding · F. N. M. Nusa
School of Civil Engineering, College of Engineering, Universiti Teknologi MARA, 40450 Shah Alam, Selangor, Malaysia
e-mail: hamidahar@uitm.edu.my

M. Kamaruding
e-mail: musmuliadi@uitm.edu.my

F. N. M. Nusa
e-mail: fatinnajwa@uitm.edu.my

C. M. Mat Isa (✉)
Center of Civil Engineering Studies, Universiti Teknologi MARA Pulau Pinang Branch, Permatang Pauh Campus, 13500 Pulau Pinang, Malaysia
e-mail: chema982@uitm.edu.my

F. N. M. Nusa
Malaysian Institute of Transport, Universiti Teknologi MARA, 40450 Shah Alam, Selangor, Malaysia

U. R. Hashim et al. (eds.), *Green Infrastructure*,
https://doi.org/10.1007/978-981-99-7003-2_3

and the scales of subcontracting management were analysed using SEMPLS software. The established performance measurement criteria will provide a reference to drive continual improvement and break down silos of thinking by performance measurement criteria of subcontracting management domains.

Keywords Performance measurement criteria · Subcontracting · Construction project

1 Introduction

The construction industry has many tribulations because of the complex nature of the operation (Kim et al., 2021; Vo et al., 2021). This industry is comprised of many occupations, professions, and organisations. They are occupied in dissimilar phases of a construction project, which include the pre-construction phase, construction phase, and post-completion phase (Bavafa et al., 2018; Carvajal-Arango et al., 2019).

The parties involved in a construction project, such as the client, consultant, contractor, and subcontractors, have the responsibility for delivering a quality project. A breakdown of any of the parties will critically influence the quality of the ultimate project (Noorzai, 2022). In all projects, contractors must plan, manage, and monitor their work and that of their workers, as well as check the competence of all their appointees and workers, train their employees, provide information to their workers, and ensure that adequate welfare facilities are available for their workers (Adinyira et al., 2020). Therefore, it is particularly essential for main contractor-subcontractor relationships to develop long-term relationships and to enjoy their working relationship (Nuyen et al., 2018).

A similar trend can be seen in the Malaysian construction industry. The construction organisation plays a key role in the construction industry. It establishes buildings and infrastructure works required for social-economic development, which contributes to the overall economic growth. Therefore, the main contractor and subcontractor are becoming increasingly important in project construction; it is uncommon for subcontractors to handle up to 90% of a project (Florence & Khoo, 2016).

Based on previous research, some potential problems with constraints in subcontracting management have been identified. Poor financial management in this context refers to the subcontractor's inability to complete the assigned task within the allocated budget, as well as the inability to avoid penalties and damages, as well as an increase in overhead and operational costs (Co et al., 2018; Gunduz & Al-Naimi, 2021; Silva et al., 2017).

Low quality of work is also an issue that is concurrent with construction projects. This is due to the adversarial relationships that frequently occur between the construction project actors, which diverts the goal of delivering the best projects to a mere task that needs to be finalised as soon as possible solely for the sake of the pay (Co et al., 2018; Goh & Loosemore, 2017; Small et al., 2017; Tan et al., 2017).

Omotayo et al. (2022) conducted a study to investigate the causes and factors of distrust among the actors in a construction project involving subcontractors and contractors. The study shows that subcontractors have limited knowledge of the clauses in contracts into which they are entering. Hence, subcontractors should carefully understand their contractual obligations and payment arrangements (Omotayo et al., 2022).

The performance of the site is also determined by the safety and health practices of the subcontractor. However, due to the adverse influence on the safety and health of the subcontractors, the performance of the project may suffer (Che Ibrahim et al., 2022). This is due to the impact that safety and health will contribute to the minimisation of damage in organisations, which will prevent financial losses due to accidents (Adeyemo & Smallwood, 2017; Small et al., 2017).

Planning and scheduling remain the key elements of success in a construction project. This includes the selection of subcontractors during the initial stages, which will have a significant impact on the subcontracting performance (Biruk et al., 2017; Small et al., 2017). Without precise planning and scheduling, there will be a possible delay.

Construction projects are works that require ultimate cooperation from all of the actors in the project team as they deal with large-scale projects (Huang et al., 2008; Jastanyah & Sidawi, 2011; Tarzijan & Brahm, 2014). This is especially important in the communication between contractors and subcontractors, as the relationship between the parties will lead to the success of the project (Martin & Benson, 2021).

Based on studies conducted in the construction (Hidiroglu, 2019; Penaloza et al., 2020; Patrucco et al., 2020) and non-construction (Pedersen et al., 2021) fields related to performance measurement criteria, there is no specific framework established on subcontracting management. The previous research covered the adaptation of the new framework in measuring the project performance and examined the benefits of quality standards, safety performance measurement systems, and supplier performance measurement.

The awareness of the importance of subcontracting management in the construction industry is quite low (Florence & Khoo, 2016; Jelodar et al., 2016). This factor has become an obstacle to completing a successful construction project. The factors that influence the subcontracting management are especially important to be studied in order to determine the solutions or ways to improve the relationships. In addition, a complete and coherent process that can provide clear guidelines on how the framework of performance measurement criteria for subcontracting management can maximise the construction performance is still lacking in the body of literature (Hidiroglu, 2019; Penaloza et al., 2020). Thus, the proposed study will address the identified gaps in the extant literature on the performance measurement criteria for subcontracting management.

2 Literature Review

2.1 Definitions

Every party involved in the construction industry has responsibilities. The main parties are clients, developers, consultants, architects, contractors and subcontractors, and suppliers. According to Wood et al. (2002), the ability to build sustainable relationships is compulsory in industries where inter-organisational relationships are becoming a gradually important part of the business. To achieve sustainable relationships, each party has to develop from a low trust/low ethics level to a high trust/high ethics level in their relationship. They are referring to good language and appropriate actions which will develop metaphors, mutual aid, and trust in the inter-organisational relationships (Florence & Khoo, 2016; Yong & Mustaffa, 2012).

The Oxford Dictionary defined a relationship as "how two or more people or things are connected, or the state of being connected, or how two or more people or groups regard and behave towards each other" (Oxford Digital Dictionary, 2003). Both parties must deliver their work and behave in such a way that they benefit each other. In business, especially in the construction industry, a relationship known as the business relationship is a normal phenomenon when the interaction between the main contractor and subcontractor or another organisation has economic consequences in a single product transaction. The process of merging happens when the transaction between both business parties is affected by their previous and future dealings with each other (Co et al., 2018). Agreement and Conditions of PAM Sub-Contract (2006) defined that:

> Contractors means the party named in the Articles of Agreement and includes the Contractor's legal successors or personal representatives or any person to whom the rights and obligations of the Contractor have been transferred with the agreement of the Employer and
>
> Sub-contractor means the party named in the second part of the Articles of Agreement and includes the Sub-Contractor's legal successors or personal representatives or any person to whom the rights and obligations of the Sub-contractor have been transferred with the agreement of the Contractor.

The same document stated that a subcontract is entered between the parties as a contractor on one side and as a subcontract on the other side under Clause 27.0 of the Main Contract Condition (Agreement & Conditions of PAM Sub-Contract, 2006). Therefore, subcontractors play a very important role in the construction industry. Subcontractors are specialty contractors who are hired to execute tasks on a project (Co et al., 2018).

According to Small et al. (2017), subcontractors are vital for the successful completion of construction projects. A subcontractor is a construction organisation that has contracted with the main contractor to carry out some of the main contractor's work. In most cases, the main contractor will execute the basic operation and subcontract the remainder to a variety of specialty contractors (Small et al., 2017).

2.2 An Overview of the Construction Supply Chain

The construction supply chain typically involves different parties supplying materials and components in the high demand of construction activities (Co et al., 2018). Construction is a fragmented industry, with complex supply networks and unlimited opportunities for integration. Restructuring in the industry showed that the creation of the current situation, which includes problems such as low-skilled workers, poor and labour productivity, began in the mid-1970s (Hui & Tsang, 2006).

Akintan and Morledge (2013) proved that increased demand for partnering, and procurement can extend down the supply chain from the client and main contractors to subcontractors and suppliers. Regarding the argument, it will delimit opportunities for perfect collaboration of processes in the various chains and provide opportunities for the industry to accelerate its rate of improvement in terms of performance (Strategic Forum for Construction, 2002). However, Hui and Tsang (2006) stated that construction partnering sometimes has been criticised for lacking transparency. The supply chain extended in partnering and collaborating to reduce a lack of trust among supply chain partners. Main contractors frequently failed to share strategic information in terms of two-way communications, and often give poor responses to these partners, who are suppliers. The worst part is that the main contractor frequently made late payments to subcontractors and suppliers. Briscoe et al. (2001) also investigated the common barriers that caused poor collaboration between contractors and subcontractors.

Cheung et al. (2003) found that collaborative working is low in the construction industry, and no parties noted the industry's emphasis on relationship exchange. Childerhouse et al. (2003) have proved that the UK house building sector, defaulted by starting its business process from a poor level and responding late in the day to the compelling need for change, is capable of developing its supply chains to embrace the concept of integration. Similarly, Aouam and Kumar (2019) have a new approach by the main contractor to establish qualified partnering approaches between contractors and subcontractors to make it easy in the supply chain to produce significant improvement in collaborative working.

Lately, the landscape for construction development delivery has been rapidly changing, with an emphasis on partnering, joint ventures, public/private partnerships, and strategic alliances (Akintan & Morledge, 2013). One of the few studies was conducted by Akintan and Morledge (2013), who investigated the issues in subcontracting practice. The few studies conducted since their studies have investigated individual issues rather than subcontracting practice in its totality. This study was conducted to obtain not only the issues in subcontracting but also the readiness for collaboration between main contractors and subcontractors in the construction project.

2.3 A Review of the Performance Measurement Criteria in Construction and Non-construction Fields

There have been few studies conducted related to the performance measurement criteria in the construction and non-construction fields. Recently, a few studies in the construction field have been conducted on the impact of a systematic approach to supplier performance measurement on project performance (Patrucco et al., 2020), safety performance measurement system (Penaloza et al., 2020), and the quality of construction organisation by applying the EFQM excellence model on processes and customer stages (Hidiroglu, 2019). Based on these papers, they considered three dimensions of project performance, namely time, cost, and quality as dependent variables.

Based on Table 1, the most important independent variables related to PMC research are quality of work, customer satisfaction, and financial status, followed by production efficiency, utilisation of resources, market share/selection, planning and scheduling, and good implementation of HSE management. The least cited are the understanding of contract and good communication. Table 1 contains a summary of the studies.

Based on previous studies, a conceptual framework for determining the performance measurement criteria of subcontracting management in the Malaysian construction supply chain was proposed. A summary of aforementioned studies is shown in Table 2.

Although the majority of these studies made significant progress in assessing performance measurement criteria in various fields, they mainly attempt to reveal evidence of project performance elements and the factors contributing to the relationships. Despite extensive studies on construction firms, the performance measurement criteria for subcontracting management appear to be lacking.

This gap may be due to the complex nature of the industry process involving construction firms, as stated by Noorzai (2022), which are related to the nature of the industries, organisational culture, human factors, and other unquantifiable factors. This study provides an extended analysis framework on performance measurement criteria of subcontracting management to provide practical information at the subcontractor company level by improvising their performance and relationship in the construction supply chain.

2.4 The Issues in Subcontracting Management

The construction industry has a bad reputation for poor quality and services, poor safety, and a history of broken promises and sharp practices. Its earlier adversarial relationships often created zero-sum games for those involved, and poor performance on cost, delivery, quality, and time for its clients (Co et al., 2018; Co et al., 2018; Silva et al., 2017). The most frequently discussed institutional form of cooperative

Table 1 Performance measurement criteria in construction and non-construction fields

No	Authors	Focus of the study	C or NC	Dependent variables			Independent variables related to PMC research									
				C	Q	T	PE (4)	UR (4)	FS (5)	MS (4)	QW (6)	CS (6)	GC (2)	PS (4)	UC (3)	HSE (4)
1	Takim et al. (2003)	An approach includes financial and nonfinancial indicators across project phases as a framework for measuring construction project performance	C	√	√	√	√	√	√	√	√	√				√
2	Azfar et al. (2014)	New conceptual framework for measuring sustainable supply chain performance	NC	√	√	√	√	√	√	√	√	√		√		
3	Ali and Rahmat (2010)	The benefits of implementing the ISO 9000 standard for construction companies and the criteria used for measuring project performance	C	√	√	√				√	√	√		√	√	√

(continued)

Table 1 (continued)

No	Authors	Focus of the study	C or NC	Dependent variables			Independent variables related to PMC research									
				C	Q	T	PE (4)	UR (4)	FS (5)	MS (4)	QW (6)	CS (6)	GC (2)	PS (4)	UC (3)	HSE (4)
4	Grigoroudis et al. (2012)	Strategic performance measurement in a healthcare organisation: A multiple criteria approach based on balanced scorecard	NC	√	√			√	√			√	√			
5	Hidiroglu (2019)	The quality of construction organisation is improved by applying the EFQM excellence model to processes and customer stages	C	√	√	√	√		√		√	√	√	√	√	√

(continued)

Table 1 (continued)

No	Authors	Focus of the study	C or NC	Dependent variables			Independent variables related to PMC research									
				C	Q	T	PE (4)	UR (4)	FS (5)	MS (4)	QW (6)	CS (6)	GC (2)	PS (4)	UC (3)	HSE (4)
6	Penaloza et al. (2020)	The contribution of the safety performance measurement system	C	√	√	√	√	√	√		√			√		√
7	Patrucco et al., (2020)	The impact of a systematic approach to measuring supplier performance on project performance	C	√	√	√				√	√	√			√	

Legend Construction (C) or Non-Construction (NC), C = Cost, Q = Quality, T = Time, PE = Production Efficiency, UR = Utilisation of Resources, FS = Financial Status, MS = Market Share/ Selection, QW = Services Quality/Quality of Work, CS = Customer Satisfaction/Perceived Quality, GC = Good Communication, PS = Planning and Scheduling, UC = Understanding on Contract, HSE = Good Implementation of HSE Management

Table 2 Summary of previous studies on performance measurement criteria of subcontracting management

Dimensions	Sub-dimensions	Authors
Financial status	Financial viability, liquidity, profitability, issued bond values, market share, revenue growth, a decrease in debts, and financial savings	Grigoroundis et al. (2012), Hidiroglu (2019), Penaloza et al. (2020), Stakeholders et al. (2011)
Quality of work	Quality assessment, deliver quality construction project; implement quality; management system; offer a quality of service; consistent level of quality; quality standards; high-quality products/services; quality specifications; skills and experiences; quality of workmanship	Hidiroglu (2019), Patrucco et al. (2020), Penaloza et al. (2020), Nguyen et al. (2022)
Good communication	A communication plan; improve internal and external communications; good interrelationships can develop a good subcontracting networking; communication tools	Abbasian-Hosseini et al. (2017), Hidiroglu (2019), Faisol (2010)
Planning and scheduling	Project running smoothly as scheduled; process controls and development; administration activities; planning and control process	Ali and Rahmat (2010), Hidiroglu (2019), Penaloza et al. (2020), Stakeholders et al. (2011)
Understanding on contract	Specified as in contractual agreement; good contract administration; contract review; contract execution; contract ongoing; contractual arrangements	Ali and Rahmat (2010), Hidiroglu (2019), Patrucco et al. (2020), Nguyen et al. (2022)
Good implementation of HSE management	Measured through injury statistics; provide information on the status of safety and health management process; functional company; balance of needs; safety problems and remedial actions; best practices; safety and health programmes	Takim et al. (2003), Hidiroglu (2019), Dale et al. (2020), Penaloza et al. (2020)
Experience on field	Foresee possible problems; higher chance of survival; ability to prevent disputes, and proper risk assessment	Na Ayudha & Kunishima (2017), Silva et al. (2017), Konno (2019)

behaviour in construction is relationships. The essence of a relationship is a single source and long-term relationship (Yong & Mustaffa, 2012). The main problems in the main contractor and subcontractor relationship have been identified as financial management, quality of work, contractual issues, poor site management, improper planning and scheduling, and a lack of communication.

2.4.1 Financial Management

The contractor may face financial problems as a result of poor cost estimates, poor management, or payments delayed by the owner (Na Ayudhya & Kunishima, 2017). This may delay the progress payment given by the contractor to his subcontractor (Co et al., 2018). Getting timely payment is one of the most serious issues in building up a better long-term relationship between the contractor and his subcontractor (Co et al., 2018). There remains a lack of trust between them, and both parties are overtly suspicious in all business dealings. In some cases, the contractor is perceived as a poor paymaster, and this will complicate the relationship even more (Silva et al., 2017).

The contractor, or subcontractor(s), will plan their work schedule and payment to vendors and labourers according to the expected cash flow of the contract progress payments. The owner may delay payment to the contractor. The contractor will therefore be forced to delay payment to the subcontractor(s), disrupting his plans. In such a case, a conflict will develop between the two parties (Co et al., 2018). Silva et al. (2017) concluded that payment problems often bother subcontractors. The retain age withheld from the main contractor is a higher percentage than that withheld by the owner from the general contractor.

Studies by Na Ayudna and Kunishima (2017) stated that a delay in interim payment can cause an eruption of construction progress, which will cause a decline in quality and work performance. Such a case will interrupt the existing work schedule, in which purchasing of materials is disrupted causing errors in the planned supply chain. Final payments are often delayed considerably, often approaching or exceeding one year after project completion (Oti-Sarpong et al., 2022). Such practices of general contractors permit them to finance their operations to a significant degree with the funds earned by but not returned to the subcontractors.

Financial management plays a very important role in making sure a certain project runs smoothly. Failure to manage a company's finances will upset both the contractor and the subcontractor. Consequently, a delay in the completion of a project will take place.

2.4.2 Quality of Work

Once the main contractor has been awarded the construction contract, they will perform the main part of the construction work, while the subcontractor(s) will undertake the remainder of the work. If either of the two parties performs substandard work in this part of the construction work, it may affect the work standard of the other party, resulting in an interface problem between them. The more experienced the contractor, the fewer mistakes they are likely to make during construction. Parties with experience in the field can foresee possible problems and risks that might occur, allowing them to plan their objectives more effectively (Borvorn & Masahiko, 2017). If the contractor or any subcontractors make a mistake in the execution of his construction work, it may affect the work of the other parties. Consequently, it will create problems between the two parties (Konno, 2018; Co et al., 2018). Small et al. (2017) concluded that all interface problems are caused by deficient experiences, leading to poor flexibility in adapting to a new environment.

Clear work-drawings and specifications are important for the effective execution of the construction work. If the work-drawings or the specifications are incomplete or unclear, it will create problems in their interpretation, affecting the quality of the project and causing problems between the contractor and his subcontractor. Enshassi et al. (2012) found that incomplete drawings have an impact on construction productivity. Low productivity leads to interface problems between the main contractor and the subcontractor.

The owner may approve a revision on the detail drawings when there is a need to add, delete, or modify the original work-drawings and the specifications. If the revision affects the scope of work on that section of the work carried out by the subcontractor, a problem may arise between the contractor and his subcontractor over the cost of carrying out the work specified in the revision (Hidiroglu, 2019). Zhan et al. (2020) indicated that the drawings and specifications alteration during execution affect productivity.

The quality of work of a main contractor and a subcontractor mainly depends on the experience of an individual. The more experienced main contractor or subcontractor will lead the less experienced one. If this kind of leadership happens in the construction process, mistakes will slightly happen. After all, greater flexibility throughout the whole process can lead to increased productivity of work and the quality of work can be controlled.

2.4.3 Contractual Issues

For one reason or another, there may be legal disputes among the parties involved in the construction project. For example, a dispute between the owner and the contractor, or between the contractor and the subcontractors, or among the subcontractors. Such a dispute may affect the work performance of the contractor or subcontractor, resulting in a conflict between them (Co et al., 2018). Silva et al. (2017) stated that the construction industry in the United States is increasingly involved in legal disputes, and much has been written about alternative means of reaching agreement without resorting to the courts.

Improper project management by either the main contractor or his subcontractors may lead to waste in material and labour costs, consequently causing problems between the two parties (Co et al., 2018). In addition, a lack of identification of responsibilities and proper records of work carried out by the subcontractors may cause disarray and confusion on the site (Silva et al., 2017).

Aslam et al. (2019) identified several contracting problems that appeared during contract execution, such as "unclear details in the drawing," "incomplete contract," and "design change." These problems usually occur when the involved parties make or execute contracts. Many subcontracts are awarded without any formal discussion taking place between the contractor and the subcontractor. This may increase the probability of a conflict after construction work has begun (Co et al., 2018).

The subcontractor may skip or neglect implementing some conditions of the agreed contract between him and the general contractor. If the contractor becomes aware of this neglect, a dispute will arise between him and the subcontractor (Co et al., 2018). The contractual issues happen primarily because of disputes between contractors and subcontractors. Sometimes changes are made by one party without getting the mutual agreement of the other. The manipulation of the agreement will cause massive problems in their construction process.

2.4.4 Safety, Health, and Environment Management

Small et al. (2017) stated that if either the contractor or his subcontractor fails to implement safety regulations and standards on the work site, there may be an injury or a loss of life among the labourers. It is unclear who is to blame for this. Dale et al. (2020) mentioned that the injury rate among subcontractors and their workers increases on large and complex projects, particularly those requiring a large number of subcontractors.

According to the contracts, the subcontractor must submit shop drawings or sample materials to the contractor for approval. The contractor, due to ineffective management, may delay the approval of the submitted materials. In such a case, a problem may arise between the two parties regarding the delays in the execution of the work and who is to blame (Dale et al., 2021).

The availability of construction materials is very important for the continuity of the work progress. If there is any shortage of construction materials in the market that are needed by either the contractor or his subcontractor, it will delay the progress of the work, resulting in conflict between the two parties (Small et al., 2017). Aslam et al. (2019) indicated that material shortage affects productivity. The low productivity leads to interface problems between the main contractor and the subcontractor.

Tarzijan and Brahm (2014) stated that labourers are a fundamental resource needed to complete the construction project. To complete the work professionally and of good quality, one requires skilled labour. A shortage of this important resource will affect the quality of the finished work, which may not be acceptable to the owner, resulting in a conflict between the contractor and his subcontractor.

Aslam et al. (2019) indicated that a lack of labour experience affects productivity. The low productivity leads to interface problems between the main contractor and the subcontractor. Aslam et al. (2019) also stated that when a subcontractor finishes a section of his work, he must submit it for approval to the contractor, who will then submit it to the owner. The owner must approve this work before the subcontractor can proceed with the remainder of his work. This process is often lengthy, and the subcontractor may be held back from proceeding with his work. This kind of delay may cause a problem between the contractor and his subcontractor. Zhan et al. (2020) indicated that the inspection affects productivity. The low productivity leads to interface problems between the main contractor and the subcontractor.

The owner or contractor may designate restricted areas on the construction site, where only authorised personnel are allowed. The subcontractor personnel may find it very difficult to obtain authorisation to use the site, even if it is necessary for their work. On the other hand, the subcontractor may impose restrictions on contractor personnel visiting and inspecting his work. In these cases, access problems may arise between the contractor and his subcontractor (Manu et al., 2013).

Poor site management, such as unclear drawings, late approval, unavailability of materials, labour shortages, and a lack of labour experience, can create uncertainty for the construction project to be completed on time. Simply put, if the main contractor and subcontractor can manage the problem on-site properly, there will be no uncertain consequences.

2.4.5 Poor Planning and Scheduling

It is well known that the cost of both materials and labour represents the largest element of any construction project. If either the contractor or subcontractor makes a mistake in the cost estimation and pricing of both material and labour, or if the prices of either material or labour have escalated beyond their estimates, the contractor or subcontractor may incur a loss rather than a profit (Small et al., 2017).

It is possible that during the project schedule, there will be scheduling conflicts between the contractor and his subcontractor due to poor scheduling of the construction activity's interface. On top of that, schedule conflicts can also occur among the subcontractors. In this case, the contractor will interfere to resolve the conflict, which may cause problems between the two parties (Small et al., 2017). Many schedule delays occur because the contractors did not recognise, in detail, the sequences of a sub's operation and realistically reflect them in determining the project schedule. Conversely, a sub may not always recognise the needs of other subs that may conflict with their operations.

Once the construction project is awarded, its time duration is identified. The awarded contractor will schedule his construction activities and those of his subcontractor(s) to meet the identified project duration. If any party delays the execution of his scheduled construction activities, it will consequently delay the progress of the activities of the other party (Small et al., 2017). In general, a high degree of subcontracting leads to a high risk of delays, which leads to inefficiency in the construction industry (Al-Emad & Rahman, 2021).

Once the contract is signed between the owner and the contractor, the completion time of the project is defined and included in the contract. If the contractor, during the construction phase, is expected to be unable to meet the agreed-upon project time limit, the main contractor may blame his subcontractors, or vice versa. Consequently, a conflict may occur between the contractor and his subcontractors (Small et al., 2017).

Poor planning and scheduling will cause a delay in the completion of a certain project. The main contractor is in charge of the major planning. The planning must be comprehensive enough, especially for the sub-operation. The proper time duration must be given to each sub-operation. Sometimes the sub-operation will happen after the previous sub-operation has been completed. Delays in the previous sub-operation will affect the next sub-operation. Therefore, the planning and scheduling must be comprehensive and effective for everyone involved in the construction project.

2.4.6 Lack of Communication

Proper communication between the contractor and subcontractors is very important and crucial to the success of the project's completion (Gamiljj & Abd Rahman, 2021). Normal communication among the construction parties could be either verbal, such as face-to-face or phone calls, or written, such as normal mail, facsimile, or other means. Poor communication between the two parties may delay work progress (Lestari et al., 2020).

Sometimes, late orders and not allowing sufficient time for both preparation and execution of a project can build up pressure on subcontractors, resulting in products that are not of the highest quality, or even meet the desired requirements. Poorly communicated information by the contractor to the subcontractor may lead to incorrect pricing. Pressure is regularly applied by the contractor on the subcontractor to reduce prices while essential information is held back, making it almost impossible to allow for proper pricing and working (Gamiljj & Abd Rahman, 2021). Zhan et al. (2020) identified communication problems that could lead to serious inefficiencies, such as poor planning and scheduling and a lack of a management system that updates new information. He also emphasised how important coordination is to the quality of the project.

If the subcontractor performs substandard construction work or violates the contract agreement between him and the contractor, the contractor may interrupt or terminate his subcontractor's work. If the subcontractor does not agree with the action of the contractor, the subcontractor may sue the contractor in court. Consequently, lengthy litigation will develop, resulting in a conflict between the two parties (Tarzijan & Brahm, 2014).

A lack of proper communication exists when the main contractor gives wrong instructions, explicitly transmits them, or does not fully share information with the subcontractor. These can take the form of project objectives, milestones, and the urgency of the project. If this matter is not properly discussed, one party may drag the other to court. This conflict can be easily resolved if they sit together and discuss it professionally.

2.5 *Proposed Conceptual Framework*

In general, the literature review reveals that the empirical studies on how to improve subcontracting management in the construction supply chain are still lacking. Having conducted a literature review on the scenario of the construction supply chain and the problems in subcontracting management, a problem was identified as "there is the need to establish a framework of performance measurement criteria of subcontracting management in Malaysian construction supply chain."

Establishing such a framework will provide subcontracting management with a systematic structured approach to measure their performance. The value of the framework of performance measurement criteria of subcontracting management will be measured based on the performance of the project. Project performance is a function-dependent variable, which includes time, cost, and quality of the project.

Based on Table 2, there are seven dimensions/factors of performance measurement criteria listed. However, the experience of work dimension was not considered in the development of the framework because it has the same interpretation as the quality of work dimension. Therefore, six independent variables of the factors contributing to subcontracting management are good communication, proper planning and scheduling, good implementation of safety, health, and environmental management, understanding of the contract, financial management, and quality of work.

The following equation implies the research question of whether independent variables (Xi) can cause dependent variables (Y) in a hypothesised manner. Y (performance of the project) = F [X1 (good communication), X2 (proper planning and scheduling), X3 (good implementation of safety, health, and environmental management), X4 (understanding the contract), X5 (financial management), X6 (quality of work)]. The theoretical framework is illustrated.

Based on the literature review, the research proposes the following hypotheses and conceptual model, as shown in Fig. 1. The hypotheses for this research are listed below:

1. Good communication has a positive impact on subcontracting management performance.

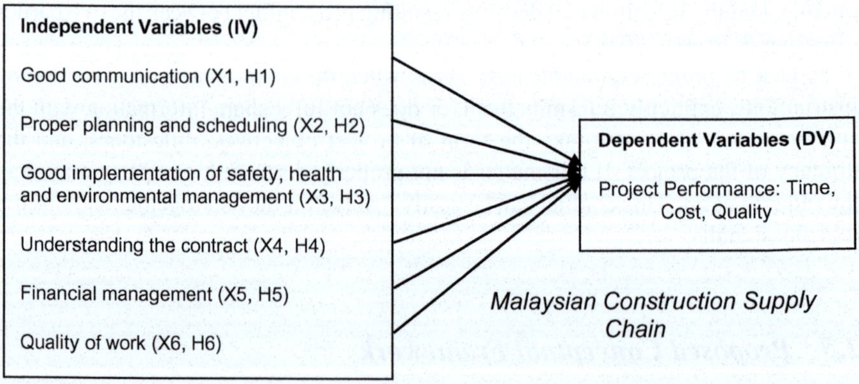

Fig. 1 Conceptual framework on performance measurement criteria of subcontracting management in Malaysian construction supply chain

2. Proper planning and scheduling have a positive impact on subcontracting management performance.
3. Good implementation of health, safety, and environment has a positive impact on subcontracting management performance.
4. Understanding of contract has a positive impact on subcontracting management performance.
5. Good financial management has a positive impact on subcontracting management performance.
6. Quality of work has a positive impact on subcontracting management performance.

3 Methodology

This study consists of six main phases as summarized in Fig. 1: (1) problem formulation; (2) determination of research objectives and scope; (3) building the theoretical framework; (4) data collection and data analysis; (5) establishment of a performance measurement criteria of subcontracting management framework; and (6) validation of research findings. Qualitative and quantitative research methods (mixed methods research) were adopted for this study. Starting with observations and theoretical insights derived from the literature, research gaps and objectives were first established, and then a theoretical framework was developed and verified as the research progressed.

3.1 Research Methods

The research methods applied include (1) literature review and content analysis; (2) questionnaire survey; (3) empirical; and (4) face-to-face semi-structured in-depth interviews. Figure 1 depicts the activities in the research process in each phase to fulfil the objectives of the study. The process was divided into six (6) phases, which are explained below:

Phase 1 focused on problem formulation that involved reviewing current knowledge and literature and identifying gaps in subcontracting management in Malaysia's construction supply chain.

Phase 2 involved the identification of objectives and defining the scope and limitations of the research related to the focus on subcontracting management in Malaysia's construction supply chain.

Phase 3 entailed developing a theoretical framework by identifying key determinants and determining relationships between determinants of subcontracting management in Malaysia's construction supply chain. A review of the literature was conducted on various fields related to subcontracting management problems, followed by an overview of construction supply chain issues and the factors contributing to good subcontracting management.

Phase 4 involved developing a research approach and instrumentation, justification for research paradigm, design, data collection methods, processes, and analysis. The theoretical review served as the basis of information for developing the preliminary survey questionnaire addressing the objectives of the research. The questionnaire, which formed the basis of this study, is based on the framework of the research area used to answer the research questions regarding subcontracting management in Malaysia's construction supply chain. The questionnaire was designed, refined, and reviewed by piloting the initial instrument and enhancing the instrument by carrying out the final check. Based on the feedback and reviews, the instrument was refined and improved. Changes were made in terms of the rewording and contents of the questionnaires. The results and comments from the pilot survey were analysed and assessed accordingly, to further enhance the scope and improve the research. The preliminary study on the key determinants of subcontracting management practices was carried out using supporting literature reviewed beforehand for validity and reliability purposes. Data from the validation process was used to statistically evaluate the psychometric properties of the questionnaire, which contain a factor analysis to guide suitable factor development and reliability of scales to demonstrate sufficient uniformity between individual key determinants. Information obtained in the preliminary study, as well as the literature review, were included in the development of the reliable questionnaire. Subsequently, a new questionnaire was developed to ensure that the assessment was measuring its intended dimensions and that the key determinants were being measured consistently.

Phase 5 involved analysing data and discussing results and findings. The findings from previous phases were consolidated and used to structure the framework. The use of a proprietary spreadsheet software was adopted, providing an automated way of collecting, retrieving, and graphically presenting data to assist in managing and monitoring the subcontracting management in Malaysia's construction supply chain.

Phase 6 entailed validating subcontracting management in Malaysia's construction supply chain framework as the final stage of the research process in order to test the reliability and practicability of the developed framework in construction projects. This involved the workability of the framework, the key determinants, its measurement and performance scales, suggestions for improvement, and so on. The interviewees were selected based on their agreement to serve as subject matter experts during the questionnaire survey phase. If only a small number of participants agree to be interviewed, a snowball sampling would have been used. Saunders et al. (2009) described that a snowball sampling emerges when the initial elements are selected and who identify further elements of the population. The latter identifies more elements, and so on, resulting in a snowball effect in the recruitment and size of the sample.

3.2 Data Analysis Techniques for the Questionnaire Survey

Various types of statistical analyses were performed on the data collected from the questionnaire, including descriptive analysis, mean ranking, and non-parametric and parametric tests, using the IBMSPSS window version 27.0. After the basic analysis was completed with IBMSPSS, the data was exported to other software to develop the framework.

This study used the Structural Equation Modelling (SEM) as a statistical technique. This statistical technique can be a powerful tool for developing the theoretical framework for main contractor and subcontractor management. The SEM is a statistical technique for concurrently testing and calculating approximate causal relationships among several independent and dependent constructs (Hair et al., 2011). Currently, there are two general approaches to SEM: (1) covariance-based structural equation modelling (CBSEM), as implemented in LISREL, AMOS, EQS, SEPATH, and RAMONA; and (2) is the PLS component-based approach. Figure 2 shows the research methodology framework for this study.

The PLS algorithm minimises the variance of all the dependent variables instead of explaining the covariation. Consequently, PLS places lower demands on measurement scales, sample size, and residual distributions (Hair et al., 2011). Therefore, for this research, the PLS approach was chosen because of its advantages over the covariance approach.

3.3 Data Analysis of the Interviews

The interviews were conducted in line with the guidelines established by Saunders et al. (2009). This guideline pertains to the appropriateness of the interviewer's appearance at the interview, approach to questioning, use of appropriate language, and other precautions are taken to ensure that the quality of the date, demonstration of interviewing competence, and avoidance of interviewer or interviewee bias. The interviews were conducted in one form which was face-to-face interviews.

The analysis of the qualitative data generated in the interviews was conducted using NVivo 10, which handles rich-text-based information and allows for deep levels of analysis on both small and large volumes of data. NVivo 10 provides a very good solution because it removes many of the manual tasks associated with the analysis, like classifying, sorting, and arranging information, giving the researcher more time to explore trends, build and test theories, and ultimately arrive at answers to questions.

It also has the capability of querying data with a powerful state-of-the-art search engine and graphically displays new ideas, connections, and findings in real time using frameworks. NVivo 10 is ideal for anyone who needs to examine or make sense of information. The software is used by researchers, academics, forensic scientists, psychologists, tourism managers, sociologists, and students around the world (NVivo

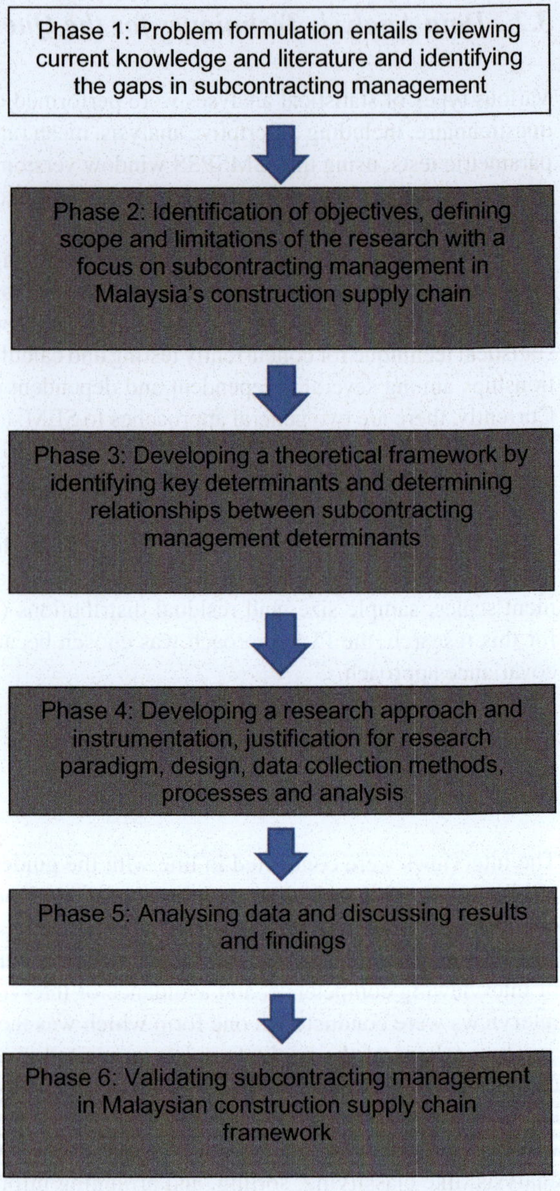

Fig. 2 Research methodology framework

10 for Windows, 2014). As a result of its efficiency, this research employed it as a qualitative data analysis tool.

4 Results

4.1 Findings

The development of Performance Measurement Criteria for Subcontracting Management in the Malaysian Construction Supply Chain is a study that has never been done before by researchers. In this regard, this study can fill the existing research gaps as well as serve as a reference in the construction supply chain, especially in the subcontracting management context. Some of the works that could lead to new findings include enhancing the understanding and knowledge of the actual ability of Performance Measure Criteria (PMC) of subcontracting management. Good communication, proper planning and scheduling, good implementation of health, safety, and environmental management, contract understanding, financial management, and quality of work all contribute to this framework and are important in subcontracting management. In addition, this framework also highlights valuable information for the Construction Industry Development Board (CIDB) and other construction players to improve the relationship between construction players and lead to successful projects.

5 Conclusion

The proposed assessment framework will enhance the understanding of the wider construction society and community on the importance of embracing the subcontracting management integration practice. The outcome of the study on the performance measurement criteria of subcontracting management in Malaysian construction supply chains will benefit the National Government and stakeholders, Construction Industry Development Board (CIDB), and the other construction players by increasing productivity, cost-efficiency, and risk-aversion. On the other hand, hiring an expert to work on one or more short-term projects can be beneficial in terms of cost, quality, and efficiency. Therefore, the findings of this research are relevant to the existing national construction supply chain policies: 12th Malaysia Plan (2021–2025), Construction 4.0 Strategic Plan (2021–2025), and 10–10 Malaysian Science, Technology, Innovation and Economy (MySTIE) Framework.

Acknowledgements Authors acknowledge the Ministry of Higher Education (MOHE) for funding under Fundamental Research Grant Scheme (FRGS) (FRGS/1/2021/TK01/UITM/02/2) and School of Civil Engineering, College of Engineering, Universiti Teknologi MARA (UiTM) for the continuous support throughout this project.

References

Abbasian-Hosseini, A. H., Liu, M., & Hsiang M. H. (2017), Social network conformity and construction work plan reliability. *Automation in Construction, 78*, 1–12. https://doi.org/10.1016/j.autcon.2017.01.004

Al-Emad, N., & Rahman, I. A. (2021). International journal of sustainable construction engineering and technology construction workers' issues from worldwide and Saudi Arabia studies. *International Journal of Sustainable Construction Engineering and Technology, 12*(4), 85–100. https://doi.org/10.30880/ijscet.2021.12.04.008

Ali, A. S., & Rahmat, I. (2010). The benefits of implementing the ISO 9000 standard for construction companies and the criteria used for measuring project performance. *Journal of Retail & Leisure Property, 9*, 25–35.

Adeyemo, O., & Smallwood, J. (2017). Impact of occupational health and safety legislation on performance improvement in the Nigerian construction industry. *Procedia Engineering, 196*, 785–791. https://doi.org/10.1016/j.proeng.2017.08.008

Adinyira, E., Agyekum, K., Danku, J. C., Addison, P., & Kukah, A. S. (2020). Influence of subcontractor risk management on quality performance of building construction projects in Ghana. *Journal of Construction in Developing Countries, 25*(2), 175–197. https://doi.org/10.21315/jcdc2020.25.2.7

Agreement and Conditions of PAM Sub-Contract. (2006). For use where the sub-contractor is nominated under the PAM contract 2006.

Akintan, O. A., & Morledge, R. (2013). Improving the collaboration between main contractors and subcontractors within traditional construction procurement. *Journal of Construction Engineering, 2013*, 1–11. https://doi.org/10.1155/2013/281236

Aouam, T., & Kumar, K. (2019). On the effect of overtime and subcontracting on supply chain safety stock. *Omega, 89*, 1–20.

Aslam, M., Baffoe-Twum, E., & Saleem, F. (2019). Design changes in construction projects—causes and impact on the cost. *Civil Engineering Journal, 5*(7), 1647–1655. https://doi.org/10.28991/cej-2019-03091360

Azfar, K. R. W., Khan, N., & Gabriel, H. F. (2014). Performance measurement: A conceptual framework for supply chain practices. *Procedia - Social and Behavioral Sciences. 150*, 803–812. https://doi.org/10.1016/j.sbspro.2014.09.089

Basu, A. K., Chau, N. H. & Soundararajan, V. (2019). Wage fairness in a subcontracted labor market. *Journal of Economic Behavior & Organization, 168*, 24–42. https://doi.org/10.1016/j.jebo.2019.09.022

Bavafa, A., Mahdiyar, A., & Marsono, A. K. (2018). Identifying and assessing the critical factors for effective implementation of safety programs in construction projects. *Safety Science, 106*, 47–56. https://doi.org/10.1016/j.ssci.2018.02.025

Biruk, S., Jaśkowski, P., & Czarnigowska, A. (2017). Minimizing project cost by integrating subcontractor selection decisions with scheduling. *IOP Conference Series: Materials Science and Engineering, 245*(7).https://doi.org/10.1088/1757-899X/245/7/072007

Camboni, R., Rondi, L. & Valbonesi, P. (2021). Temporary partnership and subcontracting: Pre-vs. post-award outsourcing in public procurement. *European Journal of Political Economy, 66*, 101950. https://doi.org/10.1016/j.ejpoleco.2020.101950

Carvajal-Arango, D., Bahamón-Jaramillo, S., Aristizábal-Monsalve, P., Vásquez-Hernández, A., & Botero, L. F. B. (2019). Relationships between lean and sustainable construction: Positive impacts of lean practices over sustainability during construction phase. *Journal of Cleaner Production, 234*, 1322–1337. https://doi.org/10.1016/j.jclepro.2019.05.216

Co, C. Y., Nguyen, T. K., Nguyen, T. N., & Tran, Q. N. (2018). The exporting and subcontracting decisions of Viet Nam's small and medium-sized enterprises. *International Review of Economics and Finance, 58*, 449–466.

Che Ibrahim, C. K. I., Manu, P., Belayutham, S., Mahamadu, A. M., & Antwi-Afari, M. F. (2022). Design for safety (DfS) practice in construction engineering and management research: A review of current trends and future directions. *Journal of Building Engineering, 52*. https://doi.org/10.1016/j.jobe.2022.104352

Cheung, S.-O., Ng, T. S., Wong, S.-P., & Suen, H. C. (2003). Behavioural aspects in construction partnering. *International Journal of Project Management, 21*(5), 333–343. https://doi.org/10.1016/S0263-7863(02)00052-2

Dale, A. M., Colvin, R., Barrera, M., Strickland, J. R., & Evanoff, B. A. (2020). The association between subcontractor safety management programs and worker perceived safety climate in commercial construction projects. *Journal of Safety Research, 74*, 279–288. https://doi.org/10.1016/j.jsr.2020.06.010

Dale, A. M., Barrera, M., Colvin, R., Strickland, J., & Evanoff, B. A. (2021). Flow-down of safety from general contractors to subcontractors working on commercial construction projects. *Safety Science, 142*. https://doi.org/10.1016/j.ssci.2021.105353

El-khalek, H. A., Aziz, R. F., & Morgan, E. S. (2019). Identification of construction subcontractor prequalification evaluation criteria and their impact on project success. *Alexandria Engineering Journal, 58*(1), 217–223. https://doi.org/10.1016/j.aej.2018.11.010

Enshassi, E., Arain, F., & Tayeh, B. (2012). Major causes of problems between contractors and subcontractors in Gaza Strip. *Journal of Financial Management of Property and Construction, 17*(1), 92–112.

Faisol, N. (2010). An investigation of relational contracting norms in construction projects in Malaysia.

Florence, Y. Y., & Khoo, L. W. K. (2016). Improving relationships in project teams in Malaysia. *Built Environment Project and Asset Management*, 6(3), 284–301.

Gamiljj, Y., & Abd Rahman, I. (2021). Studying the relationship between causes and effects of poor communication in construction projects using PLS-SEM approach. *Journal of Facilities Management*. https://doi.org/10.1108/JFM-04-2021-0039

Grigoroudis, et al. (2012, 2017), Strategic performance measurement in a healthcare organisation: A multiple criteria approach based on balanced scorecard. *Omega. 40,* (1), 104–119. https://doi.org/10.1016/j.omega.2011.04.001

Goh, E., & Loosemore, M. (2017). The impacts of industrialization on construction subcontractors: A resource-based view. *Construction Management and Economics, 35*(5), 288–304. https://doi.org/10.1080/01446193.2016.1253856

Gunduz, M., & Al-Naimi, N. H. (2021). Construction projects delay mitigation using integrated balanced scorecard and quality function deployment. *Engineering, Construction and Architectural Management*.https://doi.org/10.1108/ECAM-12-2020-1082

Hidiroglu, D. (2019). Self-assessment performance measurement in construction companies: An application of the EFQM Excellence Model on processes and customer stages. *Procedia Computer Science, 158*, 844–851.

Hair, J. F., Sarstedt, M. C., Ringle, M., & Mena, J. A. (2011). An assessment of the use of partial least squares structural equation modelling in marketing research. *Journal of the Academy of Marketing Science, 40*(3), 414–433.

Huang, R. Y., Huang, C. T., Lin, H., & Ku, W. H. (2008). Factor analysis of interface problems among construction parties a case study of MRT. *Journal of Marine Science and Technology, 16*(1), 52–63.

Hui, E. Y. Y., & Tsang, A. H. C. (2006). The inter-organizational relationship in a multi-contractor business network.https://doi.org/10.1108/13552510610685093

Jastanyah, N., & Sidawi, B. (2011). The dilemma of communications and management of remote construction projects in the kingdom of Saudi Arabia (2000, pp. 395–406).

Jelodar, M. B., Yiu, T. W., & Wilkinson, S. (2016). A conceptualization of relationship quality in construction procurement. *Internation Al Journal of Project Management, 34*(6), 997–1011.

Kim, D., Oh, W., Yun, J., Youn, J., Do, S., & Lee, D. (2021). Development of key performance indicators for measuring the management performance of small construction firms in Korea. *Sustainability (Switzerland), 13*(11). https://doi.org/10.3390/su13116166

Konno, Y. (2018). Relationship between construction performance evaluation and contractor characteristics in Japan. *Cogent Business and Management, 5*(1), 1–10. https://doi.org/10.1080/23311975.2018.1486169

Lestari, F., Sunindijo, R. Y., Loosemore, M., Kusminanti, Y., & Widanarko, B. (2020). A safety climate framework for improving health and safety in the Indonesian construction industry. *International Journal of Environmental Research and Public Health, 17*(20), 1–20. https://doi.org/10.3390/ijerph17207462

Manu, P., Ankrah, N., Proverbs, D., & Suresh, S. (2013). Mitigation the health and safety influence of subcontracting in construction: The approach of main contractors. *International Journal of Project Management, 31*, 1017–1026.

Martin, L., & Benson, L. (2021). Relationship quality in construction projects: A subcontractor perspective of principal contractor relationships. *International Journal of Project Management, 39*(6), 633–645. https://doi.org/10.1016/j.ijproman.2021.05.002

Noorzai, E. (2022). Evaluating lean techniques to improve success factors in the construction phase. *Construction Innovation.* https://doi.org/10.1108/CI-05-2021-0102

Na Ayudhya, B. I., & Kunishima, M. (2017). Risks of Abandonment in Residential Projects Caused by Subcontractors. *Procedia Computer Science, 121*, 232–237. https://doi.org/10.1016/j.procs.2017.11.032

Nguyen, D. T., Le-Hoai, L., Basenda Tarigan, P., & Tran, D. H. (2022). Tradeoff time cost quality in repetitive construction project using fuzzy logic approach and symbiotic organism search algorithm. *Alexandria Engineering Journal, 61*(2), 1499–1518.https://doi.org/10.1016/j.aej.2021.06.058

NVivo 10 for Windows. (2014). http://www.qsrinternational.com/products_nvivo.aspx

Omotayo, T. S., Danvers-Watson, O., & Oyegoke, A. S. (2022). Subcontractor trust issues on payment and valuation practices in UK private projects. *Journal of Financial Management of Property and Construction.* https://doi.org/10.1108/JFMPC-07-2021-0048

Oti-Sarpong, K., Pärn, E. A., Burgess, G., & Zaki, M. (2022). Transforming the construction sector: An institutional complexity perspective. *Construction Innovation, 22*(2), 361–387. https://doi.org/10.1108/CI-04-2021-0071

Patrucco, A. S., Moretto, A., & Knight, L. (2020). Does relationship control hinder relationship commitment? The role of supplier performance measurement systems in construction infrastructure projects. *International Journal Production Economics.*

Pedersen, A. L., Utkilen, S., Wondimu, P., & Torp, O. (2021). The risk of involving foreign contractors in the Norwegian infrastructure market. *Procedia Computer Science, 181*, 1196–1206. https://doi.org/10.1016/j.procs.2021.01.317

Penaloza, G. A., Saurin, T. A., & Formoso, C. T. (2020). Monitoring complexity and resilience in construction projects: The contribution of safety performance measurement systems. *Applied Ergonomics, 82*, 102978.

Silva, D. G., Kosmopoulou, G., & Lamarche, C. (2017). Subcontracting and the survival of plants in the road construction industry: A panel quantile regression analysis. *Journal of Economic Behavior & Organization, 137*, 113–131.

Small, E. P., Ayyash, L., & Al, H. K. (2017). Benchmarking performance of TQM principals in electrical subcontracting in Dubai: A case study. *Procedia Engineering, 196*, 622–629.

Stakeholders, M., Saleem, F., & Umer, M. (2011). Financial health of the construction industry of Pakistan—a perspective of major Stakeholders, 61–68.

Saunders, M., Lewis, P., & Thornhill, A. (2009). Research methods for business students.

Tan, Y., Xue, B., & Cheung, Y. T. (2017). Relationships between main contractors and subcontractors and their impacts on main contractor competitiveness: An empirical study in Hong Kong. *Journal of Construction Engineering and Management, 143*(7), 05017007. https://doi.org/10.1061/(asce)co.1943-7862.0001311

Tarzijan, J., & Brahm, F. (2014). Subcontracting in project-based firms: Do you follow the same pattern across your different projects? *International Journal of Project Management, 32*, 995–1006.

Takim, R., Akintoye, A., & Kelly, J. (2003). Performance measurement systems in construction. In D. J. Greenwood, (Ed.), 19th Annual ARCOM Conference, 3–5 September 2003, University of Brighton. *Association of Researchers in Construction Management, 1,* 423–32.

Vo, K. D., Pham, C. P., Phan, P. T., Vu, N. B., Duong, M. T. H., Le, L. P., & Nguyen, Q. L. H. T. T. (2021). Critical factors of subcontractor evaluation and selection: A case study in Vietnam. *Journal of Asian Finance, Economics and Business*, *8*(3), 297–305. https://doi.org/10.13106/jafeb.2021.vol8.no3.0297

Wood, G., McDermott, P., & Swan, W. (2002). The ethical benefits of trust-based partnering: the example of the construction industry. *Business Ethics: A European Review, 11*(1), 1–13.

Yong, Y. C., & Mustaffa, N. E. (2012). Analysis of factors critical to construction project success in Malaysia. *Engineering, Construction and Architectural Management*, *19*, 543–556.

Zhan, W., Pan, W., & Chen, L. (2020). Construction project productivity evaluation framework with expanded system boundaries. *Engineering, Construction and Architectural Management, 28*(4), 863–885. https://doi.org/10.1108/ECAM-12-2019-0691

Building Information Modelling Implementation Framework (BIMIF) for Government Building Construction Among Civil and Structural Engineering Consultants in Malaysia

Mohd Rashid Ya'acob, Che Maznah Mat Isa, Siti Hamidah Abdull Rahman, and Salmaliza Salleh

Abstract Implementing Building Information Modelling (BIM) into part of the design process is beneficial for civil and structural (C&S) engineering consultants involved in construction projects to ensure the project completion is within the stipulated time, cost, and quality of product delivery. However, there is still a lack of a systematic BIM process in Malaysia. Recent studies indicate that the BIM process has significantly influenced BIM implementation alongside other factors. This paper presents the initial stage of establishing a BIM implementation framework (BIMIF) for government building construction projects based on the conventional contract approach and focuses on developing the BIM process from a chosen case study. With the availability of BIMIF, the C&S consultants will have clear guidelines on the BIM process in government projects using conventional contracts to improve BIM implementation in Malaysia.

Keywords BIM · Civil and structural engineering consultants · Government building construction

1 Background of Study

Incorporation of Building Information Modelling (BIM) into part of the design process is beneficial for engineering consultants who are involved in construction projects in terms of time consumed, cost estimation, and quality of product delivery. The digitalization of the design process in BIM allows all information on design

M. R. Ya'acob · C. M. Mat Isa (✉) · S. H. Abdull Rahman
School of Civil Engineering, College of Engineering, Universiti Teknologi MARA, Shah Alam, Selangor, Malaysia
e-mail: chema982@uitm.edu.my

S. Salleh
MTC Engineering Consultancy Sdn. Bhd., Petaling Jaya, Selangor, Malaysia

U. R. Hashim et al. (eds.), *Green Infrastructure*,
https://doi.org/10.1007/978-981-99-7003-2_4

work to be stored as part of project deliverables in three-dimensional (3D) modeling. Therefore, any collaboration work between stakeholders in projects can easily be discussed, coordinated, and monitored through a real-time online system in Common Data Environment (CDE) such as BIM360 by Autodesk and eCOMs by *Jabatan Kerja Raya* (JKR) Malaysia.

BIM in the Malaysian context is defined as a modeling technology and associated set of processes to produce, communicate, analyze, and use digital information models throughout the construction project life cycle (CIBDB, 2018). The BIM process will involve all phases in a construction project from planning, design, construction, and maintenance until the building or infrastructure is proposed for renovation, rehabilitation, or demolition. BIM has been introduced in line with Construction Industrial Development Board (CIBDB) Malaysia strategic plan for 5 years from 2015 to 2020 under Construction Industry Transformation Programme (CITP) to raise productivity levels across the Malaysian construction industry.

Coherent to the inspiration of the Malaysian Government to promote Industrial Revolution (IR 4.0) in the country, digitalization become one of the main contributions toward the modernization of construction. CIDB continues the programme with new strategic 5 years plan in Construction 4.0 Strategic Plan (CSP) starting from 2021 until 2025 with objectives to improve the construction industry performance, reduce the negative impacts on the environment and promote high-income jobs for Malaysians.

CIDB is responsible for driving the implementation of BIM toward broader adoption of BIM by 2020 (CIBDB, 2013). According to Malaysia BIM Roadmap, by the year 2020, construction projects must integrate BIM technology into the construction industry in Malaysia. CIDB has recommended the mandatory use of BIM in private-sector projects by 2020.

BIM Implementation started in Malaysia's government project in the year 2007 through the Design and Build (D&B) contract for the National Cancer Institute building construction at Putrajaya which is owned by *Kementerian Kesihatan Malaysia* and implemented by JKR. Based on the success of this project, JKR as the main technical government department in Malaysia widened the implementation of BIM in other types of contracts including conventional consultant contracts. JKR, through its Strategic Plan 2021–2025 has set the adoption of the BIM mechanism to reach 50% by the year 2021 and 80% by 2025. JKR will ensure that 50% of the government projects worth RM10 million and above will use the mechanism, with a 10% increase rate in the subsequent years before 2025 (JKR, 2021).

However, Zahrizan et al. (2013) reported that Malaysian construction industry players are having difficulties implementing BIM because they do not know where1, when, and how to start as there is no national BIM standard and guideline for them to follow. In addition, Yasser and Idris (2020) conducted a survey that revealed that most construction companies lack awareness about BIM technology in Malaysia. The research by Idris and Al-Ashmori (2021) from 268 responses received revealed that only 13% of the participants from both public and private sectors are using BIM in their organization and this is a negative sign where Malaysia is still far away from the position it is supposed to be in the CIDB BIM Roadmap. The research listed a lack of awareness, implementation costs, slow adaptation, and unavailability of a clear guideline to assist organizations and policymakers toward BIM implementation as causes of the slow implementation of BIM in Malaysia.

Based on the findings from the above research, it was found that the usage of BIM in Malaysia is still low. It is expected that engineering consultants such as Civil and Structural, Mechanical, and Electrical Firms who are currently using conventional work processes will face more challenges in using BIM compared to Architects Firms where they have been comfortably practicing 3D modeling in their design processes.

Hence, this study decided to focus on Civil and Structural (C&S) engineering consultation because it is one of the important scopes for development projects as they carry out design work and approval submission to authorities. It involves complex processes which include building structures and external works such as earthwork, roads, and drainage which existence of the BIM framework will help to ease the process.

2 Literature Review

2.1 Introduction

Unit Perancang Ekonomi (UPE, 2021), *Jabatan Perdana Menteri* reported in 2021 the total achievement Gross Domestic Product (GDP) in Rancangan Malaysia ke-11 (RMKe11) from 2016 until 2020 contributed about 2.7% average rate yearly. However, economic growth for the construction industry decreased by 0.7% yearly due to Covid-19 pandemic in 2020. The government of Malaysia will continue 5 years plan from 2021 until 2025 in *Rancangan Malaysia ke-12* (RMKe12), which is the first part of the implementation of *Wawasan Kemakmuran Bersama* 2030 (WKB30). In this plan, the Government has projected that the country's GDP will rise between 4.5% and 5.5% yearly compared to RMKe11 GDP.

In this regard, the transformation in the construction industry needs to be further enhanced to help economic growth in the construction sector effectively. Construction Industrial Transformation Program (CITP) 2016–2020 under CIDB has been set to be the national agenda to transform the construction sector to become more sustainable and cost-effective. BIM has been introduced in construction as part of this CITP transformation as one of the beneficial technologies in the industry. BIM is also one of the 12 important technologies that have been listed in the Revolution of Construction 4.0 to enhance current and future technologies for the construction industry to achieve higher productivity, better safety, and toward a more sustainable approach—incorporating whole life cycle analysis (UBIM, 2021). CIDB set up MyBIM in 2017 to take the role as the main reference body for the construction industry to provide support services and training skills to expedite BIM acceptance in Malaysia. With the introduction of BIM in the construction industry, it is reported that the overall productivity of labor in 2020 was 1.1% and this result is projected to increase by 3.6% in RMKe12 (UPE, 2021). In addition, CIBDB (2020a, 2020b) reported that construction productivity increased by 60% in CITP and will continue the success with the new 5 years CIDB Strategic Plan 2021–2025 in Construction 4.0 Strategic Plan (CSP).

Meanwhile, JKR as the biggest technical department in a government agency has the responsibility to ensure successful construction of government projects for building and infrastructure in Malaysia. According to *Arahan Perbendaharaan* 182 (MOF, 2008), all other non-technical departments should get services from JKR or *Jabatan Pengairan dan Saliran* (JPS) only to execute the development project. In line with the Industrial Revolution 4.0 (IR 4.0), JKR has started BIM systematically since 2007 to transform construction projects in JKR with technology for better product delivery value. Therefore, in JKR Strategic Plan 2021–2025 (JKR, 2021), the body has already set the adoption of BIM in projects started with a value of RM 10 million above with 50% of total projects in RMKe12 will implement BIM and predicted to increase by 10% yearly.

2.2 Registration of Engineering Consultant in Malaysia

Professional engineering organizations are the primary channels by which engineers working in various technical disciplines, or otherwise possessing common interests, share technical knowledge, regulate professional practice, influence public policy, and maintain the traditions and reputation of the profession (Encyclopedia.com, 2022). In Malaysia, the registration of Engineering Consultants is handled by the Board of Engineers Malaysia (BEM).

BEM is the statutory body that oversees engineers and consulting engineering practices in Malaysia. It was formed on 23 August 1972 under the Registration of Engineers Act 1967. However, in 2015 the Registration of Engineers Act 1967 and the Registration of Engineers Regulations 1990 have been amended and known as the Registration of Engineers Act 1967 (Revised 2015) and Registration of Engineers Regulations 1990 (Revised 2015).

The primary role of BEM is to facilitate the registration of Engineers, Engineering Technologists, Inspectors of Works, Sole Proprietorships, Partnerships, and Bodies Corporate providing professional engineering services and to regulate the professional conduct and practice of registered persons in order to safeguard the safety and interest of the public (BEM, 2022).

According to BEM annual report (BEM, 2020), 3930 engineering consultancy practices were registered in the year 2020 by distinct categories which were Body Corporate, Body Corporate with Multi-Disciplinary Practice, Partnership, and Sole Proprietorship as shown in Table 1.

In 2022, the total sum of engineering consultancy practices involved in the Civil and Structural field was 1905 companies based on BEM registration directory data (BEM, 2022) as shown in Table 2.

Table 1 BEM engineering consultancy registration

Category	The year 2020
Body corporate	1621
Body corporate with multi-disciplinary	69
Partnership	356
Sole proprietorship	1884
Total	**3930**

Table 2 Civil and structure engineering consultants registered with BEM

Category	Civil	Structural	Total
Body corporate	832	0	832
Multi-discipline	44	0	44
Sole proprietorship	935	5	960
Partnerships	89	0	89
Total	**1900**	**5**	**1905**

3 Civil and Structural Engineering BIM Process

This study decided to select C&S engineering consultant conventional contract in government projects because currently only 5 out of 20 BIM projects in RMKe11 has been executed in conventional contract compared to 15 projects using D&B contract due to unclear BIM process and guidelines.

The BIM method was successfully implemented in the D&B contract due to BIM processes being led and done by the Main Contractor from the beginning of a project until completion. The D&B contractor is responsible for the conceptual design Level of Details (LOD) 100 until it is published as an as-built model LOD500. Furthermore, the D&B contractor was aware of any BIM cost and can incorporate it into the project cost. However, no additional BIM cost is allowed in the consultant fee structure in a conventional contract. The lack of skills, experience, and no clear BIM process and framework has caused C&S engineering consultants to appoint external modelers instead of doing the BIM themselves, which will result in the reduction of consultant fees. Therefore, this research will focus on developing a structural framework and process for C&S engineering consultants to implement BIM which will be delivered as conventional tender in a government project. The study will cover the planning stage until handing over the project to the client which involved the appointment of consultant engineering services stages based on the Consultancy Service Agreement (CSA) by the *Kementerian Kewangan Malaysia* (KKM). The stages are Preliminary, Design, Tender, Construction, and Defect Liability Period.

4 Introduction of Case Study A

Pusat Pendaftaran Rekod Negara (PRPN) construction was selected as Case Study A. It is one of the 5 JKR's BIM conventional contract projects selected for this study by semi-structured interview. It is also the BIM pioneer project implemented by conventional contract by the consultant. Apart from that, Case Study A has gone through all the project phases as stipulated in the CSA except the Defect Liability Period. In this regard, the information received from this project is helpful to be the main reference source for the proposed BIMIF and C&S BIM process.

4.1 Pusat Pendaftaran Rekod Negara

The *Bahagian Pusat Pengurusan Rekod (BPRN)* is one of the units at the *Jabatan Pendaftaran Negara (JPN)* headquarters under the *Kementerian Dalam Negeri*. This special BRPN building aims to provide appropriate and quality structural, security, space, and equipment requirements to manage the storage, maintenance, transfer, and reproduction of documents that include birth, death, adoption, identity card, citizenship, and marriage registration records or divorce (non-Muslims) of the Malaysian population. The records are stored in the form of physical documents, microfilms, and digital records centrally. The development of the *Pusat Pendaftaran Rekod Negara* (PRPN) construction project under the JPN is in Lot PT 12,012, Mukim Dengkil, Sepang, Selangor. This project has received approval under the provisions of the Second Rolling Plan (RP2) of the RMKe11 and has been issued by the JKR. The JKR is the project manager and implementer of the PRPN construction project until it is completed. The development of the PRPN is 3.19 acres. The PRPN building is 7 stories high and consists of an office building, repository, and parking lot and is a pioneer BIM project executed in a conventional contract by JKR illustrated in a 3D model as shown in Fig. 1.

5 Building Information Modelling Implementation Framework (BIMIF) and Process in Case Study A

5.1 1Initial Stage of BIMIF

Due to the limitation of related studies of the BIM Framework for Civil and Structural Engineering Consultants, this study came out with a conceptual framework as the initial stage of BIMIF as shown in Fig. 2 for the implementation of C&S Engineering BIM in conventional contracts. The initial stage of BIMIF is based on the level of work of the engineering consultant in the development project from planning, design, tender, and construction to the period of defect liability. Based on previous studies,

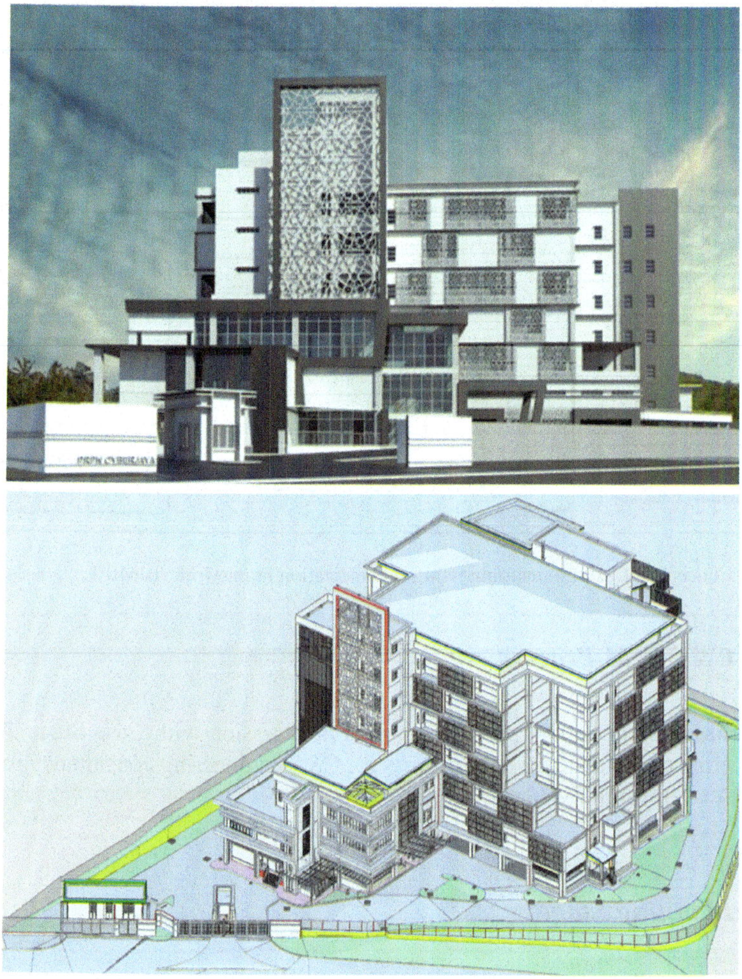

Fig. 1 3D model PRPN building

no existing framework has incorporated BIM into existing work processes by any engineering consultant. Therefore, this framework will incorporate BIM elements in all current C&S engineering work processes by integrating employer information requirements, multidisciplinary model coordination, design review, model level of detail, model integrity, and submission to local authorities and model reviewers.

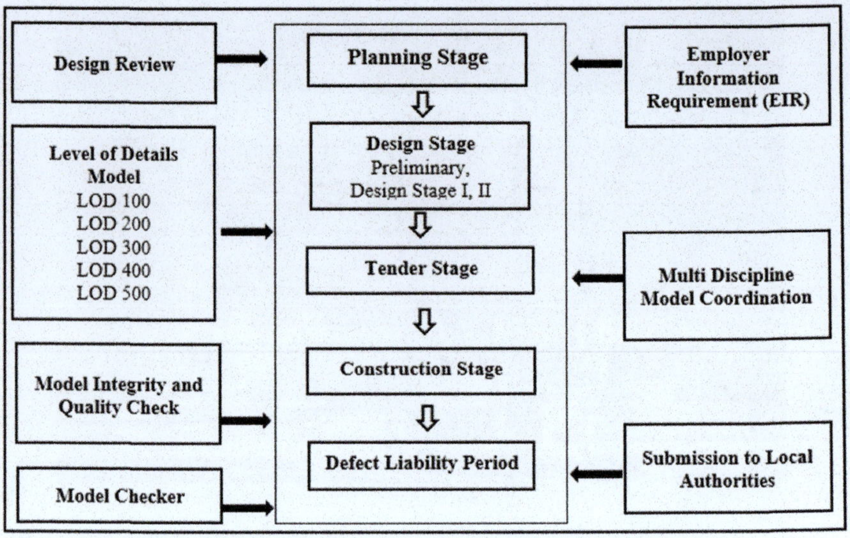

Fig. 2 Conceptual C&S Engineering BIM implementation Framework (BIMIF)

5.2 *C&S BIM Process*

Based on observation and preliminary interview session with Case Study A C&S engineering consultant, the BIM process for C&S engineering consultancy for Case Study A is shown in Fig. 3.

5.3 *Planning Stage*

The most important information needed to implement a project in a conventional BIM consultant contract is the Employer Information Requirements (EIR) document from the client/ project implementer. EIR is an important document to ensure that the project has met the requirements of BIM where the final version of this model will be submitted to the client. EIR should be a mandatory BIM document that needs to be included in the tender document and terms of reference in the consultant's contract agreement. In Case Study A, there are no specific EIR nor BIM requirements guidelines for conventional consultant contracts other than the terms of reference on model quality. However, based on input from the C&S engineering consultant, Case Study A was using the JKR Building Information Modeling (BIM) Requirements for Design & Build Projects (UBIM, 2021) as the unofficial EIR as follows:

- BIM Objective
- BIM Deliverables
- Platform and Software

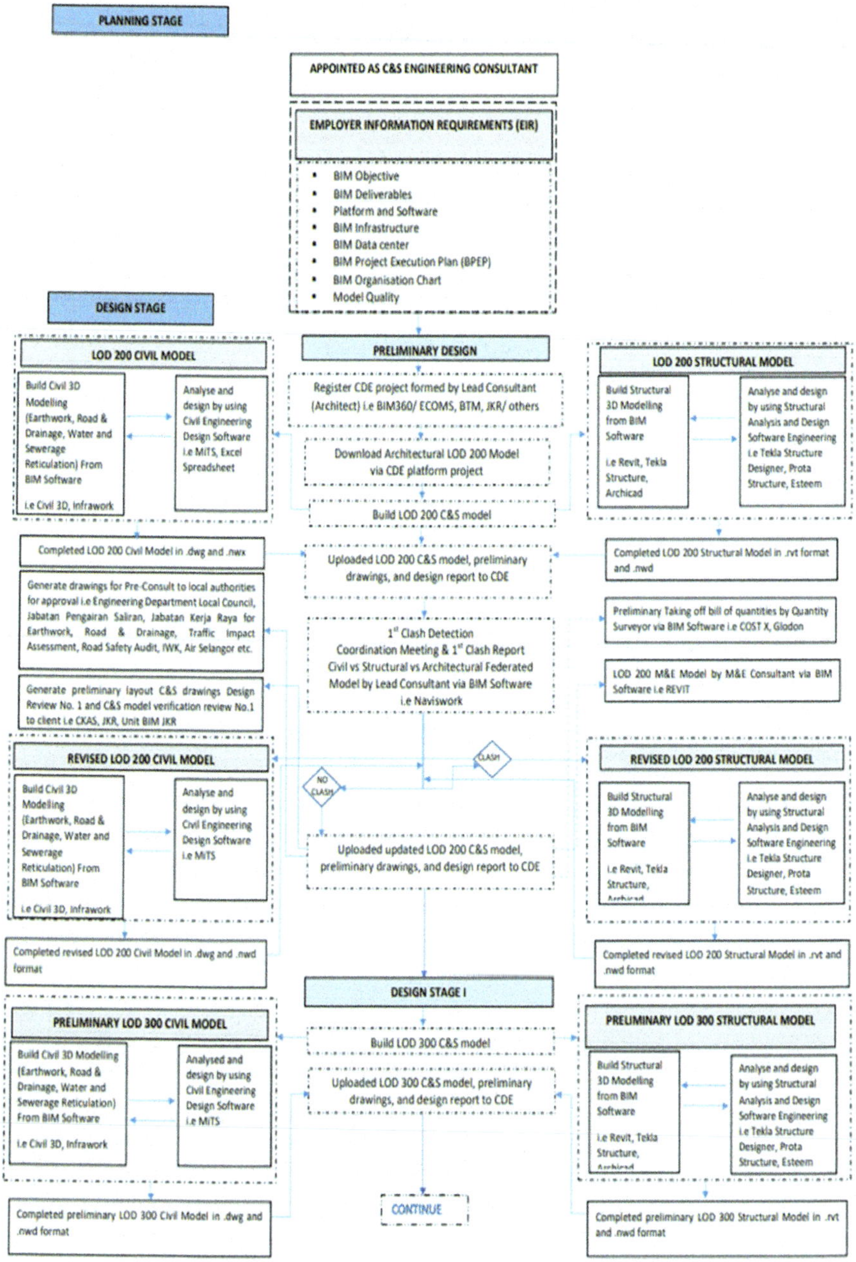

Fig. 3 C&S engineering consultant BIM process in case study A

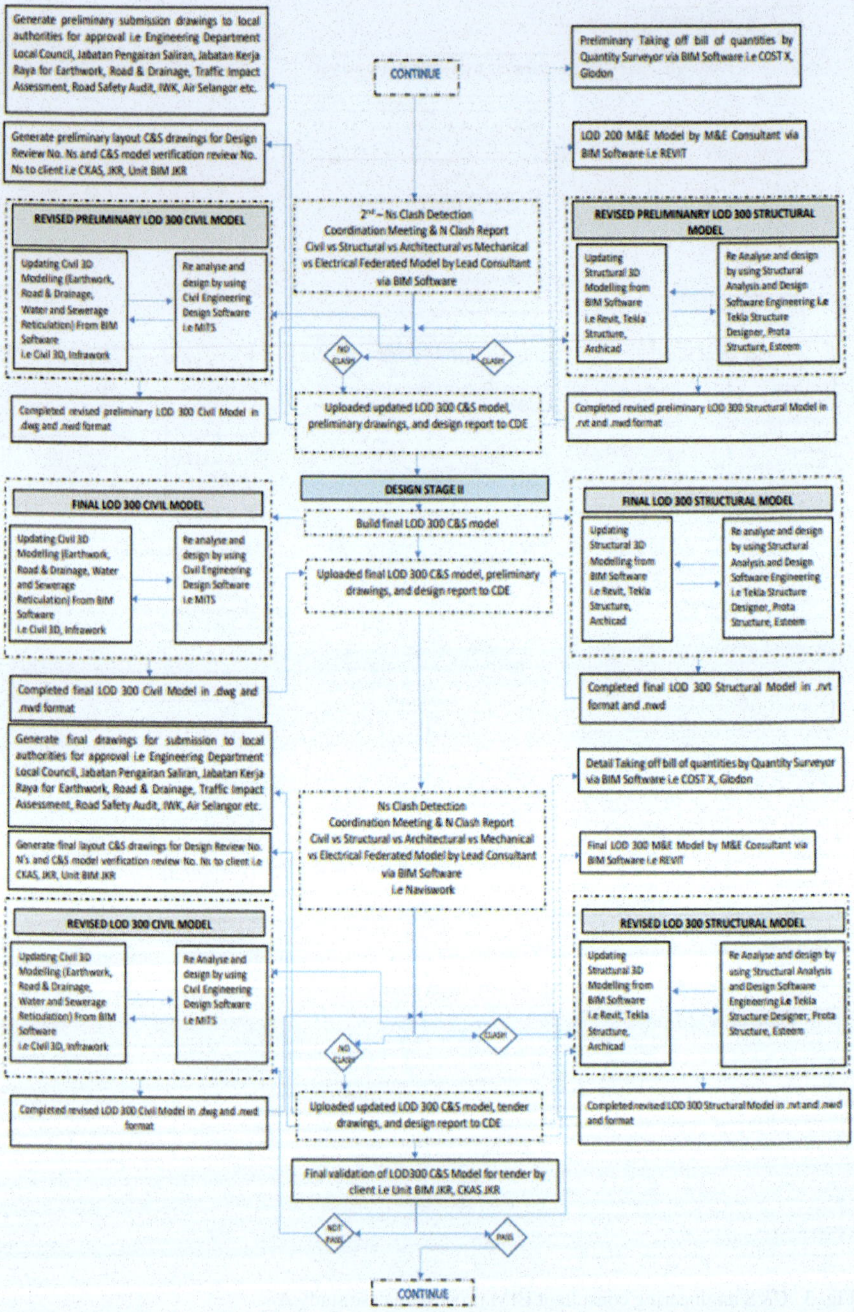

Fig. 3 (continued)

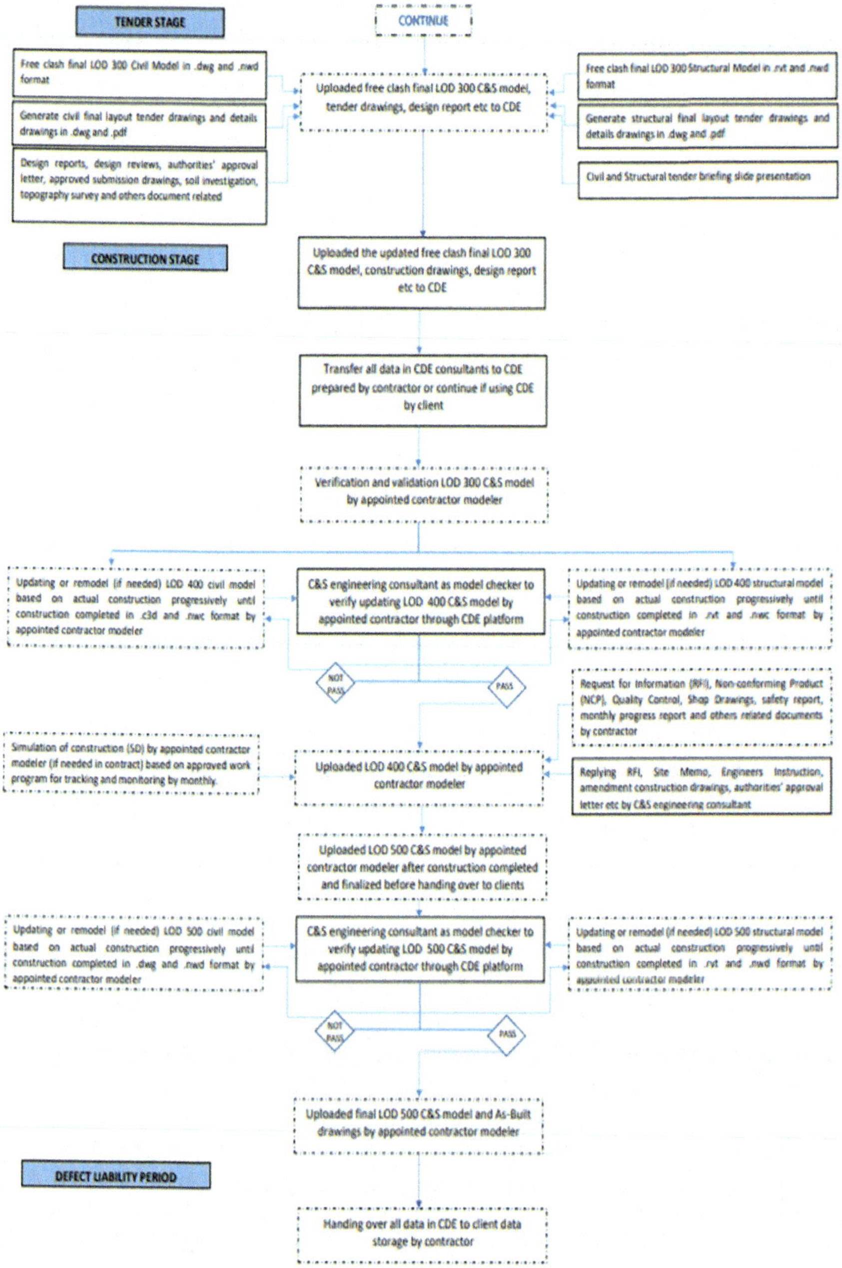

Fig. 3 (continued)

- BIM Infrastructure
- BIM Datacenter
- BIM Project Execution Plan (BPEP)
- BIM Organization Chart
- Model Quality.

5.4 Design Stage

The design stage is the most important scope of work by C&S engineering consultants in developing BIM models. The design phase is divided into 3 stages: Preliminary Design, Design Stage I, and II.

5.4.1 Preliminary Stage

The Common Data Environment (CDE) platform was created using a commercial paid database BIM360 by Autodesk for the purpose of better coordination between multi-discipline consultants and document storage such as models, drawings, design reports, and others in a cloud-based system. The Architect Consultant, generally known as the Lead Consultant in the building projects, is responsible for registering the project CDE for use by all stakeholders involved in the project, namely the consultant, JKR, and the client.

The Architect consultant will develop model LOD100-200, and the model will be uploaded to the CDE so that it can be uploaded by all users. The LOD100 model is a conceptual model before the preliminary design model LOD200 will be built. C&S engineering consultants will download the LOD200 Model Architecture via CDE to be used to build the LOD200 Civil Model and LOD200 Structure Model. LOD200 Civil Model needs to be developed using Civil 3D BIM software to build infrastructure models including Earthwork, Road & Drainage, Water, and Sewerage Reticulation. In Case Study A, the earthwork model also involves a retaining wall. The construction of each LOD200 for the Civil model requires design input either using an Excel spreadsheet or civil engineering software implemented simultaneously in the model development process. The 3D Model must follow the JKR Integrity Model and Quality Checklist and be saved in.dwg (Civil 3D) and.nwc (Navisworks) format and then uploaded to BIM360.

The development of the LOD200 Structural Model also went through the same process after obtaining the LOD200 Architect model through BIM360. Structural engineers can design and build models using the engineering software Tekla Structure Designer and Revit software from Autodesk. This process is done bilaterally until the completed LOD200 structural model and saved in.rvt and.nwc formats and uploaded in a file dedicated to the C&S folder in BIM360.

After completing the LOD200 models, Clash Detection Meeting and Clash Report must be issued between the LOD200 architectural model and the structural model using Navisworks software by Autodesk which is led by an Architect Consultant. Whether the structural and civil models need to be amended or not depends on the agreed technical discussions with the architects. If the model needs to be modified, it must go through the same process of designing and modifying the model again until it is ready to be re-uploaded in the BIM360. LOD200 architect and C&S model can be downloaded by Quantity Surveyor (QS) consultant to take off using Cost X software and a Mechanical and Electrical engineering consultant to design and model it in Revit software.

The working drawings stored in the BIM360 can be used for pre-consultation with local authorities such as the Engineering Department of *Majlis Perbandaran Selayang, Jabatan Pengairan dan Saliran, Air Selangor* submission. The LOD200 Civil and Structural model needs to go through the process of design review and model quality check by *Cawangan Kejuruteraan Awam dan Struktur (CKAS), JKR* and *Unit BIM, JKR*.

5.4.2 Design Stage

Development of the LOD300 model at the design involved stages I and II are based on details deliverables work engineering consultant in CSA. This stage will adopt the same BIM process workflow which involves the design, modeling, submission, model quality check, design review, and others as the preliminary stage in Fig. 2. The process of coordinating all models and clash reports will be done continuously between Civil versus Structure versus Architect versus Mechanical versus Electrical by using Naviswork and then uploaded into BIM360. However, if there is still a clash between the inter disciplines the design and model process need to go through the same process until it is completed, and no clash is identified before the final drawing and tender all models LOD300 are stored in the dedicated folder BIM360 for tender purpose. The LOD300 Civil and Structural Model for Case Study A is shown in Figs. 4, 5, 6, and 7.

5.4.3 Tender Stage

All tender drawings and models prepared and finalized bills of quantities will be provided by the QS consultant through the Cost X software for tendering purposes. The tender submission process for Case Study A still maintains the hardcopy submission method and no online submission. The whole process in the tender stage will be handled by the QS consultant.

5.4.4 Construction Stage

After the contractor is awarded, the contractor needs to set up a modeler team to continue updating and developing LOD400-500 models. Submission of the LOD300

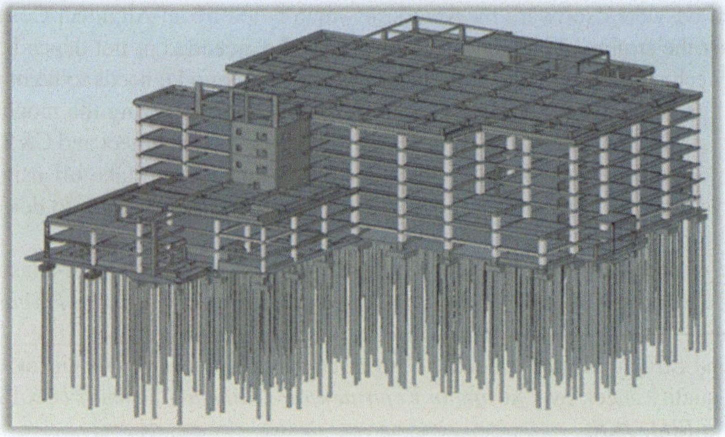

Fig. 4 LOD300 structural model

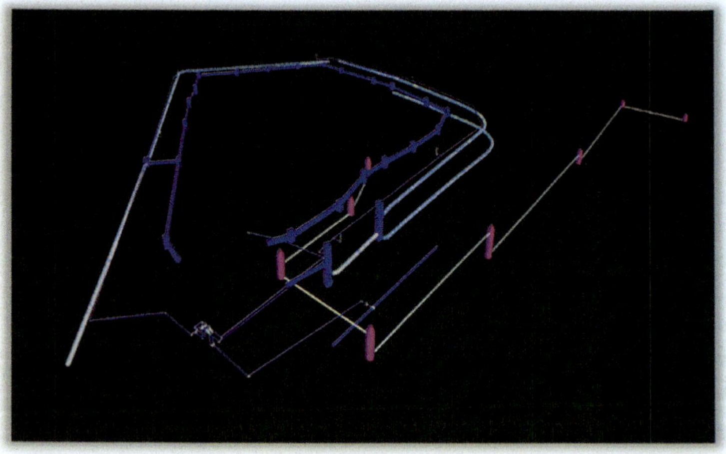

Fig. 5 LOD300 combine infrastructure model

model to the newly created platform CDE contractor needs to be established by the contractor for coordination purposes. Before all LOD300 files from BIM360 by the architect are migrated to the new BIM360 platform the contractor needs to be verified and validated by the contractor modeler team. The development of the LOD400-500 model is based on the actual site progress until it is fully completed. If there are any modifications of the model based on the design change or additional scope of works at the site due to the contractor must be amended or redone by the contractor modeler. All queries by the contractor and responses answered will use BIM360 as an alternative platform for coordination although the manual process is still used in

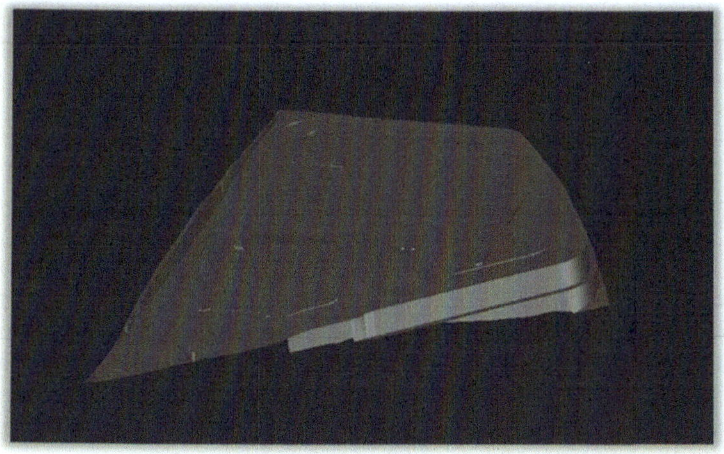

Fig. 6 LOD300 earthwork model

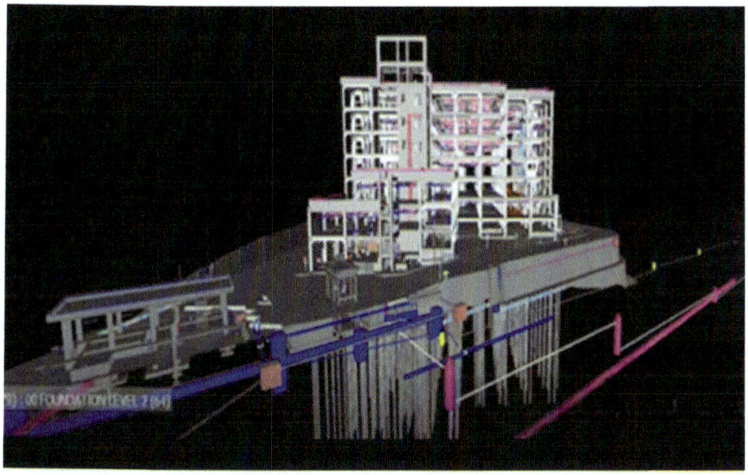

Fig. 7 The combination of the final LOD300 civil and structural model

the construction stage. When the construction of the project has been completed, the LOD500 models should be finalized by the contractor and reviewed and validated by the C&S consultant before they can be recommended to be approved by JKR.

5.4.5 Defect Liability Period

At this stage, all the data in the BIM360 has been fully distributed with the LOD500 model along with the data and other information in the BIM360, the contractor must be submitted to the client platform to complete the project, especially BIM. During this DLP period, the contractor is still responsible for updating the model if there is any obstruction and instructions from the JKR or client until the full DLP period expires.

6 Conclusion

The creation of BIMIF and BIM process for C&S engineering consultants can provide a contribution to help understand the process of implementing BIM in conventional consultant contract projects for government building construction in general. This study also can be a BIM reference to private projects and academic researchers. Indirectly, the implementation of BIM in the current C&S work process will improve the BIM adoption rate in Malaysia in the future.

Acknowledgements The authors would like to express their appreciation to the MTC Engineering Consultancy Sdn. Bhd. (MTCEC) as the C&S engineering consultant for Case Study A for their contribution in sharing the information.

References

Board of Engineers Malaysia (BEM). (2020). Annual report board of engineers Malaysia 2020. BEM, Malaysia. http://bem.org.my/web/guest/professional-engineer

Board of Engineers Malaysia. (2022). BEM register directory. BEM, Malaysia. Retrieved June 15, 2022, from http://www.bem.org.my/web/guest/body-corporate2

Construction Industrial Building Development Board Malaysia (CIBDB). (2013). BIM building information modeling roadmap for Malaysia's construction industry workshop report (Series 2). CIDB, Malaysia.

Construction Industrial Building Development Board Malaysia (CIBDB). (2018). BIM guide 1 awareness. CIDB, Malaysia.

Construction Industrial Building Development Board Malaysia (CIBDB). (2020a). Malaysia building information report 2019. CIDB, Malaysia.

Construction Industrial Building Development Board Malaysia (CIBDB). (2020b). Construction 4.0 strategic plan (2021–2025). CIDB, Malaysia.

Encyclopedia.com. (2022). *Professional engineering organizations*. An Elite Cafemedia Publisher. Retrieved June 15, 2022, from https://www.encyclopedia.com/

Idris, O., & Al-Ashmori, Y. (2021). The level of building information modelling (BIM) implementation in Malaysia. *Ain Shams Engineering Journal, 12*, 455–463.

Jabatan Kerja Raya Malaysia (JKR). (2021). Pelan strategik JKR 2021–2025. JKR, Malaysia.

Kementerian Kewangan Malaysia (MOF). (2008). Arahan perbendaharaan. Kerajaan Malaysia. Retrieved July 15, 2022, from https://www.mof.gov.my/ms/arahan-perbendaharaan

Unit Building Information Modelling (UBIM). (2021). BIM: Piawaian JKR. Bahagian Pengurusan Projek Kompleks, Cawangan Perancangan Aset Bersepadu, Jabatan Kerja Raya Malaysia. Malaysia.

Unit Perancang Ekonomi (UPE). (2021). Rancangan Malaysia kedua belas. Jabatan Perdana Menteri, Malaysia.

Yasser, Y. A., & Idris, O. (2020). BIM benefits and its influence on the BIM implementation in Malaysia. *Ain Shams Engineering Journal, 11*(4), 1013–1019.

Zahrizan, Z., Ali, N. M., Haron, A. T., Marshall-Ponting, A., & Hamid, Z. A. (2013). Exploring the adoption of building information modelling (BIM) in the Malaysian construction industry a qualitative approach. *IJRET: International Journal of Research in Engineering and Technology*. EISSN: eISSN pISS, 2319–1163.

Integrating Value Management: Determine Project Management Knowledge—Addressing *Theory–Practice* Gap

Mohd Hilmi Malek, Che Maznah Mat Isa, and Aini Jaapar

Abstract Value management (VM) is a discipline in ensuring that the projects planned to be executed and implemented can meet the project's objectives, especially against cost, schedule, and quality. Although VM has been around for decades, the theory–practice gap has constrained the successful implementation of value management, especially in the technique used when conducting VM. On the other hand, project management (PM) practices are aligned with the implementation of VM to ensure that the projects intended to be implemented the goals and meet the project performance. Nevertheless, PMBOK is the standard for PM that can guide a project team throughout the PM process and safeguard it from shattering failure. Therefore, this study was conducted to identify the PM knowledge areas that can be adopted in value management methodology. An interview was conducted with 5 Malaysian Construction Industry experts to identify significant VM activities and determine the relationship between PM processes that can be adopted into VM activities. Findings revealed that there is a relationship between the integration of PM processes into the VM process construction industry professional applied PM. This study provides evidence of the advantages of adopting PM knowledge into VM methodology.

Keywords Value management · Project management · PMBOK · Planning · Construction industry

---

M. H. Malek
School of Civil Engineering College of Engineering, Universiti Teknologi MARA, Shah Alam, Selangor, Malaysia

C. M. Mat Isa (✉)
Center for Civil Engineering Studies, Universiti Teknologi MARA, Permatang Pauh, Pulau Pinang, Malaysia
e-mail: chema982@uitm.edu.my

A. Jaapar
Centre of Studies for Quantity Surveying, College of Built Environment, Universiti Teknologi MARA, Shah Alam, Selangor, Malaysia

U. R. Hashim et al. (eds.), *Green Infrastructure*,
https://doi.org/10.1007/978-981-99-7003-2_5

1 Introduction

Value management (VM) involves improving an organization's understanding of its customer's needs, improving internal communication and increasing its knowledge of its success factors. Value Management (VM) can be described as a systematic, function-oriented, and multidisciplinary team approach. Implementation of value management enhances better, better solutions to its customers' needs (Lin et al., 2022). VM consists of the integration of proven and structured decision-making and problem-solving techniques known as value methodologies, combined with other management techniques (Thiry, 2013). SAVE (2007) defines VM as an organization applying value methodology to achieve strategic value improvement. Alsolami (2022) described VM as a valuable tool for resolving difficulties, such as restricted assets, tight schedules, costs, and sustainable alternatives in the building construction industry. Also, VM adoption in the building construction sector has proved beneficial in both developed and developing countries (Alsolami, 2022).

Value Management (VM) is a technique for increasing project competitiveness by proposing an alternative way to add value. It is a structured, organized team approach to identifying the functions of a project, product, or service with recognized techniques and providing the necessary functions to meet the required performance at the lowest possible overall cost (Abdel-Raheem et al., 2018; Elsonoki & Yunus, 2020; Latif & Ghazali, 2019; Miraj et al., 2019) Furthermore, applying the VM results in discovering the best alternative or solution via a thorough study of all items that may influence the selection choice (Mahdi et al., 2015) The construction industry is one of several industries that have embraced this idea and derived advantages from it. Examples, Building Information Modelling (Li et al., 2021; Wei & Chen, 2020), Construction Technology (Husin, 2019; Suwandi et al., 2020; Yap, 2016), Geotechnical Engineering (Ting Jude et al., 2014), Material Engineering (Elseknidy et al., 2020; Husin & Kussumardianadewi, 2018; Labuan & Waty, 2020; Latif & Ghazali, 2019), Also, previous experience has indicated that VM would bring about cost-saving up to 20% in construction projects (Mahdi et al., 2015).

Even though VM has been implemented for more than a decade, the implementation faces numerous challenges, and the construction industry's contribution is far behind other sectors (Yan, 2012). Moreover, despite the tremendous effort to promote VM, Othman et al., (2020) described VM as a non-starter for the construction industries, and the implementation of VM, still, there is no approachable solution to addressing the theory–practice gap (Gohil & Patel, 2018; Hwang et al., 2013; Latif & Ghazali, 2019; Mahinkanda et al., 2019; Yan, 2012) Furthermore, researchers highlighted that VM implementation is much more dependent on the stakeholders' commitment, the organization's readiness, the size of projects, and expert experience in VM. Aghimien et al., (2018) revealed a moderate knowledge of VM among construction professionals.

Although many researchers addressed their concerns by promoting VM knowledge and awareness, the application of VM as a viable method is frequently overlooked (Lin et al., 2022). Previous works paid less attention to VM implementation practices and activities, and most did not examine the implementation and its activities in developing countries (Kineber & Othman, 2022). Most of the related past works identify an insufficient number of factors that hinder the application of VM (Kim et al., 2016). For example, many researchers addressed/ discussed the lack of awareness of VM methodology and its application. However, many researchers failed to identify what are the critical knowledge gaps. Value management has evolved, resulting in the development of numerous tried and true methodologies. The value manager or study leader is not constrained by value management guidelines in selecting the most appropriate technique to utilize during the study (BS EN 12973, 2000).

The modern concept of PM can be defined as applying technical skills, knowledge, tools, and methods to project activities to manage PM effectively. This definition is in accordance with the modern definition of PM, which comprises a sequence of processes in which various tools, skills, and methods are implemented to achieve the project's goals (Abdulnasser & Abdulmajid, 2020; PMI, 2017). The researchers highlighted several challenges in exercising PM techniques and tools. Takagi and Varajão (2022) stressed that despite comprehensive coverage of PM knowledge areas, the standards are believed to have no specific processes focused on evaluating success. Further, the absence of these processes can limit the vision of managers and their teams on what most contributes to the success of a project. Demirkesen and Ozorhon (2017) explained that project managers nowadays seek PM tools and techniques, and integrating engineering methods, principles, and technologies into projects, products, and processes improves the performance of an organization.

Accordingly, the study's objectives outlined in this paper are (1) to identify value management activities by the construction industry practitioner and (2) to determine the project management process and its association when conducting value analysis. This study explores VM techniques to improve the VM application and its knowledge areas and bridge the theory–practice gaps.

2 Literature Review

2.1 Value Management

There are two main advantages of VM for a project client, and ultimately for asset owners, that indicate the project team has a role to play on projects regardless of size. The ratio between the benefit obtained from a course of action and the expense or work necessary to attain it may be considered value. Applying the notion of value to a project might be extremely wide. Regular project meetings should include VM discussions to improve design and construction (RICS, 2017).

Value Management (VM) is a style of management that emphasizes the importance of motivating people, developing skills, and promoting synergies. Some benefits of implementing value management include clear business decisions, the ability to offer better solutions to customers by looking at their actual needs, improved understanding by the organization's participants regarding the organization's goals, and improved internal communication (SAVE, 2007). VM is a systematic strategy for delivering a project that performs the needed functions at the lowest cost possible without sacrificing quality, performance, or reliability (Lin et al., 2022; Jaapar et al., 2009). Value management identifies a project, product, or service with recognized techniques and provides the necessary functions to meet the required performance at the cheapest overall cost (Kineber & Othman, 2022; Othman et al., 2020). VM considerably reduces building costs and enhances schedule performance. VM application broadly demonstrated that the entire construction duration could be lowered, the potential of cost-saving to the construction project cost (Brahmane & Bachhav, 2020; Danso & Osei Kwadwo, 2020; Jaapar et al., 2018).

Value management (VM) is a technique for increasing project competitiveness by proposing an alternative way to add value (Abdel-Raheem et al., 2018; Elsonoki & Yunus, 2020; Latif & Ghazali, 2019; Miraj, Dofir, Andreas, Berawi, & Abd Karim, 2019). Value Management (VM) is more than just cost reduction; it encompasses a wide range of activities (Mahdi et al., 2015). There are many definitions of concepts and definitions of value management.VM in projects must be a continual activity linked with official research targeting specific problems. At each stage of the project, VM attests to the fulfilment of expectations, enabling advancement to the following level. Formal studies should be conducted at project milestones to inform the review or approval process or at other times appropriate to the situation. The formal studies' goals will alter as the project develops (Kelly et al., 2015).

2.2 VM Activities

VM are approaches, principles, practices, procedures, and processes of understanding how a construction project's required functionality, or part of a project, can be achieved through a technical design or construction process that avoids the necessary cost (RICS, 2017). Value management has evolved, so several tested methods have been developed. BS 12973:2000 does not restrict the value manager or study leader in choosing the most appropriate technique to employ during the study. VM study involves applying one or more methods and techniques to a specific subject. A competent study leader or facilitator will lead it with defined objectives, and the issues or subjects may be strategic or technical (BS EN 12973, 2000). A value study is a discrete intervention during the project's inception, evolution, development, and delivery. It comprises all activities necessary to ensure that the project will use the least amount of resources in delivering the required functional benefits while requiring the minimum sacrifices to be made (Kelly et al., 2015).

Alternative and supplementary viewpoints on the value challenge are brought together, discussed, and investigated during the value workshop phase. Logical solutions are then evaluated and recorded. The project moves on to the next phase with the solutions that best satisfy the functional criteria and benefit the client the most.

A thorough workshop report with an action plan is one of the optimal methods to move forward with options that, on the surface, appear to give the best value (Kelly et al., 2015). Table 1 shows VM implementation activities are identified from the literature, and the activities are integrated with various VM guidelines and literature.

A Value Study (SAVE, 2007) is defined in the standard as the formal application of a value methodology to a project to improve its value. According to the standard, the value methodology encompasses value analysis (VA), Value Engineering, Value

Table 1 VM activities in the construction industry (RICS, 2017; Kelly et al., 2015; EPU, 2011; SAVE, 2007)

Phases	Codes	VM activities
Pra—Lab (PL)	PL1 PL2 PL3 PL4 PL5	Conduct a site visit Provide relevant documents about project deliverables Provide relevant documents about the financial status Define the project time and scope Define the project cost estimation
Information (IP)	IP1 IP2 IP3 IP3 IP4	Lab process briefing Engage responsibilities to construction stakeholders Simplify related project information and knowledge among stakeholders Share project information and knowledge among the stakeholders
Function (FP)	FP1 FP2 FP3 FP4 FP5 FP6 FP7	Make the owner specifically express the scope and goals of the project Focus on the functions that the project must fill Setting the essential functions Identify functions that can be performed through alternative methods Identify added features that do not benefit or contribute to essential functions Presentation of project constraints by stakeholders Recognize the aims and responsibilities of the project
Creativity (CP)	CP1 CP2 CP3 C4	Looking for alternatives to increase the effectiveness of the function Create brainstorm ideas and new alternatives to accomplish the anticipated functions Classify brainstormed ideas as theoretically adequate
Evaluation (EP)	EP1, EP2, EP3 EP4	Identify the life cycle cost for each alternative Analyse brainstormed solutions and ideas to perform anticipated functions Investigate the criteria for the alternative action plan Preparing conceptual/schematic drawings and general specifications
Development (DP)	DP1 DP2 DP3	Develop an alternative action plan Compiling the evaluated ideas Researching the ideas studied that need to be dropped due to the appropriateness of implementation
Presentation (PP)	PP1 PP2	Lab report preparation Lab proposal results presentation

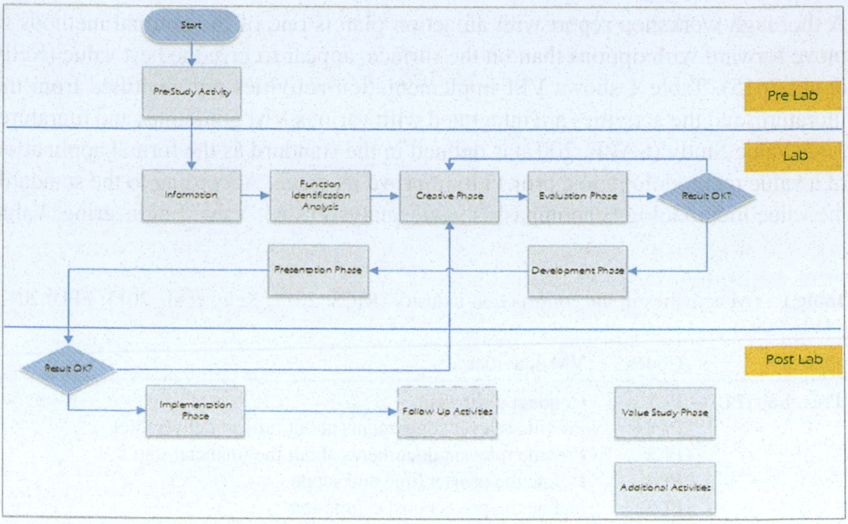

Fig. 1 The value study process flow (*Source* SAVE (2007))

Management, and Value Planning (VP). These are terms that are often used interchangeably around the world. As a result, the standard is diverse in encompassing discipline nomenclatures. According to the standard, this structured process underpins the methodology and performs function analysis as part of the overall process. Other value improvement processes can qualify as value studies if they follow the standard's Job Plan.

Value Management Job Plan is a systematic way to implement the Value Methodology that consists of the eight steps listed below: (1) Preparation, (2) Information, (3) Function Analysis, (4) Creativity, (5) Evaluation, (6) Development, (7) Presentation, (8) Implementation (SAVE, 2020). The VM workshop job plan is a part of the systematic problem-solving approach, overcoming the difficulties associated with inadequate project rationale and justification (Jaapar et al., 2009). In Malaysia, public sector projects use Implementation Guidelines for Value Management in Government Programmes/Projects (EPU, 2011) as the reference to conduct value management. The EPU guidelines adopt Value Standards and Body of Knowledge (SAVE, 2007) as a Value Management Methodology practice. Figure 1 shows the value study process flow.

2.3 Project Management

According to PMI (2017), project management (PM) is the application of knowledge, skills, tools, and techniques to project activities to meet the project's requirements. PM is not a recent development. It has a history that goes back many hundreds of years. The organization's best efforts to apply PM (PMI, 2017) described PM as using knowledge, skill, tools, and techniques to project activities to meet the project requirement. The project manager works with the project team and other stakeholders to determine and use the appropriate recognized good practices for each project. The project team will choose the proper combination of processes, inputs, tools, techniques, and life cycle phases to manage a project.

Several standards and guides, including PM methodologies, are available today for practitioners and researchers. Some of these standards and guides describe management through processes, organizing them into "process group", "subject group", "life cycle phases", or "knowledge areas". Several of these concepts are similar in the guides, although described differently. Each has its own approach to defining and organizing the concepts and good practices. The objective of standards and guides is to describe processes and related aspects, as well as to converge terminology and concepts between stakeholders and PM teams. Different institutions maintain several of these project management standards and guides. Some examples are PMBOK (PMI, 2017), ISO 21500 (ISO, 2012), Project Management Methodologies (PM2, 2016), and PRINCE2 (AXELLOS, 2017).

PM is the process by which projects are defined, planned, monitored, controlled, and delivered such that agreed benefits are realized from an investment. Projects are unique, and a temporary endeavour was undertaken to achieve desired outcomes. The project brings about change, and PM is recognized as the most efficient way of managing such change. Value studies within the PM process facilitate understanding the project's primary purpose, its constituent parts, and the value system criteria by which the project benefits are judged to be successfully delivered.

In this research, PMBOK Guide is a foundation upon which organizations can build methodologies, policies, procedures, rules, tools, techniques, and life cycle phases to practice project management.

2.4 Project Management Body of Knowledge, PMBOK

The Project Management Body of Knowledge (PMBOK) guide is based on Standard Project Management, and it is a document established by the authority, custom, or general consent as a model or example. It was developed using a process based on consensus, openness, due process, and balance. The Standard for Project Management is a foundational reference for PMI's professional development programmes and project management practice. The standard identifies the processes considered good practice on most projects and specifies the inputs and outputs usually associated with those processes. In other words, PMBOK provides details about critical concepts, emerging trends, considerations for tailoring the project management process, and information on how tools and techniques are applied to projects (PMI, 2017).

Jamali and Oveisi (2016) described the PMBOK as a group of processes and knowledge fields generally accepted in the PM discipline. PMBOK represents the result of a global effort that recognizes and recognizes the application of knowledge, processes, skills, tools, and techniques to impact project success significantly. It identifies ten areas of project knowledge management, the dual nature of the project process, and five groups of PM processes. PMBOK can be classified into ten (10) knowledge areas which are: (1) integration management, (2) scope management, (3) schedule management, (4) project cost management, (5) quality management, (6) risk management, (7) resource management, (8) procurement management, (9) stakeholder management, and (10) communication management. Table 2 shows the description of PM knowledge areas.

Table 2 PM knowledge areas (PMI, 2017)

Management knowledge areas	Description
Integration (PM1)	Project Integration Management identifies, defines, combines, unifies, and coordinates PM process group processes and activities
Scope (PM2)	Project scope management includes the techniques to guarantee that a project is efficient enough to complete the required work to achieve the desired product, service, or outcome
Schedule (PM3)	The project schedule outlines how and when the project will deliver the goods, services, and objectives defined in the project scope
Cost (PM4)	Project Cost Management is primarily concerned with the cost of the resources needed to complete project activities
Quality (PM5)	Project Quality Management addresses the project's management and the project deliverables
Risk (PM6)	Project Risk Management identifies and manages unaddressed hazards
Resource (PM7)	Project Resource Management include identifying, procuring, and managing the resources required to finish the project effectively. These procedures guarantee that the appropriate resources are accessible to the project manager and team at the proper time and location
Procurement (PM8)	Project Procurement Management encompasses the procedures required to obtain items, services, or outcomes from sources other than the project team
Stakeholder (PM9)	Project Stakeholder Management encompasses the processes required to identify the individuals, groups, or organizations who may have an impact or be impacted by the project, analyse stakeholder expectations and their impact on the project, and develop appropriate management strategies for effectively engaging stakeholders in project decisions and execution
Communication (PM10)	Project Communications Management encompasses the procedures required to guarantee that the project's and its stakeholders' information demands are satisfied via the creation of artefacts and the execution of actions aimed to enable successful information sharing

3 Methodology

The exploratory character necessitated this study using a qualitative research method. It focused on identifying significant VM activities and determining the relationship between PM processes that can be adopted into VM activities. In-depth semi-structured interviews were conducted with five (5) experienced executives representing various project stakeholders who manage and supervise the construction industry and VM activities. Data from the literature were used to identify the PM processes related to VM practice in the construction industry. The interview questions were drawn from several associated studies focusing on the ten (10) PM knowledge areas (see Table 2). During the interview session, to ensure confidence and trust, the respondents were fully informed regarding the aim and objectives of this survey. Confidentiality and the integrity of the respondents were strictly respected, and codes were assigned to each of them. The respondents include the Project Director (P1), Project Manager (P2) and (P3), Construction Manager (P4), and Architect (P5). They are highly involved in the construction industry and familiar with the Value study and the PM process, which aims to achieve project performance and a cost-quality schedule.

This research method is designed to get in-depth information regarding the PM process and detect irrelevant questions. Also, this research enabled the literature, especially non-academic material, to be validated and solutions to real problems to be elicited. The interviews were recorded and transcribed verbatim to classify and formulate data for analysis. The transcribed copy was reread to understand the ideas better and linked to the aim and objectives. The interviews enabled the literature, especially non-academic material, to be validated and PM practice to be prompted. Purposive sampling was used to determine the construction experts experienced in implementing VM and PM. The interview took approximately an average of one hour to complete the interview session. Recorded interviews were transcribed and thematically analysed.

4 Result and Discussion

The relationship between VM activities and PM activities is recognized, and the VM process to bridge the VM theory–practice gap has been discovered through the interview session. A semi-structured interview was conducted to gather qualitative data among the construction industry practitioners. This section discusses findings from the interview regarding identifying the critical PM knowledge areas that can be adopted in value management methodology.

Table 3 General characteristics of the respondents

Panel	Position	Experience	Project type	Contract amount	Organization
P1	Project director	24	Housing	More than 50 million	Developer
P2	Project manager	16	Housing	More than 50 million	Consultant
P3	Project manager	20	Infrastructure	More than 50 million	Contractor
P4	Construction manager	16	Housing	Less than 10 million	Contractor
P5	Architect	26	Housing	More than 50 million	Consultant

4.1 Demographic Study

Most respondents have more than ten (10) years of working experience and are in their organizations' middle to top management ranks. Most of them are experienced and involved in various projects, especially housing and infrastructure. Table 3 below indicates the level of experience that differs among the respondents.

All the respondents are construction experts within their organizations. They are considered exemplary practitioners in discussing value study, PM process and tools, and their implication towards project performance. Most respondents also experienced managing a project with a contract amount between MYR 10—185 million, except P4, which manages a contract with less than MYR 10 million.

4.2 VM Activities and PM Processes

After thoroughly analysing and synthesizing the employers' interview transcripts, we constructed a thematic map to visually illustrate the themes that emerged from the data. Table 4 summarises the findings between VM activities and the PM process. Based on Table 4, the respondents identified the activities during the implementation of VM and significant PM processes that can be adopted in the VM activities.

4.2.1 Integration Pre-Lab Activities with PM Activities

The pre-lab phase is a significant phase before conducting a value management programme. This phase is to plan and execute value analysis studies, including enlisting the support and commitment of top management and stakeholders to the need for research to be done, in addition to planning and deciding on the distribution of responsibilities and tasks of the participants involved in the study. Based on the interview with five (5) panels, five activities were involved in the pre-lab phase,

Table 4 Finding on VM activities and PM process

Phases	Codes	Panel						PM knowledge areas (PM)									
		P1	P2	P3	P4	P5		1	2	3	4	5	6	7	8	9	10
Pra—Lab (PL)	PL1	√	√	√	√	√	P1	√	√								√
	PL2	√	√	√	√	√	P2	√	√								√
	PL3	√	√	√	√	√	P3		√								√
	PL4	√					P4										√
	PL5						P5										√
Information (IP)	IP1	√	√	√	√	√	P1	√	√								√
	IP2	√	√	√	√	√	P2	√	√								√
	IP3	√				√	P3		√								√
	IP4						P4										√
							P5										√
Function (FP)	FP1	√	√	√	√	√	P1		√	√	√	√	√			√	√
	FP2	√	√	√	√	√	P2		√	√	√	√	√			√	√
	FP3	√	√	√	√	√	P3		√	√	√	√	√				√
	FP4	√	√	√	√	√	P4		√	√	√	√	√				√
	FP5		√	√	√	√	P5		√		√		√				√
	FP6		√	√	√	√											
	FP7																
Creativity (CP)	CP1	√	√	√	√	√	P1	√	√	√	√	√	√	√	√		√
	CP2	√	√	√	√	√	P2	√	√	√	√	√	√	√	√		√
	CP3	√	√	√	√	√	P3	√	√	√	√	√	√	√	√		√
							P4		√	√	√	√	√	√	√		√
							P5		√			√	√	√	√		√
Evaluation (EP)	EP1,	√	√	√	√	√	P1		√	√	√	√	√	√	√		√
	EP2,	√	√	√	√	√	P2		√	√	√	√	√	√	√		√
	EP3	√	√			√	P3		√	√	√	√	√	√	√		√
	EP4	√					P4		√	√	√	√	√	√	√		√
							P5			√		√	√		√		√
Development (DP)	DP1	√	√	√	√	√	P1									√	√
	DP2	√	√	√	√	√	P2									√	√
	DP3	√	√	√	√	√	P3									√	√
							P4									√	√
							P5									√	√
Presentation (PP)	PP1	√	√	√	√	√	P1									√	√
	PP2	√	√	√	√	√	P2									√	√
							P3									√	√
							P4									√	√
							P5									√	√

and the interview results showed that all panellists approved of the PL1 and PL2 activities. P1 states that PL1 activities are significant before the VA is carried out because, as a result of this site visit, the project team can plan how the project can be implemented effectively. P2, P3, and P4 say that PL1 activities are a necessity to obtain deliverable project input and among the things that need to be emphasized are matters such as work scope, management and logistic building materials, soil surface conditions and neighbourhoods, site readiness, appropriate technology and machinery, and flexibility to get workers to perform the work. P5 stated that the requirements of PL1 activities should be implemented to assess the constraints on the overall project. As the lead consultant, ensuring that the project is implemented in accordance with the main targets and objectives is significant, as it involves the management and administration of contracts.

Planning as a pre-construction activity has a significant impact on ensuring the project performance can be achieved (Amini et al., 2022). In the PM context, the planning phase can be defined as processes performed to determine a project. P2 and P3 state that the PM activities present in the pre-lab phase are project scope management. Among the processes available in scope management is preparing a scope management plan that contains how the project's scope will be defined, validated, and controlled. In addition, the process involved in planning is to collect information relevant to the project's needs. P1 pointed out that before the value analysis programme is implemented, the top management will look at the scope management plan first to take a comprehensive look at the scope of work to be carried out, such as project deliverables, financial readiness of the company, project team, obstacles and constraints, budget and estimated construction costs. P5 said all findings should be recorded and documented to facilitate the agenda and objectives of value management. In addition, P3 and P5 agree that the lesson-learned report is an essential supporting document as a reference before the value analysis phase continues.

4.2.2 Integration Lab Activities with PM Activities

Value analysis has been defined as a proven method that improves product or service effectiveness and increases an organization's competitiveness. Value analysis involves six (6) phases of work known as job plans, including the information, function, creative, evaluation, and development phases. Based on the value management guidelines retrieved from the literature, each phase of VA activities has its own goals and objectives.

The findings showed that all panellists agreed that IP3 and IP4 activities significantly ensure that VM can be carried tangibly, substantially, and quantifiably. P1 outlines that the project team needs to anticipate, realize, and understand the main objectives and requirements of a project pursued. A site visit not only lets the project team do site analysis, but it is also a significant activity, aiming the project team can transpire with design solutions that fit the needs and features. P2 and P3 insist that the ream of data that has been collected should be justified, wherefore, it is recommended that the project team summarize the project details. Necessitous data should be translated into simple, accurate information, especially the description of the law and regulation. On the contrary, P5 describes an architect, who is responsible for ensuring the product or project requirement and specifications, insists intrinsic information in particular of the law and regulation, should be translated into simple and accurate information.

IP1 and IP2 activities are inconsiderable from the respondents P1, P2 and P3. However, P1 and P5 are favourable in their purpose for the information phase activities. P2 states that IP2 activities are uncompelled for the information phase as the pre-eminent objective of the information phase is to ensure that the prominent driver of value analysis is attained, and P3 believes the VM process has a sense of its practicability, where various methods, techniques, and tools can be embraced, utilized, and designated, to achieve the goals and objectives of value analysis. P1 and P5 suggested that IP1 and IP5 activities are necessary as the value study required briefings explaining the VM process. The value team leader or value facilitator delineates the responsibilities of each project team to ensure the succession and smooth running of the VM programme.

Referring to the findings from Table 4, all respondents agreed that PM9 and PM10 are decisive and significant PM knowledge areas at the function phase. P1 and P5 suggest that communication is the essential element for the success of value analysis. Moreover, effective communication can explain the primary goal of why value analysis should be held. All the respondents agreed that the function, creativity, and evaluation phases are the three functions that need to be integrated. P2 and P3 agree that when performing a value study, with the availability of supporting and reference documents, the project team will list the elements and functions of each requirement during the pre-lab phase. The findings show that FP2, FP3, FP4, and FP5 activities agreed by all respondents and CP1, CP2, and CP3 activities should be emphasized during the creative phase. P1 and P5 state that brainstorming techniques are adequate to ensure value analysis, especially on alternative methods, implementation effectiveness, and risk management. In addition, the findings also showed that all respondents agreed that EP1, EP2, and EP3 activities were significant activities for the evaluation phase. Activities in the development phase, which include DP1, DP2, and DP3 activities, and activities in the delivery phase, are selected by all respondents. The respondents agreed that the value analysis results and findings should be recorded, documented, and presented to top management.

4.3 Determining Project Management Process

Project management is the application of knowledge, skills, tools, and techniques to project activities to meet the project requirements. PM enables organizations to execute projects effectively and efficiently. PM is accomplished by appropriately applying and integrating the project management processes identified (PMI, 2017).

Furthermore, the majority agreed that all ten (10) knowledge areas that PMBOK has asserted could be applied during the conduct of value analysis programmes, especially during the initial stage of the project and planning. P2 and P3 state that integration management (PM1) is vital because value management involves various expertise, knowledge, and discipline areas. For example, in reaching an agreement on an element and function, the project team, comprising architects, engineers, quantity surveyors, and contractors, need to work together, especially when providing input on the relevant parameters of products and services. PM1 is about ensuring that the deliverable time completion of the product, service, and benefits management plan is aligned by providing a PM plan to achieve the project objectives. Similarly, it ensures the creation and use of the appropriate knowledge and making decisions regarding fundamental changes impacting the project (PMI, 2017). The same applies to project scope management (PM2). P5 states that the availability of a scope management plan assists in defining the objectives, targets, goals, and information, and it should be translated into documentation. P2 and P3 stated that gathering data and information on project needs can be done systematically and focus on VM accomplishment, and the project team can define and verify the work scope when the VA is conducted. Constantly, according to (PMI, 2017), requirements have always been a concern in PM, and trends and emerging practices for PM2 focus on collaborating with business analysts to: (1) identify problems and identify a business need, (2) identify and recommend viable solutions for meeting the needs, (3) elicit, document, and manage stakeholder requirements to meet business and project objectives, and (4) facilitate the successful implementation of the programme's product, service, or result.

P1 and P5 say project cost management (PM4) and project quality management (PM5) activities used during the VA allow the project team to determine the overall cost estimate of the construction in line with the justification of product and service quality. On top, according to PMI (2017), PM4 is primarily concerned with the cost of the resources required to accomplish project activities, and it should examine the impact of project decisions on the following recurring cost of utilizing, maintaining, and supporting the project's product, service, and result. While, PM5 addresses the management of the project and the project deliverables, which it applies to all projects, regardless of the nature of the deliverables. PM5 approaches seek to minimize variation and deliver results that meet defined stakeholder requirements by understanding, evaluating, defining, and managing the requirements. Failure to meet the quality requirement can seriously harm the project stakeholders (PMI, 2017).

In addition, P4, a construction manager, states that schedule management (PM3) is a continuation of the information available from the scope of construction. For example, during defining activities, things such as work sequence, time and duration, productivity rates, costs, and resources should be considered because complete final information can be translated into work programmes during the presentation of VA findings.

The panel also agreed that risk management (PM6), resource management (PM7), and procurement management (PM8) could ensure that VA processes can be implemented more effectively and effectively. PM1 states that applying risk identification techniques to the elements and functions of a product can be effectively analysed, where the weaknesses and advantages of a product can be determined. Additionally, PMI (2017) describes PM6 as aims to identify and manage risks that are not addressed by the other PM processes. It should start as soon as a project is envisioned and end early in the project. As a result, the project's success is strongly associated with PM6's effectiveness. The unmanaged risk could cause the project to diverge from its original course and fall short of its intended goals. A coordinated strategy may provide alignment and coherence in risk management at all levels. As a result, offering the greatest total value for a given risk exposure increases risk efficiency in the structure programmes.

Moreover, P2 and P5 state that, with the availability of resources management (PM7), the project team can evaluate the current situation against material availability and logistics. Regardless, the project team can estimate the overall costs associated with a product. P2, P3, and P4 state that selecting contractors, vendors, and specialists can be made systematically and effectively. Notwithstanding, various aspects should be considered, such as capability and capacity, price, quality, and management system. PM8 includes processes necessary to purchase or acquire products, services, or results from outside the project team. Therefore, procurement management should be applied when implementing the VA. There are several emerging practices for PM8, such as using advanced techniques and tools, conducting risk management, and changing the contracting process (PMI, 2017).

The respondents also agreed that stakeholder management (PM9) and project communications management (PM10) significantly impacted the realization of the VA programme. P1 claims that with the availability of stakeholder management during the information phase, the project team can understand each other's roles and responsibilities and the client's needs and requirements can be translated into documentation. Finally, all respondents agreed that communication management could ensure that the VA can be conducted effectively and effectively. For example, P5 states that with systematic management, consensus between the project team can be achieved, and conflict of interest, unethical decisions, and poor decision-making can be avoided. Nevertheless, PMI (2017) defined PM10 as the processes required to ensure that the information needs of the project and its stakeholders are met through development and implementation activities designed to achieve information exchange and suggested communication, which develops the relationships essential to a successful project outcome.

5 Conclusion

Many researchers agreed on the terms used to define VM. VM is expressed as an organization applying value methodology to achieve strategic value instruments (SAVE, 2007). VM can be defined as a structured, systematic, multidisciplinary effort that increases project competitiveness by focusing on analysing the functions of projects using a systematic, logical procedure and technique under the direction of a knowledgeable value practitioner to achieve the best value at the lowest total life cycle project cost by proposing an alternative way to add value.

For decades, VM has been introduced to the construction industry, and many researchers agreed on the tremendous benefits of applying its application. However, the application of VM is still beneath the expectation. Most researchers have pointed out and found undifferentiated findings on the challenges that hinder the application of VM. Lack of VM knowledge and poor awareness among construction industry practitioners are the critical factors that encumber the implementation of VM in the construction industry (Mahinkanda et al., 2019; Aghimien et al., 2018; Tanko et al., 2018). Further, despite the knowledge barriers, Mahinkanda et al. (2019) found that the construction industry practitioner encounters the term value management. However, the VM concepts do not follow the guidelines, and the value team study focuses on reducing costs. Though, BS EN 12973 (2000) suggest that there are no absolute techniques and tools for conducting the VM, and RICS (2017) expressed that VM should not be seen as isolated or stand-alone and suggest the VM should not be viewed as a stand-alone or independent. It should be integrated with project management areas.

This study administered a qualitative approach intended to identify the relationship between VM activities within the areas of project management. Five respondents are selected based on focus groups, who have experience in the value study and managing and supervising construction projects with different backgrounds of organization. The study identifies significant VM activities and determines the relationship between PM processes that can be adopted into VM activities. The findings show that PM knowledge areas are significantly integrated into VM activities. Correspondingly, the result indicates that the respondent agrees that the ten (10) knowledge areas, that PMBOK recommends, apply to the value study. It is a tool and technique frequently used by the project team. However, the study shows that the application of the VM process is highly dependable on the VM phases and suited to the phases of the VM. Furthermore, this study found that the functional, creativity and evaluation phases should be integrated. The finding aligns with the guideline proposed by RICS (2017).

The findings of this study can be used as the milestones for bridging the theory–practice gap in VM practice. Some of the VM governing bodies are too rigid and fail to acknowledge the existence of numerous techniques, especially the project management process. For example, less affordable research addresses the construction industry practitioner's frequent techniques and strategies for implementing VM. In the bargain, this study can be used to promote VM application and the implementation by the construction industries, and bodies governing the standard, such as SAVE, IVMM, and the academician, can be used as future references.

Acknowledgements The authors would like to thank the editor and anonymous reviewers for their insightful comments and suggestions, which significantly improved the quality of the paper.

References

Abdel-Raheem, M., Burbach, V., Abdelhameed, A., Sanchez, G., & Navarro, L. (2018). Value engineering and its applications in civil engineering. In: construction research congress 2018: Infrastructure and facility management—selected papers from the construction research congress 2018, 2018-April (January) (pp. 263–272). https://doi.org/10.1061/9780784481295.027

Abdulnasser, K., & Abdulmajid, N. (2020). Factors affecting the application of project management knowledge guide (PMBOK ® guide) in construction projects in Yemen. *International Journal of Construction Engineering and Management, 9*(3), 81–91. https://doi.org/10.5923/j.ijcem.20200903.01

Aghimien, D. O., Oke, A. E., & Aigbavboa, C. O. (2018). Barriers to the adoption of value management in developing countries. *Engineering, Construction and Architectural Management, 25*(7), 818–834. https://doi.org/10.1108/ECAM-04-2017-0070

Alsolami, B. M. (2022). Identifying and assessing critical success factors of value management implementation in Saudi Arabia building construction industry. *Ain Shams Engineering Journal, 13*(6). https://doi.org/10.1016/j.asej.2022.101804

Amini, S., Rezvani, A., Tabassi, M., & Malek Sadati, S. S. (2022). Causes of cost overruns in building construction projects in Asian countries; Iran as a case study. *Engineering, Construction and Architectural Management*. https://doi.org/10.1108/ECAM-05-2021-0445

AXELLOS. (2017). *Managing successful projects with PRINCE2* (6th ed.). The Stationery Office, TSO.

Brahmane, N. S., & Bachhav, S. S. (2020). Implementation of value engineering in construction project to reduce time of the project, 7–9.

BSI. (2000). BS EN 12973:2000: Value management, British Standard Institute, UK

Danso, H., & Osei Kwadwo, R. (2020). Assessment of value engineering implementation in the Ghanaian construction sector. *Journal of the Institution of Engineers (India): Series A, 101*(1), 7–17. https://doi.org/10.1007/s40030-019-00423-6

Demirkesen, S., & Ozorhon, B. (2017). Measuring project management performance: Case of construction industry. *EMJ—Engineering Management Journal, 29*(4), 258–277. https://doi.org/10.1080/10429247.2017.1380579

Elseknidy, M. H., Ali, S., Saad, A. H., Kineber, F., & I, N. S. (2020). VE application in material selection assessment and performance of aluminum dross as cement replacement. Department of chemical and environmental engine.

Elsonoki, M. M., & Yunus, R. (2020). Value engineering practices in the Libyan construction industry: a preliminary study. In IOP conference series: Earth and environmental science (Vol. 498(1)). https://doi.org/10.1088/1755-1315/498/1/012109

European Commission Centre of Excellence in Project Management, PM2. (2016). *The PM2 project management methodology guide.*

Gohil, P., & Patel, S. (2018). Review of value engineering in Indian construction industry. *8*(Iii), 1080–1085.

Husin, A. E. (2019). Implementation value engineering in diaphragm wall at high rise building implementation value engineering in diaphragm wall at high rise building. *8*(February), 16–23.

Husin, A. E., & Kussumardianadewi, B. D. (2018). Cost performance review on value engineering optimized floor cover finishing work of high rise office building. *International Journal of Engineering and Advanced Technology, 8*(2), 146–154. https://doi.org/10.13140/RG.2.2.29165.33764

Hwang, B., Zhao, X., & Ong, S. Y. (2013). Value management in Singaporean building projects: Implementation status, critical success factors, and risk factors. *Journal of Management in Engineering*. https://doi.org/10.1061/(ASCE)ME

Jaapar, A., Azmi, N., Bari, A., & Takim, R. (2009). The impact of value management implementation in Malaysia. *Journal of Sustainable Development, 2*(2).

Jaapar, A., Maznan, N. A., & Zawawi, M. (2018). Current state of value management implementations in Malaysian public projects. *Asian Journal of Environment-Behaviour Studies, 3*(8), 71–78. https://doi.org/10.21834/aje-bs.v3i8.280

Jamali, G., & Oveisi, M. (2016). A study on project management based on PMBOK and PRINCE2. *Modern Applied Science, 10*(6), 142. https://doi.org/10.5539/mas.v10n6p142

Kelly, J., Male, S., & Graham, D. (2015). *Value management of construction projects* (2nd ed.). John Wiley & Sons.

Kim, S.-Y., Lee, Y.-S., Nguyenand, V. T., & Luu, V. T. (2016). Barriers to applying value management in the Vietnamese construction industry. *Journal of Construction in Developing Countries, 21*(2), 55–80. https://doi.org/10.21315/jcdc2016.21.2.4

Kineber A. F., & Othman, I. (2022). Modelling the relationship between value management's activities and critical success factor for sustainable building. *Journal of Engineering, Design and Technology*, *20*(2), 414–435. Emerald Publishing Limited 1726–0531. https://doi.org/10.1108/jedt-03-2021-0159

Labuan, F. L., & Waty, M. (2020). Study of selecting floor cover by using the value engineering method in the housing project. In *IOP conference series: materials science and engineering* (Vol. 1007(1)). https://doi.org/10.1088/1757-899X/1007/1/012076

Latif, M. A., & Ghazali, Z. A. (2019). Value management/value engineering (VM/VE) application in Malaysian public construction projects: Application of VM/VE study improved roads project sustainability. In *IOP conference series: Materials science and engineering* (Vol. 512(1)). https://doi.org/10.1088/1757-899X/512/1/012046

Li, X., Wang, C., & Alashwal, A. (2021). Case study on BIM and value engineering integration for construction cost control. *Advances in Civil Engineering, 2021*.https://doi.org/10.1155/2021/8849303

Lin, X., Mazlan, A. N., Ismail, S., & Durdyev, S. (2022) Status of value management implementation in small and medium construction projects in Malaysia. *Buildings, 12*(5), 658. https://doi.org/10.3390/buildings12050658

Mahdi, I. M., Heiza, K. M., & Elenen, N. E. A. (2015). State of the art review on application of value engineering on construction projects: High rise building. *International Journal of Innovative Research in Science, Engineering and Technology*, *4*(5), 2742–2753. https://doi.org/10.15680/ijirset.2015.0405013

Mahinkanda, M. M. M. P., Sandanayake, Y. G., & Ekanayake, B. J. (2019). Bridging the theory-practice gap in value management in Sri Lankan construction industry. In *Proceedings of the 8th World Construction Symposium* (pp. 147–157). https://doi.org/10.31705/WCS.2019.15

Malaysian Economic Planning Unit, EPU. (2011). *Panduan perlaksanaan pengurusan nilai dalam program/projek kerajaan.*

Miraj, P., Dofir, A., Andreas, A., Berawi, M. A., & Bari, S. (2019). Systematic review of value engineering research studies in Indonesia. *CSID Journal of Infrastructure Development*, *2*(2), 192. https://doi.org/10.32783/csid-jid.v2i2.93

Othman, I., Kineber, A. F., Oke, A. E., Khalil, N., & Buniya, M. K. (2020). Drivers of value management implementation in building projects in developing countries. *Journal of Physics: Conference Series, 1529*(4). https://doi.org/10.1088/1742-6596/1529/4/042083

PMI. (2017). A guide to the project management body of knowledge (PMBOK guide) (6th ed.). Project Management Institute.

RICS. (2017). Value management and value engineering, RICS, UK.

SAVE. (2007). Value methodology standard. SAVE International, Mount Royal.

Society of American Value Engineer, SAVE. (2020). Value methodology glossary of terms. In *VM guide: A guide to the value methodology body of knowledge.*

Suwandi, P. A. P., Husodo, I. T., & Suryadi, M. W. (2020). Value engineering in the implementation of KumKang formwork (Case study: the Alton apartment project in Semarang Indonesia). *Journal of Physics: Conference Series, 1625*(1). https://doi.org/10.1088/1742-6596/1625/1/012012

Takagi, N., & Varajão, J. (2022). ISO 21500 and success management: An integrated model for project management. *International Journal of Quality and Reliability Management, 39*(2), 408–427. https://doi.org/10.1108/IJQRM-10-2020-0353

Tanko, B. L., Abdullah, F., Mohamad Ramly, Z., & Enegbuma, W. I. (2018). An implementation framework of value management in the Nigerian construction industry. *Built Environment Project and Asset Management, 8*(3), 305–319. https://doi.org/10.1108/BEPAM-09-2017-0078

Thiry, M. (2013) A framework for value management practice. In *Project Management Institute* (2nd ed.). Project Management Institute.

Ting Jude, M. H., Afshar, N. R., Selaman, O. S., & Taib, S. N. L. (2014). Application of value engineering in slope stabilization. *1*(10), 211–216.

Wei, T., & Chen, Y. (2020). Green building design based on BIM and value engineering. *Journal of Ambient Intelligence and Humanized Computing, 11*(9), 3699–3706. https://doi.org/10.1007/s12652-019-01556-z

Yan, L. (2012). Value engineering application in construction project management. In *Proceedings of the 2012 International Conference on Industrial Control and Electronics Engineering, ICICEE 2012* (pp. 1265–1268). https://doi.org/10.1109/ICICEE.2012.336

Yap, C. L. (2016). Application of value management by IBS contractors in Malaysia. Doctoral dissertation, UTAR. Universiti Tunku Abdul Rahman.

Strategies of Carbon Reduction Management in Construction Operations

Verona Ramas Anak Joseph, Nur Kamaliah Mustaffa, and Che Maznah Mat Isa

Abstract There is a significant demand in curbing the generation of carbon emissions globally due to the detrimental effects of climate change all over the world. The construction industry was among the largest contributors to carbon emissions. However, the previous efforts of reducing carbon emissions were only focused on quantifying embodied carbon from building operations, while less study focused on the construction stage. Therefore, this study aims to investigate the current carbon emissions management practices and key strategies for reducing emissions effectively. The study was conducted by using a brief questionnaire survey of 42 targeted construction stakeholders. Relative important index analysis was used in this study to rank the criteria according to their relative importance as an indicator of the carbon reduction strategies. This study contributes to the body of knowledge on the current Malaysian construction industry's carbon practices and provides a reference to the industry in managing its carbon reduction.

Keywords Carbon emissions · Construction · Management

1 Introduction

As global climate change causes annual temperature increases, there is increased interest in decreasing carbon emissions from building projects. This is because the primary enabler of greenhouse gas is carbon emissions which trap heat in the atmosphere (United States Environmental Protection Agency, 2021). There is an increase of emissions by the building and construction sector amounted to 11Gt CO_2 for a

V. R. A. Joseph · N. K. Mustaffa (✉)
School of Civil Engineering, College of Engineering, Universiti Teknologi MARA, 40450 Shah Alam, Selangor, Malaysia
e-mail: nurkamaliah@uitm.edu.my

C. M. Mat Isa
Center of Civil Engineering Studies, Universiti Teknologi MARA Pulau Pinang Branch, Permatang Pauh Campus, 13500 Pulau Pinang, Malaysia
e-mail: chema982@uitm.edu.my

U. R. Hashim et al. (eds.), *Green Infrastructure*,
https://doi.org/10.1007/978-981-99-7003-2_6

total of 39% of global energy-related emissions (Malaysia Ministry of Environment and Water, 2020) As a result, the need for society to decrease carbon emissions has grown, notably in the construction sector, which is one of the major worldwide producers to carbon emissions. As a result, there has been a significant increase in construction emission research publications.

Malaysia now has the highest daily peak temperatures, according to Malaysia's third biannual report to the UNFCC (Malaysia Ministry of Environment and Water, 2020). This is because CO_2 emissions in 2016 amounted to 263,577 Gg CO_2, an annual increase (Malaysia Ministry of Environment and Water, 2020). Consequently, in 2015, Malaysia agreed to a 45% reduction in emissions under the Paris Agreement. Furthermore, although carbon emissions have increased every year and with the Paris Agreement in 2015, the construction industries still depend significantly on the traditional building technique that contributes greatly to emissions levels according to Wen et al. (2015), Ishak et al. (2017), Balasbaneh and Ramli (2020) and Hafizan et al. (2021).

Most of the research on emissions brought on by building activities has solely looked at the choice of materials, techniques, or instruments to minimize construction emissions. Since most publications have concentrated on quantifying embodied carbon throughout the building lifetime, thorough research on carbon emissions in construction, particularly for onsite construction, is lacking (Acquaye & Duffy, 2010; Avetisyan et al., 2012; Bilec et al., 2010; Ye et al., 2017; Li & Zheng, 2020; Wolf et al., 2017; Zhang et al., 2019). This is also true for Malaysia as the current research trend is focused on emission reduction from the pre-construction phase and operation phase (Omar et al., 2014; Wen et al., 2015), while little attention has been paid to the construction phase, the design phase has received considerable attention. Therefore, this paper's goal is to examine current emissions management practices in construction operations as well as the crucial strategies that influence how effectively emissions are reduced during the operation of a construction projects.

2 Background of Study

2.1 *Overview of Carbon Emissions in Malaysia*

Carbon dioxide fluxes in Malaysian construction are growing as a result of the use of unsustainable energy sources to plan, build, and operate structures (Ahmed Ali et al., 2020). Malaysia's unprecedented carbon increase combined with business-as-usual practices might lead to unsustainable development (Zaid et al., 2015). In addition, Malaysia has no reliable statistics or analysis to demonstrate the level of carbon emissions produced during building construction (Hafizan et al., 2021; Lim et al., 2017; Omar, 2018). This is because according to Annuar et al. (2014), there is no consensus on the standardized approach to evaluating carbon emissions and that these practices remain at the early stage. Therefore, the findings of this study will provide information regarding the current practices and reduction strategies used in the Malaysian construction industry.

2.2 *Studies of Previous Literature*

Table 1 list the plans and present practices that have been explored in past research to reduce carbon emissions. Decision-makers are crucial in promoting sustainable development across all sectors because they understand the significance of sustainable practices. Government and policy makers have launched several initiatives to cut emissions in the building industry. Unquestionably, numerous initiatives in Malaysia to cut carbon emissions have centred on the construction of green buildings. The goal of this research is to explore the reduction tactics and methods that will be utilized as criteria for carbon reduction management strategies in the questionnaire survey.

3 Methodology

Based on the content of the questionnaires, the analysis was divided into two sections: demographic and relative importance analysis. Through literature review, the questionnaire was designed using current practices for reducing carbon on site and proposed carbon reduction strategies that can highly influence lowering the generation of carbon emissions during construction operations.

The validity and reliability of the analysis are intended to lead the questionnaire questions and to assess the accuracy of the designed questionnaire. The questionnaire design was divided into four parts as follows:

- Section 1: The information on the respondent profile includes the type of construction company, designation, experience, academic qualification, company operation age, and type of most involved projects.
- Section 2: Knowledge of carbon emissions in construction operations. Respondents were asked to define their level of knowledge on a Likert scale question based on their familiarity and understanding of the concept of carbon emissions.
- Section 3: Current practices on carbon emissions management on site. This section is designed to gain insight in current practices from the participants experience in the field of construction. This section allows the respondent to have multiple answers instead of a Likert scale.
- Section 4: strategies for a way forward on low-carbon initiatives in construction. Respondents were asked to rate which strategies is the most important to be adopted for construction operation by using a five-point Likert scale question from strongly disagree to strongly agree.

Table 1 Current practices and strategies from previous studies

Carbon reduction strategies	Current carbon reduction practices	References
Education and training	Conducting carbon emissions awareness training or environmental awareness training	Wei et al. (2020), Jackson and Kaesehage (2020), Rasdorf et al. (2012), Kim et al. (2012), Zhang et al. (2019)
	Training drivers for fuel-efficient techniques	
Low-carbon designs and materials	Performing construction modular process to reduce impacted green space for storage and onsite works	Wei et al. (2020), Liu et al. (2017), Zhang et al. (2019), Shi and Bai (2020)
	Progressively switching from heavy fuel oil (HFO) to less carbon-intensive natural gas	
	Promoting electronic communication channels by reducing the use of paper	
	Using light emitting diode (LED) and compact fluorescent lamps during construction	
Recycling material and waste	Reducing the wastage of building material	Giustozzi et al. (2015), Liu et al. (2017), Hossain and Poon (2018), Wei et al. (2020)
	Using 4R approach—reuse, reduce, recycle and replace	
Logistic • Proper site logistic planning • Reduce distance of transportation	Green procurement practices	Nasab et al. (2019), Luo et al. (2019), Itoya et al. (2012)
Equipment and Machinery • Energy efficient • Reduce machine idling time • New machine/ upgrade machine • Proper maintenance • Proper selection of equipment	Applying emissions-control equipment maintenance schedule on the projects	Wei et al. (2020), Wu et al. (2019), Gottsche et al. (2016), Nasab et al. (2019), Luo et al. (2019), Liu et al. (2017), Rasdorf et al. (2012), Kim et al. (2012), Zhang et al. (2019), Szamocki et al. (2019), Frey et al. (2010)
	Optimizing the usage of equipment for energy efficiency	
	Reducing the idling time of construction stationary plants, machineries and equipment used	
	Reducing unnecessary fuel consumption	
	Regular and scheduled maintenance of construction equipment and machinery	
	Site monitoring and supervision with the recording of the usage of equipment and machinery	

(continued)

Table 1 (continued)

Carbon reduction strategies	Current carbon reduction practices	References
Management • Task planning, operational planning, activity planning • Control strategy, crash strategy, catch up strategy • Implementing weekly emission cap • Linking emission and fuel use to the project schedule	Applying a lifecycle perspective in the projects from the design and planning stage to the end product and disposal stage to reduce waste generation	Rasdorf et al. (2012), Tang et al. (2013), Wong et al. (2013), Szamocki et al. (2019)
	Assessing performance data for energy utilization—electricity consumption	
	Assessing performance data for energy utilization—fuel consumption (diesel/petrol)	
	Reducing electricity consumption	
	Turning off site lighting at night (not security lighting) over the course of the project	
	Using emission control and monitoring systems	
Construction method • Prefabrication method • Stabilization of in situ soils • Use less intensive construction method	The implementation of IBS at project sites	Itoya et al. (2012), Giustozzi et al. (2015), Hong et al. (2015), Du et al. (2019), Luo et al. (2019), Li and Zheng (2020)
Renewable energy	Installing solar-powered CCTV for security and solar powered lighting at project site	Qi et al. (2014), Zhang and Wang (2016), Guo et al. (2017), Padilla-rivera et al. (2018

3.1 Sample Size and Target Population

The respondents are construction stakeholders with civil engineering background who work as project managers, project coordinators, construction managers, site managers, site engineers, supervisors, and so on. Since this study is a preliminary study for developing a carbon reduction management framework for construction operations, the study opted for a simple random sampling. 50 questionnaire survey was distributed randomly to construction companies near UiTM Shah area that are listed in CIDB to get a glimpse of current practices and preferred carbon reduction strategies used in the Malaysia Construction Industry. However, only 42 responses were received.

3.2 Descriptive Analysis

Demographic and project involvement particulars of all respondents was first to be extracted and tabled before taking place into the more sophisticated analysis. Pallant (2007) identified these clear-cut variables were examined by computing their occurrences. Outcomes then presented participants' information such as their position, type of organization, the number of projects they were involved in, and so on.

3.3 Cronbach's Alpha

In addition, questionnaire data reliability in this study is conducted by using Cronbach's Alpha in the SPSS software to measure the internal consistency of the questionnaire. Cronbach's alpha is often used when the Likert-type scale is used in the questionnaire to assess the internal consistency of the questionnaire. According to Gliem and Gliem (2003), the coefficient for Cronbach's Alpha typically ranges from 0 to 1. Whereby, the closer the coefficient is to 1, the internal consistency of the questionnaire is greater. It is also noted that 0.8 is a reasonable goal.

Since this is a preliminary study for the strategies of carbon reduction management in Malaysian construction operations, a Cronbach's Alpha analysis is conducted. As shown in Table 2, the result is 0.924 which means the question for the level of familiarity and understanding of carbon emissions is reliable to be included in the questionnaire survey.

Table 3 presents the result of Cronbach's Alpha for current practices listed in the questionnaire which is 0.891 for 23 lists of current practices. It is also noted that 0.8 is a reasonable goal.

Furthermore, the Likert Scale questions were validated as shown in Table 4 and the result was 0.972 which is closer to 1. Therefore, the 20-item Likert scale is reliable for this questionnaire.

Table 2 Cronbach's Alpha for the level of participant's familiarity and understandings

Reliability statistics		
Cronbach's Alpha	*Cronbach's Alpha based on standardized items*	*N of items*
0.923	0.924	4

Table 3 Cronbach's Alpha for current practices

Reliability statistics		
Cronbach's Alpha	*Cronbach's Alpha based on standardized items*	*N of items*
0.891	0.890	23

Table 4 Cronbach's Alpha for carbon emissions reduction strategies listed in the questionnaire

Reliability statistics		
Cronbach's Alpha	*Cronbach's Alpha based on standardized items*	*N of items*
0.972	0.972	20

3.4 Relative Importance Index

Relative Importance Index was used for the analysis because it best fits the purpose of this study. Relative Importance Index or weight is a type of relative importance analysis. Relative index analysis was selected in this study to rank the criteria according to their relative importance. RII fosters realizing the influence a particular variable makes on the prediction of a criterion variable both by itself and in combination with other predictor variables (Johnson and LeBreton, 2004). The responses from the respondents were analysed using the Microsoft Excel application. The following formula is used to determine the relative index:

$$\text{RII} = \Sigma \text{w}/(\text{A} \times \text{N}) \tag{1}$$

where w is the weighting as assigned by each respondent on a scale of one to five with one implying the least and five the highest. A is the highest weight and N is the total number of the sample. The relative importance index (RII) was used to rank multiple items within each section ($0 \leq \text{RII} \leq 1$). An item achieving a higher RII score would rank higher than those with lower RII values.

4 Results and Discussion

This section presented the descriptive statistics, reliability, and validity analysis of the questionnaire survey conducted using SPSS version 26 and Microsoft Excel application.

4.1 Demographic Background

Table 5 summarized the demographic background of 42 respondents from various types of construction stakeholders listed by CIDB in the Klang Valley area. Demographic characteristics include the type of organizations, their function within the company, their level of experience and education, the number of projects they are primarily involved in, and their years of operation.

Table 5 Demographic background of participants

Demographic characteristic	Frequency	Percentage
Type of construction company		
Client	7	16.7
Consultant	5	11.9
Contractor	26	61.9
Developer	2	4.8
Others	2	4.8
Total	42	100.0
Role in the company		
Engineer	22	52.4
Quality/environmental/safety officer	5	11.9
Senior/project engineer	2	4.8
Site supervisor	3	7.1
Resident engineer	2	4.8
Assistant project manager	3	7.1
Project director	1	2.4
Project executive	2	4.8
Purchasing coordinator	1	2.4
Construction manager	1	2.4
Total	42	100.0
Year of experience		
<10 years	36	85.7
10–15 years	1	2.4
>20 years	5	11.9
Total	42	100.0
Academic qualification		
Bachelor's degree	32	76.2
Master/PhD	10	23.8
Total	42	100.0
Company's operation years		
<10 years	14	33.3
10–15 years	8	19.0
16–20 years	6	14.3
>20 years	14	33.3
Total	42	100.0
Type of construction project		
Residential	12	28.6
Commercial/office	5	11.9

(continued)

Table 5 (continued)

Demographic characteristic	Frequency	Percentage
Infrastructure	17	40.5
Social amenities	5	11.9
Others	3	7.1
Total	42	100.0

4.2 Familiarity and Understanding of the Concept of Carbon Emissions in Construction

Figure 1 shows the level of the participant's familiarity with the concept of carbon emissions in construction, especially in construction sites. 38% of the respondents responded that they have a moderate familiarity with the knowledge of carbon emissions. As shown in Fig. 1, 42 participants are most likely indicated to be on a lower level of familiarity. This is a result of the construction industry's failure to recognize the importance of reducing carbon emissions. However, in 2016 the government has set a mandatory adoption of Malaysian Carbon Reduction and Environmental Sustainability Tools (MyCREST) for all valued 500 million and above for government projects. However, since most companies concentrate on private building projects rather than public ones, most construction stakeholders are unfamiliar with this instrument.

Figure 2 presented the participants' level of understanding of the concept, negative effects, and facts and information related to construction emissions in Malaysia and globally. As shown in Fig. 2, the trend of the participants' level of understanding is moderately familiar for the concept of carbon emissions is 57%, for the negative

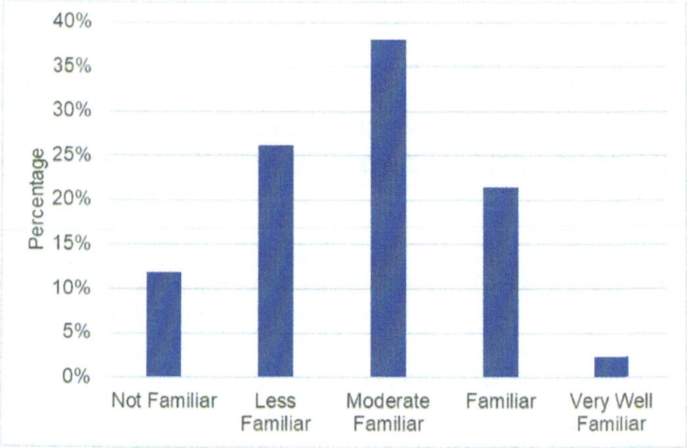

Fig. 1 Level of participant's familiarity

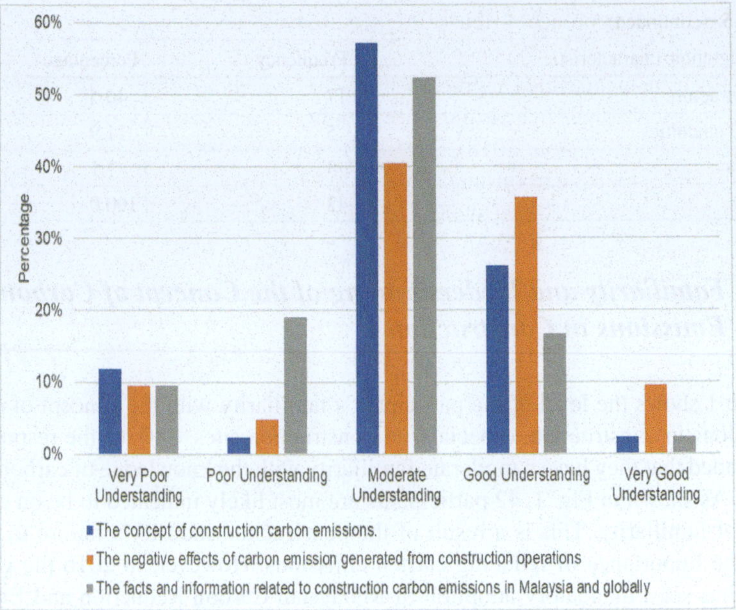

Fig. 2 Level of participants' understanding

effects of carbon emissions generated from construction operations is 40% and for the facts and information related to construction emissions in Malaysia and globally is 52%. This indicated that most of the participants have quite a moderate understanding of the knowledge on carbon emissions. However, in Fig. 1, most of the participants are more inclined to a lower level of familiarity. This confirms that most of these construction companies do not fully adopt the idea of reducing carbon emissions on site.

4.3 *Current Practices of Carbon Emissions Management in the Construction Industry Specifically in Construction Site*

The percentage frequencies of current Malaysian construction carbon reduction practices are presented in Table 6. Thirty carbon reduction practises were obtained from previous studies shown in Table 1. However, it is noted that this list was overwhelming to the participants as there are comments on the length of this questionnaire. Therefore, only the top 10 of this current practice listed in Table 6 will be designed in the final questionnaire survey. However, in Table 6 top ten of the current practices are mainly focused on reducing emissions directly on site which conforms to the aim of this research to focus on reducing construction operation on sites. Among the top

10, using the 4R approach—Reuse, Reduce, Recycle, and Replace—was ranked first because it is quick and inexpensive to implement. For instance, it can decrease material waste by using recycled aggregate from old concrete in new batches of concrete or by reusing steel panels in hoarding construction, which lowers carbon dioxide emissions (Hussain and Poon, 2018). Changing gradually from carbon-intensive heavy fuel oil (HFO) to less carbon-intensive natural gas was listed last on the list. Further investigations concluded that switching to less carbon-intensive natural gas requires more cost because construction stakeholders will have to buy new machinery that can use the low-carbon natural gas instead of regular machineries.

4.4 Strategies of Carbon Reduction Management in the Construction Industry Specifically in Construction Site

Table 7 shows the ranking of carbon reduction strategies according to the relative importance index. The overall RII shows that the most important strategy to reduce carbon emissions is improving the efficiency of equipment usage with RII = 0.833. Kim et al. (2012) claim that decreasing idle machinery and appropriate equipment maintenance will optimize fossil fuel savings and increase machinery efficiency. Building in situ piles was essential in the present case study to achieve the goal of a 2.54% reduction in GHG emissions (Luo et al., 2019). The second important strategy is implementing planned waste management through the concept of reduce, reuse, and recycle with RII = 0.824. Reusing building materials, according to Padilla-rivera et al. (2018), may save a lot of resources in addition to other environmental advantages including a decrease in the amount of waste dumped in landfills and the energy used to generate virgin materials. The degree to which materials are recyclable and how well a recycling plan is implemented may affect a building's embodied carbon in several ways.

Next is, the selection of sustainable construction methods such as the prefabrication method with RII = 0.814. This shows that using sustainable building techniques is vital since traditional techniques are often to blame for high carbon emissions methods owing to their usual ways. Yanli and Chao (2021) claim that prefabricated element CO_2 emissions and production costs might be significantly reduced by rationally enhancing the manufacturing process of prefabricated components, optimizing product performance, and minimizing waste throughout the manufacturing process. Nevertheless, the participants for this preliminary study ranked site layout planning as the least important strategy to reduce carbon emissions with RII = 0.724. Yes, surprisingly, none of the 20 tactics described in this early survey score over the index's midpoint for significance. This is very clear the importance of carbon reduction strategies in developing a carbon reduction management framework.

Table 6 Percentage Frequencies of current carbon reduction practices conducted in construction site

No	Item	Total selected current practices	Percentage of frequency (%)
1	Using 4R approach—reuse, reduce, recycle, and replace	34	8.72
2	Site monitoring and supervision with the recording of the usage of equipment and machinery	24	6.15
3	Regular and scheduled maintenance of construction equipment and machinery	22	5.64
4	Reducing electricity and fuel consumption on site	21	5.38
5	The implementation of IBS at project sites	19	4.87
6	Green procurement practices	18	4.62
7	Optimizing the usage of equipment for energy efficiency	16	4.10
8	Reducing the idling time of construction stationary plants, machineries and equipment used	16	4.10
9	Applying a lifecycle perspective in the projects from the design and planning stage to the end product and disposal stage to reduce waste generation	15	3.85
10	Assessing performance data for energy utilization—electricity consumption	15	3.85
11	Promoting electronic communication channels by reducing the use of paper	14	3.59
12	Turning off site lighting at night (not security lighting) over the course of the project	13	3.33
13	Assessing performance data for energy utilization—fuel consumption (diesel/petrol)	12	3.08
14	Using light emitting diode (LED) and compact fluorescent lamps during construction	12	3.08

(continued)

Table 6 (continued)

No	Item	Total selected current practices	Percentage of frequency (%)
15	Using emission control and monitoring systems	10	2.56
16	Applying emissions-control equipment maintenance schedule on the projects	9	2.31
17	Conducting carbon emissions awareness training or environmental awareness training	9	2.31
18	Conducting emissions and monitoring reporting for the projects	9	2.31
19	Installing solar-powered CCTV for security at project site	8	2.05
20	Conducts periodic site visits, facilitates fortnightly Green Building coordination meetings and prepares Green Building monthly progress reports	7	1.79
21	Performing construction modular process to reduce impacted green space for storage and onsite works	4	1.03
22	Training drivers for fuel-efficient techniques	4	1.03
23	Progressively switching from heavy fuel oil (HFO) to less carbon-intensive natural gas	1	0.26
		312	100.00

4.5 Discussion

According to the 42 questionnaires that were gathered, 26 of the respondents were workers for contractors, while 7 were clients, 5 were from consulting firms, 2 were from developing enterprises, and the last group was not from private businesses. In addition, most of the respondent were engineers in their respective companies which mean that these respondents were involved in construction on-site practices. However, 76.2% of the respondents have bachelor's degree as their highest academic qualification and exhibit their knowledge of carbon reduction. Based on Sect. 2 on the knowledge level of participants on construction carbon emissions, the trend of

Table 7 Relative importance index for the strategies of carbon emission reductions for construction operations

No	Items	Weighted total	Total number (N)	RII (relative important index)	Rank
1	Improve the efficiency of equipment usage	175	42	0.833	1
2	Implement planned waste management through the concept of reduce, reuse and recycle	173	42	0.824	2
3	Selection of sustainable construction methods such as prefabrication method	171	42	0.814	3
4	Conduct propaganda and training for carbon emissions reduction scheme	170	42	0.810	4
5	Increases operator expertise	170	42	0.810	4
6	Proper maintenance of equipment	170	42	0.810	4
7	Usage of low-carbon material	169	42	0.805	7
8	Reduction of machinery idling time	168	42	0.800	8
9	Optimisation of energy, electricity, gas & water and material utilization in every construction operational activity	167	42	0.795	9
10	Linking emissions and fuel usage to the project schedule	170	43	0.791	10
11	Selection of energy efficient equipment	166	42	0.791	11
12	Reduction of equipment and material transportation distance	160	41	0.781	12
13	Site logistic planning	162	42	0.771	13

(continued)

Table 7 (continued)

No	Items	Weighted total	Total number (N)	RII (relative important index)	Rank
14	Utilization of renewable energy such as solar energy panel	162	42	0.771	13
15	Applying carbon assessment practices onsite by analysing and evaluating energy performance	161	42	0.767	15
16	Uniform and relevant policies specifically for carbon emissions in construction	161	42	0.767	15
17	Implementing weekly emission cap	158	42	0.752	17
18	Supervision and inspection of energy consumption during construction operations	158	42	0.752	17
19	Using alternative fuels for equipment such as biodiesel fuel or low sulphur fuel	154	42	0.733	19
20	Site layout planning	152	42	0.724	20

understanding the concept of carbon was more on having a moderate and understanding of carbon emissions while the trend for the level of familiarity was inclined to having less familiarity towards carbon emissions. This implies that most of them must have not applied any carbon reduction management system in their projects. This is anticipated since there are no laws requiring building projects to decrease carbon emissions, except for government projects costing more than 500 million ringgits.

Figure 3 shows that the 4R strategy, which stands for reuse, reduce, recycle, and recycle of building materials, is at the top of the list of current practices. This practice is mostly used because this is one of the cheapest practices that can be done in construction not only can it curb emissions levels, but this practice can also reduce the consumption of materials and unnecessary waste. When choosing components for low-carbon strategies, the recyclability of materials and the implementation of a recycling plan should be considered since they might have different impacts on the embodied carbon of buildings. According to Hossain and Poon's (2018) study, recycling waste steel into steel scraps and using steel panels in new hoarding construction can reduce emissions by up to 90% while recycling concrete can create recycled aggregate for making new concrete, which can reduce emissions by 12% overall. While the most ideal strategy rated by the respondents was to improve the efficiency of equipment usage in construction projects. Considering this, the respondents' most recent practices put this ideal technique sixth. Fleet managers should evaluate the impact of fuel use and pollutants to estimate the usable life and failure rate of the equipment when selecting which equipment must be employed for a project and even when removing, repairing, or upgrading obsolete equipment, according to Frey et al. (2010). However, these strategies would require continuous efforts from higher management levels having decision-making roles.

Besides that, the second rank falls to site monitoring and supervision with a recording of the usage of equipment and machinery. Ahn et al. (2013) claimed that by estimating and benchmarking during the pre-construction stage and then monitoring throughout the construction stage, it is possible to capture the opportunities for reducing emissions from construction operations. Nevertheless, according to Ahn et al. (2013), efforts to continually monitor the environmental performance of operations during construction are still in their infancy because there isn't a workable monitoring method. These assessments do not provide any details on how long or how effectively the equipment was used; only whether it was used. The second ideal strategy is to implement planned waste management through the concept of reuse, reduce, and recycle which is the top rank current practice for reducing carbon on site by the respondents. This shows that the construction players are implementing their preferred ideal strategies.

Moving on, regular and scheduled maintenance of construction equipment and machinery is the third current practice and the six ideal strategies whereby Kim et al. (2012) revealed that proper equipment servicing will maximize the saving of carbon fuels and improve equipment efficiency. This practice supports that the construction players are in their growing potential to reduce emissions. The third ideal

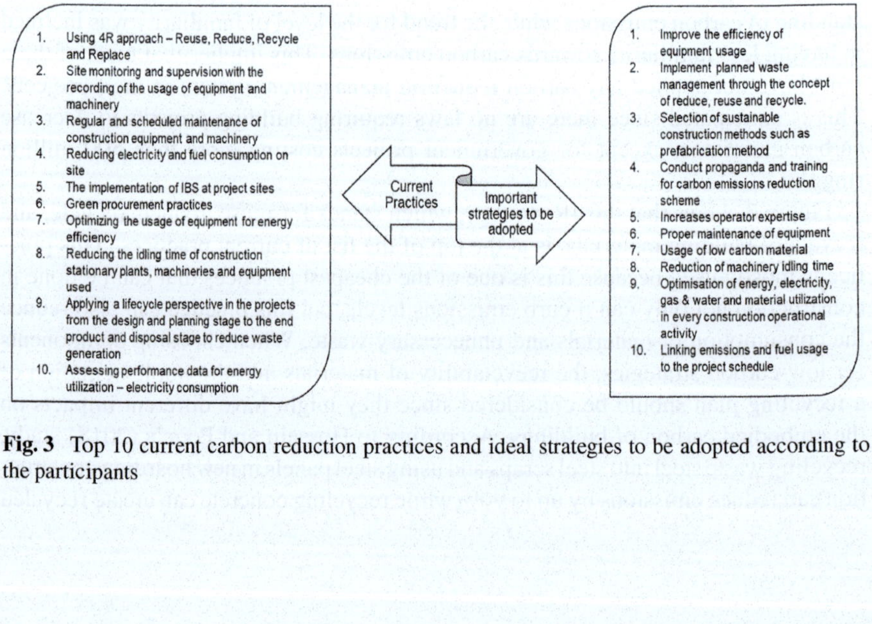

Fig. 3 Top 10 current carbon reduction practices and ideal strategies to be adopted according to the participants

strategy was selecting sustainable construction methods such as prefabrication was also the fifth current practice where the respondents mostly implemented IBS at their project sites. Prefabricated structures emit less CO_2 emissions at the construction site than traditional buildings, according to research by Du et al. (2019). Prefabrication technology thus opens even additional opportunities for the building industry to reduce carbon emissions.

Subsequently, reducing electricity and fuel consumption on site is the fourth top current practice that is also greatly linked with the ninth top ideal strategy which is the optimization of energy, electrical, gas, and water, and material utilization in every construction operational activity. While the respondent opts for advertising and training for carbon emissions reduction schemes. Shi and Bai (2020) recommended that the government and the building sector support professional education that raises awareness of the potential advantages of low-carbon public buildings. In addition, the fifth ideal strategy is also profoundly linked to the training for carbon emissions scheme as an increase in operator expertise will also reduce the impact of carbon emissions. Equipment drivers should get the proper training, according to recommendations made by Rasdorf et al. (2012) and Kim et al. (2012), to reduce the burning of fossil fuels.

Thereafter, the sixth top current practice is green procurement practices. Itoya et al. (2012) highlighted that effective procurement is essential for reducing the industrial and environmental hazards connected to road maintenance. Additionally, Yanli and Chao (2021) promoted efficient logistics transportation plans that link the length of the operation, the form of transportation, the route used, and the method used to load the components. This is because when the distance between the site the material processing plant and the waste recycling facility rises, the emissions rate connected to the supply of construction supplies and trash collection from the site increases. The usage of low-carbon material is the seventh ideal strategy to reduce emissions. According to Zhang et al. (2019), it is best to employ less materials in buildings to reduce energy consumption and pollution. The strength of a building's material usage may be significantly impacted by changes in its structural form.

Next, reducing the idling time of construction stationary plants, machineries and equipment used is the eighth top current practice and the eighth ideal strategy rated by the respondents. In the example studies that were undertaken, Szamocki et al. (2019) claimed reducing carbon emissions by 53% by doing away with idling. Rasdorf et al. (2012) expressed the same thing, stating that contractors might reduce idle time while equipment is not in use to be more fuel-efficient.

The ninth and tenth of the top current rankings are employing a lifecycle approach to projects from the design and planning stage to the final product and disposal stage to limit waste creation and evaluating performance data for energy usage notably electricity consumption. Whereas the last top-rated ideal strategy is linking emissions and fuel usage to the project schedule. According to Rasdorf et al. (2012), tying the project timeline to fuel use would minimize emissions, make costs apparent, and provide contractor management savings.

In summary, this study revealed that most of the respondents have quite high levels of familiarity in contrast when the respondent was asked to rate their level of familiarity and understanding was trending to be moderate level. Some of the responders have already started using some of the ideal tactics. However, most of these strategies will only be able to optimize through the continuous effort of higher management levels and from the local authorities to apply stricter regulations in reducing emissions.

5 Conclusion

This study described the development of a set of strategies and mitigation measures criteria and current practices in developing a framework for the management of carbon reduction in construction operations in Malaysia. Responses of forty-two respondents. A total of twenty (20) strategies and mitigation criteria and twenty-three current practices were identified on a thorough literature review and discussion with selected expertise in assessing carbon emissions in construction projects. Relative importance index analysis was used to determine the ranking of each criterion and it was revealed that all twenty strategies were perceived as very important since the RII number was closer to 1. In general, as demonstrated from the data obtained, there is an untapped potential and growing awareness for construction stakeholders to reduce carbon emissions in their respective projects. Reviewing both previous literature and the responses from construction stakeholders, to develop a set of mitigation criteria for carbon reduction framework, higher level of construction stakeholders have a decision-making role in the construction projects that helps to optimize the maximum capabilities in reducing emissions. This is because of the higher management level in selecting suitable carbon reduction methods. A shift from traditional construction management to more sustainable practices such as maximum usage of highly efficient equipment, implementing waste management, selection of low-carbon construction methods, etc. in conclusion, this study through preliminary study highlighted the most current practices of carbon reduction where more research needs to be done specifically in identifying the causes and the barriers of the adoption of these ideal strategies. The outcomes of this study will be the next step towards developing a more improved carbon reduction management framework in Malaysia. This study advances our understanding of how carbon emissions are managed in the construction sector and the best practices for implementing carbon reduction mitigation strategies in Malaysian construction operations projects.

Acknowledgements The authors would like to express their gratitude to the Ministry of Higher Education Malaysia and Universiti Teknologi MARA, Shah Alam, for supporting this publication via research grant FRGS/1/2019/ TK10/UITM/02/4 and the gratitude to the School of Civil Engineering, College of Engineering, Universiti Teknologi MARA, Shah Alam, and construction stakeholders for participating in this research.

References

Acquaye, A. A., & Duffy, A. P. (2010). Input–output analysis of Irish construction sector greenhouse gas emissions. *Building and Environment, 45*(3), 784–791. https://doi.org/10.1016/j.buildenv.2009.08.022

Ahn, C., Lee, S., Mora, F. P., & Abourizk, S. (2010a). Toward environmentally sustainable construction processes: The U.S. and Canada's perspective on energy consumption and GHG/CAP emissions. *Sustainability*, 354–370. https://doi.org/10.3390/su2010354

Ahn, C., Xie, H., Lee, S., Abourizk, S., & Mora, F. P. (2010b). Carbon footprints analysis for tunnel construction processes in the preplanning phase using collaborative simulation. *Construction Research Congress, 2010*, 1538–1546.

Ahn, C. R., Lewis, P., Golparvar-Fard, M., & Lee, S. (2013). Integrated framework for estimating, benchmarking, and monitoring pollutant emissions of construction operations. *Journal of Construction Engineering and Management, 139*(12). https://doi.org/10.1061/(asce)co.1943-7862.0000755

Ahmed Ali, K., Ahmad, M. I., & Yusup, Y. (2020). Issues, impacts, and mitigations of carbon dioxide emissions in the building sector. *Sustainability, 12*(18), 7427. https://doi.org/10.3390/su12187427

Annuar, N. M., Osmond, P., & Prasad, D. K. (2014). An exploratory review of life cycle assessment activities in the Malaysian an exploratory review of life cycle assessment activities in the Malaysian building industry. In *Life cycle thinking and policy: Towards a sustainable society*. Wellington, New Zealand

Avetisyan, H. G., Miller-hooks, E., & Melanta, S. (2012). Decision models to support greenhouse gas emissions reduction from transportation construction projects. *Journal of Construction Engineering and Management, 138*(5), 631–641. https://doi.org/10.1061/(ASCE)CO.1943-7862.0000477

Balasbaneh, A. T., & Ramli, M. Z. (2020). A comparative life cycle assessment (LCA) of concrete and steel-prefabricated prefinished volumetric construction structures in Malaysia. *Environmental Science and Pollution Research.*

Bilec, M. M., Ries, R. J., & Matthews, H. S. (2010). Life-cycle assessment modeling of construction processes for buildings. *Journal of Infrastructure Systems, 16*(3), 199–205. https://doi.org/10.1061/(ASCE)IS.1943-555X.0000022

Chen, J., Xu, C., Managi, S., & Song, M. (2019). Energy-carbon performance and its changing trend: An example from China's construction industry. *Resources, Conservation & Recycling, 145*(October 2018), 379–388. https://doi.org/10.1016/j.resconrec.2019.03.014

Du, Q., Bao, T., Li, Y., Huang, Y., & Shao, L. (2019). Impact of prefabrication technology on the cradle to site CO_2 emissions of residential buildings. *Clean Technologies and Environmental Policy, 21*, 1499–1514. https://doi.org/10.1007/s10098-019-01723-y

Frey, H. C., Rasdorf, W., & Lewis, P. (2010). Comprehensive field study of fuel use and emissions of nonroad diesel construction equipment. *Transportation Research Record: Journal of the Transportation Research Board, 2158*, 69–76. https://doi.org/10.3141/2158-09

Giustozzi, F., Flintsch, G. W., & Crispino, M. (2015). Environmental impact analysis of low-carbon road-foundation layers. *International Journal of Sustainable Transportation, 9*(1), 73–79. https://doi.org/10.1080/15568318.2012.738355

Gliem, J. A., & Gliem, R. R. (2003). Calculating, interpreting, and reporting Cronbach's Alpha reliability coefficient for Likert-type scales. In *Midwest research-to-practice conference*, pp 82–88. The Ohio State University, Columbus

Gottsche, J., Kelly, M., Taggart, M., Kelly, M., & Taggart, M. (2016). Assessing the impact of energy management initiatives on the energy usage during the construction phase of an educational building project in Ireland. *Construction Management and Economics*, June, 6193. https://doi.org/10.1080/01446193.2016.1162317

Guo, H., Liu, Y., Meng, Y., Huang, H., Sun, C., & Shao, Y. (2017). A comparison of the energy saving and carbon reduction performance between reinforced concrete and cross-laminated timber structures in residential buildings in the severe cold region of China. *Sustainability, 9*, 1426. https://doi.org/10.3390/su9081426

Hafizan, C., Hussein, N., & Noor, Z. Z. (2021). Life cycle assessment framework application in Malaysia. *IOP Conference Series: Materials Science and Engineering, 1051*(1), 012101. https://doi.org/10.1088/1757-899x/1051/1/012101

Hong, Y., Jaillon, L., Chu, P., & Poon, C. S. (2015). Comparing carbon emissions of precast and cast-in-situ construction methods—A case study of high-rise private building. *Construction and Building Materials, 99*, 39–53. https://doi.org/10.1016/j.conbuildmat.2015.08.145

Hossain, U., & Poon, C. S. (2018). Global warming potential and energy consumption of temporary works in building construction: A case study in Hong Kong. *Building and Environment, 142*(June), 171–179. https://doi.org/10.1016/j.buildenv.2018.06.026

Ishak, N. I., Mustafa Kamal, E., & Yusof, N. (2017). The green manufacturer's compliance with green criteria throughout the life cycle of building material. *SAGE Open, 7*(3), 1–12. https://doi.org/10.1177/2158244017725446

Itoya, E., Hazell, K., Ison, S., El-hamalawi, A., & Frost, M. W. (2012). Framework for carbon emission evaluation of road maintenance. *Transportation Research Record: Journal of the Transportation Research Board*. https://doi.org/10.3141/2292-01

Jackson, D. J., & Kaesehage, K. (2020). Addressing the challenges of integrating carbon calculation tools in the construction industry. *Business Strategy and the Environment*, (May), 2973–2983. https://doi.org/10.1002/bse.2551

Johnson, J., & LeBreton, J. (2004). History and use of relative importance indices in organizational research. *Organizational Research Methods, 7*, 238–257. https://doi.org/10.1177/1094428104266510

Kim, B., Lee, H., Park, H., & Kim, H. (2012). Greenhouse gas emissions from onsite equipment usage in road construction. *Journal of Construction Engineering and Management, 138*(August), 982–990. https://doi.org/10.1061/(ASCE)CO.1943-7862.0000515

Li, X., & Zheng, Y. (2020). Using LCA to research carbon footprint for precast concrete piles during the building construction stage: A China study. *Journal of Cleaner Production, 245*, 118754. https://doi.org/10.1016/j.jclepro.2019.118754

Lim, P. Y., Yahya, K., Aminudin, E., Zakaria, R., Haron, Z., Zin, R. M., & Redzuan, A. A. H. (2017). Carbon footprint of construction using industrialised building system. *IOP Conference Series: Materials Science and Engineering (IOP Publishing)*. https://doi.org/10.1088/1757-899X/271/1/012107

Liu, Y., Wang, Y., & Li, D. (2017). Estimation and uncertainty analysis on carbon dioxide emissions from construction phase of real highway projects in China. *Journal of Cleaner Production, 144*, 337–346. https://doi.org/10.1016/j.jclepro.2017.01.015

Luo, W., Sandanayake, M., & Zhang, G. (2019). Direct and indirect carbon emissions in foundation construction—Two case studies of driven precast and cast-in-situ piles. *Journal of Cleaner Production, 211*, 1517–1526. https://doi.org/10.1016/j.jclepro.2018.11.244

Malaysia Ministry of Environment and Water. (2020). *Third Biennial update report to the UNFCC*. https://unfccc.int/sites/default/files/resource/MALAYSIA_BUR3-UNFCCC_Submission.pdf

Nasab, T. J., Monavari, S. M., Jozi, S. A., & Majedi, H. (2019). Assessment of carbon footprint in the construction phase of high rise constructions in Tehran. *International Journal of Environmental Science and Technology*. https://doi.org/10.1007/s13762-019-02557-3

Omar, W. W., Doh, J., & Panuwatwanich, K. (2014). Variations in embodied energy and carbon emission intensities of construction materials. *Environmental Impact Assessment Review, 49*, 31–48. https://doi.org/10.1016/j.eiar.2014.06.003

Omar, W. M. S. W. (2018). A hybrid life cycle assessment of embodied energy and carbon emissions from conventional and industrialised building systems in Malaysia. *Energy & Buildings, 167*, 253–268. https://doi.org/10.1016/j.enbuild.2018.02.045

Padilla-rivera, A., Amor, B., & Blanchet, P. (2018). Evaluating the link between low carbon reductions strategies and its performance in the context of climate change: A carbon footprint of a

wood-frame residential building in Quebec, Canada. *Sustainability, 10*, 2715. https://doi.org/10.3390/su10082715
Pallant, J. F. (2007). *SPSS survival manual: A step by step guide to data analysis using SPSS* (3rd ed.). Open University Press.
Qi, X., Peng, L., & Lou, J. (2014). Comparative study of new construction and renovation project based on carbon emission. *Advanced Materials Research, 1008–1009*, 1269–1273. https://doi.org/10.4028/www.scientific.net/AMR.1008-1009.1269
Rasdorf, W., Lewis, P., Marshall, S. K., Arocho, I., & Frey, H. C. (2012). Evaluation of on-site fuel use and emissions over the duration of a commercial building project. *Journal of Infrastructure Systems, 18*(June), 119–129. https://doi.org/10.1061/(ASCE)IS.1943-555X.0000071
Shi, Q. J., & Bai, C. (2020). Carbon emission efficiency in the construction industry and its carbon emission control measures: A case study of Henan Province, China. *Nature Environment and Pollution Technology, 19*, 791–797. https://doi.org/10.46488/NEPT.2020.v19i02.036
Szamocki, N., Kim, M., Asce, M., Ahn, C. R., Asce, M., Brilakis, I., & Asce, M. (2019). Reducing greenhouse gas emission of construction equipment at construction sites: Field study approach. *Journal of Construction Engineering and Management, 145*(9). https://doi.org/10.1061/(ASCE)CO.1943-7862.0001690
Tang, P., Cass, D., & Mukherjee, A. (2013). Investigating the effect of construction management strategies on project greenhouse gas emissions using interactive simulation. *Journal of Cleaner Production, 54*, 78–88. https://doi.org/10.1016/j.jclepro.2013.03.046
United States Environmental Protection Agency. (2021, November 19). Overview of greenhouse gases. US EPA; United States Environmental Protection Agency. https://www.epa.gov/ghgemissions/overview-greenhouse-gases
Wei, L., Tam, V. W. Y., & Du Heng Chen, L. (2020). A holistic review of research on carbon emissions of green building construction industry. *Engineering, Construction and Architechtural Management, 27*(5), 1065–1092. https://doi.org/10.1108/ECAM-06-2019-0283
Wen, T. J., Siong, H. C., & Noor, Z. Z. (2015). Assessment of embodied energy and global warming potential of building construction using life cycle analysis approach: Case studies of residential buildings in Iskandar Malaysia. *Energy and Buildings, 93*(2015), 295–302. https://doi.org/10.1016/j.enbuild.2014.12.002
Wolf, C. D., Pomponi, F., & Moncaster, A. (2017). Measuring embodied carbon dioxide equivalent of buildings: A review and critique of current industry practice. *Energy and Buildings, 140*, 68–80. https://doi.org/10.1016/j.enbuild.2017.01.075
Wong, J. K. W., Li, H., Wang, H., Huang, T., Luo, E., & Li, V. (2013). Toward low-carbon construction processes: The visualization of predicted emission via virtual prototyping technology. *Automation in Construction, 33*, 72–78. https://doi.org/10.1016/j.autcon.2012.09.014
Wu, P., Song, Y., Zhu, J., & Chang, R. (2019). Analyzing the influence factors of the carbon emissions from China's building and construction industry from 2000 to 2015. *Journal of Cleaner Production, 221*(2019), 552–566. https://doi.org/10.1016/j.jclepro.2019.02.200
Yanli, S., & Chao, L. (2021). Measurement and analysis of carbon emissions from prefabricated buildings under the transition of new and old kinetic energy measurement and analysis of carbon emissions from prefabricated buildings under the transition of new and old kinetic energy. *IOP Conference Series: Earth and Environmental Science, 813*, 012007. https://doi.org/10.1088/1755-1315/813/1/012007
Ye, H., Hu, X., Ren, Q., Lin, T., Li, X., Zhang, G., & Shi, L. (2017). Effect of urban micro-climatic regulation ability on public building energy usage carbon emission. *Energy and Buildings, 154*, 553–559. https://doi.org/10.1016/j.enbuild.2017.08.047
Zaid, S. M., Myeda, N. E., Mahyuddin, N., & Sulaiman, R. (2015). Malaysia's rising GHG emissions and carbon "lock-in" risk: A review of Malaysian building sector legislation and policy. *Journal of Surveying, Construction & Property, 6*(1), 1–13. https://doi.org/10.22452/jscp.vol6no1.1
Zhang, X., & Wang, F. (2016). Hybrid input-output analysis for lifecycle energy consumption and carbon emissions of China's building sector. *Building and Environment, 104*, 188–197. https://doi.org/10.1016/j.buildenv.2016.05.018

Zhang, Y., Yan, D., Hu, S., & Guo, S. (2019). Modelling of energy consumption and carbon emission from the building construction sector in China, a process-based LCA approach. *Energy Policy, 134*(December 2018), 110949. https://doi.org/10.1016/j.enpol.2019.110949

Green Infrastructure Development in Malaysia: A Review

Nur Shuhada Nor Shahrudin, Nur Kamaliah Mustaffa, and Che Maznah Mat Isa

Abstract Infrastructure development is critical to the country's economic integration goal, and sustainable infrastructure is a vital enabler of economic, social, and environmental development. Several assessment frameworks and tools have emerged in response to growing concerns about sustainability. Despite this, a comprehensive study has yet to examine the key characteristics of developing a sustainable assessment tool for infrastructure projects. This study investigates the Malaysian government's top-down initiatives to encompass all attempts to transform Malaysia into a nation with green infrastructure development. The paper highlights the development of green development in Malaysia over the previous three decades by incorporating fiscal, institutional, legislative, and regulatory mechanisms into the nation's national plan. Malaysia's progress towards constructing green infrastructure and becoming a climate-resilient nation is assessed holistically as a result of the detailed study of incorporating green techniques into current assessments. This paper found that most studies on assessment tools have focused on identifying potential indicators to measure the sustainability of infrastructure projects for decision-making purposes. This paper provides information on the sustainable infrastructure attributes to facilitate sustainability development goals and the impact on infrastructure development directions and strategies, which can be a reference for future sustainable infrastructure development studies.

Keywords Green development · Infrastructure · Sustainable infrastructure · Construction · Malaysia

N. S. Nor Shahrudin · N. K. Mustaffa (✉)
School of Civil Engineering, College of Engineering, Universiti Teknologi MARA, 40450 Shah Alam, Selangor, Malaysia
e-mail: nurkamaliah@uitm.edu.my

C. M. Mat Isa
Center of Civil Engineering Studies, Universiti Teknologi MARA Pulau Pinang Branch, Permatang Pauh Campus, 13500 Pulau Pinang, Malaysia

U. R. Hashim et al. (eds.), *Green Infrastructure*,
https://doi.org/10.1007/978-981-99-7003-2_7

1 Introduction

Global warming has become a major international concern due to its negative consequences on every country's environment, social development, and economic growth. Greenhouse gas (GHG) emissions, which have increased dramatically as a result of industrialisation and urbanisation, are commonly believed to be a major contributor to global warming. According to the OECD in 2017, socio-economic activities in urban areas account for around 70% of worldwide CO_2 emissions. Sustainable development (SD) has emerged as an approach to minimising the effects of climate change and global warming. The United Nations General Assembly established the Sustainable Development Goals (SDGs) in 2015 to lead countries' efforts towards sustainable development.

Infrastructure is a major concern in both emerging and developed nations. Sustainable infrastructure is essential since it directly affects all sustainability measures. Since it is necessary for every region and economy, the sewage infrastructure system is important in developing and wealthy nations. To further balance the economic, social, and environmental components of sustainable development, a sustainable infrastructure may be especially helpful for developing countries. Having a project organised such that it achieves its aim to the best of its abilities throughout the remainder of its life, protecting users' quality of life at the lowest practicable cost, is the concept meant by Diaz-sarachaga et al. (2016). However, designers of such systems face several challenges and potential threats to long-term functioning, including age, degradation, underfunding, disruptive events, and population growth (Upadhyaya et al., 2016). It also influences the success of infrastructure construction projects (Krajangsri & Pongpeng, 2017).

The infrastructure sector earned the highest public sector development investment in Malaysia in the 12th Malaysia Plan. According to 12MP, the RM400 billion allocation is the highest in history. Half of it is for basic development expenditure, which funds development initiatives like building schools, hospitals, and roads. Assuming that 2020s containment measures are not repeated, DOSM (2020) predicted that the industry would have a considerable comeback in 2021. Investments in transportation and energy projects would have supported growth. Between 2022 and 2025, the industry is estimated to rise by 6.0–6.6% annually, with 2021 seeing a 9.8% increase. An improvement in economic conditions and investments in infrastructure, renewable energy, and residential, telecommunications, and water infrastructure projects will all contribute to the industry's growth over the projection period. The state governments of Sabah and Sarawak received a total of MYR9.6 billion (US$2.2 billion) in the 2021 budget from the federal government to develop road, electricity, and water infrastructure projects and enhance health and educational facilities. Adapting to low-carbon growth, energy conservation, and climate-resilient development using green technology is challenging in Malaysia because of the country's rapidly expanding industrial and urban sectors. The government's policies, previous studies on green infrastructure, assessments to support sustainable development goals, and the impact on infrastructure development directions and strategies are all examined in this paper.

The findings can be used as a guide for future research on the development of sustainable infrastructure. This paper opens the door for future research on sustainable infrastructure development, particularly in Southeast Asia, by providing insight into present green development policies and programmes.

2 Green Infrastructure in Malaysia

Malaysia has started sustainable development initiatives in response to widespread environmental concerns and the need to preserve resources for future generations. As a result, the CIDB formed a technical committee in June 1999 to establish good environmental practices in the construction industry. The Construction Industry Master Plan 2006–2015 (CIMP, 2007) was released in June 2006 to outline the future of Malaysian construction. The Construction Industry Master Plan (2005–2015) also emphasises the significance of sustainability and green building issues for the Malaysian construction industry. The National Green Technology Policy (NGTP) was introduced in 2009 to show how seriously the government was taking the nation's "green" initiatives. Among them is the acceleration of green technology research and innovation to commercialisation, promotion, and public awareness. The Malaysian Green Building Index (GBI) was created in 2009 by the Association of Consulting Engineers Malaysia (ACEM) and the Malaysian Institute of Architects (Pertubuhan Arkitek Malaysia or PAM) to promote sustainability in the built environment. However, most construction work in Malaysia still uses conventional techniques that are frequently not sustainable, despite the industry's rapid advancement and emphasis on directing construction towards sustainability. The distinctive qualities of IBS have the potential to help the construction industry successfully address the sustainability issue. The application of IBS, according to research, benefits adopters in terms of cost and time certainty, better construction quality and productivity, lowering risk related to occupational safety and health, resolving the issue of skilled workers and dependence on manual foreign labour, and achieving.

Malaysia has already seen major negative consequences from climate change, including increasing annual surface temperature and rainfall, sea level rise, and other extreme weather occurrences (Ho & Tang, 2019). Due to its rapid growth, the country is confronted with depleting natural resources and rising greenhouse gas (GHG) emissions. Furthermore, some Malaysian states had continuous heavy rain in late 2021 and early 2022 due to the monsoon season and high tides (Abdullah et al., 2017). This event caused exceptional flooding in several regions, including Klang Valley, South Malaysia, the East Coast, Sabah, and Sarawak. The Malaysian government launched a survey to assess the impact of floods in the affected states through the Department of Statistics Malaysia (DOSM). This study aims to calculate the total losses and damage to housing, vehicles, businesses, and industrial properties. Data from relevant government agencies were used to calculate the cost of damage to the agriculture industry, public assets, and infrastructure. The flood that struck this country in late 2021 and early 2022 damaged living quarters, vehicles, business

Table 1 Value of flood losses by type of damage (DOSM, 2020)

Type of damage	Value of loss
Living quarters	RM1.6 billion
Business premises	RM0.5 billion
Vehicles	RM1.0 billion
Agriculture	RM90.6 million
Manufacturing	RM0.9 billion
Public assets and infrastructure	RM2.0 billion

premises, the manufacturing and agriculture sectors, public assets, and infrastructure. Overall losses from the floods totalled RM6.1 billion, or 0.40 % of the nominal Gross Domestic Product. According to Table 1, living quarters losses amounted to RM1.6 billion, business premises losses totalled RM0.5 billion, vehicle losses totalled RM1.0 billion, agricultural losses totalled RM90.6 million, and manufacturing losses totalled RM0.9 billion. Public assets and infrastructure losses totalled RM2.0 billion.

Both the environment and the economy benefit from green infrastructure. It can generate prosperity by increasing competitiveness, productivity, and job opportunities; expanding the national electric grid's reach, reliability, and efficiency without polluting the environment; broadening the economic base; creating new markets, and providing inclusion and connectivity throughout Malaysia.

3 Types of Infrastructure in Malaysia

Transportation, communication, sewage, water, electricity supply, and other essential systems and amenities comprise infrastructure. Infrastructure development is critical for a country's economic progress and poverty reduction (Fig. 1).

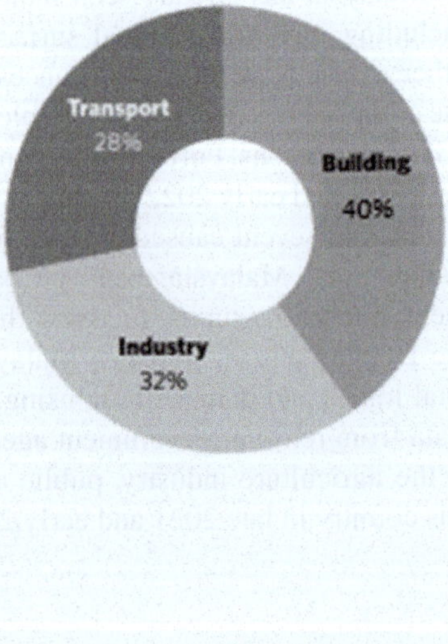

Fig.1 Energy consumption by sector (Energy Malaysia, 2017)

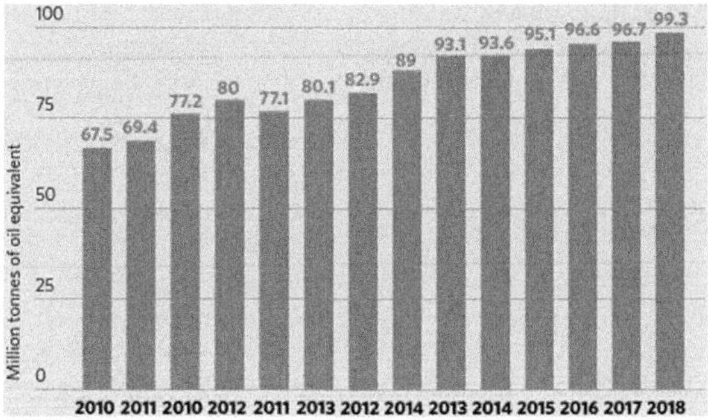

Fig.2 Primary energy consumption in Malaysia is on the rise (Davidson et al., 2020)

Globally, the buildings and construction sector has been a significant generator of CO_2 emissions, accounting for around 39% of global CO_2 emissions (Fan & Friedmann, 2021). Buildings absorb almost the most energy in Malaysia, and the most energy is consumed by electricity (Hassan et al., 2020) which is also the largest CO_2 emitter. As a result, embracing low-carbon development options for buildings is critical to Malaysia becoming a low-carbon, climate-resilient green economy. In its 2030 Sustainable Goals, the country seeks to cut 25% of total power consumption from the building sector while promising to cut up to 40% of CO_2 emissions per GDP by 2020 (Fig. 2).

Malaysia is a rapidly developing country with increasing industry and urbanisation. The country has relied substantially on energy to fund and propel its development. Malaysia's primary energy consumption increased from 67.5 million to 99.3 million tonnes of oil equivalent between 2005 and 2018. According to the World Energy Markets Observatory (WEMO) (2017) report, Malaysia's energy usage is projected to increase by an annual rate of 4.8% up to 2030. Malaysia currently has the greatest CO_2 emissions per capita in ASEAN, with approximately 7.27 tonnes, more than doubling Thailand's 3.64 tonnes and exceeding China's 6.59 tonnes (data as of 2015) (Munir et al., 2020). As a result, low-carbon mobility solutions are required to propel Malaysia's sustainable urban transportation. The recently proposed National Transport Policy, 2019–2030 has established aims and measures for increasing the country's economic competitiveness while lowering the negative environmental impact of the transportation industry (National Transport Policy, 2019). As a result, by 2030, the GHG intensity of GDP will be reduced by 45%. Malaysia has made significant advances in the field of sustainable transportation innovation. For example, it is an all-electric bus service that can serve up to 1000 passengers each hour. Rapid KL buses and MRT/LRT integration are operating successfully, boosting inner-city transit connectivity. MRT Phase One successfully removed 9.9 million cars in 2017, and an additional 62–89 million cars are expected to be removed between 2020 and 2030 (Davidson et al., 2020) (Fig. 3).

Malaysia pledged at the 2015 Paris Climate Conference and Conference of Parties (COP21) to reduce CO_2 emissions per unit of GDP by 45% by 2030 compared to 2005 levels (Kee et al., 2021). This 45% aim is also included in the intended Nationally

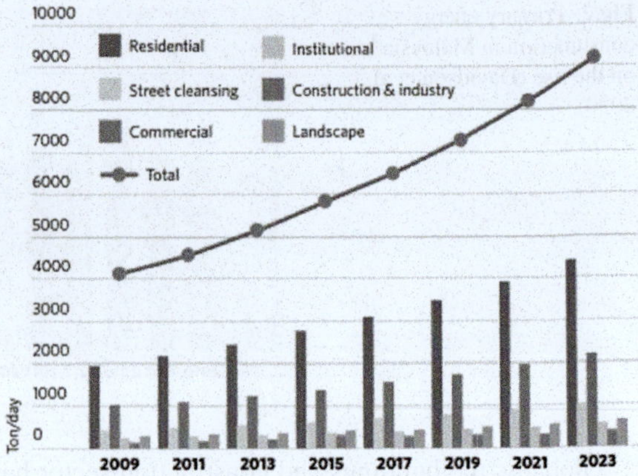

Fig.3 Proportion of waste generation in Kuala Lumpur (Fauziah & Agamuthu, 2003)

Determined Contribution (NDC), of which 35% is unconditional, and the remaining 10% is with foreign assistance (Veng et al., 2019). Thus, the country's authorities recognise the importance of waste management in reducing GHG emissions in its NDC to reach this aim.

Water management projects could also involve water capture and collection, water storage, water treatment (including methane emissions treatment), flood protection, drought protection, stormwater management, and ecological restoration/management. Malaysia has made significant progress in terms of clean water accessibility. According to the WHO/Unicef Joint Monitoring Programme report in 2000–2017, in 2017, 93% of Malaysians had access to properly managed water services 2017, and 87% had access to safely managed sanitation services.

4 Overview of Past Research on Green Infrastructure Development in Malaysia

This study has considered four key criteria for green infrastructure based on a review of the literature on green criteria in various fields (Table 2). Every green criterion has a defined scope and definition, particularly regarding infrastructure. When all the criteria for green infrastructure have been considered and carefully thought out, the function could be 100% green. As a result, there might be direct and indirect connections between the green infrastructure criteria. To analyse the effectiveness and interrelationships among the green infrastructure criteria, research development and procedures with specific methods, such as the decision-making trial and evaluation laboratory, are required to consider these relationships.

Table 2 Key criteria of green infrastructure incorporates in previous research designs

Key criteria	Sub-criteria	Author (Year)	Descriptions
Economic	Innovation	Begum et al. (2010), Shamsuddin et al. (2012), Rooshdi et al. (2021)	Innovative design and initiatives include industrialised building systems (IBS), building information modelling (BIM) or software
	Policy and requirements	Yeo et al. (2022a, 2022b), Danjaji and Ariffin (2017)	Policies, plans, acts, guidelines, and non-governmental organisations (NGOs) plans or reports contribute to green infrastructure planning and establishment
	Life cycle cost	Goh and Yang (2014)	Costs involving construction completion, along with the effect of cost decisions on using, maintaining, and supporting the infrastructure
Environment	Ecology and biodiversity	Yeo et al. (2022a, 2022b), Lourdes et al. (2022)	Mitigation and adaptation to climate change, minimising the risks of natural disasters and supporting biodiversity conservation by cooling, noise reduction and air filtration of pollutants
	Energy and carbon emissions	Kanniah et al. (2014), Zakaria et al. (2013)	Establish minimum energy efficiency performance to reduce energy consumption in buildings, thus reducing CO_2 emission to the atmosphere
	Renewable energy	Umar et al. (2020)	Consideration of using renewable energy in construction project
	Sustainable material source	Yadollahi et al. (2014)	Consideration and implementation of responsible sourcing of sustainable materials
	Waste disposal method	Reza et al. (2017)	Specific documented mechanisms for managing waste and identifying and dealing with all waste arising from the civil engineering work
	Pollution control	Amiril et al. (2014)	Creation and implementation of an environmental pollution control plan to specify actions to prevent and mitigate pollution to air, land, and water during construction
	Water preservation	Othman and Ameer (2010)	Protection or usage of existing water features from degradation or physical damage by the construction plant and processes
Social	Safety and health	Ahmad et al. (2017)	Reduction of the risks to the worker and public safety and health to acceptable levels and receipt of approval from the appropriate public health officials

(continued)

Table 2 (continued)

Key criteria	Sub-criteria	Author (Year)	Descriptions
	Project governance and strategic management	Muhaimin et al. (2021)	Appropriately skilled personnel was commissioned to implement the management plan, monitor the establishment, and review the objectives and management prescriptions
	Social and cultural impact due to the project	Cai et al. (2017)	Consideration of the wider social impacts of the project during construction and operation and the effects of the completed project on the human environment
	Job opportunity	Masrom et al. (2015)	The project creates many new jobs during its design, construction, and operation
Resilient	Adaptability of infrastructure	Haris et al. (2016, Ahmad et al. (2020)	Infrastructure adaptability is the quality of retaining long-term functional value based on safety, validity, and durability during the duration of a project and the urbanisation process
	Recovery ability of infrastructure	Sharifi and Adnan (2022)	Ability to efficiently restore the infrastructure systems and services to support a viable, sustainable community and improves resilience to and protection from future hazards
	Vulnerability to the climate threat	Khaimi and Perera (2013)	Planning and designing climate-resilient infrastructure

4.1 Economic

Shamsuddin et al. (2012) mentioned that IBS promotes construction sustainability from a controlled production environment, minimisation of waste generation, extensive usage of energy-efficient building material, and effective logistic and long-term economic stability, which can contribute to better investment in the construction industry. Meanwhile, Yeo et al. (2022a, 2022b) discovered that Malaysia still needs green infrastructure policies and regulations. Apart from that, life cycle cost is a method consisting of estimating the total cost of constructing the project, taking into account the whole life cycle of the construction and the direct and external costs (Fathollahi & Coupe, 2021). Goh and Yang (2014) developed practical life cycle cost analysis tools to evaluate highway investment decisions and reach an optimum balance between financial viability and sustainable deliverables. Schulz et al. (2012) used life cycle costings (LCCs) to identify the key issues in a streamlined sustainability assessment tool for the urban water industry. The planning phase should bring many different staff from various departments into the discussion. It is important to include operations personnel in the asset review during the planning, design, and construction phases because they will oversee the operation and maintenance of the asset once it is installed.

4.2 Environment

Ecology and Biodiversity where mitigation and adaptation to climate change minimise the risks of natural disasters and support biodiversity conservation by cooling, noise reduction, and air filtration of pollutants. Yeo et al. (2022a, 2022b) outline the conceptual framework of green infrastructure planning and establishment in Kuala Lumpur. Meanwhile, Kanniah et al. (2014) use a Landsat Thematic Mapper satellite image to derive information about the land surface temperature and vegetation and to analyse the effect of vegetation. They have been used in green (plants) and blue bodies (lakes and rivers) in moderating temperatures in Iskandar, Malaysia. In addition, Umar et al. (2020) evaluate the sustainability of Malaysia's grid-connected oil palm biomass renewable energy industry and propose a policy framework and industry roadmap. The factors investigated include resource supply, the efficiency of waste-to-energy conversion technology used in the existing plants, and the attractiveness of the electricity interconnection scheme in encouraging exports of excess power from the participating mills to the main grid.

Apart from that, Yadollahi et al. (2014) stated that the appropriate use of materials could significantly impact the surrounding environment by reducing the use of energy, reducing greenhouse gas emissions, and decreasing landfill volume, all of which are beneficial side effects. The application of green materials will positively affect the cost and energy saving, causing less forest degradation, less rotting, and corrosion during the life cycle of the project. For example, the daily temperature swings of phase-change materials can effectively provide low-energy techniques to improve occupant comfort. Meanwhile, Reza et al. (2017) identify key principles and strategies for developing integrated construction and demolition waste management. In contrast, Amiril et al. (2014) analyse how factors including low maintenance costs, reduced operational expenditures, safe building practices, and pollution prevention affect project outcomes. Lastly, Othman and Ameer (2010) research analyses the annual environmental protection disclosures of palm oil firms in Malaysia, which have major effects on environmental sustainability and public health.

4.3 Social

Safety and health are some of the sub-criteria that reduce the risks to the worker and the public on acceptable levels and receive approval from the appropriate public health officials. In Malaysia, Ahmad et al. (2017) identify critical phases for safety cost allocation throughout the construction project's life cycle, issues, and the importance of safety and health cost allocation. Meanwhile, Muhaimin et al. (2021) identify factors influencing the governance practices in a rural road project in Malaysia. According to Cai et al. (2017), illegal dumping is a major problem in many areas, which raises concerns regarding safety, property value, and quality of life in the community. Dumping waste at illegal sites in Malaysia is a common practice, particularly industrial and construction wastes. Therefore, their study explores illegal dumping activities and discusses the contributory factors in Malaysia. Besides that, Masrom et al. (2015) mentioned that the construction industry and the private sector play important roles in generating wealth and improving the quality of life for Malaysians. It can be achieved by translating the government's socio-economic policies into social and economic infrastructures and buildings. The construction industry also provides job opportunities to approximately 800,000 people, according to the CIDB in 2008.

4.4 Resilient

As the project lifespan and urbanisation process progress, it is important for infrastructure to retain its functional value based on safety, validity, and durability. Haris et al. (2016) discuss software and its functionality, accessibility, characteristics, and components in the quantitative analysis of the hydrological design software to be implemented in infrastructure projects. Meanwhile, Sharifi and Adnan (2022) assessed the initiatives taken by smart cities to decrease GHG emissions, as well as the role of the Internet of Things (IoT) and telecommunication networks in smart cities. The local development planning system in Malaysia is evaluated based on how well it has met the demands of civil society in terms of reducing vulnerability and increasing resilience in the face of climate change-induced flooding (Khaimi & Perera, 2013). According to Reddy et al. study, increased occurrence of climate change-related events and impacts have challenged the function of engineered systems and their ability to achieve sustainable development, forcing policymakers and stakeholders to consider resilience in engineering designs and projects. Resilience and sustainability are inseparable, as an engineering system can only be sustainable if it is resilient; moreover, only some tools and frameworks integrate resilience and sustainability assessments at a project level. Even so, gaps still impede the proper adoption of sustainable and resilient infrastructure, particularly in the construction industry.

5 Previous Research on Green Infrastructure Development in Malaysia

A more environmentally conscious society has brought green infrastructure, also known as sustainable infrastructure, to the forefront of built environmental research. These effects will be reduced if the infrastructure is built resilient and sustainable, and the affected communities will reap the benefits in the long run. To future societies, it is crucial to be aware of the value of infrastructure and how it can be built sustainably. A summary of prior research on green development in the Malaysian context is provided by dividing the period into three-period dimensions.

Before 2013, the majority of the published works on the subject of green infrastructure development in Malaysia centred on presenting the concept of green infrastructure evaluation approaches. Tools for evaluating green infrastructure were the focus of numerous studies during this period (i.e., indicator, criteria, sub-criteria, weightage, and certification process). Yen et al. (2008) created energy indicators to track the progress of Malaysia's energy industry. Meanwhile, the government of Malaysia is working on a framework for sustainable indicators (Hezri, 2004). Urban pollution, heat islands, erosion, and flooding are all made worse by the lack of vegetation and green space in Malaysian cities, according to a study by Lilian et al. (2002).

More analytically sound research that refined the approach for evaluating sustainable infrastructure emerged between 2013 and 2017. Case studies across the country have been studied in depth to better understand the problem and its causes, and a potential solution has been developed. Green building adoption studies involving the government and builders were considered (Chan et al., 2014; Amiril et al., 2014). To find the sweet spot between economic viability and sustainability, Goh and Yang (2014) suggest a platform for highway project stakeholders to develop instruments for doing just that. As well, the efforts of Malaysia to determine what factors encourage or discourage the use of IBS in green and sustainable building projects have been described (Shamsuddin et al., 2012; Khoshnava et al., 2014). They've brought to light possible factors that could aid in overcoming challenges and obstructions to enacting green practices. Later, further contexts (such as comparison studies and assessment criteria) were taken into account in the literature to better understand the elements driving better green practises in Malaysia (Balubaid et al., 2015; Masrom et al., 2015; Yadollahi et al., 2014; Zakaria et al., 2013).

After 2017, new avenues of inquiry opened up. Advanced studies focus on developing methods for incorporating robust infrastructure that is also sustainable (economically, environmentally, and socially) at every stage of a project's life cycle. For instance, sustainable construction is committed to promoting environmentally friendly building practices by developing a management structure, raising knowledge of relevant concepts, establishing regulatory procedures, and developing conceptual models to guide practice (Danjaji & Ariffin, 2017; Umar et al., 2020). Research on the creation of environmentally friendly policies and programmes by the government and agencies in Malaysia can be found in (Danjaji & Ariffin, 2017). Meanwhile, a

Table 3 Overview of past research on green infrastructure development in Malaysia

Key aspect	Sub-period <2013	Sub-period (2013–2017)	Sub-period >2017
Main topic	Concept, initiative and indicator of green infrastructure assessment methods	Green development, initiatives, indicators, attributes, perceptions, challenges, and drivers	Green infrastructure performance, assessment, benefits, strategies and measures
Level of study	International level, national level	National level, project level	National level, project level
Methodological framework	Review paper and mixed method	Review paper, interview, case study and questionnaire	Review paper, the survey, case study

conceptual framework and model for the planning and establishment of green infrastructure in Kuala Lumpur are being established (Lourdes et al., 2022; Yeo et al., 2022a, 2022b).

To yet, research has only resulted in high-level summaries of existing assessment procedures, which itself span the entirety of a project's life cycle in the construction industry. Because of this, a close look at the dynamics and trends in resilient and sustainable infrastructure research will yield invaluable insights. However, additional research into the ways in which businesses commit to utilising green practices and the effects of green infrastructure on the industry as a whole is necessary. The results of previous research on green infrastructure in Malaysia are summarised in Table 3.

6 Discussion

The success of a project can be determined when all the sustainable pillars are met. However, a construction project must consider other aspects of sustainable infrastructure. The prerequisite is to work towards each of the sustainability pillar features in construction projects, which will assist stakeholders in understanding the areas in which they should consider successfully implementing sustainable construction. This line of reasoning was used to identify any features connected to improving sustainable infrastructure in various construction projects. This study summarised four (4) key sustainability criteria and the seventeen (17) sub-criteria for sustainable infrastructure.

Infrastructure developments' effects on the environment and environmental criteria are well understood. Environmental factors are typically or significantly related to environmental health. Environmental criteria, in more detail, primarily reflect the concern with minimising pollution on the one hand and maximising preservation on the other, necessary and complementary actions in terms of the environmental health of a city or region. Addressing environmental sustainability has become more crucial due to the widespread environmental degradation and the severe pressures on the global commons. Additionally, it works to lessen environmental disruption by preserving areas with significant ecological value, biodiversity, and ecosystem functions.

Economic criteria make an effort to make sure that financial and economic factors are properly addressed. Transparent and efficient regulatory frameworks should be in place, along with enforcement measures to guarantee the project's smooth operation. Prices must be established and adjusted to preserve commercial sustainability and social affordability. To guarantee the long-term effectiveness and profitability of the assets, adequate design and operating requirements, as well as pricing and other incentives, should be considered. Thus, infrastructure projects have a direct impact on the environment and public health, a sense of community, the economies of neighbouring cities and regions, and urban sustainability, which increases the need for indicators of infrastructure project sustainability. Indicators are crucial in assessing stakeholder education and infrastructure sustainability. Numerous frameworks for infrastructure indicator development and sustainability emerged in this context. These frameworks link indicators to create an international index. In line with the findings of other researchers studying the topic, economic indicators are still contentious topics. They have yet to be extensively studied in infrastructure projects' context of sustainability indicators.

According to social criteria, infrastructure projects should be socially sustainable throughout their lifetimes. It also aims to prevent discrimination and ensure fairness, equality of opportunity, and gender inclusivity in project design. The fact that the sub-criteria have a direct connection to the workers or public health shows how crucial it is to consider how infrastructure projects affect the health of those whom they will most directly impact: residents and project workers. The other social indicators include one representing project governance concerns and two relating to the project's social responsibility to the affected community, highlighting the necessity of strong project leadership for sustainability.

Aside from that, there are few tools for evaluating how environmental, economic, and social factors interact. Malaysia is under pressure to create a second generation of sustainable development indicators to serve as a basis for integrated sustainable development systems (Dizdaroglu, 2017). Developing a more robust model for evaluating sustainability performance is necessary to ensure that infrastructure projects contribute to the overall development. To increase the probability of achieving sustainable development goals, sustainability assessment and reporting techniques must be developed to update stakeholders on the goals' progress (Krajangsri & Pongpeng, 2017).

There is a need for immediate and preventive action through new partnerships and collaborations to implement ambitious current ideas to mitigate climate change. Yusof (2021) stated that in the context of Malaysia, Malaysia will need to develop smart infrastructure by planning, constructing, and operating infrastructure in ways that will promote and maintain sound economic development while simultaneously safeguarding our essential natural resources and environment. He thinks resilient, inclusive, and sustainable infrastructure can improve our quality of life. Infrastructure development must promote more effective and efficient use of financial resources, consideration of carbon footprint, social cohesion, and stewardship of natural ecosystems throughout its life cycle.

7 Conclusion and Recommendations

Green infrastructure offers enormous potential to revive economic growth, tackle societal difficulties, and safeguard natural resources for future generations. The purpose of this study is to investigate the Malaysian government's top-down measures for integrating all reform efforts in Malaysia into long-term infrastructure development. The paper compiles comprehensive data on Malaysia's green development efforts and initiatives. This study looks at how green and sustainable development has evolved in Malaysia over the last 30 years and how it has been included in the national plan, policies, fiscal, institutional, legislative, and regulatory systems. The evaluation emphasised green and sustainable policies, infrastructure projects in Malaysia, and green assessment techniques. Based on the policy review study, a framework based on the strategic mechanism defined as credible policy support, scaling up green investment, technological innovation, and capacity building has been established. This report examines Malaysia's progress towards sustainable development and climate resilience.

There are several consequences for Malaysian green major influencers. This research assists stakeholders in understanding the current state of green and sustainable advancement, as well as what needs to be considered for preparing for a sustainable ecosystem. It helps to clarify the roles and activities of important participants in green practices. Second, the findings contribute to identifying gaps and issues, advancing sustainability literacy, developing more research evaluations of significant efforts, and reconstructing development goals for a better and greater nation. The study findings could be utilised to build stronger solutions and policy orientations for Malaysia, as well as to strengthen capacities at all levels to promote green growth and ensure the implementation of the Twelfth Malaysia Plan 2021–2025 and the 2030 Agenda.

This study includes the findings of a top-down data collection on green and sustainable initiatives in Malaysia. However, the following limitations are recognised. No data was obtained because the focus of this study was exclusively on policy review. The obvious problem is to connect policy to the practical implementation of functional climate infrastructure; thus, future research should include bottom-up viewpoints or internal elements imposed by people or organisational behaviour participating in the sustainability agenda.

Acknowledgements The authors would like to thank the School of Civil Engineering, College of Engineering, Universiti Teknologi Mara, Shah Alam, Selangor, Malaysia, for the research grant (600-RMC/GIP 5/3 (120/2021), supporting the research work and the building stakeholders who have involved in the case studies.

References

Abdullah, H., Aznin, N., Bakar, A., Razani, M., Jali, M., & Ibrahim, F. W. (2017). The current state of Malaysia's journey towards a green economy: The perceptions of the companies on environmental efficiency and sustainability. *International Journal of Energy Economics and Policy, 7*(1), 253–258.

Ahmad, A. C., Mara, U. T., Iskandar, S., & Khairunnisa, N. N. (2017). Preliminary survey of safety and health cost framework for urban rail infrastructure project in Malaysia. https://doi.org/10.21834/e-bpj.v2i5.712

Ahmad, E. F., Nianti, I., Zin, M., & Alauddin, K. (2020). Criteria of resilience infrastructure in flood-prone areas in Kelantan. *7*(1), 173–190.

Amiril, A., Nawawi, A. H., Takim, R., Nur, S., & Ab, F. (2014). Transportation infrastructure project sustainability factors and performance. *Procedia—Social and Behavioral Sciences, 153*, 90–98. https://doi.org/10.1016/j.sbspro.2014.10.044

Balubaid, S., Bujang, M., Nur Aifa, W., Kian Seng, F., Raja Muhammad Rooshdi, R. R., Hamzah, N., Mohd Yazid, Y. S., Abd. Majid, M. Z., Mohamad Zin, R., Zakaria, R., Hainin, M. R., Yaacob, H., & Ismail, H. H. (2015). Assessment index tool for green highway. *16*(Phase 5), 99–104.

Begum, R. A., Satari, S. K., & Pereira, J. J. (2010). Waste generation and recycling: Comparison of conventional and industrialized building systems. https://doi.org/10.3844/ajessp.2010.383.388

Cai, B., Tan, X., Li, Q., Lin, Q., Soterroni, A. C., & Mosnier, A. (2017). Construction waste generation in Malaysia construction industry: Illegal dumping activities. https://doi.org/10.1088/1757-899X/271/1/012040

Chan, Y. H., Lee, B. C. T., & Lee, J. C. (2014). Sustainability in the construction industry in Malaysia: The challenges and breakthroughs. *International Journal of Economics and Management Engineering, 8*(4), 1218–1222.

Construction Industry Master Plan 2006–2015 (CIMP 2006–2015). (2007). *Construction Industry Development Board (CIDB) Malaysia*, December 2006, Kuala Lumpur.

Danjaji, A. S., & Ariffin, M. (2017). Green infrastructure policy for sustainable urban development. July 2018. https://doi.org/10.1504/IJESD.2017.10004010

Davidson, K., Nguyệt, P. T. M., Gunawan, N., & Boulle, B. (2020). Green infrastructure investment opportunities. https://doi.org/10.18235/0002638

Department of Statistic Malaysia (DOSM). (2020). *Gross Domestic Product (GDP) By State 2020.* https://www.dosm.gov.my/. Accessed 17 Feb 2022.

Dizdaroglu, D. (2017). The role of indicator-based sustainability assessment in policy and the decision-making process: A review and outlook. *Sustainability, 9*(6), 1018.

Diaz-sarachaga, J. M., Jato-espino, D., Alsulami, B., & Castro-fresno, D. (2016). Evaluation of existing sustainable infrastructure rating systems for their application in developing countries. *Ecological Indicators, 71*, 491–502. https://doi.org/10.1016/j.ecolind.2016.07.033

Fan, Z., & Friedmann, S. J. (2021). Review low-carbon production of iron and steel: Technology options, economic assessment, and policy. *Joule, 5*(4), 829–862. https://doi.org/10.1016/j.joule.2021.02.018

Fathollahi, A., & Coupe, S. J. (2021). Life cycle assessment (LCA) and life cycle costing (LCC) of road drainage systems for sustainability evaluation: Quantifying the contribution of different life cycle phases. *Science of the Total Environment, 776*, 145937. https://doi.org/10.1016/j.scitotenv.2021.145937

Fauziah, S. H., & Agamuthu, P. (2003). Municipal solid waste management: A comparative study on selected landfill in Selangor. In *Proceedings of Environment* (pp. 434–437).

Goh, K. C., & Yang, J. (2014). Managing cost implications for highway infrastructure sustainability. *International Journal of Environmental Science and Technology, 11*(8), 2271–2280. https://doi.org/10.1007/s13762-014-0572-5

Harris, H., Chow, M. F., Usman, F., Sidek, L. M., Roseli, Z. A., & Norlida, M. D. (2016). Urban stormwater management model and tools for designing stormwater management of green infrastructure practices. https://doi.org/10.1088/1755-1315/32/1/012022

Hassan, J. S., Zin, R. M., Majid, M. Z. A., Balubaid, S., & Hainin, M. R. (2020). Building energy consumption in Malaysia: An overview. *Jurnal Teknologi, 7*(2014), 33–38.

Hezri, A. A. (2004). Sustainability indicator system and policy processes in Malaysia: A framework for utilisation and learning. *Journal of Environment, 73*, 357–371. https://doi.org/10.1016/j.jenvman.2004.07.010

Ho, K., & Tang, D. (2019). Science of the total environment climate change in Malaysia: Trends, contributors, impacts, mitigation and adaptations. *650*, 2018–2020.

Kanniah, K. D., Sheikh, A., & Kang, C. S. (2014). Exploring the role of green and blue infrastructure in reducing the temperature in Iskandar Malaysia using a remote sensing approach. *IOP Conference Series Earth and Environmental Science*, 0–6. https://doi.org/10.1088/1755-1315/18/1/012156

Kee, K., Lim, Y. S., Wong, J., & Chua, K. (2021). Impact of nonintrusive load monitoring on CO_2 emissions in Malaysia. *Bulletin of Electrical Engineering and Informatics, 10*(4), 1803–1810. https://doi.org/10.11591/eei.v10i4.2979

Khaimi, D., & Perera, R. (2013). Mainstreaming disaster resilience attributes in local development plans for the adaptation to climate change induced flooding: A study based on the local plan of Shah Alam City, Malaysia. *Land Use Policy*, *30*(1), 615–627. https://doi.org/10.1016/j.landusepol.2012.05.003

Khoshnava, S. M., Rostami, R., Ismail, M., & Lamit, H. (2014). Obstacles and drivers in steering IBS towards green and sustainability. *Research Journal of Applied Sciences, Engineering and Technology, 8*(14), 1639–1647.

Krajangsri, T., & Pongpeng, J. (2017). Effect of sustainable infrastructure assessments on construction project success using structural equation modeling. *Journal of Management in Engineering, 33*(3), 04016056. https://doi.org/10.1061/(asce)me.1943-5479.0000509

Lilian, T. Y. C., Ho, C. S., & Ismail, S. (2002, April). Some planning consideration of garden city concept towards achieving sustainable development. In R. Omar, Z. Ali Rahman, M. T. Latif, T. Lihan, & J. H. Adam (Eds.), *Proceedings of the Regional Symposium on Environment and Natural Resources* (pp. 261–269).

Lourdes, K. T., Hamel, P., Gibbins, C. N., Sanusi, R., Azhar, B., & Lechner, A. M. (2022). Planning for green infrastructure using multiple urban ecosystem service models and multicriteria analysis. *Landscape and Urban Planning*, *226*(September 2021), 104500. https://doi.org/10.1016/j.landurbplan.2022.104500.

Masrom, M. A. N., Rahim, M. H. I. A., Mohamed, S., Chen, G. K., & Yunus, R. (2015). Successful criteria for large infrastructure projects in Malaysia. *Procedia Engineering, 125*, 143–149. https://doi.org/10.1016/j.proeng.2015.11.021

Muhaimin, A., Latiff, A., Jaapar, A., Maznah, C., & Isa, M. (2021). Factors influencing governance practices in rural public infrastructure projects: A case study of rural road project in Malaysia. *International Journal of Sustainable Construction Engineering Technology, 12*(4), 35–45.

Munir, Q., Hooi, H., & Smyth, R. (2020). CO_2 emissions, energy consumption and economic growth in the ASEAN-5 countries: A cross-sectional dependence approach. *Energy Economics, 85*, 104571. https://doi.org/10.1016/j.eneco.2019.104571

Othman, R., & Ameer, R. (2010). Environmental disclosures of palm oil plantation companies in Malaysia: A tool for stakeholder engagement. *62*(November 2009), 52–62.

Policy, N. T. (2019). National transport policy.

Reza, M., Halog, A., & Rigamonti, L. (2017). Strategies for minimising construction and demolition wastes in Malaysia. *Resources, Conservation & Recycling, 120*, 219–229. https://doi.org/10.1016/j.resconrec.2016.12.014

Rooshdi, R. R. R. M., Ismail, N. A. A., Sahamir, S. R., & Marhani, M. A. (2021). Integrative assessment framework of building information modelling (BIM) and sustainable design for green highway construction: A review. *Chemical Engineering Transactions, 89*(August), 55–60. https://doi.org/10.3303/CET2189010

Schulz, M., Short, M., & Peters, G. (2012). A streamlined sustainability assessment tool for improved decision-making in the urban water industry.

Shamsuddin, S. M., Zakaria, R., Mohamed, S. F., & Mustaffar, M. (2012). Drivers and challenges of industrialised building system (IBS) in sustainable construction.
Sharifi, A., & Adnan, Y. M. (2022). Envisioning sustainable and resilient petaling Jaya through low-carbon and smart city framework. https://doi.org/10.1007/978-3-030-95037-8
Umar, M. S., Urmee, T., & Jennings, P. (2020). A policy framework and industry roadmap model for sustainable oil palm biomass electricity generation in Malaysia. *Renewable Energy, 128*(2018), 275–284. https://doi.org/10.1016/j.renene.2017.12.060
Upadhyaya, J., Kwan, E., Tam, L., & Biswas, N. (2016). Infrastructure challenges and way forward: An innovative framework for sustainability assessment applied to stormwater systems, June 2012.
Veng, V., Suryadi, B., Pranadi, A. D., & Shani, N. (2019). Review renewable energy development and its policy under nationally determined contributions in ASEAN. 149–161. https://doi.org/10.12720/sgce.9.1.149-161
WEMO. (2017). World energy markets observatory.
World Health Organization. (2017). Progress on household drinking water, sanitation and hygiene 2000–2020: Five years into the SDGs.
Yadollahi, M., Ansari, R., Zaimi, M., Majid, A., & Yih, C. H. (2014). Multicriteria analysis for bridge sustainability assessment: A case study of Penang Second Bridge, February 2015, 37–41. https://doi.org/10.1080/15732479.2014.893002.
Yen, C. H., Sopian, K., Ali, B., & Mokhtar, M. B. (2008). Energy indicators for sustainable development in Malaysia. *European Journal of Social Sciences, 6*(2), 213–231.
Yeo, O. T. S., Mohd Yusof, M. J., Maruthaveeran, S., Shafri, H. Z. M., & Saito, K. (2022a). Green infrastructure conceptual framework for Kuala Lumpur. https://doi.org/10.1088/1755-1315/1053/1/012002
Yeo, O. T. S., Yusof, M. J. M., Sreetheran, M., Saito, K., & Kasim, J. A. (2022b). Green infrastructure transitional management sphere analysis of policies and regulations in Kuala Lumpur, Malaysia. *Planning Malaysia, 20*(2), 61–73.
Zakaria, R., Seng, F. K., Zin, R. M., Hainin, M. R., Puan, O. C., Derin, N., Ainee, F., Hamzah, N., Balubaid, S. O., Mazlan, A. N., Ismail, M. A., Yazid, S., Rafidah, R., Mohd, R., & Moayedi, F. (2013). Energy efficiency criteria for green highways in Malaysia. *Jurnal Teknologi, 3*, 91–95.

Why Current Procurement Systems Require Modifications to Suit the Natures of Malaysian Pre-fabricated Construction

Ahmad Abd Jalil, Mastura Jaafar, Fadhilah Md Fazil, Nurina Nawi, and Mohd Amir Shazwan Hashim

Abstract The Malaysian government has put many efforts into improving the performance and quality of local projects and one of the steps was using prefabrication construction. Worldwide, this concept has gradually overtaken the traditional construction and as evidence, many prefabricated companies have emerged with plenty of innovative products, making the prefabricated method a preferred choice. Various advantages of using this method, as it changes from wet in situ construction into applying a manufacturing style that is simpler, safer, faster, and better quality. It brings more than just cost savings as this method will take less than half of the time compared to using conventional cast in situ, and it accelerates projects toward achieving sustainable construction and fulfilling the green concept regulations. However, in Malaysia, this concept hardly achieves its maximum benefits because prefabricated projects still use current procurement system which is not developed to suit prefabricated characteristics. When prefabricated projects apply unsuitable procurement, it exposes them to claim and payment hurdles, warranty disputes, weak supply chain, market monopoly, rigid project flow process, flawed understanding, and rivalry relationships. As a result, the prefabricated projects ended with higher costs, more risk, and unfair treatment to certain parties despite having contributed significantly. Much research has proved that the current procurement

A. Abd Jalil (✉)
Department of Quantity Surveying, Faculty of Built Environment, Universiti Malaysia Sarawak, Kota Samarahan, Sarawak, Malaysia
e-mail: ajahmad@unimas.my

M. Jaafar
Department of Quantity Surveying, School of Housing, Building and Planning, Universiti Sains Malaysia, Penang, Malaysia

F. Md Fazil
Department of Law, Universiti Koperasi Polis (UNIKOP), Cyberjaya, Malaysia

N. Nawi
School of Science and Technology, Wawasan Open Universiti Malaysia, Ipoh, Perak, Malaysia

M. A. Shazwan Hashim
Faculty of Engineering and Quantity Surveying, INTI International University, Nilai, Negeri Sembilan, Malaysia

U. R. Hashim et al. (eds.), *Green Infrastructure*,
https://doi.org/10.1007/978-981-99-7003-2_8

systems do not welcome the crucial role of prefabricated companies. Besides, it disallows prefabricated companies to be involved since the detail design stage and neglects the supervision during and after installation, causing prefabricated projects more exposed to quality disputes, defects in installation, and warranty conflicts. Current procurement also treats the prefabricated companies as a minority despite their role constituting 70–90% of the building's structure. Since IBS Roadmap (2007), there has been urgency to develop suitable procurement that is unswerving and can cater good flow of the supply chain, empowering the roles of the prefabricated design team, codifying skills transfer, proper logistics, and increase higher productivity. This research aims to highlight the necessity and nature of prefabricated projects in Malaysia, and why the current procurement system needs to be adjusted so that it can help the projects to reap maximum benefits from the prefabricated concept.

Keywords Modification · Procurement · Pre-fabricated · Construction

1 Introduction

As the modern era developed, all industries had embedded the latest technology advancement causing the methods they used either in production or construction to be improved. This includes the method to construct buildings, where a prefabrication concept which is similar to manufacturing, had been applied in construction projects. This concept originally began by imitating the style of modern manufacturing into construction, where this strategy transformed the construction of buildings from using traditional materials that were built in situ to using building components (Nawi et al., 2015). The building components are actually fabricated in the factories, using specialized automated machines where various materials are joined together to form the prefabricated components, which are then transported to the site for assembling and finally, erected as buildings (Hamid et al., 2011; Shukor et al., 2016; Tatum et al., 1986). Lately, the concept of a mobile prefabrication plant has been implemented that allows prefabricated components to be manufactured on-site, and no longer needs to be fabricated in a permanent factory (Azman et al., 2011).

Din et al. (2012) and Goodier and Gibb (2007) explained that prefabricated components are made using modern precise technology, so their quality and build-up process is better and simpler compared to traditional construction. This concept has been applied in many countries under different names but carries the same purpose such as Off-site Construction Method (OSM) in Australia; Modern Method Construction (MMC) in the United Kingdom and Industrialized Building System (IBS) in Malaysia (Azman et al., 2012b; Mahbub, 2016).

Many overseas projects had applied the prefabrication concept and were successfully completed with numerous advantages, thus proving that the prefabrication had performed better compared to conventional. As a result, it has attracted the attention of many developing countries as they also want to reap the benefits from this adaptation of modern manufacturing style construction (Blismas & Wakefield, 2009). Much research has been conducted on the application of the prefabrication concept and their findings revealed that it has been exponentially demanding especially for countries that aim to cater to the urgent needs for more homes and commercial buildings (Abd Jalil et al., 2016; Ibrahim et al., 2010; Pan et al., 2005).

In Malaysia, the prefabrication concept has been implemented since 1960s and was greatly encouraged by the government. It embedded modern manufacturing techniques that were product-based, where building components systems are manufactured in a factory and later transported to the site for installation (CIDB, 2010). The prefabricated components are manufactured either in a permanent factory which is called off-site, or near to site which is called on-site and mobile IBS plant. Then the components will be transported to the site, positioned to their intended location, and installed with minimal site work (IBS Roadmap, 2011). Chung and Kadir (2007) underlined that the Malaysian prefabricated concept involves mass production of building components, made using factory styles under strict specifications that produce standard form, shapes, and dimension components, and carried to the site for installation. If conducted with success, this concept surely can simplify the construction process, thus reducing the on-site activities and changing the site conditions from wet to dry (Najuwa et al., 2016; Shaari & Ismail, 2003).

2 Research Background

To transform from using the traditional construction into prefabricated construction will certainly require major adaptations. Prefabricated construction needs heavy investment in resources, equipment, plant, machinery, and technology as it uses modern facilities, automated machines, cranes, casting beds, special mold, and other measurement equipment (Jalil et el., 2015; Kamar et al., 2009a). These are necessary whether the prefabricated components are made either in a permanent factory or in the temporary mobile plant which is set up near to project site (Abd Shukor et al., 2009; IBS Roadmap, 2011). Both plants require an extensive logistic plan, skilled workers, a casting yard, and curing and storing facilities.

Not only that, but the prefabricated concept also needs labors with specific skill and knowledge. Syazwana et al. (2017) explain this concept consists of at least four major tasks which are design, fabrication of components, delivery, and installation. Each task will be conducted by different teams who are specialized in their detail works to ensure quality products, on-time completion and avoid wastage and errors during and after installations.

Kamar et al. (2011) revealed that this concept demands total integration of construction subsystems, components, and building elements into one overall system that involves industrialized production, assembling, mobilization, and erection on site. Nawi et al. (2011) illustrate that it is an innovative construction method in which components are manufactured either in a factory or on-site, monitored under strict procedures with detailed fabrication and controlled process toward producing the construction desired products. It simplifies the construction process by replacing most construction activities from build in situ, into build in the factory. It brings faster completion, economical cost, and higher quality as the components are made using automated machines with precise techniques, produced under strict surveillance, and monitored and installed by skilled laborers (Shukor et al., 2011a, 2011b). However, these benefits could not be achieved if the needs toward a successful prefabricated concept, are not well taken care of. The major conclusion can be made is that most of the processes in prefabricated construction are dissimilar to traditional construction, thus the system and all supported mechanisms for prefabricated must be suitable to meet its own natures because prefabrication depends more on components and skills while conventional depends on raw materials and manual labor (Faizul, 2006; Fournery, 1997; Ibrahim et al., 2010).

3 Problem Statement

In Malaysian projects history, prefabricated construction was often associated with a negative image due to its past failures (Kamar et al., 2009b; Nawi et al., 2011). Previous research showed that many contractors perceived prefabricated as having a lack of flexibility in design, facing logistical barriers, claim and payment hurdles, warranty disputes, hard decoration finishes, lack of skilled labor, being unfriendly to certain shapes or patterns, and applying unfamiliar materials and techniques (Abdullah & Nasir, 2017; Najuwa et al., 2016; Nawi et al., 2014). These issues embolden many clients to have less confidence in adopting the prefabricated concept unless they were forced by regulations or to gain tax exemption (Rahim et al., 2012).

To ensure success, the manufacturers of the prefabricated components need to play significant roles as they develop the prefabricated design, fabricate the components, logistic to the site, and are responsible for installation which accounts for between 70 and 90% of the project structure. However, their involvement was lacking in terms of official visibility, thus they could not interfere much in the decision, especially on supply chain, quality of skilled labor, quality control, logistical arrangement, time scheduling, warranty, and cost efficiency. Previous researches had covered these issues and they concluded that these problems originated due to prefabricated projects applying unsuitable procurement systems.

Many agreed that current procurement systems are not suitable to be used in prefabricated projects because they are designed to suit traditional construction methods, not prefabricated methods. As a result, when these current procurements are applied to prefabricated projects, many problems will inadvertently occur leading to unavoidable bad effects (Abd Jalil et al., 2021). The problems include design errors, disruption of components fabrication, coordination barriers, installation defects, disputes on payment, schedule interruption, and unclear warranty and liability. In fact, these problems which were initiated by using unsuitable procurement caused stereotypes among industry players that believe the prefabricated concept would cost more and difficult than the traditional method. However, Kamar et al. (2009b) argued that those projects could be completed with lower cost and better performance if they adopted the right approach and applied some modifications to the current procurement. He stressed that the reason why prefabricated projects likely face problems is because the projects have often been treated the same way as traditional projects, neglecting the prefabricated natures.

To discover more, this research was conducted where the information was collected directly from the professional respondents involved in the prefabricated construction in Malaysia. The aim is to analyze the details of the prefabricated natures and characteristics, especially on the real issues and challenges that relate to the current procurement system. Through the findings, this research proves that there are many strong reasons why prefabricated projects must apply different procurement system that can meet the prefabricated natures. By doing this, the benefits of prefabricated concepts can be optimized and the project targets can be improved significantly.

4 Methodology

Bazeley (2004), Creswell and Plano (2011) explained for research investigating construction issues involving many parties, the best methodology is to use a mixed method that combines questionnaire and interview. After strengthening the literature review, this research conducted a pilot study with eight experienced industry people in this area and proposed a sample questionnaire and interview questions. After receiving feedback, corrections were made and then the real data collection began. The Respondents for both the pilot study and real data collection consist of prefabricated companies, consultants, developers, prefabricated installers, academicians, main contractors, and mechanical and electrical (M&E) contractors.

The questionnaire consists of two sides, first respondents must rate the frequency of procurement issues in the prefabricated projects that they had experienced, and second to rate their agreement level toward the proposed steps on solving the procurement issues. Software SPSS 17.0 was used to tabulate the questionnaire results, and then the data were verified through other deep interviews with twelve experts that represent every party. After completing the interviews with the experts, the inputs were filtered using the thematic analysis technique. This step is suggested by Driscoll et al. (2007) who conclude to ensure the research carries stronger data and includes real causes and reasons, the result from the quantitative method should be verified and checked by the qualitative method which is the interviews.

Overall, a number of 260 questionnaires were sent to various companies that are involved in prefabricated projects around Malaysia and the return is 118, which is 45%. The feedback comes from the main contractor and M&E contractor (30.5%), prefabricated installers and prefabricated companies (27.1%), consultants comprising architect, QS, M&E, and C&S (23.7%), and lastly housing developers (18.7%). The detail of the respondents' group is tabulated in Table 1.

Table 1 Detail of the respondents' group

Respondent	Distributed	Returned	Percentage (%)	Response rate (%)
Housing developers	49	22	18.7	45
Main contractors and M&E contractors	70	36	30.5	51
Prefabricated installers/ companies	61	32	27.1	52
Consultants	80	28	23.7	35
Total	260	118	100	45

5 Analysis and Discussion

After receiving the feedback from questionnaires and conducting deep interviews with expert respondents, the below essence had been summarized together with their sources and literature.

5.1 *Characteristics of Prefabricated Construction Compared to Traditional Construction*

Prefabricated construction applies the use of off-site production, manufacturing construction, and mass production of building components (Lessing et al., 2005; Najuwa et al., 2016; Rahman & Omar, 2006; Shaari, 2022; Thanoon et al., 2003; Warszawski, 1999). IBS Roadmap (2003–2010) page 12 defines the prefabricated concept as 'a new construction method which components are manufactured in a controlled production (on or off site), transported, positioned and assembled to form buildings with minimal site works.' Below are the characteristics of prefabricated construction in Malaysia:

1. Reduce on-site activity

Prefabrication components were specifically designed to reduce the on-site activities, thus saving much cost as it shortens the period of hiring the machinery, cranes, site operation, loan commitment, insurances, and minimizes the interest payment. In addition, most of the construction activities have been simplified, thus avoiding injuries or fatal accidents (Mahbub, 2016).

2. Components made under strict process

Components are made under strict process with detailed measurements using automated machines (Abd Jalil et al., 2016; Din et al., 2012). They are fabricated precisely following the project specifications on structural, loads, safety, and standard tolerances. Once completed, the components will be tested before they are allowed to be sent to the site. Any crack or defect can be detected because the checking process is made using special tools and computerized machines. Besides, the components' design is made using advanced software that can trace any errors or faults before production starts.

3. On-site activities are transferred to the factory

Due to most of the on-site activities being transferred to the factory, the project site will become the place to install the components, not conduct building works (Azman, 2012b). As a result, this can avoid wastages at the site, reduce environmental impact, and change the site condition from wet construction to dry construction. It automatically reduces the amount of dumping cost, as there are not many unused construction materials on site (Finn, 1992).

4. High initial cost

The cost to set up prefabrication plant is expensive as it involves high technology, skilled labors, casting beds, specific molds and automated machines (Abd Shukor et al., 2009; Jalil et el., 2015; Nawi et al., 2011). However, it can be economical if the prefabricated plant can manufacture the components in mass production, which can be achieved when the components are highly in demand and easily marketed (Shukor et al., 2011b).

5. Precise quality checks

Prefabricated components are sensitive to water leakage problems where the water may seep into the building through some imperfect joint space or connection, especially between wall and beam (Hamid et al., 2011). This usually happens several years after component installation. It will decay the concrete and cause corrosion to metal materials such as wiring and ducting cables, thus causing danger and affecting the building M&E services.

6. Requires trained labor and sophisticated machines

Prefabricated components heavily depend on trained and skilled laborers who must be qualified and authorized to conduct components' production and installation. Besides, it requires sophisticated machines operated by trained operators. The machines and cranes must be well maintained to avoid operation problems such as mechanical failure or breakdown especially when the production is operating for long hours (Din et al., 2012).

7. Transportation barriers

There are always transportation barriers whether the prefabricated components are produced either off-site (factory) or on-site (mobile plant). According to Abd Jalil et al. (2021) and Nadim and Goulding (2010) if the components are produced in the factory, they need to be transported to the site but if the components are produced on-site (mobile plant), the prefabrication production facilities need to be set up on the temporary plant. So, either using the factory or mobile plant, the projects will still face transportation barriers, which between transporting the components (using factory), or transporting the production equipment (using the mobile plant). This can be worsened if the site is hard to access or it is located in the city center where the routes are facing traffic difficulties (Shukor et al., 2011b).

8. Storage

The prefabricated components need to be stored at site for several reasons such as the installer team cannot do the installation works due to sudden bad weather, too many components arriving at the site at one time or the cranes not functioning (Nawi et al., 2011). In addition, some sites are already congested, so the components cannot be placed there, but need to be stored at proper storage that can protect them from vandalism, away from danger, and avoiding them from blocking the on-site activities.

5.2 *Major Differences Between Prefabricated Construction vs Traditional Construction*

According to Najuwa et al. (2016), Kamaruddin et al. (2022), and Hamid et al. (2011), the traditional construction method requires many stages and the process is driven by manual on-site activities. Besides, it requires the construction works from foundation to structure, roof, etc. to be labor intensive. As a result, the traditional method needs a long time to finish such as for the work of brick laying, plastering, concreting, and other in situ works. However, by using the prefabrication method, the prefabricated components will replace all these manual and tiring works, and the components can start to be installed once the foundation works are completed.

Azman et al. (2012b) explained that through prefabrication the contractors only need to place the components and fix them to their right position, without waiting for concreting or plastering. This can reduce much of the construction time as most of the building activities have been simplified. Due to this simplification, the installation of prefabrication components can be completed within a short period, saving labor costs, better safety, and minimizing other risk such as weather, pandemics and fluctuation of price. Below are the detail stages for prefabricated construction projects.

5.2.1 Design Stage

In most prefabricated construction, the project starts after the project design has been completed that include the architectural design, structural design, and M&E design but without the prefabrication components' design (Kamar et al., 2011). The prefabrication manufacturers are not involved during the design stage as they are not yet being appointed until the main contractors appoint them which happens after construction starts. According to Nawi et al. (2014), due to no participation from components manufacturers in the design team, the project design is first made in conventional design, then converted into a prefabricated design. As a result, problems always arise due to two different designs for the same work, making it hard to optimize prefabricated adoption. Besides, the two different designs cause clash of drawings and cannot adopt much of the prefabricated concept and both need to be re-designed the second time to make it adhere to the prefabricated concept. There are also disputes on the verification of the drawings because the prefabricated design is made by the components' manufacturer while the structural design is made by the C&S consultant (Nawi et al., 2011). The problem happens when the C&S consultant is reluctant to verify the prefabricated design despite the design and components being made through a strict process. They argued that the prefabricated designs are not produced by the C&S consultant and, thus, not under their verification. Some components are designed by different designers, so when they are combined and fixed together, the connections cannot be joined properly due to miss-match design and insufficient tolerances (Goodier & Gibb, 2007; Nawi et al., 2011). To make matters worse, some prefabrication manufacturers are using different design systems with special measurements, causing their components to be different from the other manufacturers or suppliers. This leads to errors during installation and weakens the building's structure and appearance.

5.2.2 Tendering Process to Select the Main Contractor

When all designs are completed, the client will appoint the main contractor through a tendering process. Usually, the main contractor will be responsible for construction including the prefabrication works. So, their duty includes appointing the prefabricated components' manufacturer but this happens after initial construction starts (Shukor et al., 2011b). At the same time, the main contractor will also start to appoint other sub contractors for M&E, foundation, external works, etc., and to procure the required materials for the project.

5.2.3 Construction Start

When the project enters the stage of construction, most of the on-site activities are managed using a conventional approach, which cannot meet the needs and natures of the prefabrication concept (Abd Jalil et al., 2016). In terms of on-site labor, the majority are trained to construct using traditional methods, while only a few have prefabrication experience or at least have attended prefabricated classes or courses. Due to the projects involving mixed construction, the main contractors must first identify which part adopts prefabrication concept and which part uses traditional method. Both constructions run concurrently, except for foundation or initial construction works.

5.2.4 Procuring IBS Components

Due to most main contractors do not have their in-house prefabrication manufacturers, the prefabricated components are procured from external manufacturers (Din et al., 2012). Research showed that it will be cheaper if the components are procured from external manufacturers rather than the main contractors setting up their own prefabrication plant (Hamid et al., 2011). Once appointed into the project, the manufacturer will prepare the prefabrication design which will be the basis for components production. Besides, through the prefabrication design, the main contractor will have clear direction on which part will adopt prefabrication and which part will use traditional. However, due to late appointment in joining the project, the prefabrication manufacturers always have insufficient time to conduct proper works that include preparing prefabricated design, manufacturing of the components, supervision, and monitoring, delivery, and precise installations at the site. Prefabricated manufacturers have always been appointed during the middle of project progress, not since the design stage.

5.2.5 Installing Prefabricated Components

Before finishing the installation, the components will be checked to ensure their position is correct and there is enough space for M&E services such as for the ducting, piping, wiring sockets, plumbing, etc. (Chung & Kadir, 2007). If there is an installation error, the repair work will cause additional time, and cost and might affect the other structures (Hussein, 2007). To avoid installation errors, it is better to engage a prefabrication installer team who are trained for the work (Hamid et al., 2011). However, there are some main contractors who choose to install by using their on-site laborers to save the cost, but they engage a prefabrication supervisor who will be stationed at the site, to monitor the installation works. However due to the projects involving mixed construction between prefabricated and traditional, the C&S consultant is always reluctant to take responsibility for supervising the prefabricated work as they stressed that this is not part of their scope and they are not experts in prefabrication (Azman et al., 2012a).

5.2.6 Finishing Works

After installing the prefabricated components into their final position, the finishing work can be started. Among the finishing works are painting the walls and fixing the windows and doors. Besides, the concrete wall or slabs which have surface differences will also be plastered to smoothen their surfaces, so that they can be covered with wallpaper or flooring tiles (Shaari & Ismail, 2003).

6 Issues and Difficulties of Current Procurement System in Fulfilling the Natures of Malaysian Pre-fabricated Construction

Much research proved that there are many constraints on the current procurement system that is being used in Malaysian prefabrication projects. These issues have slowed down the project efficiency, reduced the optimization of prefabrication, and distracted the project focus which eventually hard to maximize the benefits (Kamar et al., 2009b; Rahman & Omar, 2006; Shukor et al., 2011a). Below are the procurement issues in Malaysian prefabrication projects.

6.1 Payment

At present, the payment for prefabricated components manufacturers is made by the main contractor, not by the clients (Faizul, 2006). When the main contractors have agreed to procure the prefabricated components, they are required to pay 30% upfront and this money will be used by the manufacturers to buy materials and for their production cost (Rahman & Omar 2006). After the completion of the fabrication process, the main contractors are required to settle the whole payment before the prefabricated components can be sent to the site. This regulation also matches the requirement of the 'Material on site' clause under the Malaysian standard form of contract. However, many main contractors claim that this payment clause is not fair to them because they need to wait until the components have reached the site before they can claim to the clients (Nawi et al., 2011; Shukor et al., 2016). To lessen the burden, some manufacturers take lenient measures by sending the prefabricated components despite not receiving full payment from main contractors (Din et al., 2012). This decision also benefits them as they will not store the components for a long time and the components cannot be sold to other projects because they are specifically designed to suit one particular project. Due to many main contractors facing difficulties in paying for the components, prefabrication manufacturers cannot demand strict payment as they do not want to jeopardize their future contracts deals. Most manufacturers find it hard to secure good prospects with the industry if they demand strict payment (Shukor et al., 2011b). However, sending the components without receiving full payment has caused difficulties for them because the main contractors are always delaying in settling their full payment to manufacturers. To demand the delayed payment, some manufacturers had taken their own action by delaying or stopping the delivery of the components, which eventually affected the project's schedule (Hamid et al., 2011). The main contractor, on the other hand, justifies his delay in paying the cost of the components by claiming that they need to wait for the payment from the client before they can pay to the manufacturers.

6.2 *Fragmentation Within Prefabrication Projects*

Fragmentation in construction projects is defined as the segregation or isolation of work among the different parties in the same project (Nawi et al., 2014). Due to the current procurement system not designed to cater to prefabrication needs, coupled with strict duration for manufacture, delivery, and installation, the parties involved spend less time to integrate as most are focusing to their own task (Abd Jalil et al., 2016; Saad et al., 2002). Besides, each party has a different specialization, works in different locations, and carries their own liability (Abdul Rashid et al., 2006). Without integration, it is hard to share project information, progress updates, changing ideas, and effective discussions (Nawi et al., 2014). Fragmentation in prefabrication projects happens because most projects do not encourage early integration, thus the parties only fulfill their duty and only communicate when necessary (Azman et al., 2012a). They have difficulties coordinating with others, have less connection, hard to maintain cooperation and good working environment. As a result, many disputes and conflicts occur such as design clashes, misunderstandings, communication barriers, miss-schedule on delivery of components, and late response (Abadi, 2005).

6.3 *Procuring Prefabricated Components*

Prefabricated components are manufactured under a strict monitoring process as each component will have direct impact on the building's final structure (Faizul, 2006). Every component is made based on a specific design, which can only suit one particular project (Jacqueline, 1999). There is also an issue with the warranty especially for prefabrication installation. If the components' manufacturer carries the duty for prefabrication design until installation work, they will give a warranty for the components and their installation. However, if the manufacturer is only responsible for producing and delivering the components without installation, they will not give a warranty for the installation (Kamar et al., 2009b). Some projects request the manufacturer to include the installation work while some other projects choose to install by themselves, to save the budget (Nadim & Goulding, 2010). Without a warranty on prefabrication installation, the main contractors will bear the repair cost if defects on installation occur. The other issue with procuring prefabricated components is there are not many manufacturers who can produce large quantities of components at one time (Nawi et al., 2011). In addition, the manufacturers have always been appointed late into the project, so they have insufficient time to prepare the prefabricated design, manufacture the components, deliver, and install (Abd Jalil et al., 2021; Kamar et al., 2009b). Besides, the late orders from the main contractor will also cause late components delivery, thus delaying the project.

6.4 Components Production

Prefabrication manufacturers and main contractors are different parties, and they are not operating under the same organization or in one roof (Kamar et al., 2011; Shaari & Ismail, 2003). They usually work without close coordination until the time of component delivery (Rahman & Omar, 2006). Besides, the manufacturer is appointed in the middle of the project's progress, so they have limited time to produce prefabricated drawings, fabricate the components, and plan for installation (Chung & Kadir, 2007). Due to time constraints, they have less time to develop effective coordination with the main contractors and other consultants, causing them to be isolated without much communication with other parties (Blismas & Wakefield, 2009). Most manufacturers only employ C&S engineers who are responsible for monitoring the components' structure, but they do not employ M&E engineers who can monitor the M&E works for the components (Goodier & Gibb, 2005). Due to this, some prefabricated components lack on providing enough space for M&E services, thus they need to be modified or fixed at the site. This is common especially when the project involves vast prefabricated components and the time to manufacture the components is short (Shaari & Ismail, 2003).

6.5 Logistics

The components are huge, hard, long span, heavy and carry a massive load. They are not suitable to be placed on site for a long time as this will cause site congestion and danger to workers (Kamar et al., 2011). Placing them not in their right place can expose them to cracks and damages (Blismas & Wakefield, 2009). There are many logistic barriers to transporting and unloading the components such as weather conditions, safety for other road users, road regulations, hard to access the project site and traffic congestion (Lessing et al., 2005). Each delivery must have a detail plan to ensure the components are placed near their intended location. This is to avoid 'double handling' work where the components are moving multiple times before they are placed in their final position. The time to deliver the components must be punctual and the transporters need to consider various obstacles because the installation team needs the components to arrive on schedule, not too early and not too late (Azman et al., 2012a).

6.6 Installation and Supervision

For components' installation, the main contractor can choose either to install by themselves or hire an installer team. If the main contractors choose to install using their own labors, they will not get any warranty, but if they choose to hire an installer team, the installer will provide a warranty for the installation (Kamar et al., 2009b). Some main contractors choose to install because they want to save the budget (Nadim & Goulding, 2010). To ensure the installation is correctly done, the main contractor will request 'supervision work' from the components' manufacturer (Kamaruddin et al., 2022; Nawi et al., 2007). However, the 'supervision team' will only be responsible for monitoring the installation work, but they do not carry any liability if there is any error. Many argue about the effectiveness of the 'supervision work' because it is not enough to ensure the quality of the on-site labors, and whether can conduct the installation work properly (Shaari & Ismail, 2003). To further reduce the cost, some main contractors conduct the installation works even without proper supervision from the manufacturers (Trikha & Ali, 2004). They only refer to the method statement and installation drawings provided by the manufacturers, which many said are too brief, do not detail, and are uncomprehensive for untrained laborers. As a result, their installation is vulnerable to errors and defects (Nadim & Goulding, 2010). In some other projects, the C&S engineers are reluctant to take part in the components' installation because they claim that their responsibility is only for the conventional part, not including the prefabricated part (Din et al., 2012). For mixed construction that combines prefabrication and traditional method, there are always disputes on who will be held responsible if defects happen, either the main contractor who construct in situ structures, the manufacturer who is responsible for prefabricated part, or the project C&S consultant who is responsible for C&S design.

6.7 *No Standard Regulation or Prefabricated Form of Contract*

Generally, the components' installation process has not been properly recorded, cataloged, documented, or compiled to become guides or references for future projects (Faizul, 2006). Besides, there is no standard manual that provides the details on how the installation work should be conducted, except for the method statement which many complain is too brief and not comprehensive (Kamar et al., 2009b). Many believe that prefabrication is only about shifting to different methods, but do not realize it also carries major differences as it involves specific design, precision in components and installation, detailed measurement and tolerances, tight schedule on planning, and close coordination between parties (Chung & Kadir, 2007). There are many issues that evolve around payment, roles and liabilities of prefabricated manufacturers, components quality assessment, design disputes, insufficient time for prefabricated design work and warranty, etc. Until now, there is no specific form of contract that is designed to suit prefabricated natures and needs (Abd Jalil et al., 2015a, 2015b; Shukor et al., 2011b). The government has encouraged prefabrication adoption and made it compulsory, but they do not provide a good environment for the prefabrication industry to grow, even the rights of components' manufacturers are not well protected (Nawi et al., 2011).

6.8 Design

In current practice, the initial design for prefabricated is made in conventional format, before they are converted into prefabricated design (Kamar et al., 2009b). Many suggest that the design should not be converted later, instead, they should be made in prefabricated format at the beginning. However, this can only happen if the manufacturers have been involved since the design stage. As for current practice, the manufacturers are only involved after the design has been completed, thus they cannot participate in design stage and cannot consult other designers on the needs of prefabrication (Nawi et al., 2014). Without the involvement of manufacturers, the designers may not know how to optimize the prefabrication concept (Hamid et al., 2008). As a result, errors are likely to happen because the initial design and prefabricated design are not made at the same time. This causes difficulty in achieving high prefabrication scores, cannot fully utilize prefabricated in many areas and the achievement of the prefabricated adoption is less than the original targets. The other issue on design is the insufficient time for design checking especially between prefabricated design and M&E designs (Edge et al., 2002). As a result, the manufacturers do not have enough time to match their drawings with other drawings, causing mismatches and design clashes. These problems may also cause the components hard to install because their size, measurement, and load do not match with other designs (Blismas & Wakefield, 2009). Due to most prefabricated projects involve mixed methods of construction, so there are always disputes on the design responsibility where the C&S consultant will only be responsible for C&S design and the prefabrication manufacturers only responsible for prefabricated design (Hallowell & Toole, 2009). If defects occur, it is hard to find who will be held liable. Some claimed that due to the components being designed by the manufacturer, the C&S consultant is reluctant to be involved in the components' checking process. On the components' certification issue, the components are guaranteed by their manufacturers without verification from the C&S consultant. The C&S engineers compromise by trusting the quality checks made by engineers from the manufacturers, assuming they had followed the quality standard and the design already complied with the required strength, load, and structure (Hussein, 2007).

6.9 Challenges of the Prefabrication Manufacturers

Currently, the roles of prefabricated manufacturers are significant where they contribute 70–90% of the building structures and bear the greatest liability (Jacqueline, 1999). However, their rights are not secure, and they face difficulties in surviving including payment, coordination with the site team, unfair contracts, communication barriers and hard to match with other drawings and the design team (Din et al., 2012; Kamar et al., 2009b; Shukor et al., 2011b). The manufacturers also do not have a direct contractual relationship with the clients; thus they cannot highlight their problems or complaints to demand greater rights and better treatment (Dulaimi et al., 2002). This causes them to feel unheard, demotivated, less attention and find it hard to receive quick responses except when the components are about to be delivered (Nawi et al., 2011).

6.10 Late Appointment of IBS Manufacturers

At present, the appointment of prefabricated manufacturers is made by the main contractors and it happens in the middle of the project's progress (Shukor et al., 2011b). Due to this late appointment, the manufacturer cannot join or consult the design team, thus many problems arise such as hard in increasing prefabrication adoption and misconception about the constructability of the project (Nawi et al., 2011). The late involvement also gives them less time to integrate with other consultants, difficult to communicate or share project information, and limiting their contribution to the project designers regarding the best method to construct the project (Barlow, 1999).

6.11 Barriers to Maximize Prefabricated Components

The contracts for prefabricated manufacturers in the projects are only for the short term, so they find it hard to share their technology especially when they cannot have a direct relationship with the clients (Rahman & Omar, 2006). As a result, not all prefabricated technologies are exposed to developers and consultants. The main contractors also worry about the higher usage of prefabricated components because the more prefabricated is adopted, the lower margin and profit that the main contractors can achieve (Azman et al., 2012a). This happens because most main contractors depend on in situ works to gain profit, so when the in situ works are replaced with prefabricated components, their job scope will be reduced and their profit will be less. Meanwhile, that replacement creates jobs for prefabricated manufacturers and they take away the main contractor's profit (Ball, 1999). In addition, due to the short-term relationship between clients and manufacturers, the transfer of prefabricated technology and skills to the clients is hardly to be continuously developed and this will decelerate prefabrication adoption in future projects (Webster, 1993).

7 Conclusion

The implementation and process of Malaysian prefabrication projects have been discussed in detail in this chapter, including the uniqueness of prefabrication compared to traditional methods. After revealing how Malaysian industry treats the prefabrication concept, it can be concluded that it is hard for Malaysian projects to reap the optimization of prefabrication benefits because the procurement they use does not suit the prefabrication needs.

Since Malaysia started prefabrication, there is still no specific regulations or standard guidelines for contract or procurement system to cater on prefabrication's special characteristics. The only available references are only limited to technical aspects such as Precast Concrete Building Components for Residential Buildings (2012), Modular Coordination Implications—Building By-Laws and Regulations, and Joints and Tolerances for Building Construction (2014).

Jaafar and Radzi (2013) stressed that when there are changes in the method of construction, there is a need to adopt new suitable procurement that meets the projects' characteristics. They stressed that using unsuitable procurement would not only expose the projects to delay, payment disputes, less efficiency, and design clashes but it would also cause misconceptions and coordination problems that later lead to cost increases, time consumption, and quality defects. This was supported by Shukor et al. (2016) who emphasized that a new procurement should be introduced for prefabrication projects because their nature and needs are different. He added, that despite many efforts to promote the prefabricated concept, the procurement issues still became a major obstacle. If the method of construction has changed to prefabrication, the procurement must also be reviewed to suit the new differences. If no procurement that suits prefabrication is made, the industry will find difficulties in maximizing the benefits and they surely will revert back to traditional construction that is labor intensive, wastage, and cannot respond to huge demand within a short time.

References

Abadi, M. (2005). *Issues and challenges in communication within design teams in the construction industry: Investigation into the use of virtual teams and information and communication technologies (ICTs) in the UK construction industry*. Doctoral dissertation, University of Manchester.

Abd Jalil, A. B., Nuruddin, R., Jaafar, M., & Mydin, A. O. (2015a). A new procurement method for housing projects implementing IBS modular system. In *International Conference on Advances in Civil and Environmental Engineering (Penang)* (pp. 1–8).

Abd Jalil, A. B., Nuruddin, A. R., Jaafar, M., & Mydin, M. A. O. (2015b). Challenges for developers implementing IBS in Malaysian housing projects. *Journal of Construction in Developing Countries, 2*(3), 1–10. ISSN 1823-6499. http://eprints.usm.my/30177/

Abd Jalil, A. B., Nuruddin, A. R., Jaafar, M., & Mydin, M. A. O. (2016). Barriers and drivers to speed up prefabricated housing in Malaysian housing projects. *Journal of Construction in Developing Countries (JCDC), 10*(1), 1–14. ISSN 1823-6499.

Abd Jalil, A. B., Mohamed Shaari, S., & Nawi, N. (2021). IBS in residential projects; dissolving the problems of on-site component installation. In *Asset Management Business Solution International Conference Universiti Teknologi Malaysia UTM (AMBiS 1.0 2021)* (pp. 31–38).

Abd Shukor, A. S., Mohammad, M. F., Mahbub, R., & Halil, F. (2009). Integration of issues and problems of construction supply chain management in industrialised building system (IBS). In *2nd Construction Industry Research Achievement International Conference (CIRAIC)* 3rd-5th November 2009, Kuala Lumpur.

Abdullah, W. Z. W., & Nasir, S. R. M. (2017). Supply chain integration issues and challenges in industrialised building system (IBS) construction projects in Malaysia. *Malaysian Construction Research Journal (MCRJ), 22*(2), 73–89.

Abdul Rashid, R., Mat Taib, I., Ahmad, W., Basiron, W., Nasid, M., Wan Ali, W. N., & Mohd Zainordin, Z. (2006). Effect of procurement systems on the performance of construction projects. In *International Conference on Construction Industry*, 21–25th June, Padang, Indonesia.

Azman, M. N. A., Ahamad, M. S. S., Majid, T. A., & Hanafi, M. H. (2011). Permanent and mobile industrialised building system manufacturing plant based on Malaysian case study. In *2nd World Conference on Information Technology (WCIT 2011), 23–27th November, Antalya, Turkey.*

Azman, M. N. A., Ahamad, M. S. S., & Hilmi, N. D. (2012a). The perspective view of Malaysian industrialized building system (IBS) under IBS precast manufacturing. In *The 4th International Engineering Conference-Towards Engineering of 21st Century, 15–16th October, Faculty of Engineering, The Islamic University of Gaza.*

Azman, M. N. A., Ahamad, M. S. S., & Hussin, W. M. A. W. (2012b). Comparative study on prefabrication construction process. *International Surveying Research Journal, 2*(1), 45–58.

Ball, M. (1999). Chasing a snail: Innovation and housebuilding firms' strategies. *Housing Studies, 14*(1), 9–22.

Barlow, J. (1999). From craft production to mass customisation. Innovation requirements for the UK housebuilding industry. *Housing Studies, 14*(1), 23–42.

Bazeley, P. (2004). Issues in mixing qualitative and quantitative approaches to research. *Applying Qualitative Methods to Marketing Management Research, 141*, 156.

Blismas, N., & Wakefield, R. (2009). Drivers, constraints and the future of offsite manufacture in Australia. *Construction Innovation, 9*(1), 72–83.

Chung, L. P., & Kadir, A. M. (2007). *Implementation strategy for industrialized building system.* PhD thesis, Universiti Teknologi Malaysia (UTM), Johor Bahru.

Creswell, W., & Plano Clark, V. L. (2011). *Designing and conducting mixed methods research* (2nd ed.). Sage Publications.

CIDB. (2010). Industrialised building system (IBS) Info, Issue 3/2010, IBS Centre Malaysia Publication, Construction Industry Development Board Malaysia (CIDB), Kuala Lumpur, Malaysia.

Din, M. I., Bahri, N., Dzulkifly, M. A., Norman, M. R., Kamar, K. A. M., & Hamid, Z. A. (2012). The adoption of industrialised building system (IBS) construction in Malaysia: The history, policies, experiences and lesson learned. In *ISARC. Proceedings of the International Symposium on Automation and Robotics in Construction* (Vol. 29, pp. 1–8). Vilnius Gediminas Technical University, Department of Construction Economics & Property, Lithuania.

Driscoll, D. L., Appiah-Yeboah, A., Salib, P., & Rupert, D. J. (2007). Merging qualitative and quantitative data in mixed methods research: How to and why not. *Ecological & Environmental Anthropology, 3*(1), 19–28.

Dulaimi, M. F., Y. Ling, F. Y., Ofori, G., & Silva, N. D. (2002). Enhancing integration and innovation in construction. *Building Research & Information, 30*(4), 237–247.
Edge, M., Craig, A., Laing, R., Abbott, L., Hargreaves, A., Scott, J., & Scott, S. (2002). Overcoming client and market resistance to prefabrication and standardisation in housing. *Robert Gordon University, Aberdeen.*
Faizul, N. A. (2006). Supply chain management in IBS industry. *Malaysia International IBS Exhibition (MIE 06) 21–24th November 2006*. Akademi Binaan Malaysia, Jalan Chan Sow Lin, Kuala Lumpur, *2*(5), 15–28.
Finn, D. W. (1992). Towards industrialized construction. *Construction Canada, 34*(3), 25–26.
Fournery, M. (1997). *Prefabricated housing in Argentina and U.S*, Book Published by Editorial Primera Linea SA. Buenos Aires, Argentina.
Goodier, C., & Gibb, A. (2005). Barriers and opportunities for offsite in the UK. In *11th Joint CIB International Symposium*, 13–16 June 2005 (pp. 148–158).
Goodier, C., & Gibb, A. (2007). Future opportunities for offsite in the UK. *Construction Management and Economics, 25*(6), 585–595.
Hallowell, M., & Toole, T. M. (2009). Contemporary design-bid-build model. *Journal of Construction Engineering and Management, 135*(6), 540–549.
Hamid, Z. A., Kamar, K. A. M., Ghani, M. K., Zura, M., Zain, M., & Rahim, A. H. A. (2011). Business strategies of contractors in industrialised building systems (IBS) construction. *Vertilite Asia, 1*(3), 18–30.
Hamid, Z., Kamar, K. A. M., Zain, M., Ghani, K., & Rahim, A. H. A. (2008). Industrialized building system (IBS) in Malaysia: The current state and R&D initiatives. *Malaysia Construction Research Journal, 2*(1), 1–13.
Hussein, J. (2007). Industrialised building systems: The challenge and the way forward. In *Keynote address at Construction Industry Research Achievement International Conference, CIDB, PWTC, Kuala Lumpur.*
Ibrahim, A., Roy, M. H., Ahmed, Z., & Imtiaz, G. (2010). An investigation of the status of the Malaysian construction industry. *Benchmarking: An International Journal, 17*(2), 294–308.
Industrialized Building System (IBS) Roadmap 2011–2015, Construction Industry Development Board (CIDB), Malaysia, 2011.
Jaafar, M., & Radzi, N. M. (2013). Level of satisfaction and issues with procurement systems used in the Malaysian public sector. *Australasian Journal of Construction Economics and Building, 13*(1), 50–63.
Jacqueline, G. (1999). The future for precast concrete in low-rise housing. *Precast Housing Feasibility Study Group, UK.*
Kamar, K. A. M., Hamid, Z. A., Azman, M. N. A., & Ahamad, M. S. S. (2011). Industrialized building system (IBS): Revisiting issues of definition and classification. *International Journal of Emerging Sciences, 1*(2), 120–133.
Kamar, K. A. M., Alshawi, M., & Hamid, Z. (2009a). Industrialised building system: the critical success factors. In *9th International Postgraduate Research Conference (IPGRC), University of Salford, Salford*, 29–30th January 2009 (pp. 29–30). Salford Quays, Greater Manchester, United Kingdom.
Kamar, K. A. M., Alshawi, M., & Hamid, Z. (2009b, January). Barriers to industrialized building system (IBS): The case of Malaysia. In *Proceedings of the BuHu 9th International Postgraduate Research Conference (IPGRC), Salford, UK* (pp. 29–30).
Lessing, J., Stehn, L., & Ekholm, A. (2005). Industrialised housing: definition and categorization of the concept. In *Annual conference of the International Group for Lean Construction: 18/07/2005–21/07/2005* (pp. 471–480). International group for lean construction.

Mahbub, R. (2016, April). Framework on the production and installation of industrialized building system (IBS) construction approach in Malaysia. In *Proceedings of International Conference on Architecture and Civil Engineering (ACE 2016)* (pp. 25–26).

Nadim, W., & Goulding, J. S. (2010). Offsite production in the UK: The way forward? A UK construction industry perspective. *Construction Innovation, 10*(2), 181–202.

Najuwa, M. N., Mohd Nasrun, M. N., Mohd Kamarul, I. A. R., Yusni, A., & Tapa, A. (2016). A review of delay factors in Malaysian industralised building system (IBS) construction project. *ARPN Journal of Engineering and Applied Sciences, 3*(3), 18–29.

Nawi, M. N. M., Mydin, M. A. O., Elias, M. E., Shaharanee, I. N. M., & Yusoff, M. N. (2015). Suitability of IBS formwork system in Malaysian construction industry. *Australian Journal of Basic and Applied Sciences, 9*(7), 89–91.

Nawi, M. N. M., Lee, A., Azman, M. N. A., & Kamar, K. A. M. (2014). Fragmentation issue in Malaysian industrialised building system (IBS) projects. *Journal of Engineering Science and Technology, 9*(1), 97–106.

Nawi, M. N. M., Lee, A., & Nor, K. M. (2011). Barriers to implementation of the industrialised building system (IBS) in Malaysia. *The Built & Human Environment Review, 4*(2), 34–37.

Nawi, M. N. M., Nifa, F. A. A., Abdullah, S., & Yasin, F. M. (2007, November). A preliminary survey of the application of industrialised building system (IBS) in Kedah and Perlis Malaysian construction industry. In *Conference on Sustainable Building South East Asia* (Vol. 5, No. 7).

Pan, W., Gibb, A., & Dainty, A. (2005). Offsite modern methods of construction in housebuilding: perspectives and practices of leading UK housebuilders. *Loughboroough: Buildoffsite.*

Rahim, A. A., Hamid, Z. A., Zen, I. H., Ismail, Z., & Kamar, K. A. M. (2012). Adaptable housing of precast panel system in Malaysia. *Procedia-Social and Behavioral Sciences, 50*, 369–382.

Rahman, A. B. A., & Omar, W. (2006, September). Issues and challenges in the implementation of industrialised building systems in Malaysia. In *Proceedings of the 6th Asia-Pacific Structural Engineering and Construction Conference (APSEC 2006)* (pp. 5–6).

Saad, M., Jones, M., & James, P. (2002). A review of the progress towards the adoption of supply chain management (SCM) relationships in construction. *European Journal of Purchasing & Supply Management, 8*(3), 173–183.

Shaari, S. N., & ISmail, E. (2003). Promoting the use of industrialised building systems (IBS) and modular coordination (MC) in the Malaysia construction industry. *Bulletin of Institute Engineers Malaysia (IEM), 14*(6), 91–112.

Shukor, A. S. A., Muhammad, M. F., Khaderi, S. S., & Halil, F. M. (2016). Supply chain integration: Establish the appropriate challenges in improving integrated SC in an innovative approach of IBS. *Environment-Behaviour Proceedings Journal, 1*(3), 79–88.

Shukor, A. A., Mohammad, M. F., & Mahbub, R. (2011a). Supply chain integration challenges in project procurement in Malaysia: IBS contractors' perspective. *Journal Management and Innovation for a Sustainable Built Environment, Amsterdam, the Netherlands, 3*(5), 77–92.

Shukor, A. S. A., Mohammad, M. F., Mahbub, R., & Ismail, F. (2011b). Supply chain integration challenges in project procurement in Malaysia: The perspective of IBS manufacturers. In *Proceedings of the 27th annual ARCOM conference,* 5–7 September 2011, *Association of Researchers in Construction Management* (pp. 495–504), Bristol, UK.

Syazwana T.A., Yunus, R., Mohammad, H., & Yaman, S. K. (2017). A preliminary review on economies of scale (EOS) towards industrialized building system (IBS) Manufacturer. In *MATEC Web of Conferences* (Vol. 103, p. 03008). EDP Sciences.

Tatum, C. B., Vanegas, J. A., & Williams, J. M. (1986). Constructability improvement using prefabrication, pre-assembly and modularization, Construction Industry Institute, Austin, Texas, USA.

Thanoon, W. A., Peng, L. W., Kadir, M. R. A., Jaafar, M. S., & Salit, M. S. (2003). The essential characteristics of industrialised building system. *International Conference on Industrialised Building Systems, 10*(11), 12–21, Kuala Lumpur.

Trikha, D. N., & Ali, A. A. A. (2004). *Industrialized building system* (1st ed.). Universiti Putra Malaysia Press.

Warszawski, A. (1999). *Industrialised and automated building system, Technion-Israel Institute of Technology*. Publications E & FN Spon.

Webster, A. C. (1993). Japanese building design and construction technologies. *Journal of Professional Issues in Engineering Education and Practice, 119*(4), 358–377.

A Review of Green Open Space Implementation Towards Green City Development in Developing Countries

M. Nabilaa, V. Thenmolli, and M. Z. Zarina

Abstract Green open space (GOS) has an imperative ecological act to the success of green city development (GCD). It is evident that currently many GOS are plunging and transmitted due to distinct conditions. In Malaysia and Indonesia, the maximum allocation of GOS in every city development area is 10 and 30%, respectively. However, developers' adherence to this procedure extend to decay in recent years succeeding in the minimization of GOS. Consequently, there is a compelling requirement for systemic preliminary planning to preclude the eradication of existing and future green spaces in urban areas. This study aimed to understand the advantages and implementation of GOS in GCD in Malaysia and Indonesia. Through the systematic literature review method—there are four types of ecosystem services (ES) that highlighted the advantages of GOS on GCD, while 19 services were listed for the improvement of urban ecosystem quality.

Keywords Ecosystem services · Green city · Green open space · Sustainability · Sustainable development goals

1 Introduction

The government introduced the green open space (GOS) concept to improve the quality of life of its citizens by establishing a natural social ecosystem, contributing to the sustainable progress of cities and communities and the ecosystem as a whole (Şenik & Uzun, 2022; Zhang et al., 2021). This concept is incorporated in the green city (GC) model to ensure the ideal use of natural resources (Olaniyan, 2020) and the building of environmentally friendly urban structures (European Landscape Contractors Association (ELCA), 2011; United Nations Environment Programme

M. Nabilaa · V. Thenmolli (✉) · M. Z. Zarina
Faculty of Administrative Science and Policy Studies, Universiti Teknologi MARA, Shah Alam, Selangor, Malaysia
e-mail: thenm020@uitm.edu.my

M. Z. Zarina
e-mail: zarina752@uitm.edu.my

U. R. Hashim et al. (eds.), *Green Infrastructure*,
https://doi.org/10.1007/978-981-99-7003-2_9

(UNEP), 2011) that can maintain a sustainable ecology and support urban ecosystems (Ajrina & Kustiwan, 2019). Despite its importance, GOS continues to decline due to uncontrolled urbanisation and industrialisation (Murtini et al., 2020a).

Currently, most cities in Indonesia have less than 30% of GOS (Endangsih et al., 2020; Iqbal & Jumiati, 2019), while only 10% of this green space can be found in Malaysian urban areas (Ibrahim et al., 2017; Mansor et al., 2019). Over the years, there have been attempts to establish urban green spaces, but the focus of these projects is the social interest of city dwellers while the environment is forced into the back seat (Bush, 2020; Karuppannan et al., 2014). Unfortunately, urban developers often turn a blind eye to the necessity of GOS. Therefore, this study aimed to delve deeper into the implementation and advantages of GOS towards green city development (GCD) in Malaysia and Indonesia.

2 Literature Review

2.1 Green City

The GC concept is associated with the famous Sustainability Theory by Barbier (1987); a city should be developed based on the three pillars of sustainability: economic, social, and environmental aspects (Brilhante & Klaas, 2018; Lewis, 2015) to preserve environmental functions (Díaz et al., 2015) while maintaining the quality of life of its inhabitants (Chiesura, 2004). City developers use the GC model as a guideline in structuring environmentally friendly towns (Brilhante & Klaas, 2018; ELCA, 2011), where GOS is an essential element (Brilhante & Klaas, 2018; Zheng et al., 2019) incorporated in their green infrastructure strategy (Mutini et al., 2020b; Meerow, 2019) to accomplish sustainable urban development (Kruize et al., 2019; Wikantiyoso et al., 2020).

2.2 Green Open Space

In Malaysia, the government launched the GOS initiative towards greener and sustainable development in 2010 (Malek, 2013; Malek & Nashar, 2018; Malek et al., 2018), followed by the Green City Action Plan established in 2013 to achieve the GC status (Lewis, 2015). Meanwhile, the Indonesian government announced its GCD Programme in 2011 to implement GOS towards a greener and sustainable development (Kirmanto et al., 2012). Their mission to be among the 40 cities endorsed with the GC status by 2036; a project that was officiated during the Chief Ministers and Governors Forum (CMGF) of the Indonesia-Malaysia-Thailand Growth Triangle (IMT-GT) meeting in 2012 (Centre for IMT-GT Subregional Cooperation (CIMT), 2017). Some well-known initiatives spearheading the GOS projects in Malaysia are

Pocket Parks and Green Neighbourhood (Hashim et al., 2019; Ramli et al., 2019), whereas Ruang Terbuka Hijau is at the forefront of the GOS initiative in Indonesia space (Iqbal & Jumiati, 2019).

2.3 *Ecosystem Services*

GOS preserves ecosystem functions by supporting the continuity of human life and socio-economic growth (Amir & Adlansyah, 2021). These advantages that people gain from the ecosystem are termed ecosystem services (ES) (Millennium Ecosystem Assessment (MEA), 2005). There are mainly four types of ES on GCD that are obtained via the establishment of GOS: (1) supporting, (2) regulating, (3) cultural, and (4) provisioning (Hagemann et al., 2020; Keane et al., 2014). Keane et al. (2014) reported that GOS gave rise to 19 urban ES that presented the advantages of GOS. All 19 services under the four ES were evaluated in the present study, and the most suitable ones were identified for application in Malaysia and Indonesia to achieve their GC status.

Supporting Ecosystem Service (SES) is the main ES that supports other ES (Sandhu et al., 2010; Sinnett et al., 2016), and it has become increasingly challenging in this fast-paced world (Saha, 2009). For instance, in Malaysia, the implementation of GOS in the metropolitan city of Kuala Lumpur has succeeded in increasing the number of various species of birdlife in the large urban area (Karuppannan et al., 2014; Rakhshandehroo et al., 2017). Meanwhile, Indonesia has proven that the combination of ornamental and natural vegetation in GOS maintains the species richness of flora and fauna in Pontianak City (Ratih & Febrianto, 2016).

Regulating Ecosystem Service (RES) refers to its importance in benefiting human health and well-being in terms of safety and quality of life in urban environments (Maes et al., 2013; Setiawan, 2015). For example, the coverage of green spaces in Malaysia, especially in Kuala Lumpur city managed to reduce the effect of urban heat island (Elsayed, 2018; Razali et al., 2020) and enhance air quality (Rakhshandehroo et al., 2017; Razali et al., 2020). GOS in Indonesia succeeded in reducing air pollution and eliminating noise pollution (Vety Jayanti et al., 2020). It was reported that the soft walls of greenery and trees have vital effects on noise reduction (Veisten et al., 2012; Watts et al., 2013). Indonesia also intensified its efforts to set up more GOS to eradicate urban floods (Ratnawati, 2017). The locations involved are Padang City (Iqbal & Jumiati, 2019) and Jakarta City, where plants are being grown along coastlines via the East Flood Canal Revitalization Project (Wahdaniyat, 2019).

The advantages of GOS under the cultural ecosystem service (CES) cover non-material advantages (Gonzales et al., 2018; MEA, 2005) to city dwellers through recreation, rest, relaxation, spiritual enhancement (Sinnett et al., 2016), cognitive development, reflection, aesthetic experiences (MEA, 2005), religious, and (Gonzales et al., 2018) other advantages derived from human–ecological relations (Chan et al., 2019). In Malaysia, the implementation of GOS has been one of the initiatives to promote a healthy lifestyle since 2010 due to the increase in obesity and other diseases (Hussain et al., 2016; Mansor & Harun, 2018). Meanwhile, the GOS at Bogor Regency, West Java, Indonesia, has become a community gathering centre for sports and other therapeutic activities (Dewi et al., 2018). Moreover, through the Child-Friendly City Programme by the Indonesian government, certain green spaces in Jakarta city were built as child-friendly GOS to increase the quality of nature pedagogic among people in metropolitan areas (Yuniastuti & Hasibuan, 2019a, 2019b).

Provisioning Ecosystem Service (PES) ability to supply products to people (Lundh, 2017; Sinnett et al., 2016) by extracting directly from the ecosystem like edible crops, freshwater catch, flowers, and fibres (Lundh, 2017; Haines-Young & Potschin, 2011). For instance, a study by Ridwan et al. (2017) in Bandung City, Indonesia, found that the productive GOS implementation at Taman Sari has a high potential in producing food specifically for the less fortunate. These residents have become the key players in maximising the use of GOS as an urban farming land through urban agriculture. The concept of urban agriculture not only improves food security (Ridwan et al., 2017) but is also one step closer to becoming a GC (Ramandhani et al., 2020; Ridwan et al., 2017). On the other hand, the local government has conducted urban agriculture projects in Malaysia since 2013 at several community gardens in GOS areas of Subang Jaya City, Selangor, for the advantage of the lower-income groups, contributing to their economic sustainability (Othman et al., 2018). Nevertheless, Keane et al. (2014) did not include other business types under PES. Since then, some studies have found that GOS also promoted various business activities besides food production. Open spaces or parks and recreation areas have increased the value of properties in their vicinity (Crompton, 2000; Hussain et al., 2016) and enhanced the economy of a particular area (Rostami et al., 2013). Most people are willing to pay for residential areas near parks and open spaces (Crompton, 2000), suggesting parks act as a marketing tool that has successfully boosted the tourism and business sector and become the city's landmarks and attractions (Hussain et al., 2016).

3 Research Methodology

GOS and GCD-related readings were analysed conceptually to identify the relationship between GOS implementation and its impact on economic, social, and environmental sustainability that led to the GCD. A search for relevant articles was carried out in several search engines, including Science Direct, SCOPUS, Web of Science, Sage Publications, Emerald Publishing, and Google Scholar. As referred to Fig. 1 a total of 282 articles were initially identified; however, only 56 were shortlisted after the screening process and followed by the eligibility process accordingly.

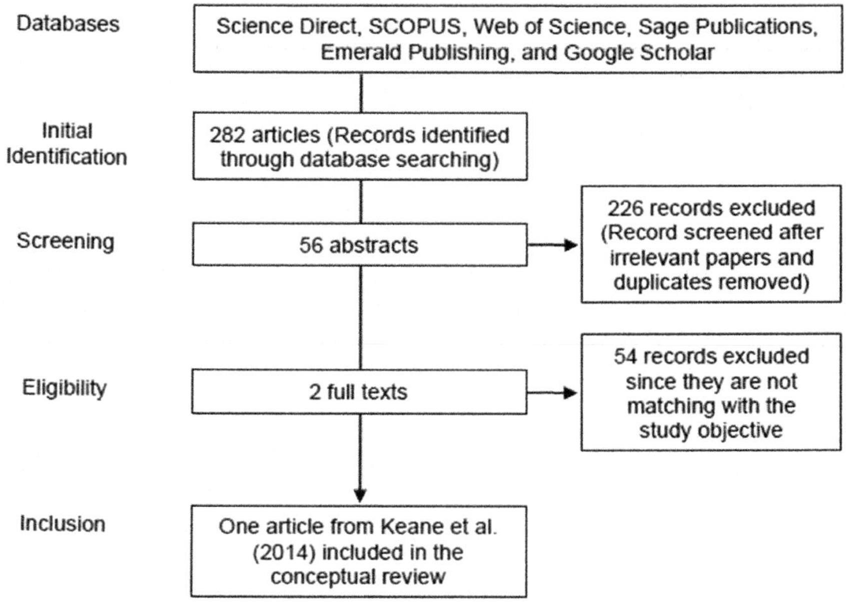

Fig. 1 Flow diagram of the process in selecting article

3.1 *Identification*

Search string activity was conducted by using reliable keywords. For example, the keyword used in SCOPUS database is TITLE-ABS-KEY (("important" OR "essential" OR "vital" OR "benefit" OR "advantage" OR "influence" OR "affect" OR "impact" OR "effect") AND ("green space" OR green open space" OR "urban green" OR "urban green space" OR "urban green open space" OR "urban park" OR "metropolitan park" OR "public park" OR "public open space" OR "recreational park" OR "pocket park" OR "ecology park" OR "eco park")).

3.2 *Screening*

Through the screening process, as many as 26 duplicate articles were eliminated and another 200 articles were removed because did not follow the inclusion criteria. The first criteria is that selected articles must be within a 5-year period (from 2016 to 2021) to get the latest information. Henceforth, the literature type must only focus on articles from journals in the final stage because consists of primary sources that involve complete empirical and non-empirical data. Therefore, publication in the form of a book, book chapter, book series, conference paper or proceedings, editorial, business article, erratum, short survey, and trade publications were excluded, and all literature reviewed must be published in the English language.

3.3 *Eligibility*

After examining two articles comprehensively in the eligibility process, the researchers found that only one article was relevant to the topic of the study and was selected to answer the research question, what are the advantages of GOS implementation towards GCD in Malaysia and Indonesia. In the eligibility process, all articles were filtered by looking at the title and abstract including the contents in order to achieve the objective of the current study. Consequently, a total of 54 articles were excluded because consisted of hard science study and did not focus on advantages of GOS on GC. Finally, one article was remained to be an adapted theory in this study which is the Urban Ecosystem Services Theory by Keane et al. (2014). This theory has been selected as a guideline to measures the relationship of GOS's advantages on GCD (refer to Table 1).

4 Results and Discussion

The indifference of city residents, especially city developers, in caring for the environment in every city development project has caused a variety of pollution that disturbs the health of the city's environment until it becomes worse not only due to extreme urbanisation but also the unregulated industry which also causes the problem to become worse every year because the absence of sufficient green areas as an effective natural urban resilience (Murtini et al., 2020a). Therefore, the awareness of urban citizens to know the advantages of environmental protection must be emphasised and applied in every municipal and industrial sector. Based on the conceptual review that has been made by identifying the advantages that can be produced by GOS for the well-being of a city and not only for the environment but also for covering social and economic development in a city, it has been found that there are 19 advantages that can be produced by GOS (refer to Table 1). Based on Keane et al. (2014) theory, four types of ES that benefited city sustainability (Pakzad & Osmond, 2015) were identified. This shows that a productive GOS can help city developers to create a successful GC. Green spaces are vital to the sustainability concept (Ahmad et al., 2017; Chiesura, 2004) and as a driver to achieve Sustainable Development Goals (SDGs) (World Bank, 2003). Although GOS brings various advantages and well-being to the city, it is still not fully used by both countries, namely Malaysia and Indonesia. The results show that there are still elements of SE that are left empty and not fully paid attention to.

In addition, the success of GCD through the implementation of GOS is not only required to touch the 4 types of SE but these 4 SEs also need to touch the 3 pillars

Table 1 The 19 advantages of GOS towards GCD

No	Urban ecosystem services theory (Keane et al., 2014)	Malaysia	Indonesia	3 Pillars of sustainability theory (Barbier, 1987)
	Support ecosystem services			Environmental sustainability
1	Biodiversity	√	√	
2	Ecological interconnections	√	√	
3	Soil fertility	x	x	
4	Habitat	√	√	
	Regulating ecosystem services			
5	Air quality	√	√	
6	Noise regulation	x	√	
7	Extreme weather protection	√	√	
8	Water quality regulation	√	√	
9	Climate adaptation	√	√	
10	Pollination	x	x	
	Cultural ecosystem services			Social sustainability
11	Health	√	√	
12	Sensory experience	√	√	
13	Social interactions	√	√	
14	Nature pedagogic	x	√	
15	Symbolic and spiritual	x	x	
	Provisioning ecosystem services			Economic sustainability
16	Food production	√	√	
17	Freshwater	x	x	
18	Material	x	x	
19	Energy	x	x	
		11 Services	13 Services	

Source Adapted from Keane et al. (2014, pp. 13–14) and Barbier (1987, p. 104)

of sustainability that have been recommended by Barbier (1987) to build a GC that is balanced with the economic, social and environmental development of the surrounding area that then can also achieve the success of the SDGs at the global level. In fact, GCD is also part of a global mission under the SDGs that has been widely mobilised under the United Nations General Assembly together with several countries around the world since 2015 (Kuklina et al., 2021). Therefore, the implementation of GOS towards GCD is considered one of the most important world missions for the eternal life of mankind.

Through the advantages produced by GOS, a productive GOS can produce a strong GC. Therefore, the concern of urban citizens and the world community of the 21st millennium can be seen with GCD which is one of the usages of GOS elements.

As seen, although the implementation of GOS in Malaysia and Indonesia has started since 10 years ago (Kirmanto et al., 2012; Malek, 2013; Malek & Nashar, 2018; Malek et al., 2018), the amount of green space in the city is still not encouraging, each of Malaysia and Indonesia only has less than 10% green area (Ibrahim et al., 2017; Mansor et al., 2019) and 30% (Endangsih et al., 2020; Iqbal & Jumiati, 2019). Although the existence of GOS has been able to provide all the advantages (refer to Table 1), city residents in Malaysia and Indonesia still ignore "green" in the efficient and comprehensive management of their city development (Endangsih et al., 2020; Ibrahim et al., 2017; Iqbal & Jumiati, 2019; Karade et al., 2017; Mansor et al., 2019; Zheng et al., 2019) then slow down the GCD process (Pakzad & Osmond, 2015; Mutini et al., 2020b). It is even worse when GOS in soil fertility, pollination, symbolic and spiritual, freshwater, material, and energy are still not emphasised in the implementation of GOS in Malaysia and Indonesia. Furthermore, in Malaysia, noise regulation and pedagogic nature are not included in their GOS planning. Besides, many research had reported good cooperation among all city stakeholders from multidimensional green governance is compulsory (Li et al., 2018; Ojha et al., 2019) to implement GOS (Mao et al., 2020). Therefore, Malaysia and Indonesia need to intensify their efforts in developing high-quality GOS for successful GC.

5 Conclusions and Recommendations

In the theory of Keane et al (2014) there are some important elements that have not been included in his theory such as tourism, business and real estate development because previous research studies found real estate areas that have productive GOS areas can increase the value of the surrounding real estate (Crompton, 2000; Hussain et al., 2016; Rostami et al., 2013) and at the same time can attract domestic and foreign tourists. For example in Kuala Lumpur, Malaysia where Taman Tasik Perdana and Taman Tasik Titiwangsa which are productive GOS have succeeded in increasing the economy of the nearest community, it has succeeded in attracting tourists to come there because of various social activities held every year such as arts and food festivals, sport events, musicals, and theatricals (Hussain et al., 2016; Sherer, 2006). While in Indonesia, the existence of their GOS has shown an increase in the small business sector (street vendors) in 5 regions of Indonesia with the existence of 12 GOS, one of which is Kartini Garden in West Java Province (Ali et al., 2021). Therefore, researchers would like to recommend Keane et al. placing elements such as tourism, business and real estate development in their theory which is located under the "Provisioning Service Ecosystem" in the hope that it can strengthen the economic sustainability of the city's citizens to achieve GC status.

Additionally, this research has identified several recommendations that are useful for future studies. Based on Table 1, out of the 19 advantages, only 11 are currently benefited by GOS in Malaysia and 13 in Indonesia. Therefore, both countries should consider designing GOS for phytoremediation purposes, especially in industrial areas for soil purification and a soil fertility strategy under the SES category. Furthermore,

Malaysia could establish GOS as an initiative in their flood management strategy and noise reduction management strategy. Moreover, both countries could utilise GOS for pollination to improve nature's quality and diversity, besides enhancing ecological interaction under the RES. Meanwhile, child-friendly GOS could be established to support nature pedagogy, and future study needs to investigate the relationship between GOS and symbolism and spirituality in Malaysia and Indonesia in the CES category. From the conceptual reviews, it is also recommended for Keane et al. (2014) to include "business" as a new product in PES since the GOS has been shown to boost ecotourism and the retail sector in the community. Most importantly, there is limited literature discussing the ability of GOS in supplying freshwater, materials, and energy. Other green methods in supplying freshwater for example that has been implemented by Japan at their underground green space. They were storing the rainwater in one huge place known as the temple water storage basin beneath Tokyo due to limited green open space and at the same time increase the resilience of the city (Admiraal & Cornaro, 2020). Therefore, it is highly recommended for future research to be conducted in these areas.

Acknowledgements The main author appreciates the support from supervisors, Dr. Thenmolli Vadeveloo and Dr. Zarina Mohd Zain from Universiti Teknologi MARA.

References

Admiraal, H., & Cornaro, A. (2020). Future cities, resilient cities—The role of underground space in achieving urban resilience. *Underground Space, 5*, 223–228. https://doi.org/10.1016/j.undsp.2019.02.001

Ahmad, P., Misni, A., Kamaruddin, S. M., & Daud, N. (2017). Green neighbourhood adaptive model for urban living: A conceptual review. *Environment-Behaviour Proceeding Journal.* In *5th AMER International Conference on Quality of Life, 2*(5), 55–63. https://doi.org/10.21834/e-bpj.v2i5.690

Ajrina, H., & Kustiwan, I. (2019). From green open space to green infrastructure: The potential of green open space optimization towards sustainable cities in Bekasi City & Regency, Indonesia. *IOP Conference Series: Earth and Environmental Science, 399*, 012130. https://doi.org/10.1088/1755-1315/399/1/012130

Ali, S., Sulistiowati, R., Wulandari, C., & Riniarti, M. (2021). Maximizing the social-economy impacts of urban green space in several cities in Indonesia. *IOP Conference Series: Earth and Environmental Science, 918*, 012007. https://doi.org/10.1088/1755-1315/918/1/012007

Amir, N., & Adlansyah, B. (2021). Implementation of Government Oversight Functions Related to Child Friendly Green Open Space in Gresik Regency. *Journal Equity of Law and Governance, 1*(1), 68–76. https://doi.org/10.22225/elg.v1i1.3247

Barbier, E. B. (1987). The concept of sustainable economic development. *Environmental Conservation, 14*(2), 101–110. https://doi.org/10.1017/S0376892900011449

Brilhante, O., & Klaas, J. (2018). Green city concept and a method to measure green city performance over time applied to fifty cities globally: Influence of GDP, population size and energy efficiency. *Sustainability, 10*(2031). https://doi.org/10.3390/su10062031

Bush, J. (2020). The role of local government greening policies in the transition towards nature-based cities. *Environmental Innovation and Societal Transitions, 35*, 35–44. https://doi.org/10.1016/j.eist.2020.01.015

Centre for IMT-GT Subregional Cooperation. (2017). *IMT-GT vision 2036*. https://www.adb.org/sites/default/files/related/72271/imt-gt-vision-2036.pdf

Chan, K., Goldstein, J., Satterfield, T., Hannahs, N., Kikiloi, K., & Woodside, U. (2019). Cultural services and non-use values. In P. Kareiva, H. Tallis, T. H. Ricketts, G. C. Daily, & S. Polasky (Eds.), *Natural capital: Theory & practice of mapping ecosystem services* (pp. 206–228). Oxford University Press. https://doi.org/10.1093/acprof:oso/9780199588992.003.0012

Chiesura, A. (2004). The role of urban parks for the sustainable city. *Landscape and Urban Planning, 68*, 129–138. https://doi.org/10.1016/j.landurbplan.2003.08.003

Crompton, J. L. (2000). *The impact of parks and open space on property values and the property tax base*. https://saugatuckdunescoastalalliance.com/wp-content/uploads/2018/06/crompton_2000.pdf

Dewi, O. C., Chairunnisa, I., Hidayat, T., Anggraini, M., & Napitupulu, A. (2018). Green open space: Awareness for health or sustainability? *IOP Conference Series: Earth and Environmental Science, 120*, 012014. https://doi.org/10.1088/1755-1315/120/1/012014

Díaz, S., Demissew, S., Carabias, J., Joly, C., Lonsdale, M., Ash, N., Larigauderie, A., Adhikari, J. R., Arico, S., Báldi, A., Bartuska, A., et al. (2015). The IPBES conceptual framework—connecting nature and people. *Current Opinion in Environmental Sustainability, 14*, 1–16. https://doi.org/10.1016/j.cosust.2014.11.002

Elsayed, I. S. M. (2018). Type of gardens that reduce the intensity of an urban heat island. *European Journal of Social Sciences, 56*(3), 297–310. http://www.europeanjournalofsocialsciences.com

Endangsih, T., Prayitno, B., & Kusumawanto, A. (2020). Sustainable hybrid village: Regeneration of settlement in Jatinegara, Indonesia. *IOP Conference Series: Earth and Environmental Science, 520*(012019), 1–7. https://doi.org/10.1088/1755-1315/520/1/012019

European Landscape Contractors Association. (2011). *Green city Europe—for a better life in European cities*. www.green-city.eu

Gonzales, A. L., Arshad, H. S. H., Rahaman, M. S., Singh, S., & Li, Y. (2018). Sustainable greening strategies for megacities in developing countries. https://prospernet.ias.unu.edu/wp-content/uploads/2018/04/ProSPER.Net-YRS-2018-Group-2-Research-Proposal.pdf

Hagemann, F. A., Randrup, T. B., & Sang, Å. O. (2020). Challenges to implementing the urban ecosystem service concept in green infrastructure planning: A view from practitioners in Swedish municipalities. *Socio-Ecological Practice Research, 2*, 283–296. https://doi.org/10.1007/s42532-020-00054-3

Haines-Young, R., & Potschin, M. (2011). Common international classification of ecosystem services (CICES): 2011 Update. https://cices.eu/content/uploads/sites/8/2009/11/CICES_Update_Nov2011.pdf

Hashim, N. I., Yusof, N., Anuar, A. N. A., & Sulaiman, F. C. (2019). The restorative environment offered by pocket park at Laman Standard Chartered Kuala Lumpur. *Journal of Hotel and Business Management, 08*(01), 1–10. https://doi.org/10.35248/2169-0286.19.8.194

Hussain, N., Kher Hussein, M., & Li Thing, L. (2016). Edu-eco park at Bukit Serindit, Melaka. *Alam Cipta, 9*(1), 47–54. http://www.frsb.upm.edu.my/dokumen/FKRSE1_p5.pdf

Ibrahim, P. H., Dali, M. M., & Haron, R. C. (2017). Open space development planning in Malaysia: Issues and problems. *Advanced Science Letters, 23*(7), 6322–6326. https://doi.org/10.1166/asl.2017.9261

Iqbal, M., & Jumiati. (2019). Pengelolaan ruang terbuka hijau berdasarkan peraturan Daerah Kota Padang no 3 tahun 2017. *Ranah Research: Journal of Multidisclipinary Research and Development, 1*(2), 154–161. https://jurnal.ranahresearch.com/index.php/R2J/article/view/36

Karade, R. M., Kuchi, V. S., & Salma, Z. (2017). The role of green space for sustainable landscape development in urban areas. *International Archive of Applied Sciences and Technology, 8*(2), 81–84. https://doi.org/10.15515/iaast.0976-4828.8.2.8184

Karuppannan, S., Baharuddin, Z. M., Sivam, A., & Daniels, C. B. (2014). Urban green space and urban biodiversity: Kuala Lumpur, Malaysia. *Journal of Sustainable Development, 7*(1), 1–16. https://doi.org/10.5539/jsd.v7n1p1

Keane, Å., Stenkula, U., Wijkmark, J., & Johansson, E. (2014). Ekosystemtjänster i stadsplanering—en vägledning. *C/O City*, 1–52.

Kirmanto, D., Ernawi, I. S., & Djakapermana, R. D. (2012). Indonesia green city development program: An urban reform. In *48th ISOCARP Congress*. https://www.isocarp.net/data/case_studies/2124.pdf

Kruize, H., Vliet, N. Van Der, Staatsen, B., Bell, R., Chiabai, A., Muiños, G., Higgins, S., Quiroga, S., Martinez-Juarez, P., Yngwe, M. A., Tsichlas, F., Karnaki, P., Lu, M., Garc, S., Jal, D., Khan, M., Morris, G., Stegeman, I., Kruizse, H., & Stegeman, I. (2019). Urban green space: Creating a triple win for environmental sustainability, health, and health equity through behavior change. *International Journal of Environmental Research and Public Health, 16*(4403), 1–22. https://doi.org/10.3390/ijerph16224403

Kuklina, V., Sizov, O., & Fedorov, R. (2021). Green spaces as an indicator of urban sustainability in the Arctic cities: Case of Nadym. *Polar Science*. https://doi.org/10.1016/j.polar.2021.100672

Lewis, E. (2015). Green City Development Tool Kit. Asian Development Bank. www.adb.org

Li, W., Xu, J., & Zheng, M. (2018). Green governance: New perspective from open innovation. *Sustainability, 10*(3845), 1–19. https://doi.org/10.3390/su10113845

Lundh, J. (2017). Indicators for ecosystem services in urban green space management. Swedish University of Agriculture Sciences. In *Uppsala Universitet*. https://www.diva-portal.org/smash/get/diva2:1128847/FULLTEXT02

Maes, J., Teller, A., Erhard, M., & Bidoglio, G. (2013). Mapping and assessment of ecosystems and their services: An analytical framework for ecosystem assessments under action 5 of the EU biodiversity strategy to 2020. In *Publications office of the European Union*. Luxembourg. https://doi.org/10.2779/12398

Malek, N. A. (2013). Use pattern analysis in green open spaces of Malaysia (Patent No. 1455). IIUM Research, Invention and Innovation Exhibition 2013.

Malek, N. A., Mohammad, S. Z., & Nashar, A. (2018). Determinant factor for quality green open space assessment in Malaysia. *Journal of Design and Built Environment, 18*(2), 26–36. https://doi.org/10.22452/jdbe.vol18no2.3

Malek, N. A., & Nashar, A. (2018). Measuring successfulness of malaysian green open spaces: An assessment tool. *Theoretical and Empirical Researches in Urban Management, 13*(2), 21–37. https://um.ase.ro/no132/2.pdf

Mansor, M., & Harun, N. Z. (2018). Health issues and green space as reinforcement for a healthy lifestyle in Malaysia. *Journal of ASIAN Behavioural Studies, 3*(9), 69–77. https://doi.org/10.21834/jabs.v5i18.87

Mansor, M., Zakariya, K., & Harun, N. Z. (2019). Challenges on use of recreational parks in Kuala Lumpur. *Journal of Construction in Developing Countries, 24*(1), 141–162. https://doi.org/10.21315/jcdc2019.24.1.8

Mao, Y., Imara, I. D., Natawiguna, I. M. P. D., Pratiwi, P. I., Oka, T., & Furuya, K. (2020). A comparison of young generation's perception regarding home garden in urban area: Case study of Indonesia and Japan. In *The 4th International Symposium of Sustainable Landscape Development—IOP Conference Series: Earth and Environmental Science, 501 012025*, 1–9. https://doi.org/10.1088/1755-1315/501/1/012025

Meerow, S. (2019). A green infrastructure spatial planning model for evaluating ecosystem service tradeoffs and synergies across three coastal megacities. *Environment Research Letters, 14*(125011), 1–14. https://doi.org/10.1088/1748-9326/ab502c

Millennium Ecosystem Assessment. (2005). Ecosystems and human well-being: Wetland and water synthesis. In *World Resources Institute*. https://www.millenniumassessment.org/documents/document.358.aspx.pdf

Murtini, S., Zain, I. M., Sutedjo, A., Lestari, A. P., & Hariyanto, B. (2020). Analysis of green open space needs in Surabaya City. In *Proceedings of the 3rd International Conference on Social Sciences (ICSS 2020) 473*, (pp. 300–303). https://doi.org/10.2991/assehr.k.201014.064

Murtini, S., Sutedjo, A., & Zain, I. (2020, January). Analysis of green open space in Krembangan, Surabaya City. In *International Conference on Research and Academic Community Services (ICRACOS 2019)* (pp. 162–164). Atlantis Press.

Ojha, H. R., Ghate, R., Dorji, L., Shrestha, A., Paudel, D., Nightingale, A., Shrestha, K., Watto, M. A., & Kotru, R. (2019). Governance: Key for environmental sustainability in the Hindu Kush Himalaya. In P. Wester, A. Mishra, A. Mukherji, & A. B. Shrestha (Eds.), *The Hindu Kush Himalaya assessment: Mountains, climate change, sustainability and people* (pp. 545–578). Springer International Publishing. https://doi.org/10.1007/978-3-319-92288-1_16

Olaniyan, O. M. (2020). Green city interventions for socio-ecological resilience of urban areas to covid-19 in Nigeria. *Lead City Journal of the Social Sciences, 5*(1), 1632–1645.

Othman, N., Mohamad, M., Latip, R. A., & Ariffin, M. H. (2018). Urban farming activity towards sustainable wellbeing of urban dwellers. *IOP Conference Series: Earth and Environmental Science, 117*, 012007. https://doi.org/10.1088/1755-1315/117/1/012007

Pakzad, P., & Osmond, P. (2015). A Conceptual framework for assessing green infrastructure sustainability performance in Australia. *The State of Australian Cities Conference*. https://apo.org.au/sites/default/files/resource-files/2015-12/apo-nid63220.pdf

Rakhshandehroo, M., Yusof, M. J. M., Arabi, R., Parva, M., & Nochian, A. (2017). The environmental benefits of urban open green spaces. *Alam Cipta, 10*(1), 10–16. https://core.ac.uk/download/pdf/153833984.pdf

Ramandhani, K., Budiarti, T., & Makalew, A. D. N. (2020). Development concept of productive landscape with urban agriculture based on potential green open space and community preference in Bogor City. In *The 4th International Symposium of Sustainable Landscape Development—IOP Conference Series: Earth and Environmental Science, 501*(012010). https://doi.org/10.1088/1755-1315/501/1/012010

Ramli, R., Omar, D., & Ahmad, P. (2019). Malaysia's green neighbourhood initiatives: Implementing and approach in Putrajaya, Selangor and Johor. *International Journal of Engineering and Advanced Technology, 8*(5C), 553–562. https://doi.org/10.35940/ijeat.E1081.0585C19

Ratih, Y., & Febrianto, J. Z. (2016). Biodiversity as part of urban green network system planning case study: Pontianak City. *Procedia—Social and Behavioral Sciences, 227*(November 2015), 583–586. https://doi.org/10.1016/j.sbspro.2016.06.118

Ratnawati, T. (2017). Potensi dan prospek lahan pekarangan sebagai ruang terbuka hijau dalam menunjang kota cerdas. *Prosiding Seminar Nasional Tahunan Matematika, Sains Dan Teknologi, 26*, 42–51.

Razali, N. D. M., Rani, S. N. S. A., Yatim, S. R. M., & Ain, N., (2020). The Effect of urban green space on particulate matter and heat level in Kuala Lumpur. *International Journal of Advanced Science and Technology, 29*(9), 1682–1691. https://www.academia.edu/download/64064799/13807-Article-Text-20584-1-10-20200506(1).pdf

Ridwan, M., Sinatra, F., & Natalivan, P. (2017). Promoting productive urban green open space towards food security: Case study Taman Sari, Bandung. *IOP Conference Series: Earth and Environmental Science, 91*(1). https://doi.org/10.1088/1755-1315/91/1/012030

Rostami, R., Lamit, H., Khoshnava, S. M., & Rostami, R. (2013). Urban green spaces and city sustainability. *Asian Journal of Microbiology, Biotechnology and Environmental Sciences, 15*(2), 441–446.

Saha, D. (2009). Factors influencing local government sustainability efforts. *State and Local Government Review, 41*(1), 39–48. https://doi.org/10.1177/0160323x0904100105

Sandhu, H. S., Wratten, S. D., & Cullen, R. (2010). The role of supporting ecosystem services in conventional and organic arable farmland. *Ecological Complexity, 7*(3), 302–310. https://doi.org/10.1016/j.ecocom.2010.04.006

Şenik, B., & Uzun, O. (2022). A process approach to the open green space system planning. *Landscape and Ecological Engineering, 18*(2), 203–219. https://doi.org/10.1007/s11355-021-00492-5

Setiawan, W. (2015). Reading the urban planning in Indonesia: A journey towards sustainable development. *Sinektika: Jurnal Arsitektur, 14*(2), 257–268. https://doi.org/10.23917/sinektika.v14i2.1446

Sherer, P. (2006). The benefits of parks: Why america needs more city parks and open space. *The Trust for Public Land*, (January 2003), 1–37.

Sinnett, D., Calvert, T., Martyn, N., Williams, K., Burgess, S., Smith, N., & King, L. (2016). *Green infrastructure*. http://ec.europa.eu/environment/nature/ecosystems/index_en.htm

UNEP & United Nations Environment Programme. (2011). Pathways to sustainable development and poverty eradication—A synthesis for policy makers. *Towards a GREEN Economy,* 52. www.unep.org

Veisten, K., Smyrnova, Y., Klæboe, R., Hornikx, M., Mosslemi, M., & Kang, J. (2012). Valuation of green walls and green roofs as soundscape measures: Including monetised amenity values together with noise-attenuation values in a cost-benefit analysis of a green wall affecting courtyards. *International Journal of Environmental Research and Public Health, 9*(11), 3770–3778. https://doi.org/10.3390/ijerph9113770

Vety Jayanti, A., Priyo Purnomo, E., & Nurkasiwi, A. (2020). Vertical Garden: Penghijauan Untuk Mendukung Smart Living Di Kota Yogyakarta. *Al Imarah: Jurnal Pemerintahan Dan Politik Islam, 5*(1), 41. https://doi.org/10.29300/imr.v5i1.2916

Wahdaniyat, H. (2019). Ruang Terbuka Hijau yang Masih Terpinggirkan di Indonesia. *Kementerian Pekerjaan Umum Dan Perumahan Rakyat*. http://ciptakarya.pu.go.id/pbl/index.php/detail_berita/565/ruang-terbuka-hijau-yang-masih-terpinggirkan-di-indonesia

Watts, G., Miah, A., & Pheasant, R. (2013). Tranquillity and soundscapes in urban green spaces—Predicted and actual assessments from a questionnaire survey. *Environment and Planning B: Planning and Design, 40*(1), 170–181. https://doi.org/10.1068/b38061

Wikantiyoso, R., Tutuko, P., Suhartono, T., Sulaksono, A. G., & Safrilia, A. (2020). Green city MIS as a sustainable urban GOS provision control implementation model. *International Review for Spatial Planning and Sustainable Development, 8*(1), 160–172. https://doi.org/10.14246/irspsda.8.1_160

World Bank. (2003). *World Development Report*. https://openknowledge.worldbank.org/handle/10986/5985

Yuniastuti, E., & Hasibuan, H. S. (2019a). Child-friendly green open space to enhance the education process for children. *IOP Conference Series: Earth and Environmental Science, 243*(1). https://doi.org/10.1088/1755-1315/243/1/012161

Yuniastuti, E., & Hasibuan, H. S. (2019b). Green open space, towards a child-friendly city (A case study in Lembah Gurame Park, Depok City, Jakarta Greater Area, Indonesia). *IOP Conference Series: Earth and Environmental Science, 328*(1). https://doi.org/10.1088/1755-1315/328/1/012016

Zhang, L., Cao, H., & Han, R. (2021). Residents' preferences and perceptions toward green open spaces in an urban area. *Sustainability, 13*(3), 1558. https://doi.org/10.3390/su13031558

Zheng, Q., Zhao, X., & Jin, M. (2019). Research on urban public green space planning based on taxi data: A case study on three districts of Shenzhen, China. *Sustainability*. https://doi.org/10.3390/su11041132

Environmental Impacts of a Forensic Unit Construction at a Teaching Hospital in Malaysia

Nur Syafiqah Nabila Shaari, Nurul Syazwani Khuzaini, Fatin Nurhanani Adenan, Nimi Dan-Jumbo, and Farah Ayuni Shafie

Abstract This study aims to understand the material waste generation, energy and water consumption, and total carbon emissions from constructing a forensic unit at a teaching hospital in Malaysia before the COVID-19 pandemic. The material waste magnitude identified using the material flow analysis approach showed that bricks have the biggest share (88.6%) in construction waste, whereas the smallest proportion is derived from sand (1.5%). Meanwhile, the highest and lowest water consumption to construct the forensic unit was 519 kL in May 2019 and 155 kL in December 2019, respectively. The annual electricity-related energy consumption at this site was 112,826.28 MJ, which was the lowest compared to other energy usages, such as operating the bulldozer, which consumed 914,191.71 MJ of energy per year. Interestingly, the energy source that emitted the lowest amount of carbon was the excavator, which contributed 5% (equivalent to 12,420.86 $kgCO_2e$) to the total emissions of the forensic unit construction, while the source with the largest carbon footprint remained the bulldozer (28%, equivalent to 64,463.04 $kgCO_2e$). Our findings will help develop strategies, policies, and rules to effectively manage waste and carbon emissions from construction activities in Malaysia.

Keywords Construction waste · Material flow analysis · Carbon emissions · Water · Electricity

N. S. N. Shaari · N. S. Khuzaini · F. N. Adenan · F. A. Shafie (✉)
Centre for Environmental Health and Safety Studies, Faculty of Health Sciences, Universiti Teknologi MARA, Puncak Alam Campus, Puncak Alam, Malaysia
e-mail: farahayuni@uitm.edu.my

N. Dan-Jumbo
School of Engineering and Built Environment, Edinburgh Napier University, Edinburgh, UK
e-mail: n.dan-jumbo@ulster.ac.uk

U. R. Hashim et al. (eds.), *Green Infrastructure*,
https://doi.org/10.1007/978-981-99-7003-2_10

1 Introduction

The construction industry is one of the major sectors that contributes to the Gross Domestic Product (GDP) of Malaysia's economy (Alaloul et al., 2021). It helps provide necessary socio-economic infrastructures such as houses, hospitals, schools, and other enhanced facilities to improve the quality of life and living standards of Malaysian citizens. It has expanded by 5.1% within the fourth quarter of 2019, considerably faster than the actual GDP growth of 4.5% overall. However, the Coronavirus (COVID-19) pandemic had significantly impacted Malaysia's construction industry (Kamarazaly et al., 2020; Syed Zakaria & Mahinder Singh, 2021). The country's GDP from various sectors fell by 17.1% in the second quarter of 2020 because of the COVID-19 containment measures, which brought construction as well as other economic activities to a halt (Department of Statistics Malaysia, 2022). Malaysia's total GDP then recovered in the third quarter of 2020 and peaked in the second quarter of 2021 as lockdown restrictions were relaxed and economic activities resumed.

While the construction industry is considered one of the pillars of economic stability, from an environmental point of view, the construction of buildings brings concern due to the enormous amounts of natural resources such as raw material, energy, and water needed for the highly complex processes (Heravi et al., 2017). Construction materials that are not fully utilized will lead to waste generation, and at the same time, natural resources will suffer from depletion (Alsheyab, 2022; Ametepey & Ansah, 2015). Apart from that, the construction sector operations lead to high energy and water consumption, which is related to the emission of greenhouse gases (GHGs) (United Nations Environment Programme, 2021). The energy at construction sites, especially electricity and diesel fuel, is responsible for the enormous amount of GHG emissions, and total GHG emissions have been increasing steadily in Malaysia from 1995 to 2017 (Latif et al., 2021; Sharrad et al., 2007). In addition, the emission of GHGs, especially carbon dioxide, is the leading cause of global warming (Zhang & Wang, 2016).

An increase in energy demand is expected as the need for development, especially among developing countries, continues to grow (Edeoja & Edeoja, 2015). However, it is challenging for the industry to sustain the environment as development can cause depletion of natural resources, change in living surroundings, unbalanced ecology, and global warming. Thus, this study explored the generation of material waste, energy, and water consumption from the construction of a forensic unit of a teaching hospital to understand their impacts on the environment. Material flow analysis was adopted to characterize the flow of material usage and disposal. The rate of waste generation and amount of carbon emissions were also calculated so that these parameters can be extrapolated and applied in future projects.

Fig. 1 Summary of the procedure in assessing the environmental impacts of UiTM Hospital's forensic unit construction

2 Materials and Methods

This study was conducted in three stages, as shown in Fig. 1. The first stage was the preparation of a checklist and the selection of one department from a Universiti Teknologi MARA (UiTM) teaching hospital, which was under construction in 2019. The forensic unit of UiTM Hospital in Puncak Alam Campus was selected. Data were collected in the second phase (from January to December 2019) and recorded in the Bill of Quantities (BOQ) documents for waste quantification and monthly water consumption. Total electricity and diesel fuel used were obtained from the construction management team of the construction site. All the recorded data were routinely examined to ensure that no tasks on the checklist were overlooked. Related documents of the forensic unit construction, such as material purchase invoices, were also reviewed. The final stage involved analyzing the obtained information to compute material construction waste and its generation rate, water consumption, and carbon emissions using specific formulas (Table 1). The flow of material usage and waste generation is depicted as a Sankey diagram using the material flow analysis approach. This method was previously used to extrapolate the production of construction and demolition debris as well as the lifetime of residential building stock (Cochran & Townsend, 2010; Condeixa et al., 2017).

3 Results

3.1 Magnitude and Rate of Material Waste Generated from the Construction Work

The quantity of construction materials used and wasted is comparable, except for sand, in which the weight of wasted sand (880 kg) is much lower than the sand used (7100 kg; Fig. 2; Table 2). Bricks were the largest amount of waste produced (88.6%) at the forensic unit construction site, with a total weight of 51,420 kg. Meanwhile, the smallest magnitude of waste generated from the construction activities was 1.5%, which belonged to sand. Besides, the amount of wasted cement was similar to wasted steels (1680 kg and 1600 kg, respectively).

After the quantity of each material waste had been identified, the waste generation rates were analyzed. The trend in the rate of construction waste generation follows

Table 1 Quantification formulae of material construction wastage, water consumption, and carbon emission

	Formula	Usage
Quantification of material construction waste	Rate of waste generation (kg/m^2) = Total weight of material waste (kg)/Total construction area (m^2)	To estimate the amount of waste from the construction area, known as generation rate calculation (GRC)
Quantification of water consumption	Total volume of water usage = Total volume of water consumption for whole construction area (kL)/Specified constructed area (m^2)	To calculate the water usage in a specific construction area
	Water usage per cement bag (L) = Water cement ratio × Weight of cement (kg)	To determine the indirect water usage in cement
	Total water usage in cement mixing process (L) = Total of water used per cement bag × Total cement bags used	To quantify the total water used in the cement mixing process throughout the construction
Quantification of carbon emission	Carbon emission = Activity data × Emission factor	To compute the carbon dioxide emission from daily activity with the emission factor as the constant[a]

[a] The emission factor for diesel fuel combustion is 2.7 $kgCO_2e/L$, while the emission factor for electricity usage is 0.667 $kgCO_2e/kWh$ (Yazdani et al., 2013; Malaysia Green Technology Corporation, 2017)

the same pattern of the magnitude of waste produced (Fig. 3). The highest rate of waste generated was 69.49 kg/m^2 by bricks, followed by wood (3.35 kg/m^2) and cement (2.27 kg/m^2).

3.2 *Direct and Indirect Water Consumption at the Construction Site*

Water was consumed six days a week in a single shift of 12 h daily. The maximum and minimum direct water consumption in a year were 519 kL (in the fifth month) and 155 kL (in the last month), respectively (Fig. 4). The water consumption was relatively high from January to May 2019, all above 400 kL, due to many construction activities conducted compared to June until December 2019. Toward the end of 2019, the project was nearly completed; hence, the usage of water was decreasing. However,

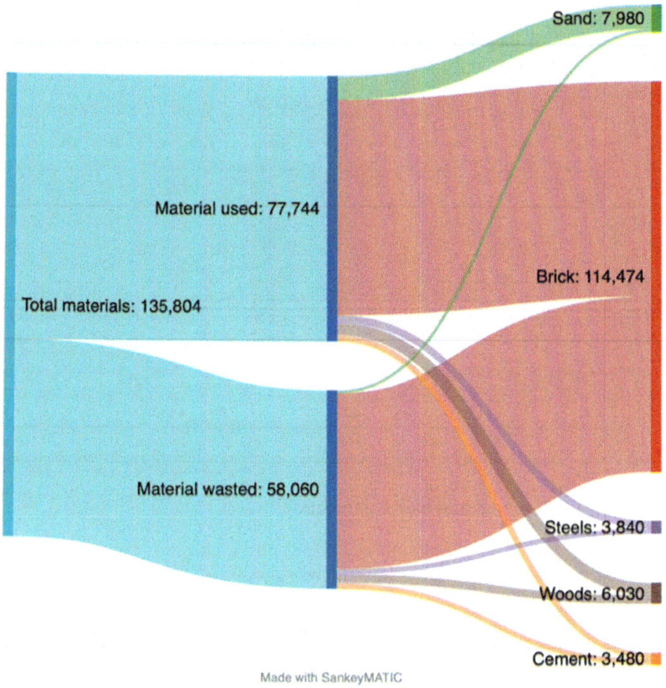

Fig. 2 The flow of materials used and wasted from the construction activities. Figure was created using Sankeymatic.com. The unit is in kg

Table 2 The quantity and magnitude of materials used and wasted at the construction site

Construction materials	Amount of material used (kg)	Amount of material waste (kg)	Magnitude of waste generated (%)
Cement	1800	1680	2.9
Sand	7100	880	1.5
Bricks	63,054	51,420	88.6
Steels	2240	1600	2.7
Woods	3550	2480	4.3
Total	77,744	58,060	100

in October 2019, the water consumption was high due to site cleaning activity. In total, direct water consumption to construct the forensic unit from January to December 2019 was 4,497 kL/m^2.

Indirect water consumption was mainly generated from cement mixing activity. The water-cement ratio ranges from 0.45 to 0.60, depending on the type of cement used. At this construction site, a ratio of 0.45 was used for the cement mixing process. This activity consumed 810 L of water, making the total embodied water used for

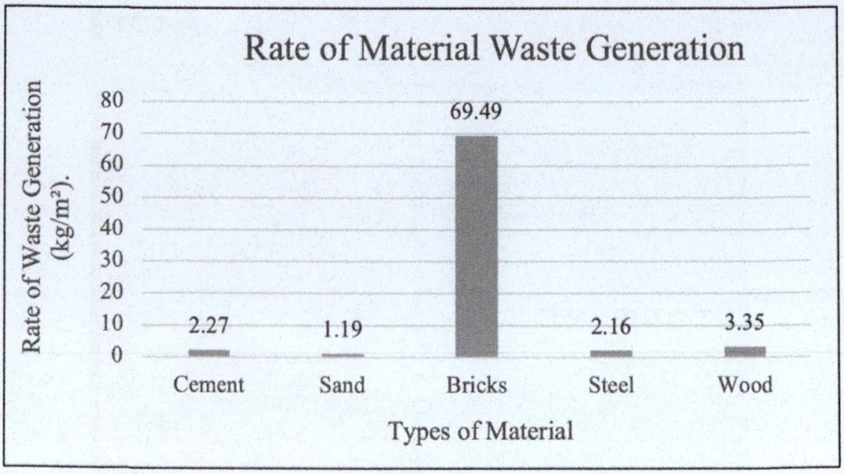

Fig. 3 Rate of construction material waste generation

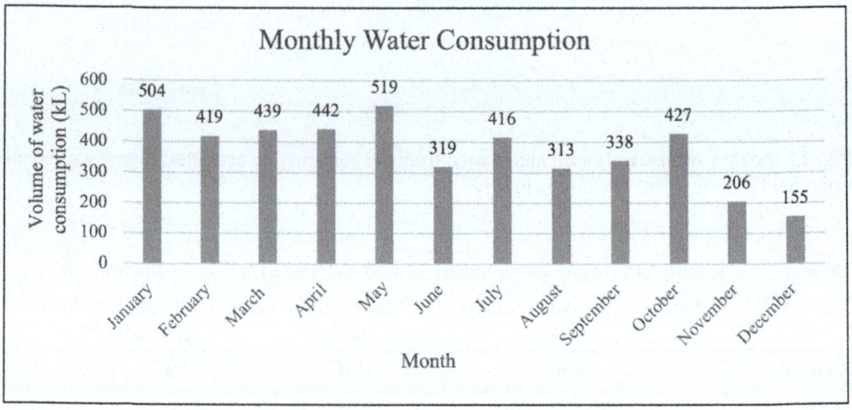

Fig. 4 Direct water consumption at the forensic unit construction site from January until December 2019

the construction works of the forensic unit 4497.81 kL in a year (Table 3). The direct water bodied was higher than the indirect water bodied due to the massive high-priority activities at the construction area, which required a fast water flow.

Table 3 The total of embodied water in construction

Indirect embodied water (kL)	Direct embodied water (kL)	Total embodied water (kL)
0.81	4,497	4,497.81

3.3 Amount of Energy Consumed for the Construction Activities

An excavator consumed 23.96 L of diesel fuel per hour, 8 h per day, and 24 days per year (Table 4). Excavator operated the least in this study as it was only used at the early stage of construction for earthwork activity. Meanwhile, a bulldozer utilized 33.16 L of diesel fuel per hour for 8 h per day and 90 days per year (Table 4). The bulldozer showed the greatest amount of diesel consumed per hour compared to other heavy construction equipment and vehicles. A crawler crane utilized the least diesel fuel per hour, which was 18 L per hour for 8 h per day. However, a crawler crane operated for 144 days a year. This made the crawler crane the most frequently used machine in this study compared to other machines. Besides, it has been observed that the on-site electricity consumption was 10.77 kWh per hour and was running for 12 h per day for various activities. Electricity continued to run for 264 days a year.

The unit for the annual energy consumption at the construction site was then standardized into megajoules (MJ) and visualized as a bar chart (Fig. 5). It has been demonstrated that the bulldozer consumed the highest energy compared to other sources (Fig. 5). This is because despite the yearly energy consumption of bulldozer was slightly lower than that of electricity as shown in Table 4 (23,875.2 L vs. 34,119.36 kWh), one liter of diesel equals to 38.29 MJ, while one kWh is equivalent to 3.6 MJ (John, 2009; Rapid Tables, 2022). Therefore, the annual energy consumption in MJ for the bulldozer was higher than electricity (914,191.71 vs. 122,826.28 MJ, respectively; Fig. 5). In contrast, the source that consumed the least energy per year compared to others was the excavator, which utilized 176,148.24 MJ of energy only. The excavator consumed quite high energy per hour, but it was

Table 4 Diesel fuel and electricity consumption for the construction of UiTM Hospital forensic unit. The energy consumption data were collected in liters (L) for diesel and kWh for electricity

Type of energy	Item	Quantity	Energy consumption (per hour)	Energy used (hours/day)	Energy used (days/year)	Total energy used in a year
Diesel (L)	Excavator	1	23.96	8	24	4600.32
	Bulldozer	1	33.16	8	90	23,875.2
	Crawler crane	1	18.00	8	144	20,736
	Heavy truck	1	22.74	8	90	16,372.8
	Truck	1	15.39	8	96	11,819.52
Electricity (kWh)	On-site electricity used	N/A	10.77	12	264	34,119.36

N/A denotes not applicable

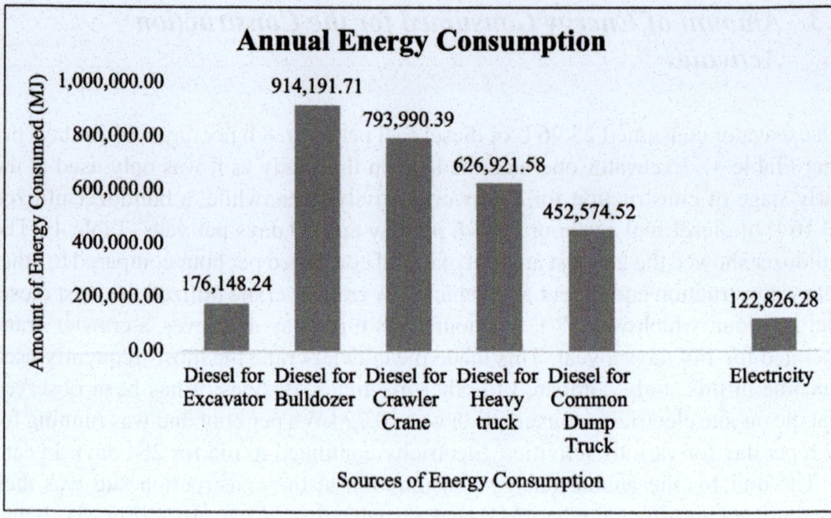

Fig. 5 The energy consumption of different sources at the forensic unit construction site

operated for 24 days only in a year, making it the machine that consumed the lowest energy.

3.4 *Total Carbon Emissions of the Forensic Unit Construction*

The largest contributor to total carbon emissions in this study was the bulldozer, followed by the crawler crane and heavy truck (Fig. 6). The bulldozer consumed 23,875.20 L of diesel fuel and emitted up to 64,463.04 kgCO_2e which is 28% of the total emission. Meanwhile, the amount of carbon emissions of the crawler crane and heavy truck were 55,987.2 kgCO_2e (equals to 24% of total emissions) and 44,206.56 kgCO_2e (shared 19% of total footprint), respectively. On the contrary, the smallest source of carbon emission was the excavator, which made up 5% of the total carbon emission and released a total of 12,420.86 kgCO_2e carbon. Surprisingly, the on-site electricity contributed to 10% of the total carbon footprint only (equivalent to 22,757.61 kgCO_2e), which is consistent with its contribution to the total annual energy consumption (Fig. 5).

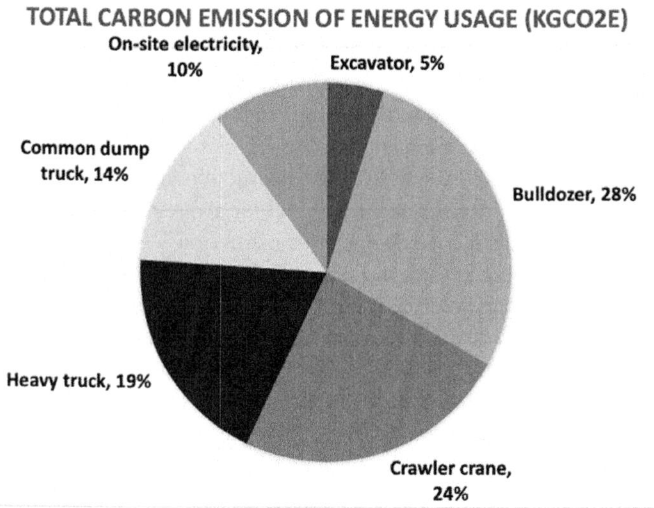

Fig. 6 Total carbon emission from the operation of heavy construction equipment, vehicles including the heavy truck and dump truck, as well as electricity consumption

4 Discussion

The construction sector undoubtedly contributes to the economic stability of a nation and improves humankind's living standards and well-being (Alaloul et al., 2021). Notwithstanding, there has been a concern over waste generation at construction and demolition sites, accounting for 25% of the total wastage worldwide (Alsheyab, 2022). Moreover, nearly 40% of the final global energy and energy-related carbon footprint comes from the building construction industry (United Nations Environment Programme, 2021). In addition, the Solid Waste and Public Cleansing Management Corporation of Malaysia stated that construction projects generate approximately eight million tons of construction waste per year (Construction Industry Development Board, 2018). This study identified the types of construction material wastage and carbon emissions from the construction of a forensic unit of a teaching hospital in Puncak Alam. According to (Eusuf et al., 2012), bricks are the most used material among others because they are used to form internal walls, fittings, and partitions of a building and have a variety of advantages such as good load-bearing capacity, long life, and strength for building construction. However, in this study, the total weight of bricks wasted was similar to that of bricks used, translating to a considerably large magnitude of construction waste produced.

Meanwhile, sand was the second most used material in this study after bricks. Ipsos Business Consulting (2017) stated that 914 sand mining licenses were issued in Malaysia in 2015. For instance, Kedah, Pahang, and Perak had the highest number of licenses, while Selangor, Johor, and Negeri Sembilan had the largest quantity of sand production, accounting for 76% of national production. Importantly, it has been shown that 40% of natural source depletion is caused by the construction and operation of buildings, and sand is one of the consumed resources (Guo & Huang, 2019; Meredith, 2021). Moreover, most of the construction materials in this study demonstrated no significant difference in their usage and wastage, indicating a rather ineffective construction materials management, such as materials over-ordering. Other factors that can affect waste generation at construction sites are damaged materials caused by mishandling or the weather and lack of awareness of construction waste (Ikau et al., 2016). Therefore, to deal with construction wastage issues, contractors should explore and implement the proper handling of daily waste, and the first step to this is by quantifying them (Wu et al., 2014). The total construction area of the UiTM Hospital forensic unit is 740 m^2. Using the generation rate calculation (GRC) method, the contractor could produce an estimation of waste produced from the forensic unit and apply it to the other departments of the teaching hospital to minimize waste generation. Other quantification methods that can also be considered are lifetime analysis, site visits, variables modeling, and classification system accumulation, in which some projects require more than one method (Wu et al., 2014). Besides, a work breakdown structure can also be applied to produce an accurate quantitative estimation of the needed materials before purchasing them, as previously described by Li et al. (2016).

On the other hand, the amount of water used monthly for construction activities may vary depending on several factors, such as the types of construction activities conducted and the awareness level among workers toward the implementation of water management practices. In October 2019, the water consumption was high due to the cleaning of the site. The workers used the spray gun with different spray patterns and left the cleaning device operating when it was not in use. The spray patterns also affect the efficiency of water consumption. The water consumption declined toward the end of the year due to the near completion of the building as targeted by the planning team; hence, fewer construction activities were conducted. Moreover, the rainy season which starts from mid to the end of the year also allowed the workers to use rainwater as a non-portable source for general site cleaning. The water consumption of rainwater was not recorded on a digital meter. Therefore, the water consumption reading was relatively low in the second half of the year. Meanwhile, the awareness of the water management practice was relatively low at the construction site due to the lack of communication between the management team and the foreign workers. Most of the foreign workers could not communicate in Malay and English, and only a few could communicate well in English. These observations indicate that water management practice needs the participation of both the management team and construction workers to effectively reduce water consumption at construction sites, apart from cutting the cost spent on the water bill monthly.

The energy at construction sites is generally provided by electricity, diesel fuel, natural gas, and gasoline. Essentially, diesel fuel is highly consumed by vehicles for the transportation of building materials and other equipment. The vehicles used in this study were one heavy and one normal truck. Usually, the heavy truck is used for the transportation of soil, prefabricated components, and waste (Mao et al., 2013). Due to the relatively small built area for the forensic unit (740 m^2), one unit of each type of machinery, equipment, and vehicle was sufficient. However, a large proportion of the total carbon footprint came from the operation of these machines, which somehow did not meet the current government's effort to develop sustainable construction practices (Abdul-Azeez, 2016; Esmaeilifar et al., 2015; Yazdani et al., 2013). Apart from that, electricity is another necessity in construction activities, as it is used to fulfill various energy needs (Esmaeilifar et al., 2015). The on-site electricity in this study was used to supply power for the lights, HVAC systems, power tools, and elevators in the forensic unit construction site, which indirectly emitted carbon dioxide and other GHGs. Unfortunately, the use of vehicles at the deconstruction stage, which transports waste to the recycling or reuse facility and the landfill for disposal, was excluded in this study due to the lack of data obtained. A previous study by Zhang and Wang (2016) stated that the difficulty in calculating carbon emissions of the transportation stage is that the average transport distance estimation is different for each building material. In addition, most of the material waste is sent to the landfill by road using garbage trucks, while some is transported via railway or sea. In this case, different vehicle speeds release different amounts of GHGs into the environment.

5 Conclusion

The construction sector is one of the crucial and productive sectors of the Malaysian economy. Energy, materials, and water are the essential resources for constructing the forensic unit of UiTM Hospital, Puncak Alam. Misusing these resources will negatively influence the environment through the generation of construction waste and carbon emissions, concomitantly depleting natural resources. By understanding the flow of material waste, its generation rate, water consumption, and total carbon emissions from different sources at the construction site, proper strategies, policies, and rules can be developed to support more sustainable construction practices in Malaysia.

Acknowledgements The authors sincerely acknowledge the information and support provided by the UiTM Hospital construction management team and Sarah Zulkifli's professional editing assistance in completing this chapter.

References

Abdul-Azeez, I. A. (2016). Measuring and monitoring carbon emission to promote low-carbon development in Johor Bahru. https://malaysiacities.mit.edu/paperadeyemi

Alaloul, W. S., Musarat, M. A., Rabbani, M. B. A., Iqbal, Q., et al. (2021). Construction sector contribution to economic stability: Malaysian GDP distribution. *Sustainability (Switzerland), 13*(9). https://doi.org/10.3390/su13095012

Alsheyab, M. A. T. (2022). Recycling of construction and demolition waste and its impact on climate change and sustainable development. *International Journal of Environmental Science and Technology, 19*(3), 2129–2138. https://doi.org/10.1007/s13762-021-03217-1

Ametepey, S. O., & Ansah, S. K. (2015). Impacts of construction activities on the environment: The case of Ghana. *Journal of Environment and Earth Sciences, 5*(3), 18–26.

Cochran, K. M., & Townsend, T. G. (2010). Estimating construction and demolition debris generation using a materials flow analysis approach. *Waste Management, 30*(11), 2247–2254. https://doi.org/10.1016/j.wasman.2010.04.008

Condeixa, K., Haddad, A., & Boer, D. (2017). Material flow analysis of the residential building stock at the city of Rio de Janeiro. *Journal of Cleaner Production, 149*, 1249–1267. https://doi.org/10.1016/j.jclepro.2017.02.080

Construction Industry Development Board. (2018). Advanced in construction site waste management (Bina Mampan Malaysia's Sustainable Construction Periodical 2, 1–29. https://www.cidb.gov.my/sites/default/files/2020-04/Bina-MAMPAN-Full-Version.pdf

Department of Statistics Malaysia. (2022). Malaysia economic performance first quarter 2022. https://www.dosm.gov.my/v1/index.php?r=column/cthemeByCat&cat=100&bul_id=Zm8xRXoyVitzKzVUbG9Cc0pPQ0s3Zz09&menu_id=TE5CRUZCblh4ZTZMODZIbmk2aWRRQT09

Edeoja, J. A., & Edeoja, A. O. (2015). Carbon emission management in the construction industry – Case studies of Nigerian construction industry. *American Journal of Engineering Research, 4*(7), 112–122.

Esmaeilifar, R., Samari, M., Mirzaei, N. F., et al. (2015). How is electricity consumption on construction sites in Malaysia related to sources of CO_2? *Advances in Environmental Biology, 9*(5), 160–163.

Eusuf, M. A., Ibrahim, M., & Islam, R. (2012). The construction and demolition wastes in Klang Valley, Malaysia. *Planning Malaysia Journal, 10*(3), 99–106. https://doi.org/10.21837/pmjournal.v10.i3.103

Guo, D., & Huang, L. (2019). The state of the art of material flow analysis research based on construction and demolition waste recycling and disposal. *Buildings, 9*(10). MDPI AG. https://doi.org/10.3390/buildings9100207

Heravi, G., Fathi, M., & Faeghi, S. (2017). Multi-criteria group decision-making method for optimal selection of sustainable industrial building options focused on petrochemical projects. *Journal of Cleaner Production, 142*, 2999–3013. https://doi.org/10.1016/j.jclepro.2016.10.168

Ikau, R., Joseph, C., & Tawie, R. (2016). Factors influencing waste generation in the construction industry in Malaysia. *Procedia - Social and Behavioral Sciences, 234*, 11–18. https://doi.org/10.1016/j.sbspro.2016.10.213

Ipsos Business Consulting. (2017). Market review of building materials in the construction industry. http://mbam.org.my/wp-content/uploads/2017/11/Market-Review-of-Building-Materials-in-the-Construction-Industry-Draft-Final-091117-v2compressed.pdf

John, S. (2009). Online converter. https://onlineconversion.vbulletin.net/forum/main-forums/convert-and-calculate/7182-diesel-generator-to-kwh

Kamarazaly, M. A., Badaruddin, A. E., Chin, S. C. A., et al. (2020). The impact of coronavirus (Covid-19) outbreak towards contractors' performance in Malaysia. *Journal of Build Environment, Technology and Engineering, 8*, 42–51.

Latif, S. N. A., Chiong, M. S., Rajoo, S., et al. (2021). The trend and status of energy resources and greenhouse gas emissions in the Malaysia power generation mix. *Energies, 14*(8). https://doi.org/10.3390/en14082200

Li, Y., Zhang, X., Ding, G., et al. (2016). Developing a quantitative construction waste estimation model for building construction projects. *Resources, Conservation and Recycling, 106*, 9–20. https://doi.org/10.1016/j.resconrec.2015.11.001

Malaysia Green Technology Corporation. (2017). 2017 CDM electricity baseline for Malaysia. https://www.mgtc.gov.my/wp-content/uploads/2019/12/2017-CDM-Electricity-Baseline-Final-Report-Publication-Version.pdf

Mao, C., Shen, Q., Shen, L., et al. (2013). Comparative study of greenhouse gas emissions between off-site prefabrication and conventional construction methods: Two case studies of residential projects. *Energy and Buildings, 66*, 165–176. https://doi.org/10.1016/j.enbuild.2013.07.033

Meredith, S. (2021). A sand shortage? The world is running out of a crucial – but underappreciated – commodity. *CNBC*. https://www.cnbc.com/2021/03/05/sand-shortage-the-world-is-running-out-of-a-crucial-commodity.html#:~:text=World%20Politics-,A%20sand%20shortage%3F,%E2%80%94%20but%20under%2Dappreciated%20%E2%80%94%20commodity&text=Sand%20is%20the%20world's%20most,ingredient%20to%20our%20everyday%20lives

Rapid Tables. (2022). Energy conversion. https://www.quora.com/How-do-you-convert-kWh-to-joules

Sharrad, A. L., Matthews, H. S., & Roth, M. (2007). Environmental implications of construction site energy use and electricity generation. *Journal of Construction Engineering and Management, 133*, 846–854. https://doi.org/10.1061//ASCE/0733-9364/2007/133:11/846

Syed Zakaria, S. A., & Mahinder Singh, A. K. (2021). Impacts of Covid-19 outbreak on civil engineering activities in the Malaysian construction industry: A review. *Jurnal Kejuruteraan, 33*(3), 477–485. https://doi.org/10.17576/jkukm-2021-33(3)-10

United Nations Environment Programme. (2021). 2021 global status report for buildings and construction. https://globalabc.org/sites/default/files/2021-10/GABC_Buildings-GSR-2021_BOOK.pdf.

Wu, Z., Yu, A. T. W., Shen, L., et al. (2014). Quantifying construction and demolition waste: An analytical review. *Waste Management, 34*(9), 1683–1692. https://doi.org/10.1016/j.wasman.2014.05.010

Yazdani, Z., Talkhestan, G. A., & Kamsah, M. Z. (2013). Assessment of carbon footprint at university technology Malaysia (UTM). *Applied Mechanics and Materials, 295–298*, 872–875. https://doi.org/10.4028/www.scientific.net/AMM.295-298.872

Zhang, Z., & Wang, B. (2016). Research on the life-cycle CO_2 emission of China's construction sector. *Energy and Buildings, 112*, 244–255. https://doi.org/10.1016/j.enbuild.2015.12.026

Mechanical Properties of Concrete Containing POFA as Cement and Sand Replacement

Arif Fahmi Baharom, Mohd Afiq Mohd Fauzi, Muhd Norhasri Muhd Sidek, and Rabitah Handan

Abstract The cement demand in the construction industry in Malaysia reached 19.49 million metric tons in 2020 (Müller J (2021) Malaysia: Cement Production 2020 | Statista. Statista), giving a perfect view that the price of the material will rise following the re-opening of the industry after the Covid-19 pandemic. The main problem with using cement is that the carbon dioxide (CO_2) emission from the cement production industry gets higher as the cement volume increases. Hence, waste material such as Palm Oil Fuel Ash (POFA) was used to reduce the use of cement. Only a few studies have used POFA in concrete applications. This research aims to analyze the mechanical properties of POFA in concrete as a partial replacement for cement and sand. The influence of POFA content in the range of 0 to 20% of cement and sand replacement on the slump test and strength characteristic level of concrete was identified. The concrete specimens were tested for strength at the age of 3, 7, and 28 days. The curing for this research is water cured. The result indicated that the compressive strength of concrete containing POFA replacement as cement and sand performs better than the normal concrete.

Keywords Palm Oil Fuel Ash (POFA) · Concrete · Sand replacement · Cement replacement · Compressive strength · Slump test

A. F. Baharom · M. A. Mohd Fauzi (✉) · M. N. Muhd Sidek
School of Civil Engineering, College of Engineering, Universiti Teknologi MARA (UiTM), 40450 Shah Alam, Selangor, Malaysia
e-mail: mohdafiq3473@uitm.edu.my; falseafiq_50230@yahoo.com

M. N. Muhd Sidek
e-mail: muhdnorhasri@uitm.edu.my

M. A. Mohd Fauzi · M. N. Muhd Sidek
Institute for Infrastructure Engineering and Sustainable Management (IIESM), Universiti Teknologi MARA, 40450 Shah Alam, Selangor, Malaysia

R. Handan
Faculty of Engineering and Life Sciences, Universiti Selangor, Bestari Jaya, Selangor, Malaysia
e-mail: rabitah@unisel.edu.my

U. R. Hashim et al. (eds.), *Green Infrastructure*,
https://doi.org/10.1007/978-981-99-7003-2_11

1 Introduction

Palm Oil Fuel Ash (POFA) on concrete is a mix of POFA as a substance to replace a percentage of a substance such as cement and sand, used to construct concrete as an element. The POFA will be part of the concrete mixture, as the palm oil waste will be burned to ashes, fly ashes, and bottom ashes, and be accumulated and sieved to get the specific size to be used as the cement and sand. The POFA contains some similar characteristics to the cement, which acts as the binder and sand acts as the filler, giving compatible strength to the concrete itself.

In the world of construction, concrete is one of the main casts of the field, which is significantly used in most construction projects. In Malaysia, concrete was widely implemented in nearly all construction projects. Concrete definitely has some properties and characteristics, which are important as it will be made as one of the elements such as beams, slabs, and columns. The mechanical properties of concrete are one of the instruments that need to be supervised when making concrete. Compressive strength is one of the important evaluations in concrete, which can be defined as the capacity of concrete to withstand loads before failure (Jaya, 2020). The cubes of concrete specimens crushed using the compression machine generally translate the concept of compressive strength test.

According to Liyana Ahmad Sofri et al. (2015), countries such as France, Germany, Italy, Japan, Netherlands, and Russia have implemented blended cement and its production for a long time, while the United States and Canada have prioritized the mineral admixture in concrete by use of waste industry. Meanwhile, in Malaysia, there were some limitations to implementing the use of waste because of the industry itself. Therefore, as moving toward the modern construction industry, Malaysia should be aware of the technology and start to conduct the use of waste to be equivalent to other countries' construction fields. More researchers also need to master waste usage, not only in the research stage but to implement new ways of having economic construction for real.

According to Statista (Müller, 2021), the production of cement during the past five years was at an average of 16 to 20 million metric tons per year which shows the high demand for cement which may increase after the economy recovered from the Covid-19 pandemic. Sinar Harian also stated that the cost of housing is increasing as the main materials, such as reinforcement bars and concrete cement have increased by up to 50% more than before (Mohd Izzatul Izuan Tahir, 2021). According to Free Malaysia Today, the chairman of Guild of Bumiputra Contractors Bhd., Md. Nasir Ibrahim mentioned that the cost of components such as sand, cement, and bitumen had also risen (Ariff, 2021).

Therefore, utilizing POFA in concrete can be one of the effective ways to reduce cement production and also stabilize the cost of houses despite applying normal concrete materials which have increased in price. The aim of this research is to determine the compressive strength of POFA in concrete as a partial replacement for cement and sand.

2 Materials and Methods

2.1 Materials Selection

In this study, the materials used were according to the specification that meets the requirement of appropriate Malaysian and British Standards.

The type of cement used to make the concrete is called Ordinary Portland Cement (OPC). The cement served as the binder in concrete, strengthening the concrete with its composition (LeBow, 2018). OPC is usually grey in color and is used widely in civil engineering tasks and work, all over the world. In this study, the cement was used in powder form despite there being various techniques in using the cement to make the concrete (liquidation). By referring to the BS EN 197-1:2000, the OPC used in this study follows the standard. The OPC contains calcareous materials such as limestone, calcium, and argillaceous materials such as alumina and laterite.

Water was used as the liquid agent of the concrete. The water utilized must be free from chemical composition and organic matter. In Malaysia, the quality of water was referred to the Public Work Department of Malaysia (PWD) and BS EN 1008:2002 for specifications of concrete.

The crushed gravel and stone were used, as they can be categorized as particle that is higher than 20 mm, produced by the disintegration process. The coarse aggregate used in the preparation of the specimen was the maximum size is 20 mm in accordance with BS EN 12620:2002.

According to BS EN 12620:2002, the uncrushed aggregates, having the size of particles less than 6 mm, the sand was used for the fine aggregates. Despite its small size of particles, sand is very important, because it serves as a filler component.

Palm Oil Fuel Ash (POFA) is the result of burning the waste from the palm oil elements. The POFA has been discovered to be a useful waste with a certain significance, with some unique characteristics as it produced impacts and it was implemented in the optimum ways. The POFA was collected from Raub Oil Mill Sdn. Bhd., Raub, Pahang Darul Makmur.

2.2 Experimental Method

In this study, two different phases of testing were done. The first phase consists of five different percentages of POFA replacing cement with the variation of 0, 5, 10, 15, and 20%. The compressive strength of the first phase was determined by the optimum percentage of the POFA as a cement replacement of the concrete. The second phase assessed the compressive strength of concrete by replacing various percentages of fine aggregate with POFA, at 0, 10, 20, and 30%, with the use of the optimum percentage of POFA replacing cement as the control. With controlling all the materials besides the cement, the test cube dimension with the dimension of

100 mm × 100 mm × 100 mm was prepared for the compression test, for 3, 7, and 28 days.

2.3 Mix Proportion

The concrete specimens used the same concept as normal concrete casting using the mix design from the British Department of Environment (DoE), however, the use of POFA as partial cement on phase 1 testing and POFA as partial of both sand and cement on phase 2 testing were implemented. The use of cement, water, fine aggregate, and coarse aggregate was designed using the design mix calculation to determine the exact usage of each of the materials. The percentage of the replacement of POFA for cement was referred to in several past studies which were 2.5, 5, 10, and 15%. Meanwhile, the control specimen for phase 1 was extracted from the normal design mix proportion with 0% of POFA replacing the cement.

In the first phase, the cement usage was reduced against each percentage of the POFA used. Other proportions of the materials were followed accordingly from the design mix. The control specimens apply accordingly from the design mix as mentioned which applies 0% of any replacement of material and other specimens consist of the sample of POFA replacing cement with 5, 10, 15, and 20%.

Next, the second phase applied the replacement of sand with the POFA by using the optimum percentage of the first phase, which is the one that produces the highest compressive strength. The replacement of sand with POFA was tested at the variation of percentages of 10, 20, and 30%. The control specimen for the second testing used the exact proportion of materials of the optimum percentage gained from the phase 1 compressive test.

As for achieving the best result, the use of a concrete mixer ensures that the mixing process is done well. Cleaning the concrete mixer was a very important step, as the exact proportion must be referred to in the calculation. The fine and coarse aggregate after the accurate weight was determined and followed by the cement after a minute of the mixture. Last but not least, the water was filled to complete the combination of the materials to make the concrete. Table 1 shows the mix proportion for this research.

2.4 Casting and Curing

For the first phase, the POFA was replaced with cement, while the other materials were used the same as the composition of conventional concrete based on the calculation of the design mix. The casting of the concrete was implemented on the 100 mm × 100 mm × 100 mm mold as it was tested on the compressive strength of the sample.

Table 1 The mix proportion of this research

Type of mixture	Cement	POFA (cement replacement)		Water	Fine aggregate	POFA (sand replacement)		Coarse aggregate
	kg/m^3	%	kg/m^3	kg/m^3	kg/m^3	%	kg/m^3	kg/m^3
N1	350.0	0	–	190	705	0	–	1120
C1	332.5	5	17.5	190	705	0	–	1120
C2	315.0	10	35.0	190	705	0	–	1120
C3	297.5	15	52.5	190	705	0	–	1120
C4	280.0	20	70.0	190	705	0	–	1120
S1	315.0	10	35.0	190	634.5	10	70.5	1120
S2	315.0	10	35.0	190	564.0	20	141.0	1120
S3	315.0	10	35.0	190	493.5	30	211.5	1120

After the first phase was done, the second phase was conducted. The optimum result of the compressive strength of the first phase was used as the control specimen, with the utilization of various percentages of POFA replacing the sand. The POFA was exactly sieved with the size of the sand and was used to replace partial sand. The other material was used as the optimum percentage of cement replacement material proportion from phase 1.

Once the concrete reached 24 h after cast, the curing process took place and the samples were cured in the water for 3, 7, and 28 days using the water curing technique whereby the sample was immersed in the water until the tested days. The curing process was conducted in accordance with BS EN 12390-2:2009.

2.5 *Test Procedures*

The workability of concrete is the ability to work with the concrete, which means the higher slump test computes the lower rate of workability of concrete. For the fresh concrete, the slump test will determine the workability of the mix. The slump test was applied to the procedure according to British Standard BS EN 206-9:2010. Using the standard as the guide, the slump test was performed based on the specifications highlighted in the procedure.

A compressive test is one method for determining the mechanical properties of concrete, as compressive strength is defined as the capacity of concrete to withstand loads before failure (Jaya, 2020). The compressive strength was compared to the design mix strength as the concrete supposedly reaches the design strength or even exceeds the supposed strength. The compressive strength can be gained using the formula in (1).

$$\text{Stress}, \sigma = \frac{P}{A} \tag{1}$$

where:

P is Load/Pressure (kN).

A is Area (mm^2).

3 Results and Discussions

3.1 X-Ray Fluorescence

The composition of the materials such as POFA cement and sand were identified by using the X-Ray Fluorescence (XRF) test. All the specimens of the materials were sent to the lab to undergo the testing procedures.

Cement plays the important role in the binding process of all the materials used to make concrete. For the POFA replacing cement and cement, cement showed higher Silicon Dioxide and Calcium Oxide to undergo the process of formation of C–S–H compound which is related to the hydration process of the concrete. Table 2 shows the chemical composition of POFA as cement replacement and cement.

Referring to Harrisson (2019), to generate the C–S–H compound for the hydration process, the original cement may have higher quality regarding the chemical compounds contained in both materials.

The C–S–H formation needs two important compounds which were C_3S and C_2S which can be formed by the equation:

$$Ca_2SiO_5 + H_2O = CaO - SiO_2 - H_2O \tag{2}$$

$$C_2S + H = C - S - H \tag{3}$$

Table 2 Chemical Composition of POFA as cement replacement and Cement

Chemical compound	POFA	Cement
MgO	14.4931	<LOD
Al_2O_3	10.4536	2.9162
SiO_2	<LOD	21.3045
P_2O_5	12.1097	<LOD
SO_3	4.4263	3.8827
K_2O	19.7535	0.2724
CaO	26.3141	67.124
TiO_2	0.945	0.1467
MnO	0.7117	0.0585
Fe_2O_3	10.793	3.5755

$$Ca_3SiO_5 + H_2O = CaO - SiO_2 - H_2O + Ca(OH)_2 \tag{4}$$

$$C_3S + H = C - S - H + CH \tag{5}$$

According to Kamaruddin et al. (2018), the chemical compound of POFA can result in the secondary C–S–H attribute, as described in the equations in (6) and (7).

$$CaO + H_2O = Ca(OH)_2 \tag{6}$$

$$Ca(OH)_2 + SiO_2 + H_2O = C - S - H \tag{7}$$

According to Chelberg (2019), the compressive strength increases as the alumina helps in the rapid rate of performing C–S–H compound on concrete which means the presence of alumina assists the high reactivity of particles. POFA replacing cement has a higher concentration of Aluminum Oxide and Iron Oxide compared to cement, which may affect the concrete as well. Alumina and Iron Oxide is known as the agent of mineralizer which increase the rate of combination between solid and liquid, which may increase the productivity of making concrete. Other than that, there were several compounds that differentiated but the Potassium Oxide of both compounds shows a huge difference between POFA replacing cement which has 19.75, and original cement which has just 0.27 in terms of concentration. Potassium Oxide has the ability to reduce thermal expansion (Zhang, 2016) by stabilizing the hydration process and increasing the rate of strength in concrete (Neville, 2009).

POFA replacing sand has a higher concentration of Magnesium Oxide, Phosphorus Oxide, Sulphur Oxide, Potassium Oxide, and Calcium Oxide compared to the prior sand (Table 3). The chemical compound may not affect the role of the filler, but it will contribute to other factors related to the hydration process and strength of the concrete. While using the POFA to replace sand, the composition of Calcium Oxide in the sand may affect the hydration process which also fulfills the role of the filler in concrete. After the sand is combined with water and other materials, the chemical effect of the sand also leads to a better hydration process due to the bonds generated by the materials to form C–S–H gel (Hashim et al., 2019). The same goes for other chemical compounds such as Magnesium Oxide, Iron Oxide, etc. All the chemical compounds will double up the effect compared to the mixture that uses POFA just for cement replacement (Chelberg, 2019).

3.2 *Fresh Properties of POFA Concrete*

The slump test was conducted for all the percentages involved in the first phase, and the results are shown in Fig. 1.

Table 3 Chemical composition of POFA as sand replacement and sand

Chemical compound	POFA	Cement
MgO	14.429	<LOD
Al_2O_3	11.7458	65.448
SiO_2	<LOD	<LOD
P_2O_5	10.5743	0.3286
SO_3	6.2546	1.3912
K_2O	25.089	13.9055
CaO	20.2931	3.0272
TiO_2	0.7859	2.7999
MnO	0.5747	0.2338
Fe_2O_3	10.2526	12.8659

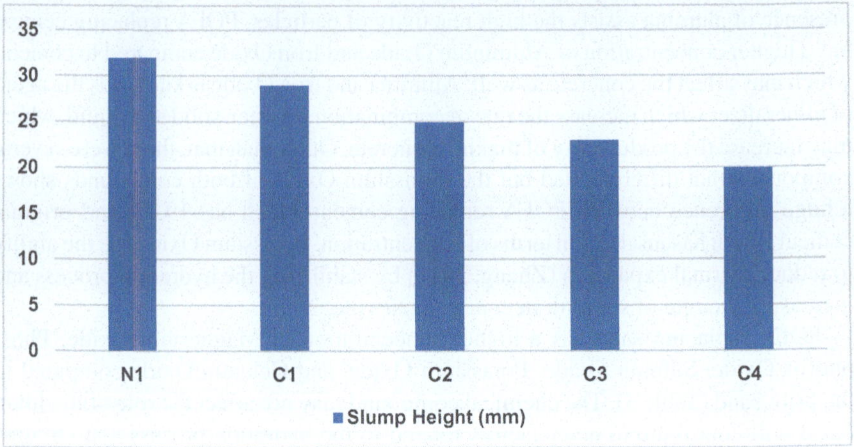

Fig. 1 The slump height for partial replacement of cement with POFA

The higher percentage of POFA used to replace cement, the lower the slump height for each batch of specimens. The use of full cement in the fresh concrete (N1) showed the highest slump which was 32 mm compared to the specimen which used POFA replacing cement. The 5% of cement replacement of POFA (C1) showed a slump height of 29 mm while 10% of cement replacement (C2) showed a 4 mm difference with 5% which was 25 mm. The 20% of POFA replacement for cement (C4) showed the lowest slump height which was 20 mm because finer particles will lead to a high rate of hydration process, therefore insufficient water for the mix will lower the slump height of the fresh concrete (Moutei et al., 2018).

As the level of POFA increases, the workability of concrete decreases. This behavior was observed, and it describes how workability decreased as the amount of replacement cement material increased (Rosseli et al., 2019).

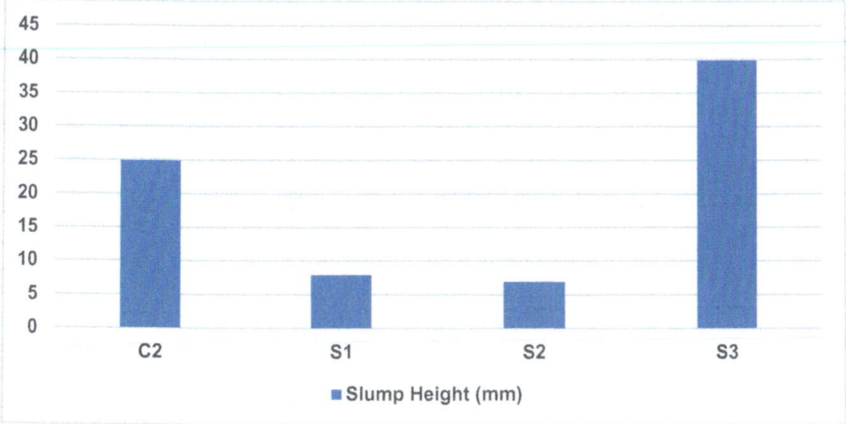

Fig. 2 The slump height for partial replacement of cement and sand with POFA

For the second phase, the slump height showed a different pattern compared to the first phase shown in Fig. 2. Generally, the first phase used POFA to replace cement only but in the second phase, POFA was used to replace both cement and sand partially.

For the 10% replacement of cement and 30% replacement of sand, the slump height showed the highest, 40 mm due to its slump condition, which was collapsed, thus the slump height could not be used. On the other hand, the slump height showed a decrement from 25 to 7 mm as the use of POFA to replace sand increased because the volume of POFA used was too much considering the water ratio was fixed. This was due to a higher volume of POFA that reduced the water with the high rate of hydration process as more C–S–H reaction first from the POFA replacing cement combined with the POFA replacing sand, increasing the reaction of POFA toward permeability of the porosity (Hasanuzzaman et al., 2016). The water absorption has been highlighted in past research whereby the POFA with high porosity tends to absorb more water (Saffuan et al., 2017). The more POFA used, the more the volume of water needed to bind and fill the porosity of the fresh concrete.

3.3 Compressive Strength Development with Age of POFA in Concrete

For the mechanical properties of the concrete, the compressive test was executed for all the specimens. For the first phase, all samples showed an increment of strength from day 3 until day 28 as shown in Fig. 3. The 10% of POFA replacing cement (C2) showed the highest compressive strength for 3 days, 7 days, and 28 days which were 19.64 kN/mm^2, 25.34 kN/mm^2, and 37.71 kN/mm^2 respectively. 5% and 15% of POFA replacement for cement also showed increment through days until they

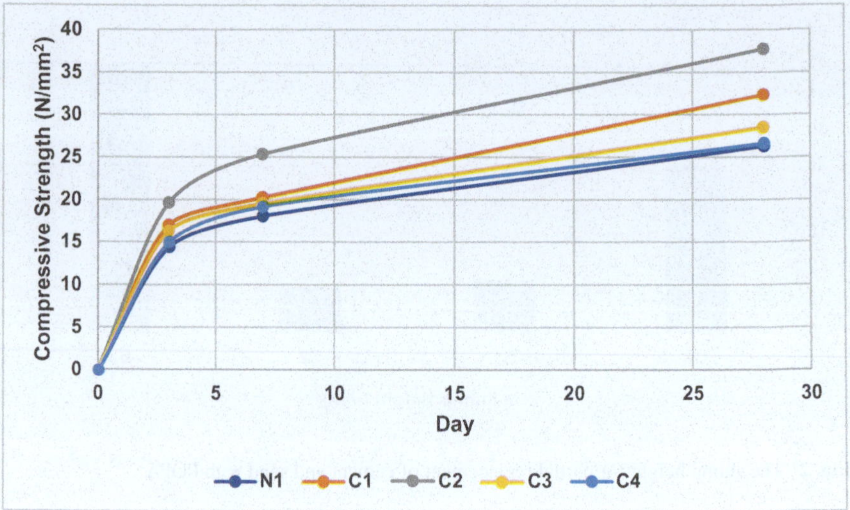

Fig. 3 The compressive strength of concrete with cement replacement of POFA

reached the compressive strength of 32.44 kN/mm^2 and 28.63 kN/mm^2 respectively after 28 days. However, 20% of POFA replacement for cement (C4) still showed higher compressive strength compared to normal concrete which applied 0% of cement replacement (N1).

The 20% of POFA replacing cement (C4) shows the lowest compressive strength among the specimen of replacement, but then the specimen still has higher compressive strength compared to the control specimen which has a difference of 0.59 kN/mm^2 for the 3 days, 1.11 kN/mm^2 on the 7 days, and 0.33 on the 28 days. Therefore, any percentage up to 20% replacement of POFA on cement shows a higher compressive strength compared to the control specimen which uses 100% cement. The strength of concrete may be higher as POFA has Potassium Oxide which enhances the strength of the concrete (Neville, 2009).

The concrete applied with POFA as cement replacement contributes to the second reaction of pozzolanic material to form an extra C–S–H compound which led to an increase in strength especially the early strength (Aiswarya et al., 2017). Therefore, more C–S–H reactions enhance the strength of the concrete. Nevertheless, the POFA replacing cement will be more effective if the water ratio used is compatible with the percentage of POFA applied in concrete (Hamid & Hamsan, 2021).

The second phase uses POFA as partial cement and sand replacement shown in Fig. 4. The 10% POFA replacing cement and 10% POFA replacing sand (S2) show the highest compressive strength on all three days compared to all specimens. For the 3 days compressive strength, 10% cement replacement and 10% sand replacement computed 17.81 kN/m^2 while 0%, 20%, and 30% of the sand replacement (10% replacement of cement for all specimens) showed 15.71 kN/mm^2, 17.03 kN/mm^2, and 5.06 kN/mm^2 respectively.

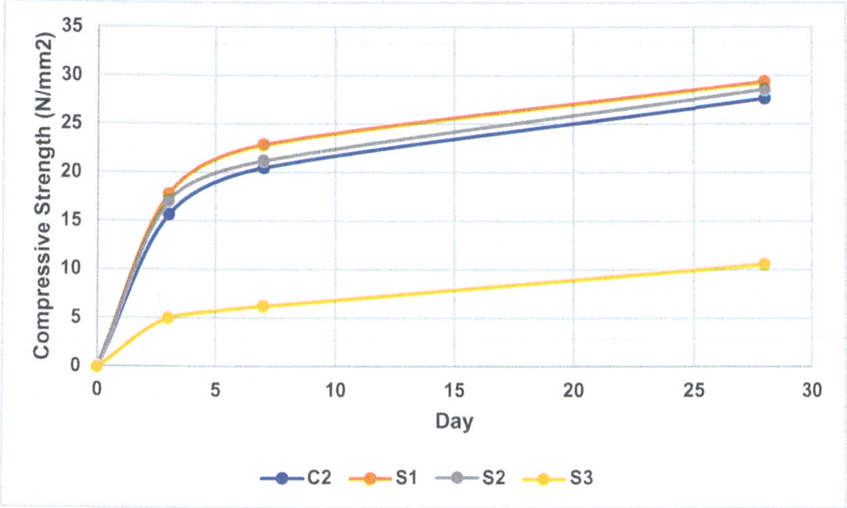

Fig. 4 The compressive strength of concrete with cement and sand replacement of POFA

Meanwhile, for the 7 and 28 days, the pattern of compressive strength resulted in the same 10% cement replacement and 10% sand replacement resulted in 22.93 kN/mm^2 for 7 days and 29.5 kN/mm^2 for 28 days. While the other replacement still showed an increment in strength through the days which somehow proved that the compressive strength kept increasing as the days went by.

The 10% cement replacement and 0% sand replacement for phases 1 and 2 also showed some differences, as the strength of the phase 1 specimen could not be reached as the second phase, which was affected by human error or the vibration process, which did not fully concentrate the specimen and caused some segregation due to the error.

The impact of more volume of POFA may be constant in the strength of concrete as more C–S–H compounds were reacted and more chemical compounds led to improve the strength of concrete due to high reaction of pozzolanic materials (Hamada et al., 2018). Not to mention that the chemical compound of POFA replacing sand also affects the strength of the concrete, making it stronger and more sustainable. The high volume of POFA in concrete specimens increases strength when the water is sufficient for the compounds to react, but insufficient water reduces compressive strength and thus the durability of the concrete (Huang et al., 2021).

3.4 . Relationship Between Strength and Age

Figure 5 describes the relationship between the percentage of POFA used as cement and the compressive strength. From Fig. 5, 10% of POFA replacement for cement

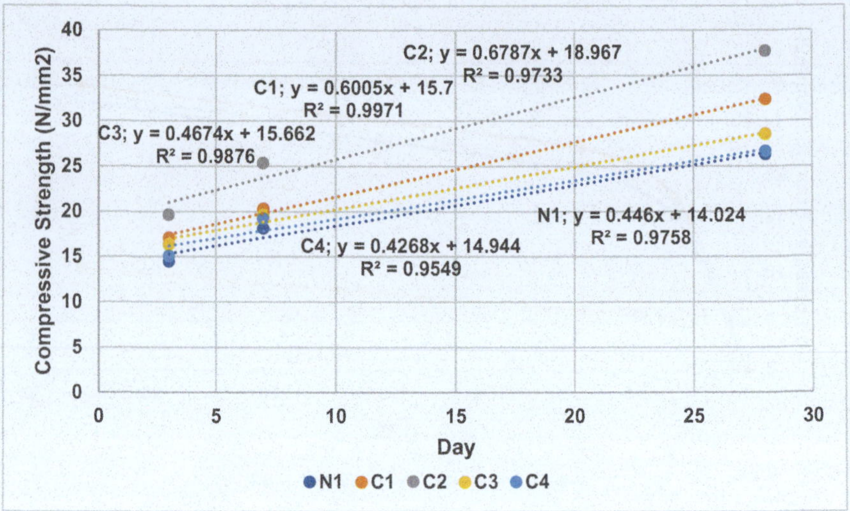

Fig. 5 The relationship between days and the compressive strength of POFA replacing cement

(C2) had the highest compressive strength at 3, 7, and 21 days. Toward 20% of cement replacement with POFA, the compressive strength decreased as the optimum percentage of POFA replacement for cement was 10%.

From the linear regression for respective percentages, the effective replacement of POFA can be defined by

$$\text{C2; } y = 0.6787x + 18.967 \tag{8}$$

whereby:

x = days.

y = compressive strength.

The higher the numbers of the day, the higher the compressive strength computed by the replacement. Other equations were generated for different percentages of replacement of POFA for cement such as follows:

$$\text{N1; } y = 0.446x + 14.024 \tag{9}$$

$$\text{C1; } y = 0.6005x + 15.7 \tag{10}$$

$$\text{C3; } y = 0.4674x + 15.662 \tag{11}$$

$$\text{C4; } y = 0.4268x + 14.944 \tag{12}$$

This figure exhibits a good correlation between day and compressive strength, in which coefficient of determination (R^2) for all concrete containing POFA as cement replacement and normal concrete has a value close to 1 as follows; where the coefficient of determination value for N1, C1, C2, C3, and C4 is 0.9758, 0.9971, 0.9733, 0.9876, and 0.9549 respectively.

All the percentages of replacement for cement showed an inclined pattern as the compressive strength increased and the number of days increased. The replacement of cement with POFA was also computed to be higher than normal concrete even though the compressive strength was not too enhanced.

Figure 6 describes the relationship between the percentage of POFA used as cement and sand and the compressive strength. From the graph above, 10% of POFA replacement for cement and 10% of POFA replacement for sand had the highest compressive strength for 3, 7, and 21 days. For the 30% of sand replacement with POFA, the compressive strength was the lowest as the optimum percentage of POFA replacement for cement and sand was 10%.

From the linear regression for respective percentages, the effective replacement of POFA can be defined by

$$\text{S1};\ y = 0.4165x + 18.137 \tag{13}$$

whereby:

x = days.

y = compressive strength.

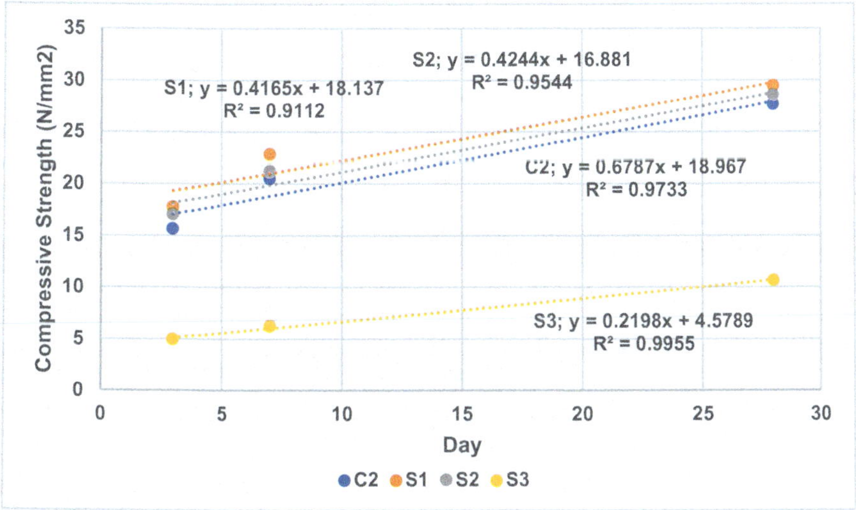

Fig. 6 The relationship between days and compressive strength of POFA replacing cement and sand

The higher the number of days, the higher the compressive strength computed by the replacement. Another equation was generated for different percentages of replacement of POFA for cement and sand such as follows:

$$C2;\ y = 0.6787x + 18.967 \quad (14)$$

$$S2;\ y = 0.4244x + 16.881 \quad (15)$$

$$S3;\ y = 0.2198x + 4.5789 \quad (16)$$

whereby:

x = days.

y = Compressive strength.

From Fig. 6, it was found that the range of coefficient of determination for concrete specimens (R^2) showed a good linear relationship between compressive strength and curing days since the value of R^2 is 0.9112–0.9955 for 28 days of curing age.

4 Conclusion

From the experiment done in this research, the main objects were successfully achieved with relevant and satisfying outcomes after a few months of casting and testing the sample. Generally, POFA can be said to be suitable to replace parts with cement and fine aggregate. When talking about suitability, the strength of concrete may be the main issue or factor that leads to the suitability, which is that 10% of POFA replacing cement show 43% higher strength compared to the normal concrete. However, a more accurate percentage may give a better improvement in concrete. The POFA seems to have pros and cons but the chemical compound and its fineness may give advantages to the enhancement of concrete strength.

As for POFA replacing partially both cement and sand, it also can be defined as a compatible replacement as the strength of concrete does not too different from the first phase. When it comes to POFA with sand, the role as filler was done perfectly because the smaller particle may occupy a small void, as it may be compacted well but the problems will be related to the hydration process, which is good to know that finer particles absorb more water than the coarser one.

Therefore, the optimum percentage of POFA replacing cement is 10%, which gives the highest compressive strength while applying the optimum percentage of 10% replacement of cement, and 10% replacement of cement shows the highest improvement in the concrete performance. However, the use of POFA-replaced sand showed lower improvement compared to the sample that just apply 10% of cement replacement.

Acknowledgements The author would like to acknowledge the Research Management Institute for providing the "Private Grant" (100-TNCPI/PRI 16/6/2 (059/2021)), Institute for Infrastructure Engineering and Sustainable Management (IIESM), and College of Engineering, Universiti Teknologi MARA for providing the facilities to accomplish the experimental work for this research.

References

Aiswarya, V. S., Beyoola, W., Harsha, V. N., & Preethi, M. (2017). Palm oil fuel ash as partial replacement of cement in concrete. *International Journal of Engineering Research & Technology, 6*(3), 544–546.

Ariff, I. (2021). Free Malaysia Today (FMT) | free and Independent. Retrieved December 10, 2021, from https://www.freemalaysiatoday.com/category/nation/2021/01/23/after-mco-prices-of-materials-have-gone-sky-high-complain-builders/.

BS EN 12390-2:2009, *Testing hardened concrete. Making and curing specimens for strength tests.* British Standard, pp 1–12

BS EN 197-1. (2000). *Cement - Part 1: Composition, specifications and conformity criteria for common cement*. British Standard, pp. 1–45.

BS EN 1008. (2002). *Mixing water for concrete - Specification for sampling, testing, and assessing the suitability of water, including water recovered from processes in the concrete industry, as mixing water for concrete*. British Standard, pp. 1–23.

BS EN 12620:2002, Aggregates for concrete, British Standard, pp 1–47

BS EN 206-9:2010, *Concrete. Additional Rules for Self-Compacting Concrete (SCC)*. British Standard, pp 1–16

Chelberg, M. (2019). *The effect of fly ash chemical composition on compressive strength of fly ash Portland cement concrete*. Doctoral dissertation, The Ohio State University.

Hamada, H. M., Jokhio, G. A., Yahaya, F. M., & Humada, A. M. (2018). Properties of fresh and hardened sustainable concrete due to the use of palm oil fuel ash as cement replacement. *IOP Conference Series: Materials Science and Engineering, 342*, 012035. https://doi.org/10.1088/1757-899x/342/1/012035

Hamid, N. A. A., & Hamsan, N. F. (2021). A Comparative study on utilization of palm oil fuel ash (POFA) as partial cement replacement in concrete. *Recent Trends in Civil Engineering and Built Environment, 2*(1), 773–782.

Harrisson, A. M. (2019). *Constitution and specification of Portland cement lea's chemistry of cement and concrete*, pp. 87–155. https://doi.org/10.1016/b978-0-08-100773-0.00004-6.

Hasanuzzaman, M., Rafferty, A., Sajjia, M., & Olabi, A.-G. (2016). Production and treatment of porous glass materials for advanced usage. *Reference Module in Materials Science and Materials Engineering*. https://doi.org/10.1016/b978-0-12-803581-8.03999-0

Hashim, N. H., Muhd Sidek, M. N., Noor, N. M., Roselli, S. R., Mohd Yusoff, M. A., & Saiful Bahari, M. F. (2019). Utilisation of palm oil fuel ash (POFA) as sand replacement for fresh and hardened concrete by using powder and liquidation methods. *Journal of Physics: Conference Series, 1349*(1), 012055. https://doi.org/10.1088/1742-6596/1349/1/012055

Huang, Q., Zhu, X., Liu, D., Zhao, L., & Zhao, M. (2021). Modification of water absorption and pore structure of high-volume fly ash cement pastes by incorporating nanosilica. *Journal of Building Engineering, 33*, 101638. https://doi.org/10.1016/j.jobe.2020.101638

Jaya, R. P. (2020). Porous concrete pavement containing nanosilica from black rice husk ash. *New Materials in Civil Engineering*, 493–527. https://doi.org/10.1016/b978-0-12-818961-0.00014-4

Kamaruddin, S., Goh, W. I., Jhatial, A. A., & Lakhiar, M. T. (2018). Chemical and fresh state properties of foamed concrete incorporating palm oil fuel ash and eggshell ash as cement replacement. *International Journal of Engineering & Technology, 7*(4.30), 350. https://doi.org/10.14419/ijet.v7i4.30.22307.

LeBow C (2018) Effect of Cement Content on Concrete Performance. University of Arkansas

Mohd Izzatul Izuan Tahir. (2021). *Harga bahan binaan naik, rumah semakin mahal*. Sinarharian; sinarharian. https://www.sinarharian.com.my/article/172393/SUARA-SINAR/Analisis-Sinar/Harga-bahan-binaan-naik-rumah-semakin-mahal.

Moutei, L., Benbrahim, Y., El ghailassi, T., Bouih, A., Labied, S., Guedira, T., & Benali, O. (2018). The effect of the addition of alumina powder on the confinement properties of a cement mortar. *MATEC Web of Conferences, 149*. https://doi.org/10.1051/matecconf/201814901055.

Müller, J. (2021). *Malaysia: Cement production 2020 | Statista*. Statista. https://www.statista.com/statistics/719188/cement-production-malaysia/

Neville, A. M. (2009). *Properties of concrete*. Prentice Hall/Pearson Education.

Rosseli, S. R., Sidek, M. N., Halim, N. I., Hashim, N. H., Hamid, M. S., & Tutur, N. (2019). Effect on consistency and performance of POFA concrete. *Journal of Physics: Conference Series, 1349*(1), 012067. https://doi.org/10.1088/1742-6596/1349/1/012067

Saffuan, W. A., Muthusamy, K., Salleh, & N. A., Nordin, N. (2017). Properties of concrete containing ground palm oil fuel ash as fine aggregate replacement. *IOP Conference Series: Materials Science and Engineering, 264*.https://doi.org/10.1088/1757-899x/264/1/012008.

Sofri, L. A., Affendi, Z., Zahid, M. Z. A. M., Ahmran, M. S. A. (2015). Performance of concrete by using Palm Oil Fuel Ash (POFA) as a cement replacement material. ResearchGate, Trans Tech Publications, Ltd. https://www.researchgate.net/publication/279181705_Performance_of_Concrete_by_Using_Palm_Oil_Fuel_Ash_POFA_as_a_Cement_Replacement_Material

Zhang, H. (Ed.). (2016). *Building materials in civil engineering*. Woodhead.

A Review of Graphene Research and Its Outputs: Waste Carbon Source and Synthesis Technique

M. Z. Nurfazianawatie, H. Omar, N. F. Rosman, N. S. A. Malek, A. N. Afaah, M. Maryam, I. Buniyamin, M. J. Salifairus, M. F. Malek, M. M. Mahat, M. Rusop, and N. A. Asli

Abstract Graphene is a novel material that emerged following the era of carbon nanotubes and is an extraordinary two-dimensional (2D) material. The consumption of materials derived from petroleum has highlighted numerous sustainability concerns. New approaches to graphene production and waste management have recently emerged. This study focuses on the characterization of graphene produced from carbon-rich waste materials using bottom-up and top-down techniques. Considering graphene is primarily made up of chemically bonded carbon atoms, starting materials employed must contain high carbon as their main element. In this study, the carbon-rich waste materials are of natural and synthetic waste categories, such as waste cooking oil, waste engine oil, coconut shells, and plastic. These carbon-rich wastes are highly potential to be converted to high-quality graphene. Using waste materials can indirectly reduce the environmental impact by lowering the quantity of abandoned waste that accumulates in landfills and reducing the toxic composition of the land. Producing graphene from abundant waste materials would be a game-changing invention, especially given how attractive graphene has been.

Keywords Graphene · Carbon · Waste carbon source · Synthesis technique · Raman spectroscopy

---

M. Z. Nurfazianawatie · H. Omar · N. F. Rosman · N. S. A. Malek · A. N. Afaah · M. Maryam · I. Buniyamin · M. J. Salifairus · M. F. Malek · M. Rusop · N. A. Asli (✉)
Centre for Functional Materials and Nanotechnology, Institute of Science, Universiti Teknologi MARA, 40450 Shah Alam, Selangor, Malaysia
e-mail: asnida1462@uitm.edu.my

M. Z. Nurfazianawatie · H. Omar · N. F. Rosman · N. S. A. Malek · A. N. Afaah · M. Maryam · I. Buniyamin · M. J. Salifairus · M. F. Malek · M. M. Mahat · N. A. Asli
School of Physics and Material Studies, Faculty of Applied Sciences, Universiti Teknologi MARA, 40450 Shah Alam, Selangor, Malaysia

M. Rusop
NANO-Electronic Centre, College of Engineering, Universiti Teknologi MARA, 40450 Shah Alam, Selangor, Malaysia

U. R. Hashim et al. (eds.), *Green Infrastructure*,
https://doi.org/10.1007/978-981-99-7003-2_12

1 Introduction

Graphene is a two-dimensional honeycomb lattice substance made up of single carbon atomic layers that are chemically bonded to one another (Koosha et al., 2021; Lavin-Lopez et al., 2019; Nurfazianawatie et al., 2023a; Nurfazianawatie et al., 2023b; Yesilbag et al., 2021). Graphene has excellent electrical, thermal, and mechanical characteristics (Das et al., 2019; Nurfazianawatie et al., 2020; Salifairus et al., 2016; Wu et al., 2018), including carrier mobility of greater than 10^4 cm^2/Vs and thermal conductivity greater than 10^3 W/mK (Jalani et al., 2016). The strength of graphene is about 100 times that of ordinary steel. Graphene has a tensile strength of 42 N/m and a thickness of around 0.335 nm (Aïssa et al., 2015; Ataç et al., 2020; Bhoria, 2019; Saeed et al., 2020). With its unique physical and chemical characteristics, including mechanical stiffness, strength, elasticity, and extraordinarily high electrical and thermal conductivity (Alemour et al., 2018; Amato, 2018; Ferrari & Basko, 2013; Klimchitskaya et al., 2014; Nilsson et al., 2008; Poniatowska et al., 2019; Selvakumar et al., 2016), 2D graphene is classified as an atomic crystal (Aïssa et al., 2015). Graphene properties have made graphene the best material for various types of electronic applications such as LEDs, solar cells, transistors, and displays due to the extremely higher transparency of graphene (Aïssa et al., 2015). The general structure of graphene is similar to that of many benzene rings linked together when hydrogen atoms are generated by carbon, and it is considered hydrophobic because of the absence of oxygen groups (Mbayachi et al., 2021).

An allotrope of carbons, called graphene, is a plane of sp^2 bonded atoms (Lavin-Lopez et al., 2019; Yesilbag et al., 2021). When multiple graphene layers are layered together, they produce graphite, a more complicated compound. It is possible to oxidize graphene and graphite to produce graphene oxide (GO) and graphite oxide, respectively, which include carbonyl (C=O), hydroxyl (OH), and carboxyl (COOH) groups (Mbayachi et al., 2021). Due to its remarkable properties and structure, graphene has also attracted the attention of several scientists and engineers over the last few decades. It is now the subject of extensive study (Bleu et al., 2019). Figure 1 shows the chemical structures of graphene, graphite, graphene oxide, and reduced graphene oxide.

Since graphene is so rare in nature, its synthesis has been a topic of much discussion. Synthesis is currently the best method for producing graphene (Bleu et al., 2019; Lisi et al., 2017). The graphene manufacturing process will require using a material source, commonly referred to as a precursor (Lisi et al., 2017). There are a wide variety of carbon-based substances or products that may be used as precursors in the synthesis of graphene. Several suggestions for these precursors are graphite, biowastes, and general garbage. Waste is substances thrown after use because they are no longer useful (Alias et al., 2018; Ramadhansyah et al., 2020; Tsai, 2019). Waste may be classified into several categories and forms, including hazardous and radioactive wastes, wastewater, and industrial by-products (Ikram et al., 2020; Pandey et al., 2019; Yacob et al., 2015). Unmanageable waste disposal contributes to numerous sorts of pollution to the environment, including water, air, and soil pollution, as well

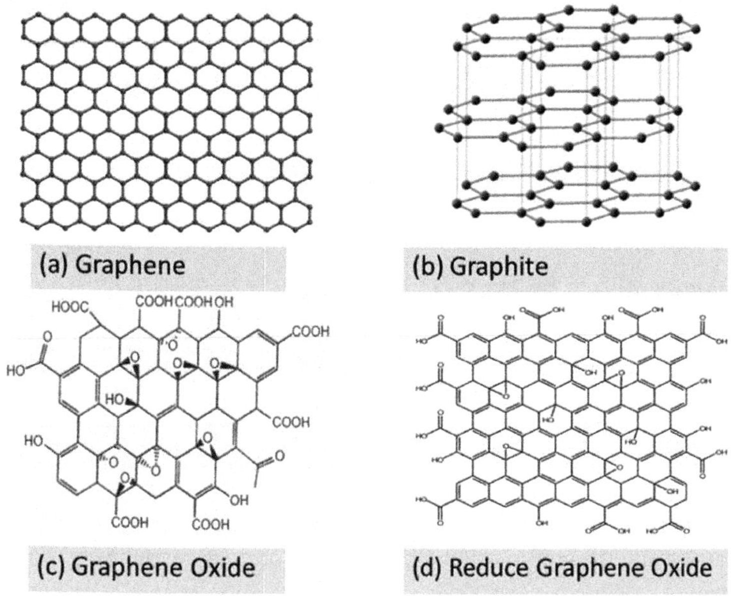

Fig. 1 Structure of **a** graphene, **b** graphite, **c** graphene oxide, GO, and **d** reduced graphene oxide, rGO

as disrupting marine ecosystems and blocking drains, all of which have a severe unfavourable influence on the environment (Lam et al., 2015; Yacob et al., 2015).

Globally, around 10 billion tonnes of solid waste are created each year, with this figure expected to rise to 27 billion tonnes by 2050 (Jahirul et al., 2022). Reuse and recycle of solid wastes and bio-wastes might help sustaining the environmental issues such as greenhouse effect and ozone layer thinning thus, would benefit the nation. Most wastes are made up of carbon atoms. For example, waste cooking oil has acids that are mostly made of carbon chains, and polymers are made of carbon polymer chains (Jahirul et al., 2022). Carbon is one of the elements that may be found in abundance on earth, and its name is derived from the Latin word for 'charcoal,' which is abundant in the world annually (Mbayachi et al., 2021; Shams et al., 2015; Zakaria et al., 2022). Research done by (Freitas et al., 2020; Nurfazianawatie et al., 2020; Ullah et al., 2014) reveals that waste cooking oil (WCO) is abundant and readily available. It was reported that in Malaysia, around 500,000 tonnes or over 453,590,000 kilos of waste cooking oil were disposed of yearly without being treated (Yacob et al., 2015). As a result, it is recommended to utilize regular waste for graphene production (Deng et al., 2016).

This study is about reviewing the graphene synthesis from various carbon sources and their synthesizing techniques. Some techniques can be used to synthesize graphene from its various precursors, including chemical vapour deposition (CVD) methods, Hummer's method, liquid-phase exfoliation (LPE), and pyrolysis. Each technique requires different precursors, undergoes different processes, and yields different products of graphene (Mbayachi et al., 2021). The summary of methods, carbon source, advantages, disadvantages, and findings of graphene synthesis is shown in Table 1. For graphene characterization, Raman spectroscopy, ultraviolet–visible spectroscopy (UV–Vis) spectroscopy, transmission electron microscope (TEM), field emission scanning electron microscopy (FESEM), atomic force spectroscopy (AFM), and X-ray diffraction (XRD) can be used.

2 Precursor

A precursor is a source that serves as a catalyst to kickstart graphene synthesis. The precursors for graphene synthesis must contain carbon atoms, and the typical carbon compound gaseous that are used for this purpose are methane, acetylene, and ethylene (Kairi et al., 2017). These rich carbon compounds gaseous can be synthesized or naturally emitted from the sources such as waste cooking oil, waste engine oil, coconut shells, and plastics through the chemical vapour deposition (CVD) method, thermal conversion method, pyrolysis method, and natural process, respectively (Ikram et al., 2020).

2.1 *Waste Cooking Oil*

About 24 million metric tons (MT) of waste oil is globally produced annually (Lam et al., 2015). Cooking oil is a substance that is often used in homes and food industries for cooking and frying(Alias et al., 2018; Rubianto et al., 2020). Commonly referred to as ‘waste cooking oil,’ or ‘used cooking oil,’ waste cooking oil is a biowaste that has a high carbon content and fatty acid content (Allah & Alexaandru, 2016; Awogbemi et al., 2019; Patel et al., 2019; Tsai, 2019; Ullah et al., 2014). Cooking oil undergoes chemical and physical changes as a result of reactions such as hydrolysis, thermal degradation, oxidation, and polymerization (Mohadesi et al., 2019; Zik et al., 2020). The WCO content of oil can vary due to thermal or other high-temperature reactions that occur during frying (Tsai, 2019). WCO has high carbon content, 76.95%, compared to refined palm oil, 68.81% (Nurfazianawatie et al., 2023b; Nurfazianawatie et al., 2020). WCO is considered to be part of the wastes generated, which is comprised of materials generated by residences and has the potential to serve as a valuable secondary raw resource if it is handled well (de Feo et al., 2020). Since many people throw away their used cooking oil, it’s important to find a way to recycle it (Mohadesi et al., 2019; Tsai, 2019), It may be reused or recycled, or even used to produce graphene in a laboratory. WCO is easy to acquire and recycle, so it is often seen as a low-cost, renewable feedstock for biodiesel and bio-based products (Allah & Alexaandru, 2016; Elkady et al., 2015; Tsai, 2019).

Thermogravimetric analysis (TGA) which is based on the observation of sample mass loss with respect to time or temperature at a specific heating rate. In producing

Table 1 Summary of methods, carbon source, advantages and disadvantages, and findings of graphene synthesis

Method	Carbon source	Advantages	Disadvantages	Findings	Ref.
CVD	Waste cooking oil (WCO), Waste Engine oil (WEO)	CVD process profile for graphene synthesis Free contamination Variety types of precursor such as solid, liquid, and gas	Production in small scale Require high-temperature in synthesis graphene	CVD techniques allow the production of high-quality material and as large-area graphene films Application: transparent conductive films such as interactive touch film touch screens, flexible LCD, smart windows, solar cells, and OLED	(Deng et al., 2019; Gi Choi et al., 2022; Kairi et al., 2017; Kumar et al., 2011; H. Liu & Liu, 2019; Mat Tahir et al., 2020; Pedrazzetti et al., 2022; Saulnier et al., 2017; 2009) Seo et al., 2017; Wang et al.,
Hummer's	Coconut shells (CS)	Produce in large quantity	Produce graphene with many unsuitable functional groups for some applications due to a few unstable structures. However, it can be improved by reducing the functional group that produces reduced graphene oxide with similar properties to graphene	High yield of graphene obtained Application: sensors, conductive thin film	(Alkhouzaam et al., 2020; Selvakumar et al., 2016; Sujiono et al., 2020; Zaaba et al., 2017)

(continued)

Table 1 (continued)

Method	Carbon source	Advantages	Disadvantages	Findings	Ref.
Liquid-phase exfoliation	Graphite, graphite oxide	Single-step functionalization and exfoliation;	Ionic liquids are costly Separation from dispersion liquid necessarily requires centrifugation or sedimentation	Functionalized graphene with excellent electrical conductivity Application: sensors, conductive thin film	(Omar et al., 2020; Bhoria, 2019; Li et al., 2019; Li et al., 2020; Liu et al., 2008)
Pyrolysis	Plastic aerosol, camphor leaves, coconut shells	The process can yield high volumes of liquid oil, up to 90 wt% at moderate temperatures of 500 °C–550 °C	The process is complex, and a system for cleaning the air is required to treat the flue gases from pyrolysis	Graphite in the form of powder Application: act as additives in composites, energy	(Indriati et al., 2019; Jahirul et al., 2022; Kong et al., 2020; Pandey et al., 2019)

graphene, the WCO goes through the chemical vapour deposition (CVD) method. The samples were prepared by varying the CVD synthesis profile parameter in obtaining optimum graphene sample (Deng et al., 2019). Before starting CVD routes, precursor was undergoing thermogravimetric analysis (TGA) to observe mass loss as a function of time or temperature. During CVD process, reactive gas from volatile precursor fed into deposition furnace, where the chemical reaction takes place and solid materials are deposited on substrate. Volatile by-products of non-reacted gases are removed by gas flow through the reaction chamber. After the synthesis process was done, the sample was left at room temperature for a few hours to cool down (Milenov et al., 2017; Omar, 2022; Asli et al., 2022; Rosman et al., 2022; Wang et al., 2009). The flow of carrier gas was continued until the furnace reached room temperature to complete the deposition and formation of graphene on substrate (Asli et al., 2022). During the synthesis, the gas is let to flow to make the ambient environment, get rid of the air in the air around the tube, and get rid of any contaminants (Asli et al., 2022). The detailed procedure for the CVD approach is described in Sect. 3.1.

2.2 *Waste Engine Oil*

Waste engine oil (WEO) is easily obtained, and available abundantly. Engine oil is mainly used for engine lubrications and other metal joining parts lubrication purposes (Abdullah et al., 2020; Kalanidhi & Nagaraaj, 2022). WEO is a high polycyclic aromatic hydrocarbon (PAHs) that contains metal contaminants. It makes WEO an environmentally harmful waste that is difficult to dispose of due to the presence of undesired species such as soot and metals (Lam et al., 2015). The flash point of fresh engine oil is 242.5 °C, meanwhile WEO has a higher flash point of 178 °C which indicates lower temperature needed to vaporize the oil (Syrmanova et al., 2017). Elemental percentages of C, H, and S in WEO are 85, 13.6, and 0.01 wt%, respectively (Ko & Kim, 2006) and high content of 85 wt% C in WEO and their chemical composition of many hydrocarbon compounds made them potential and suitable to be used as carbon sources to produce graphene. The CVD method can be used as a method in synthesis graphene (Abdullah et al., 2020). WEO was introduced as a precursor, and nickel was used as a substrate. Both in alumina boats and in a quartz tube. The samples were prepared by varying the parameter to get desired graphene properties (Deng et al., 2019). As the furnace (precursor) reaches 200 °C, the carbon is then turned into vapour by gas flow and deposited on a substrate. The sample was left at room temperature for a few hours after synthesis (Milenov et al., 2017; Nurfazianawatie et al., 2023a; Asli et al., 2022; Wang et al., 2009). Flow of carrier gas continued until the furnace achieved room temperature (Asli et al., 2022). The sample then were characterized to analyse the data. In Sect. 3.1, the CVD method's procedure is explained.

2.3 *Coconut Shells*

Coconut shells are literally the non-edible, outer covering of coconuts and are considered one of the waste products of agriculture (Somashekhar et al., 2018). Once converted into powder, coconut shells stand out among other materials due to their affordability, renewability, low density, and environmental friendliness (Somashekhar et al., 2018). Mixing coconut shell powder with epoxy resins improves its properties, resulting in new opportunities for further use (Somashekhar et al., 2018). Additionally, some researchers have turned coconut shells into charcoal and used it as filler to fend off freezing, crushing, and absorption (Ting et al., 2015). Coconut shell charcoal, which has a carbon content of roughly 74.3%, is a form of coconut shells that can be found in nature and contains hydrocarbon compounds (Sujiono et al., 2020). Graphite powder is perhaps produced via graphitization by pyrolysis of coconut shells, while graphene oxide (GO) is prepared using Hummer's technique from graphite powder (Sujiono et al., 2020). When exfoliating in aqueous solution using chemical synthesis, the generation of graphene oxide from natural graphite, which is coconut, can perform better than synthetic graphite (Aunkor et al., 2016). Activated carbon from carbonized coconut shells is also used to produce graphene (Ghosh et al., 2019; Nita et al., 2021). Natural carbon precursors were far simpler to make than synthetic ones, notably graphite, as the latter required higher temperature synthesis (Kim et al., 2016). Graphitic carbon from coconut waste contains a high graphitization degree of 84.88%, a little bit less graphitization than commercial graphite (90.23%) (Destyorini et al., 2021).

2.4 *Plastic*

Solid plastic waste can also be utilized as a precursor for graphene production and is one of the most often encountered wastes in the environment, which can help to tackle the environmental problems created by this solid waste indirectly (Pandey et al., 2019). A few nanomaterials are synthesized by these polymeric carbon-rich compound plastics wastes such as polyethylene (PE), polypropylene (PP), and polystyrene (PS) that can be used to synthesize multiple forms of carbon nano materials (CNM) like carbon spheres (CS), carbon nanofibers (CNF), and carbon nanotubes (CNT) (Pandey et al., 2019). Numerous efforts have been made to recycle plastic wastes but have yet to achieve the aspect of being environmentally friendly and cost-effective in the upcycle of waste (Kausar, 2020; Pandey et al., 2019). Plastic waste can be converted into graphene nanosheets by first finely chopping the collected plastics with a shredder, then washing with soap in a washing unit to remove the oily and greasy impurities, drying, mixing with bentonite, and finally pyrolysis, which generates the backbone for the nucleation of graphene nanosheets (Pandey et al., 2019). This procedure achieves a number of the previously listed criteria, particularly in terms of mass scalability, thanks to its precursor, which can be discovered in large quantities in the surrounding region and with the aid of local municipalities (Pandey et al., 2019). Waste plastic pyrolysis is popular because it produces a high amount of liquid oil up to 90% by volume at moderate temperatures of 500 °C–550 °C (Jahirul et al., 2022).

3 Graphene Synthesis Technique

Many technique methodologies can be used to synthesize graphene. Chemical vapour deposition, Hummer's method, and the liquid-phase exfoliation method are discussed based on their respective sections. Some techniques will instead yield graphene variations, such as graphite or graphite oxide, from which the basic form of graphene may be obtained by further processes.

3.1 *Chemical Vapour Deposition*

CVD is a chemical technique used in the semiconductor industry to produce high-quality, solid, thin-film materials (Deng et al., 2019). (CVD) is considered as a promising method for producing graphene sheets on transition metals (Jalani et al., 2016). Transition metals such as nickel (Ni), cuprum (Cu), iridium (Ir), ruthenium (Ru), and palladium (Pd) could very well be used as metal substrates in CVD (Yan et al., 2016). However, given the variety of CVD growth conditions, synthesizing a single-layer graphene, and regulating the quality graphene film on nickel are extremely challenging (Nurfazianawatie et al. 2023a). Graphene growth via CVD may be regulated by adjusting profile process parameters and substrate types. Chemical vapour deposition (CVD) is a technique that employs gas-phase mixtures and surface reactions. (Bleu et al., 2019). Carrier gases transport the volatile precursors to the reaction zone, where they either combine with other reaction gases or breakdown while still in the gaseous phase (a process known as homogeneous reaction), or they absorb on the surface's substrate and react with the reaction gases or decompose (known as heterogeneous reaction) (Ahmed et al., 2016; Azadmanjiri et al., 2020; Koosha et al., 2021). Both reactions would result in the production of a solid thin coating on the substrate and gaseous by-products (Lam et al., 2015).

The CVD method's procedure varies based on the application, but it typically includes a carrier gas and gas inlet system (to control gas flow), gas delivery system (provides precursor in gaseous state—also known as vaporizer), gas distribution system (controls the gas ratios from varying sources), energy source (provides energy for deposition reaction), and exhaust-gas cleaning system (destroys or traps any volatiles). Figure 2. illustrates a graphical procedural flow of the chemical vapour deposition technique.

Figure 2. shows the carbon source is in the gas phase, the precursor. The precursor is WCO, heated using a high energy source such as plasma or a resistivity heated coil, and an inert gas is introduced into the reaction chamber for the chemical reaction to occur (Nurfazianawatie et al., 2020). The role of gas flow in the CVD chamber on the large-scale homogeneity of graphene thickness, as well as the accurate control of gas flow for uniform and high-quality graphene growth (Deng et al., 2019). During the CVD process, the substrate is loaded with metal catalysts such as nickel or copper, which are employed since of their inert qualities and serve as the location for film deposition (Rosman et al., 2022; Alam et al., 2020; He et al., 2019; Salifairus et al., 2015). As carrier gases, hydrogen or argon gases are used to remove and

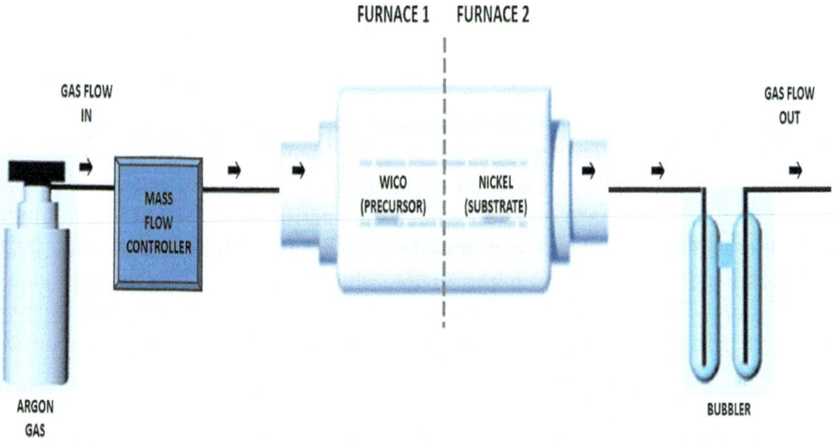

Fig. 2 A typical schematic setup for chemical vapour deposition (CVD) method

clean undesirable oxides from metal surfaces. The deposited layer on the substrate is graphene sheet. (Shams et al., 2015). The heated precursor gas is decomposed into carbon (the thin film) and a by-product gas throughout the whole process (Deng et al., 2019). The general chemical equation for the decomposition process is shown by Eq. (1):

$$\Delta \text{Precursor Gas (g)} \rightarrow \text{Carbon (s)} + \text{By - product Gas (g)} \quad (1)$$

The WCO product consists of carbon (which is our graphene layer) and hydrogen gas as a gas by-product. Equation (2) shows the process's defined chemical equation:

$$\Delta CH_4(g) \rightarrow C(s) + 2H_2(g)$$

Using cooking oil waste and nickel foil as a precursor and substrate respectively gives multilayer graphene based on FESEM and Raman analysis (Nurfazianawatie et al., 2020). Besides based on UV–Vis reflectance, the 250scc/m flow rate gives an optimum in synthesis graphene using CVD method (Nurfazianawatie et al., 2020).

In Fig. 3. the Raman spectra revealed the presence of G and 2D bands. Raman spectroscopy revealed that multilayer graphene at 350 °C had an intensity of graphene I_{2D}/I_G of 0.38, which is close to 0.5 (multilayer graphene) (Nurfazianawatie et al., 2022). Using the CVD method, multilayer graphene was effectively synthesized. The optimum temperature was successfully reached; hence it can be concluded that the experiment was a success. Chemical Vapour Deposition (CVD) processes produce thin graphene film (Nurfazianawatie et al., 2020). Because of their huge size, these films are suited for usage in applications such as touch screens, flexible LCD, smart windows, solar cells, and OLED. CVD techniques allow for the production of high-quality materials and large-area graphene films.

3.2 *Hummer's Method*

Hummer's method is one of the techniques utilized in manufacturing graphene from graphite oxide. It was developed in 1958 by William S. Hummers and Richard E. Offeman as a modified technique to the Staudenmaier method (Hummers & Offeman, 1958). Hummer's approach substitutes potassium permanganate ($KMnO_4$) for potassium chlorate ($KClO_4$), which is reported to be harmful (Alkhouzaam et al., 2020). This technique only yields graphite oxide, which requires a further step such as pyrolysis of the graphite oxide. Graphene, graphene oxide (GO), and graphene oxide reduction can all be accomplished using the synthesis process (Adetayo & Runsewe, 2019).

Sulfuric acid (H_2SO_4) and sodium nitrate ($NaNO_3$) are used as reactants in Hummer's technique, and potassium permanganate ($KMnO_4$) is mixed with graphite powder to oxidize graphite by graphite oxidation, which takes around 24 h of reaction time (Alkhouzaam et al., 2020). Hummer's technique has also gone through

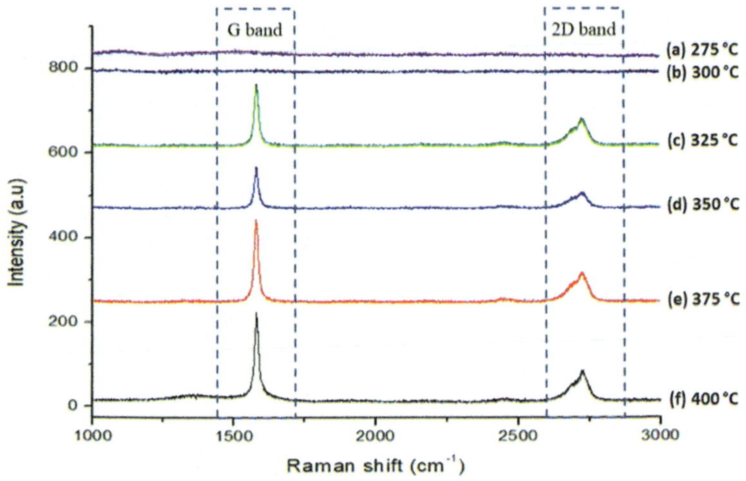

Fig. 3 Raman results of graphene using waste cooking oil as carbon source (reproduced with permission. Copyright 2023, Materials Today Proceedings) (Nurfazianawatie et al., 2023a)

a variety of modifications and variations in order to increase the yield of graphite oxide characteristics, shorten the time required for synthesis, or even fully remove the releases of toxic gases, such as sodium nitrate ($NaNO_3$), by omitting it fully from the substance pool used to prevent the production of NO_2 and N_2H_2 gases and by utilizing an entirely different set of reactants, such as H_2SO_4 and phosphoric acid (H_3PO_4) or even potassium dichromate ($K_2Cr_2O_7$) instead of $KMnO_4$ which leads to the process that only takes about 2 h of reaction time (Alkhouzaam et al., 2020; Guerrero-Contreras & Caballero-Briones, 2015; Zhu et al., 2022).

Some researchers proposed on eliminating potassium permanganate and sodium nitrate as the oxidation agent (Kudus et al., 2020). Several modified Hummer techniques have been used during the past 20 years were improved to increase yields and prevent generation of dangerous gases (such as NO_2, N_2O_4) and lingering nitrate eliminating $NaNO_3$ (Liu et al., 2018). In Hummer's method, the amount of sufficient hydrogen peroxide which is to remove manganese ion improves the less defective structure of graphene oxide (Cucu et al., 2015; Yoo & Park, 2019).

According to Alkhouzaam et al. (2020), the first step is to mix H_2SO_4 and H_3PO_4 in varied volume ratios while stirring in an ice bath for a few minutes. Next, the $KMnO_4$ and graphite powder are added slowly while stirring the solution. The mixture is then placed in an oil bath at 85 ± 2 °C for half an hour before being added 50 mL deionized water. After that, the mixture is reintroduced to an ice bath with 150 mL of deionized water and 20 mL of hydrogen peroxide ((H_2O_2) is added to terminate spontaneous reactions before it is decreased in temperature. Graphite oxide is synthesized using Hummer's method, as shown in the schematic experimental procedure in Fig. 4. Based from research done by Zhu et al., less concentrated acid sulfuric was needed to synthesis high-quality rGO which is safer and less toxic (Zhu et al., 2022). During

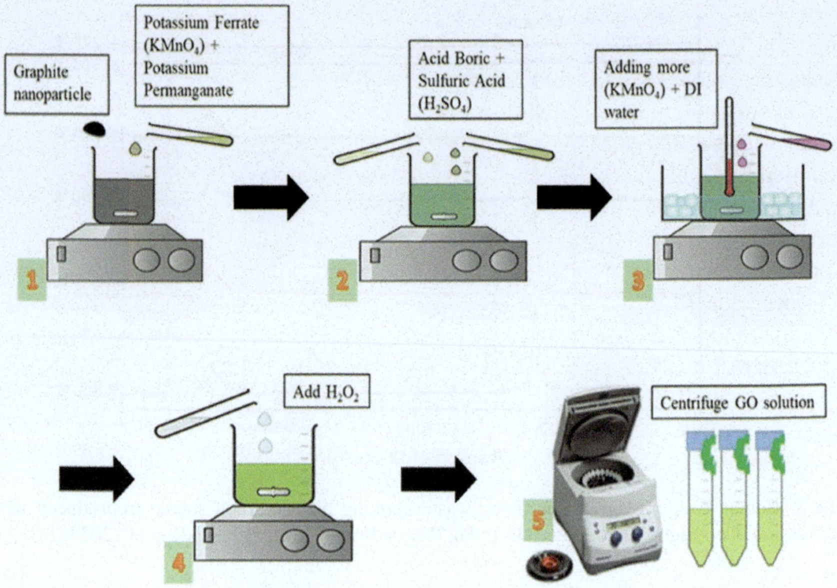

Fig. 4 Schematic of graphene oxide synthesis through Hummer's method

the oxidizing process functional groups like hydroxyl groups (OH) in graphite oxygen were added (Al-Gaashani et al., 2019).

Graphite is the primary precursor for graphene-related synthesis, and it exists in three forms in nature: 70–80% as amorphous, 90–98% as crystalline flakes, and 90–99% as crystalline lumps or veins (Sujiono et al., 2020). Graphite, which is made of several layers of graphene, is a well-known material (Kumar et al., 2021; Omar et al., 2022). Therefore, it must be feasible to reverse this process and generate graphene from graphite. There are many methods for preparing graphene, but one of them is mechanical cleavage of graphite crystals. The high crystalline kish graphite was cleaved using double-sided sticky tapes is split repeatedly until the flake becomes transparent with a thickness between 18–108 nm (Inagaki et al., 2014). The micromechanical cleavage method yields flakes with thicknesses ranging from 3–100 nm and lateral dimensions of around 2 μm. Although optically transparent, the flakes, including monolayer graphene, are visible on the SiO_2/Si substrate (Inagaki et al., 2014). The flakes produced by repeated micromechanical cleavage of graphite are then suspended in a liquid solution and attached to a micrometre-sized metallic scaffold for transmission electron microscopy (TEM) analysis of monolayer graphene (Inagaki et al., 2014). The particle size of graphite plays a significant role as it interferes with the oxidation time. Larger-particle graphite, on the other hand, only causes a partial oxidation of its structure (Zhu et al., 2022).

Graphene-based material produced from coconut shell chemical synthesis shows broad peaks at 2θ 23.18° and 42.98° showing the existence of a reflection (002) plane (Samnur et al., 2021). Moreover, wide peaks occur at 2θ 42.5° representing the

(100) planes (Suarso et al., 2021). From Raman spectroscopy, three dominance peaks were observed indicating carbon nanomaterials which at wavenumbers 1370 cm^{-1}, 1620 cm^{-1} representing D band and G band. The D band was caused by defects in crystal structure during synthesis while G band is the existence of graphite bonds which usually can be observed for all carbon nanomaterials as shown in Fig. 5 (a) as reported by Omar et al., 2022 and Fig. 5 (b) by Suarso et al., 2021 (Fig. 5).

Graphene formed from coconut shell using the hydrothermal method produced monolayer graphene as reported by Putri et al. with I_D/I_G ratio of more than 1 at an optimum calcination temperature of 1000 °C shown in Fig. 6 (Putri et al., 2019). However, the I_D/I_G for CSC itself was a relatively low value which is below 1, which is due to the porous structure that is underdeveloped obtained from Raman spectroscopy analysis (Lee et al., 2021).

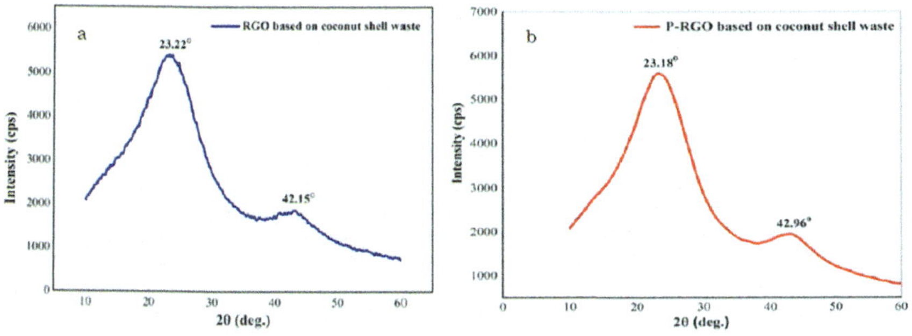

Fig. 5 XRD diffraction pattern for reduce graphene oxide based coconut shell for rGO and porous rGO

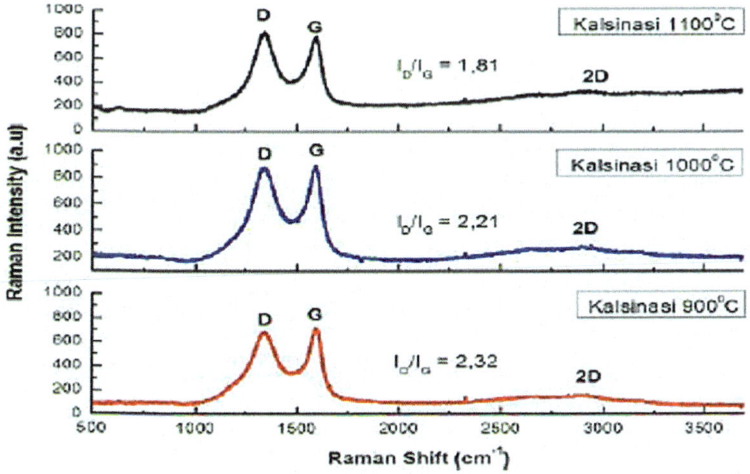

Fig. 6 Raman spectra for rGO based coconut shell calcined at different temperatures (Putri et al., 2019)

3.3 Liquid-Phase Exfoliation

One technique for producing graphene from graphite or graphite oxide, a covalent-type intercalation component of graphene, is liquid-phase exfoliation (LPE) (Inagaki et al., 2014; Li et al., 2019). Overcoming the van der Waals attraction force between the sheets of graphite or graphite oxide is the key step in the LPE process, which results in a monolayer or separated multilayers of graphene (Li et al., 2019). Ultrasonic degradation, mechanical exfoliation, and electrochemical exfoliation are further typical of LPE techniques. The LPE process typically employs natural flake graphite as a raw material (Li et al., 2019). To produce graphene from graphite, the graphite is first grounded into powder and then exposed to mechanical force operations in a liquidous solution. The sonication probe is used to sonicate the powders, which helps separate them into flakes. It is able to classify graphene and graphite layers by size after extracting the flakes and sorting them by thickness.

Electrochemical oxidation of either natural graphite or graphite intercalation compounds (GIC) is the initial step in the graphene production from graphite oxide process if graphite has not already been oxidized (Inagaki et al., 2014). A structural change from graphite oxide to graphene has been said on graphite particles of around 45 μm in size by rapid heating of 1050 °C. The exfoliation process onto graphite oxide is initiated at high temperatures at around 1000 °C, where some functional groups on the graphite oxide's surface are removed and the thin flakes tend to coalesce. Flakes of graphite are typically reduced in thickness by a reduction process after exfoliation. After 48 h of heat treatment at 500 °C in an argon and hydrogen gas combination, a monolayer with a thickness of 0.37 m was obtained (Inagaki et al., 2014).

3.4 Pyrolysis

Pyrolysis is another widely used methodology used in the graphene synthesis process to transform several materials to graphite. The term pyrolysis is derived from the Greek elements 'pyro' and 'lysis,' where 'pyro' means 'fire' and 'lysis' means 'to separate'; hence, pyrolysis means 'to separate in the presence of fire' (Mbayachi et al., 2021). Pyrolysis, in particularly, was used to most waste items in graphene production, such as aerosol, camphor leaves, coconut shells, and plastics, since the end result is generally graphite in the form of powder (Pandey et al., 2019; Sujiono et al., 2020).

According to Pandey et al. (2019), pyrolysis of plastic waste begins by shredding the collected plastics using a shredder and mixing them with bentonite (Pandey et al., 2019). After that, the sample mixture needs to undergo slow pyrolysis at 400 °C in an inert atmosphere of N_2 gas to eliminate all oils and hydrocarbons from the plastics and reduces its tendency to form waxy substances (Lam et al., 2015). Following the slow pyrolysis process, a black-coloured amorphous carbon-type charred residue is produced, which is then put in a ball milling machine to produce ultra-fine powder before undergoing the secondary heating process to optimize graphene nanosheet productivity in the system (Pandey et al., 2019; Rosman et al., 2022).

For synthesizing graphene-like carbon materials, the methods and carbon source discussed in this review serve to be fast, reliable, efficient, and cost-effective. Several industrial wastes, including cooking oil, engine oil, coconut shell, and plastic, can be used as precursors for graphene synthesis. The majority of the products are mono- or few-layered graphitic nanosheets. It is capable of mass-producing cost-effective strutted graphene with large lateral size and thin thickness on a large scale. Choosing the best synthesis routes depends on the materials and the outcome expected to achieve. Transforming from a waste carbon source to a graphene product saves the environment, reduces time-consuming and money-saving. It is strongly advised that industries devote more attention to the uses and applications of graphene and its advantages in the long run. Strongly recommended for researchers and technologists to further increase the flexibility of graphene applications, studies in the enhancement of graphene properties.

Acknowledgements The authors would like to express their gratitude to the Ministry of Higher Education Malaysia, Research Management Centre (RMC), GIP grants (600-RMC/GIP 5/3 (167/ 2021)), and Institute of Science (IOS) Universiti Teknologi MARA (UiTM).

References

Abdullah, H. B., Irmawati, R., Ismail, I., & Yusof, N. A. (2020). Utilization of waste engine oil for carbon nanotube aerogel production using floating catalyst chemical vapor deposition. *Journal of Cleaner Production, 261*(121188). https://doi.org/10.1016/j.jclepro.2020.121188.

Adetayo, A., & Runsewe, D. (2019). Synthesis and fabrication of graphene and graphene oxide: A review. *Open Journal of Composite Materials, 09*(02), 207–229. https://doi.org/10.4236/ojcm.2019.92012

Ahmed, M., Uddin, M. J., Rahman, M. A., Kishi, N., & Soga, T. (2016). Controlled Cu nanoparticle growth on wrinkle affecting deposition of large scale graphene. *Journal of Crystal Growth, 449*, 156–162. https://doi.org/10.1016/j.jcrysgro.2016.06.004

Aïssa, B., Memon, N. K., Ali, A., & Khraisheh, M. K. (2015). Recent progress in the growth and applications of graphene as a smart material: A review. *Frontiers in Materials, 2*(September). https://doi.org/10.3389/fmats.2015.00058.

Alam, K. M., Kumar, P., Manuel, A. P., Vahidzadeh, E., Goswami, A., Zeng, S., Wu, W., Mahdi, N., Cui, K., Kobryn, A. E., Gusarov, S., Song, Y., & Shankar, K. (2020). CVD grown nitrogen doped graphene is an exceptional visible-light driven photocatalyst for surface catalytic reactions. *2D Materials, 7*(1). https://doi.org/10.1088/2053-1583/ab4554.

Alemour, B., Yaacob, M. H., Lim, H. N., & Hassan, M. R. (2018). Review of electrical properties of graphene conductive composites. *International Journal of Nanoelectronics and Materials, 11*(4), 371–398.

Al-Gaashani, R., Najjar, A., Zakaria, Y., Mansour, S., & Atieh, M. A. (2019). XPS and structural studies of high quality graphene oxide and reduced graphene oxide prepared by different chemical oxidation methods. *Ceramics International, 45*(11), 14439–14448. https://doi.org/10.1016/j.ceramint.2019.04.165

Alias, N. I., Javendra Kumar, & Shahrom, M. Z. (2018). Characterization of waste cooking oil for biodiesel production. *Jurnal Kejuruteraan, 30*(2), 79–83. 10.17576.

Alkhouzaam, A., Qiblawey, H., Khraisheh, M., Atieh, M., & Al-Ghouti, M. (2020). Synthesis of graphene oxides particle of high oxidation degree using a modified Hummers method. *Ceramics International, 46*(15), 23997–24007. https://doi.org/10.1016/J.CERAMINT.2020.06.177

Allah, F. U. M., & Alexaandru, G. (2016). Waste cooking oil as source for renewable fuel in Romania. *Materials Science and Engineering, 147*, 1–8. https://doi.org/10.1088/1757-899X/147/1/012133

Amato, G. (2018). High temperature growth of graphene from cobalt volume: Effect on structural properties. *Materials, 11*(2), 1–14. https://doi.org/10.3390/ma11020257

Asli, N. A., Azhar, N. E. A., Nurfazianawatie, M. Z., Yusoff, K. M., Omar, H., Rosman, N. F., Malek, N. S. A., Akhir, R. M., Buniyamin, I., Salifairus, M. J., Husairi, F. S., Khusaimi, Z., Malek, M. F., Rusop, M., & Abdullah, S. (2022). Mechanism of vertically arrays of carbon nanotubes by camphor based catalysed in-situ growth. *Fullerenes Nanotubes and Carbon Nanostructures, 30*(4), 476–486. https://doi.org/10.1080/1536383X.2021.1958317

Ataç, D., Sanderink, J. G. M., Kinge, S., Gravesteijn, D. J., Kovalgin, A. Y., & de Jong, M. P. (2020). Large scale structures in chemical vapor deposition-grown graphene on Ni thin films. *Thin Solid Films, 709*(May), 138225. https://doi.org/10.1016/j.tsf.2020.138225

Aunkor, M. T. H., Mahbubul, I. M., Saidur, R., & Metselaar, H. S. C. (2016). The green reduction of graphene oxide. In *RSC Advances* (Vol. 6, Issue 33, pp. 27807–27825). Royal Society of Chemistry. https://doi.org/10.1039/c6ra03189g

Awogbemi, O., Onuh, E. I., & Inambao, F. L. (2019). Comparative study of properties and fatty acid composition of some neat vegetable oils and waste cooking oils. *International Journal of Low-Carbon Technologies, 14*(July), 417–425. https://doi.org/10.1093/ijlct/ctz038

Azadmanjiri, J., Srivastava, V. K., Kumar, P., Sofer, Z., Min, J., & Gong, J. (2020). Graphene-Supported 2D transition metal dichalcogenide van der Waals heterostructures. *Applied Materials Today, 19*. https://doi.org/10.1016/j.apmt.2020.100600.

Bhoria, R. S. (2019). Enhancing liquid phase exfoliation of graphene in organic solvents with additives. In *Intech*. London, United Kingdom: IntechOpen. https://doi.org/10.5772/intechopen.81462.

Bleu, Y., Barnier, V., Christien, F., Bourquard, F., Loir, A. S., Garrelie, F., & Donnet, C. (2019). Dynamics of carbon diffusion and segregation through nickel catalyst, investigated by in situ XPS, during growth of nitrogen doped graphene. *Carbon, 155*, 410–420. https://doi.org/10.1016/j.carbon.2019.08.084

Cucu, A., Catalin, C., Maria, A., & Trefilov, I. (2015). Graphite oxide post-synthesis processing protocols. *International Journal of Science Communication*, 33–37. https://doi.org/10.13140/RG.2.1.2452.9041.

Das, S., Pandey, D., Thomas, J., & Roy, T. (2019). The Role of graphene and other 2D materials in solar photovoltaics. *Advanced Materials, 31*(1), 1–35. https://doi.org/10.1002/adma.201802722

de Feo, G., di Domenico, A., Ferrara, C., Abate, S., Osseo, L. S., Feo, G. de, Domenico, A. di, Ferrara, C., Abate, S., & Osseo, L. S. (2020). Evolution of waste cooking oil collection in an area with long-standing waste management problems. *Sustainability, 2020, 12,* 8578, *12*(20), 1–16. https://doi.org/10.3390/SU12208578.

Deng, B., Liu, Z., & Peng, H. (2019). Toward mass production of CVD graphene films. *Advanced Materials, 31*(9). https://doi.org/10.1002/adma.201800996.

Deng, J., You, Y., Sahajwalla, V., & Joshi, R. K. (2016). Transforming waste into carbon-based nanomaterials. *Carbon, 96*, 105–115. https://doi.org/10.1016/j.carbon.2015.09.033

Destyorini, F., Irmawati, Y., Hardiansyah, A., Widodo, H., Yahya, I. N. D., Indayaningsih, N., Yudianti, R., Hsu, Y. I., & Uyama, H. (2021). Formation of nanostructured graphitic carbon from coconut waste via low-temperature catalytic graphitisation. *Engineering Science and Technology, an International Journal, 24*(2), 514–523. https://doi.org/10.1016/j.jestch.2020.06.011

Elkady, M. F., Zaatout, A., & Balbaa, O. (2015). Production of biodiesel from waste vegetable oil via KM micromixer. *Journal of Chemistry, 2015*.https://doi.org/10.1155/2015/630168.

Ferrari, A. C., & Basko, D. M. (2013). Raman spectroscopy as a versatile tool for studying the properties of graphene. *Nature Nanotechnology, 8*(4), 235–246. https://doi.org/10.1038/nnano.2013.46

Freitas, N. E., Araújo, M. J., Oliveira, R. L., Lanna, D. P. D., Marques, C. A. T., Torreão, J. N. C., Santos, C. B., Júnior, J. M. S., Edvan, R. L., & Bezerra, L. R. (2020). Production, composition, fatty acid profile and sensory traits of milk from goats fed crude glycerin from waste frying oils used in biodiesel production. *Livestock Science, 238*(April), 104060. https://doi.org/10.1016/j.livsci.2020.104060

Ghosh, A., Razzino, C. do A., Dasgupta, A., Fujisawa, K., Vieira, L. H. S., Subramanian, S., Costa, R. S., Lobo, A. O., Ferreira, O. P., Robinson, J., Terrones, M., Terrones, H., & Viana, B. C. (2019). Structural and electrochemical properties of babassu coconut mesocarp-generated activated carbon and few-layer graphene. *Carbon, 145*, 175–186. https://doi.org/10.1016/j.carbon.2018.12.114.

Gi Choi, M., Park, S., Lee, H., & Kim, S. (2022). Correlating surface structures and nanoscale friction of CVD Multi-Layered graphene. *Applied Surface Science, 584*(October 2021), 152572. https://doi.org/10.1016/j.apsusc.2022.152572.

Guerrero-Contreras, J., & Caballero-Briones, F. (2015). Graphene oxide powders with different oxidation degree, prepared by synthesis variations of the Hummers method. *Materials Chemistry and Physics, 153*, 209–220. https://doi.org/10.1016/j.matchemphys.2015.01.005

He, W., Geng, D., & Xu, Z. (2019). Pattern evolution characterizes the mechanism and efficiency of CVD graphene growth. *Carbon, 141*, 316–322. https://doi.org/10.1016/j.carbon.2018.09.046

Hummers, W. S., & Offeman, R. E. (1958). Preparation of Graphitic Oxide. *Journal of the American Chemical Society, 80*(6), 1339. https://doi.org/10.1021/JA01539A017

Ikram, R., Jan, B. M., & Ahmad, W. (2020). Advances in synthesis of graphene derivatives using industrial wastes precursors; prospects and challenges. *Journal of Materials Research and Technology, 9*(6), 15924–15951. https://doi.org/10.1016/J.JMRT.2020.11.043

Inagaki, M., Kang, F., Toyoda, M. Konno, H. (2014). Graphene: Synthesis and preparation. In *Advanced materials science and engineering of carbon*.https://doi.org/10.1016/B978-0-12-407789-8.00003-X.

Indriati, L., Elyani, N., Dina, S. F., Omar, M. F., Hamid, N. A. A., Maynet, W., Samsudin, E. M., & Nik, N. M. Z. (2019). Increasing yield of graphene synthesis from oil palm empty fruit bunch via two-stages pyrolysis. *Materials Science and Engineering, 543*. https://doi.org/10.1088/1757-899X/543/1/012032.

Jahirul, M. I., Rasul, M. G., Schaller, D., Khan, M. M. K., Hasan, M. M., & Hazrat, M. A. (2022). Transport fuel from waste plastics pyrolysis—A review on technologies, challenges and opportunities. *Energy Conversion and Management, 258*(October 2021), 0–2. https://doi.org/10.1016/j.enconman.2022.115451.

Jalani, D., Rahman, S. F. A., & Hashim, A. M. (2016). Defect-free mixed mono- and bi-layer graphene synthesized from refined palm oil by thermal chemical vapor deposition. *Materials Letters, 182*, 168–172. https://doi.org/10.1016/j.matlet.2016.06.112

Kalanidhi, K., & P. Nagaraaj. (2022). A green approach for synthesis of highly fluorescent carbon dots from waste engine oil: A strategy for waste to value added products. *Diamond and Related Materials, 121*(November 2021), 108724. https://doi.org/10.1016/j.diamond.2021.108724.

Kairi, M. I., Khavarian, M., Bakar, S. A., Vigolo, B., & Mohamed, A. R. (2017). Recent trends in graphene materials synthesized by CVD with various carbon precursors. *Journal of Materials Science, 5*(2). https://doi.org/10.1007/S10853-017-1694-1.

Kausar, A. (2020). Shape memory polyurethane/graphene nanocomposites: Structures, properties, and applications. *Journal of Plastic Film and Sheeting, 36*(2), 151–166. https://doi.org/10.1177/8756087919865296

Kim, T., Lee, J., & Lee, K. H. (2016). Full graphitization of amorphous carbon by microwave heating. *RSC Advances, 6*(29), 24667–24674. https://doi.org/10.1039/c6ra01989g

Klimchitskaya, G. L., Mostepanenko, V. M., & Petrov, V. M. (2014). Reflectivity properties of graphene and graphene-coated substrates. *Lecture Notes in Computer Science (Including Subseries Lecture Notes in Artificial Intelligence and Lecture Notes in Bioinformatics), 8638 LNCS*, 451–458. https://doi.org/10.1007/978-3-319-10353-2_40.

Ko, Y. G., & Kim, C. H. (2006). Confirmation of heavy metal ions in used lubricating oil from a passenger car using chelating self-assembled monolayer. *Journal of Colloid and Interface Science, 301*(1), 27–31. https://doi.org/10.1016/j.jcis.2006.04.050

Kong, X., Zhu, Y., Lei, H., Wang, C., Zhao, Y., Huo, E., Lin, X., Zhang, Q., Qian, M., Mateo, W., Zou, R., Fang, Z., & Ruan, R. (2020). Synthesis of graphene-like carbon from biomass pyrolysis and its applications. In *Chemical Engineering Journal* (Vol. 399). Elsevier B.V. https://doi.org/10.1016/j.cej.2020.125808.

Koosha, N., Karimi-Sabet, J., Moosavian, M. A., & Amini, Y. (2021). Improvement of synthesized graphene structure through various solvent liquids at low temperatures by chemical vapor deposition method. *Materials Science and Engineering B: Solid-State Materials for Advanced Technology, 274*. https://doi.org/10.1016/j.mseb.2021.115458.

Kudus, M. H. A., Zakaria, M. R., Akil, H. M., Ullah, F., & Javed, F. (2020). Oxidation of graphene via a simplified Hummers' method for graphene-diamine colloid production. *Journal of King Saud University - Science, 32*(1), 910–913. https://doi.org/10.1016/j.jksus.2019.05.002

Kumar, S., McEvoy, N., Kim, H. Y., Lee, K., Peltekis, N., Rezvani, E., Nolan, H., Weidlich, A., Daly, R., & Duesberg, G. S. (2011). CVD growth and processing of graphene for electronic applications. *Physica Status Solidi (B) Basic Research, 248*(11), 2604–2608. https://doi.org/10.1002/pssb.201100179.

Kumar, V., Kumar, A., Lee, D. J., & Park, S. S. (2021). Estimation of number of graphene layers using different methods: A focused review. *Materials, 14*(16). https://doi.org/10.3390/ma14164590.

Lam, S. S., Liew, R. K., Cheng, C. K., & Chase, H. A. (2015). Catalytic microwave pyrolysis of waste engine oil using metallic pyrolysis char. *Applied Catalysis b: Environmental, 176–177*(1), 601–617. https://doi.org/10.1016/j.apcatb.2015.04.014

Lavin-Lopez, M. P., Valverde, J. L., Ordoñez-Lozoya, S., Paton-Carrero, A., & Romero, A. (2019). Role of inert gas in the CVD-graphene synthesis over polycrystalline nickel foils. *Materials Chemistry and Physics, 222*(October 2018), 173–180. https://doi.org/10.1016/j.matchemphys.2018.09.083.

Lee, K. C., Lim, M. S. W., Hong, Z. Y., Chong, S., Tiong, T. J., Pan, G. T., & Huang, C. M. (2021). Coconut shell-derived activated carbon for high-performance solid-state supercapacitors. *Energies, 14*(15). https://doi.org/10.3390/en14154546.

Li, L., Zhou, M., Jin, L., Liu, L., Mo, Y., Li, X., Mo, Z., Liu, Z., You, S., & Zhu, H. (2019). Research progress of the liquid-phase exfoliation and stable dispersion mechanism and method of graphene. *Frontiers in Materials, 6*(December), 1–16. https://doi.org/10.3389/fmats.2019.00325

Li, Z., Young, R. J., Backes, C., Zhao, W., Zhang, X., Zhukov, A. A., Tillotson, E., Conlan, A. P., Ding, F., Haigh, S. J., Novoselov, K. S., & Coleman, J. N. (2020). Mechanisms of liquid-phase exfoliation for the production of graphene. *ACS Nano, 14*(9), 10976–10985. https://doi.org/10.1021/ACSNANO.0C03916/SUPPL_FILE/NN0C03916_SI_001.PDF

Lisi, N., Dikonimos, T., Buonocore, F., Pittori, M., Mazzaro, R., Rizzoli, R., Marras, S., & Capasso, A. (2017). Contamination-free graphene by chemical vapor deposition in quartz furnaces. *Scientific Reports, 7*(1), 1–11. https://doi.org/10.1038/s41598-017-09811-z

Liu, H., & Liu, Y. (2019). Controlled chemical synthesis in CVD graphene. *Physical Sciences Reviews, 2*(4), 1–28. https://doi.org/10.1515/psr-2016-0107

Liu, N., Luo, F., Wu, H., Liu, Y., Zhang, C., & Chen, J. (2008). One-step ionic-liquid-assisted electrochemical synthesis of ionic-liquid-functionalized graphene sheets directly from graphite. *Advanced Functional Materials, 18*(10), 1518–1525. https://doi.org/10.1002/ADFM.200700797

Liu, X., Shi, L., Jiang, W., Zhang, J., & Huang, L. (2018). Taking full advantage of KMnO4 in simplified Hummers method: A green and one pot process for the fabrication of alpha MnO2 nanorods on graphene oxide. *Chemical Engineering Science, 192*, 414–421. https://doi.org/10.1016/j.ces.2018.07.044

Mat Tahir, N. A., Abdollah, M. F. bin, Tamaldin, N., Amiruddin, H., Mohamad Zin, M. R. bin, & Liza, S. (2020). Tribological performance of the graphene synthesized from fruit cover plastic waste and oil palm fiber using a CVD method. *Industrial Lubrication and Tribology, 72*(6), 771–777. https://doi.org/10.1108/ILT-11-2019-0486.

Mbayachi, V. B., Ndayiragije, E., Sammani, T., Taj, S., Mbuta, E. R., & Khan, A. ullah. (2021). Graphene synthesis, characterization and its applications: A review. *Results in Chemistry, 3*. https://doi.org/10.1016/j.rechem.2021.100163.

Milenov, T. I., Valcheva, E., & Popov, V. N. (2017). Raman spectroscopic study of As-deposited and exfoliated defected graphene grown on (001) Si substrates by CVD. *Journal of Spectroscopy, 2017*. https://doi.org/10.1155/2017/3495432.

Mohadesi, M., Aghel, B., Maleki, M., & Ansari, A. (2019). Production of biodiesel from waste cooking oil using a homogeneous catalyst: Study of semi-industrial pilot of microreactor. *Renewable Energy, 136*, 677–682. https://doi.org/10.1016/j.renene.2019.01.039

Nilsson, J., Castro Neto, A. H., Guinea, F., & Peres, N. M. R. (2008). Electronic properties of bilayer and multilayer graphene. *Physical Review B - Condensed Matter and Materials Physics, 78*(4), 1–34. https://doi.org/10.1103/PhysRevB.78.045405

Nita, C., Zhang, B., Dentzer, J., & Matei Ghimbeu, C. (2021). Hard carbon derived from coconut shells, walnut shells, and corn silk biomass waste exhibiting high capacity for Na-ion batteries. *Journal of Energy Chemistry, 58*, 207–218. https://doi.org/10.1016/j.jechem.2020.08.065

Nurfazianawatie, M. Z., Afaah, A. N., Salifairus, M. J., Malek, M. F., Mahat, M. M., Rusop, M., & N. A. Asli. (2020). Morphological and optical properties of graphene film synthesized from waste of industrial cooking oil, AYAMAS using DTCVD method. *Science Letters, 14*(2), 24. https://doi.org/10.24191/sl.v14i2.9540.

Nurfazianawatie, M. Z., Omar, H., Rosman, N. F., S. A. Malek, N., Afaah, A. N., Buniyamin, I., Salifairus, M. J., Malek, M. F., Mahat, M. M., Rusop, M., & Asli, N. A. (2023a). Forming multilayer graphene at different precursor temperature using cooking oil waste from AYAMAS via chemical vapour deposition. *Materials Today: Proceedings*. https://doi.org/10.1016/j.matpr.2022.10.169.

Nurfazianawatie, M.Z., Omar, H., Rosman, N.F., Malek, N.S.A., Afaah, A.N., Buniyamin, I., Salifairus, M.J., Malek, M.F., Mahat, M.M., Rusop, M., Asli, N.A. (2023b). Cooking oil waste from AYAMAS as a carbon source in forming multilayer graphene films. *Energy Source Part A-Recovery Utilization and Environmental Effects*, 45, 11198–11206. https://doi.org/10.1080/15567036.2023.2252372

Omar, H., Afaah, A. N., Khusaimi, Z., Rusop, M., Abdullah, S., Asli, N.A. (2020). Structural properties of graphene coated on silicon nanowires by mechanical exfoliation (2020). *International Journal of Electroactive Materials*, 8, 1–9.

Omar, H., Abdul Malek, N. S., Mohd Zain, N., Khusaimi, Z., Abdullah, S., Rusop, M., & Asli, N. A. (2022). Potential of microparticles graphitize coconut shell charcoal with low ball milling time. *Lecture Notes in Mechanical Engineering*, 407–414. https://doi.org/10.1007/978-981-19-2890-1_39/COVER.

Pandey, S., Karakoti, M., Dhali, S., Karki, N., SanthiBhushan, B., Tewari, C., Rana, S., Srivastava, A., Melkani, A. B., & Sahoo, N. G. (2019). Bulk synthesis of graphene nanosheets from plastic waste: An invincible method of solid waste management for better tomorrow. *Waste Management, 88*, 48–55. https://doi.org/10.1016/j.wasman.2019.03.023

Patel, C., Chandra, K., Hwang, J., Agarwal, R. A., Gupta, N., Bae, C., Gupta, T., & Agarwal, A. K. (2019). Comparative compression ignition engine performance, combustion, and emission characteristics, and trace metals in particulates from Waste cooking oil, Jatropha and Karanja oil derived biodiesels. *Fuel, 236*(August 2018), 1366–1376. https://doi.org/10.1016/j.fuel.2018.08.137.

Pedrazzetti, L., Gibertini, E., Bizzoni, F., Russo, V., Lucotti, A., Nobili, L., & Magagnin, L. (2022). Graphene Growth on Electroformed Copper Substrates by Atmospheric Pressure CVD. *Materials, 15*(4). https://doi.org/10.3390/ma15041572.

Poniatowska, A., Trzaskowski, M., & Ciach, T. (2019). Production and properties of top-down and bottom-up graphene oxide. *Colloids and Surfaces A, 561*, 315–324. https://doi.org/10.1016/j.colsurfa.2018.10.049

Putri, N. P., Kusumawati, D. H., Agustina, L., & Munasir. (2019). Effect of calcination temperature on characteristics of reduced Graphene Oxide (rGO) made from old coconut shell. *Journal of Physics: Conference Series, 1171*(1). https://doi.org/10.1088/1742-6596/1171/1/012042

Ramadhansyah, P. J., Masri, K. A., Wan Azahar, W. N. A., Mashros, N., Norhidayah, A. H., Mohd Warid, M. N., Satar, M. K. I. M., & Yaacob, H. (2020). Waste Cooking Oil as Bio Asphalt Binder: A Critical Review. *IOP Conference Series: Materials Science and Engineering, 712*(1). https://doi.org/10.1088/1757-899X/712/1/012040

Rosman, N. F., Nurfazianawatie, M. Z., Omar, H., Malek, N. S. A., Buniyamin, I., Salifairus, M. J., Malek, M. F., Rusop, M., & Asli, N. A. (2022). Reducing okara size as novel carbon source graphene. *Materials Today: Proceedings*. https://doi.org/10.1016/j.matpr.2022.10.148.

Rubianto, L., Santosa, S., Sudarminto, H. P., & Udjiana, S. S. (2020). Glycerol from waste frying oil as sustainable fuel. *IOP Conference Series: Materials Science and Engineering, 732*(1). https://doi.org/10.1088/1757-899X/732/1/012066.

Saeed, M., Alshammari, Y., & Majeed, S. A. (2020). Chemical vapour deposition of graphene—Synthesis, characterisation, and applications: A review. *Molecules, 25*(17), 2–62. https://doi.org/10.3390/molecules25173856

Salifairus, M. J., Hamid, S. B. A., Alrokayan, S. A. H., Khan, H. A., & Rusop, M. (2016). The effect of synthesis time on graphene growth from palm oil as green carbon precursor. *AIP Conference Proceedings, 1733*. https://doi.org/10.1063/1.4948884.

Salifairus, M. J., Hamid, S. B. A., Soga, T., Alrokayan, S. A. H., Khan, H. A., & Rusop, M. (2015). Surface topography of synthesized graphene from green carbon source using thermal chemical vapor deposition. In *2015 IEEE Student Conference on Research and Development, SCOReD 2015, December*, pp 522–526. https://doi.org/10.1109/SCORED.2015.7449391.

Samnur, S., Azizah, N., Fahira, N., Zurnansyah, Zabrian, D., Zharvan, V., & Sujiono, E. H. (2021). Comparative Study and characterization of reduced graphene oxide (RGO) and porous reduced graphene oxide (P-RGO) based on coconut shell waste. *Journal of Nano- and Electronic Physics, 13*(6), 1–6. https://doi.org/10.21272/JNEP.13(6).06017.

Saulnier, M., Trudeau, C., Cloutier, S. G., & Schougaard, S. B. (2017). Investigation of CVD multilayered graphene as negative electrode for lithium-ion batteries. *Electrochimica Acta, 244*, 54–60. https://doi.org/10.1016/j.electacta.2017.04.101

Selvakumar, N., Pradhan, U., Krupanidhi, S. B., & Barshilia, H. C. (2016). Structural and optical properties of graphene oxide prepared by modified Hummers' method. *52*(3), 173–184. https://doi.org/10.3970/cmc.2016.052.173.

Seo, D. H., Pineda, S., Fang, J., Gozukara, Y., Yick, S., Bendavid, A., Lam, S. K. H., Murdock, A. T., Murphy, A. B., Han, Z. J., & Ostrikov, K. K. (2017). Single-step ambient-air synthesis of graphene from renewable precursors as electrochemical genosensor. *Nature Communications, 8*. https://doi.org/10.1038/ncomms14217.

Shams, S. S., Zhang, R., & Zhu, J. (2015). Graphene synthesis: A review. *Materials Science-Poland, 33*(3), 566–578. https://doi.org/10.1515/msp-2015-0079

Somashekhar, T. M., Naik, P., Nayak, V., Mallikappa, & Rahul, S. (2018). Study of mechanical properties of coconut shell powder and tamarind shell powder reinforced with epoxy composites. *IOP Conference Series: Materials Science and Engineering, 376*. https://doi.org/10.1088/1757-899X/376/1/012105.

Suarso, E., Setyawan, F. A., Subhan, A., Ramli, M. M., Ismail, N. S., Zainuri, M., Arin, Z., & Darminto, D. (2021). Enhancement of LiFePO4 (LFP) electrochemical performance through the insertion of coconut shell derived rGO-like carbon as cathode of Li-ion battery. *Journal of*

Materials Science: Materials in Electronics, 32(24), 28297–28306. https://doi.org/10.21203/rs.3.rs-298403/v1.

Sujiono, E. H., Zurnansyah, Zabrian, D., Dahlan, M. Y., Amin, B. D., Samnur, & Agus, J. (2020). Graphene oxide based coconut shell waste: Synthesis by modified Hummers method and characterization. *Heliyon, 6*.https://doi.org/10.1016/j.heliyon.2020.e04568.

Syrmanova, K. K., Kovaleva, A. Y., Kaldybekova, Z. B., Botabayev, N. Y., Botashev, Y. T., & Beloborodov, B. Y. (2017). Chemistry and Recycling Technology of Used Motor Oil. *Oriental Journal of Chemistry, 33*(6), 3195–3199. https://doi.org/10.13005/ojc/330665

Ting, T. L., Putra Jaya, R., Hassan, N. A., Yaacob, H., & Jayanti, S. (2015). A Review of Utilization of Coconut Shell And Coconut Fiber In Road Construction. *Jurnal Teknologi (Sciences & Engineering), 76*(14), 2180–3722. https://doi.org/10.11113/jt.v76.5851

Tsai, W.-T. (2019). Mandatory Recycling of Waste Cooking Oil from Residential and Commercial Sectors in Taiwan. *Resources, 8*(38). https://doi.org/10.3390/resources8010038

Ullah, Z., Bustam, M. A., & Man, Z. (2014). Characterization of waste palm cooking oil for biodiesel production. *International Journal of Chemical Engineering and Applications, 5*(2), 134–137. https://doi.org/10.7763/ijcea.2014.v5.366

Wang, X., You, H., Liu, F., Li, M., Wan, L., Li, S., Li, Q., Xu, Y., Tian, R., Yu, Z., Xiang, D., & Cheng, J. (2009). Large-scale synthesis of few-layered graphene using CVD. *Chemical Vapor Deposition, 15*(1–3), 53–56. https://doi.org/10.1002/cvde.200806737

Wu, J. B., Lin, M. L., Cong, X., Liu, H. N., & Tan, P. H. (2018). Raman spectroscopy of graphene-based materials and its applications in related devices. *Chemical Society Reviews, 47*(5), 1822–1873. https://doi.org/10.1039/c6cs00915h.

Yacob, M. R., Kabir, I., & Radam, A. (2015). Households willingness to accept collection and recycling of waste cooking oil for biodiesel input in Petaling District, Selangor, Malaysia. *Procedia Environmental Sciences, 30*, 332–337. https://doi.org/10.1016/j.proenv.2015.10.059

Yan, Md. S. A., Uddin, Md. N., Islam, Md. M., Bipasha, F. A., & Hossain, S. S. (2016). Synthesis of graphene. *International Nano Letters, 6*(2), 65–83. https://doi.org/10.1007/s40089-015-0176-1

Yesilbag, Y. O., Tuzluca Yesilbag, F. N., Huseyin, A., Tuzluca, M., Ismail, I., & Ertugrul, M. (2021). Synthesis and characterization of graphene/carbon nanotube hybrid: effects of Ni catalyst thickness and H2 flow rate on growth and morphological structure. *Journal of Materials Science: Materials in Electronics, 32*(6), 7943–7955. https://doi.org/10.1007/s10854-021-05519-z.

Yoo, M. J., & Park, H. B. (2019). Effect of hydrogen peroxide on properties of graphene oxide in Hummers method. *Carbon, 141*, 515–522. https://doi.org/10.1016/j.carbon.2018.10.009

Zaaba, N. I., Foo, K. L., Hashim, U., Tan, S. J., Liu, W. W., & Voon, C. H. (2017). Synthesis of graphene oxide using modified hummers method: Solvent influence. *Procedia Engineering, 184*, 469–477. https://doi.org/10.1016/J.PROENG.2017.04.118

Zakaria, N. Z. J., Rozali, S., Mubarak, N. M., & Ibrahim, S. (2022). A review of the recent trend in the synthesis of carbon nanomaterials derived from oil palm by-product materials. *Biomass Conversion and Biorefinery*. https://doi.org/10.1007/s13399-022-02430-3

Zhu, Y., Kong, G., Pan, Y., Liu, L., Yang, B., Zhang, S., Lai, D., & Che, C. (2022). An improved Hummers method to synthesize graphene oxide using much less concentrated sulfuric acid. *Chinese Chemical Letters, 33*(10), 4541–4544. https://doi.org/10.1016/j.cclet.2022.01.060

Zik, N. A. F. A., Sulaiman, S., & Jamal, P. (2020). Biodiesel production from waste cooking oil using calcium oxide/nanocrystal cellulose/polyvinyl alcohol catalyst in a packed bed reactor. *Renewable Energy, 155*, 267–277. https://doi.org/10.1016/j.renene.2020.03.144

Influence of Waste Paper Sludge Ash on Mechanical and Durability Properties of Self-consolidating Lightweight Foamed Concrete

Mohd Afiq Mohd Fauzi, Muhd Norhasri Muhd Sidek, Aidan Newman, Nurliza Jasmi, Muhamad Syahmi Norizan, and Muhammad Amirul Razin Roslan

Abstract Self-Consolidating Lightweight Foamed Concrete (SCLFC) is a combination of two technology of concrete which is Self-Consolidating Concrete (SCC) and Lightweight Foamed Concrete (LFC). This SCLFC offers great potential for improved ease of positioning and scaled-down cost through reduced time and labour. Generally, a large amount of waste or paper sludge is generated every day from the paper industry and at the same time brings potential and devastating harmful impacts to the surroundings. For this reason, the search into the use of waste paper sludge ash (WPSA) in the SCLFC is looked into. This research aims to study the effect of the addition of WPSA concerning workability, strength, ultrasonic pulse velocity, porosity, and water absorption characteristics of SCLFC. The effect of the addition of WPSA in the range from 0 to 15% on the workability, strength physical, and absorption characteristics of SCLFC was established and compared to the normal SCLFC. The concrete specimens were subjected to strength at the age of 3, 7, and 28 days. The curing for this research is ambient conditions. The result indicated that the WPSA tends to be adding to the beneficial workability, strength, ultrasonic pulse velocity, porosity, and absorption characteristics of SCLFC.

Keywords Waste paper sludge ash (WPSA) · Self-consolidating lightweight foamed concrete · Slump flow · Mechanical · Durability

M. A. Mohd Fauzi (✉) · M. N. Muhd Sidek · A. Newman · M. S. Norizan · M. A. R. Roslan
School of Civil Engineering, College of Engineering, Universiti Teknologi MARA (UiTM), 40450 Shah Alam, Selangor, Malaysia
e-mail: mohdafiq3474@uitm.edu.my

M. N. Muhd Sidek
e-mail: muhdnorhasri@uitm.edu.my

M. A. Mohd Fauzi · M. N. Muhd Sidek · N. Jasmi
Institute for Infrastructure Engineering and Sustainable Management (IIESM), Universiti Teknologi MARA, 40450 Shah Alam, Selangor, Malaysia
e-mail: nurlizajasmi@uitm.edu.my

U. R. Hashim et al. (eds.), *Green Infrastructure*,
https://doi.org/10.1007/978-981-99-7003-2_13

1 Introduction

In this growing area, there had been so many innovations and creations that had been made. The construction industry had become advanced through this modernization. At present time, concrete is one of the most used materials in the world. This is because concrete is extensively by means in the development of an economy. Concrete is a combination of cement, coarse aggregate, fine aggregate, and water. In the building industry, factors affecting the selection of concrete to be used in the construction industry should always desire workability, strength, and durability to keep away from the problem during construction or post-construction. In order to recognize the properties of the concrete, numbers of concrete blocks need to be produced to be tested.

Self-Consolidating Concrete (SCC) consists of cement, water, fine aggregates, coarse aggregates, additional mineral, and chemical admixtures. An example of mineral admixtures is fly ash which is the waste from industrial and is commonly used in the construction industry. Meanwhile, the chemical admixture is silica fume and acts as a superplasticizer (Siti et al., 2016). The factor required to produce SCC is limited aggregate content, low water-powder ratio, and use of superplasticizer. The ratio of aggregates for SCC is 40% for coarse aggregates and 60% for fine aggregates (Okamura & Ouchi, 2003). Superplasticizer was usually used to increase the workability of the concrete and to produce a low water-powder ratio (Kushwaha et al., 2013).

Self-Consolidating Concrete (SCC) required mineral admixtures such as fly ash and it must contain the same chemical composition as cement to allow the hydration process to occur. The main function of cement is to bind fine and coarse aggregates into a solid mass. The production of cement required a large amount of limestone which non-renewable resource and is called Ordinary Portland cement. The other examples of industrial wastes that can act as mineral admixtures are Palm Oil Fuel Ash (POFA), rice husk ash, bamboo ash, sugar cane ash, Waste Paper Sludge Ash (WPSA), and others. In addition, the usage of admixture can improve the properties of the concrete. Admixture can be from by-products or waste material from industry such as Waste Paper Sludge Ash (WPSA).

There was some research stating that the Waste Paper Sludge Ash (WPSA) contains alumina and silica material called pozzolanic material and can be used in concrete. The utilization of WPSA in the concrete can improve the hardened properties of the concrete which is compressive strength. With the concern of environmental issues, WPSA can be added to the concrete as an additive material to increase the compressive strength and can be used as a cement replacement material in the concrete (Sharipudin & Ridzuan, 2013). The usage of WPSA can give an impact on the environment which is producing greener concrete and a friendly environment.

Foamed concrete was created to produce a lightweight material for building. Foamed concrete is produced when a large volume of air is present in the cement paste which is called a chemical foaming agent. Foamed concrete can be available either in fresh foam concrete mix or precast foam concrete (Muthusamy et al., 2017). Due to the lower density of concrete, precast foam concrete is easy to handle, and high-quality the building can be produced. The chemical foaming agent is diluted with water and bubbles are created. The density of foamed concrete can be created from 600 kg/m^3 to 1600 kg/m^3 (Elrahman et al., 2019). According to Sharipudin and Ridzuan, (2013), the dead load of the building can be reduced when using the formed concrete.

Generally, proper utilization of WPSA can produce a new product of concrete. The combination of self-consolidating concrete (SCC) and foamed concrete forms a new product called Self-Consolidating Lightweight Foamed Concrete (SCLFC). This innovation can reduce the cost of labour during construction and environmental issues.

2 Materials and Method

2.1 Materials Selection

Waste paper sludge ash (WPSA) was the waste material from the paper industry. In this study, waste paper sludge ash (WPSA) was used as additional material in the concrete with several percentages which were 0, 5, 10, and 15%. Waste paper sludge ash (WPSA) that passed a 0.212 mm sieve was used. The addition of waste paper sludge ash (WPSA) was used due to determine the workability of the fresh concrete and the strength of the concrete.

Cement can be described as a powdery substance that consists of a lot of chemical compositions such as alumina, silica, iron oxide, lime, and other chemicals. Cement was the main material in concrete because the cement was acted as a binder in the concrete. Cement bound all the materials together when the cement reacted with water and solid mass concrete was produced. In this study, Ordinary Portland Cement (OPC) was used because OPC was suitable for most types of work and widely used in the world.

In this study, crushed coarse aggregates with a size of less than 10 mm were used due to produce a well-graded mix, and the aggregates were sieved using a sieving machine. According to EFNARC (2002), 40% of coarse aggregates from total aggregates were required due to an increase in the workability of the concrete. The size of the coarse aggregates was one of the factors that can affect the workability of fresh concrete. Coarse aggregates are very important in resisting the compression load. The size of the aggregates was based on the requirements defined in BS 933-1:2012.

Fine aggregates were required to fill the voids between the coarse aggregates and river sand was used in this study as a fine aggregate. Some factors that can affect the workability of fresh concrete which was the moisture content of the river sand and the size of the sand. In this study, river sand in dry conditions with a size of less than

5 mm was used. According to EFNARC (2002), 60% of coarse aggregates from total aggregates were required due to an increase in the workability of the concrete. The size of the sand had been based on the requirement in BS 933-1:2012 specification.

Tap water was used to mix the concrete in this study. The presence of water in the concrete mix allowed the hydration process to occur. The tap water must be clean of any impurities such as dust and oil. The presence of impurities in the water can affect the strength of the concrete and the quality of the water was based on BS EN 1008:2002.

Superplasticizer was used in this study to increase the workability of the fresh concrete due to produce self-consolidating concrete. Polycarboxylic Ether (PCE), the brand of Master Glenium ACE 8589 from BASF Chemical Sdn Bhd was the superplasticizer used in this study and the ratio used is 1L of superplasticizer for every 100 kg of cement. The correct ratio was needed to avoid any effect on the concrete such as segregation and bleeding effect.

A synthetic foaming agent was used for this study to produce lightweight concrete. The concrete density of 600–1800 kg/m^3 can be produced by using the foaming agent. The foaming agent was diluted with a ratio of a 1-L foaming agent with 20 L of water and stirred using a mixer until the bubble was produced.

Silica fume was known as micro silica and it is a by-product that is the production of silicon and ferrosilicon alloy. The particle size of the silica fume was less than 1 micron and about 100 times smaller than the average particle size of cement. The hydration process happens when silica fume reacts with calcium hydroxide and calcium silicate was produced. From this reaction, the strength of concrete increases, and the permeability of concrete decreases. In this study, 10% of silica fume from the amount of cement was used.

2.2 *Mix Proportion*

Self-consolidating concrete (SCC) was designed by using the DOE method and lightweight foamed concrete (LFC) was designed by using the in-house developed spreadsheet. Meanwhile, self-consolidating lightweight foamed concrete (SCLFC) was designed by using both methods. The DOE method is the method based on the British method Department of Environment. The DOE method is used to determine the weight of water content, cement content, density, coarse aggregate, fine aggregate, and water-cement ratio. Next, data from the DOE method are used for the determination of foaming agents by using the in-house developed spreadsheet. Then, the new mix proportion of self-consolidating lightweight foamed concrete (SCLFC) was produced. For waste paper sludge ash (WPSA), the amount of WPSA is determined by multiplying the percentage of WPSA with the amount of cement from the mix design that is calculated (Table 1).

Table 1 Mix proportion of SCLFC

Batch no.	Mix proportion								
	Cement	WPSA		SF	Water	FA	CA	Sp	Foaming Agent
	kg/m^3	%	kg/m^3	kg/m^3	kg/m^3	kg/m^3	kg/m^3	kg/m^3	kg/m^3
M1	509	0	–	50.9	180	666	444	5.09	234
M2	509	5	25.4	50.9	180	666	444	5.09	234
M3	509	10	50.9	50.9	180	666	444	5.09	234
M4	509	15	76.35	50.9	180	666	444	5.09	234

Note
WPSA = Waste Paper Sludge Ash
SF = Silica Fume
FA = Fine Aggregate
CA = Coarse Aggregate
Sp = Superplasticizer

2.3 *Casting and Curing*

In this study, the mixing process had been done by using a concrete pan mixer machine due to minimize the error during mixing and to produce good concrete. All the materials were measured and weighed approximately to the mix design that was calculated. Before the mixing process started, the mixer needs to be wiped with a wet cloth, and all the materials were added in sequence. The process of mixing was carried out according to BS 1881: Part 125:2013. Firstly, coarse aggregate and fine aggregate were added to the mixer for a while. Then cement, silica fume, and waste paper sludge ash (WPSA) were gradually added into the mixer and mixed for 2 min. Then, water and superplasticizer were added to the mixture. Lastly, the foam was added to the mixture and mixed for a while. The casting process was a process where the concrete mix was poured into the mould when the concrete was well blended.

Then the hardened concretes were demoulded after 24 h of casting and the samples were cured by using the ambient method for 3, 7, and 28 days before the compression test.

2.4 *Test Methods*

The concrete mixes were having the slump flow test after the mixing process. The slump flow test was to determine the ability of SCLFC to assess horizontal free flow. Firstly, the base plate is wiped with a wet cloth and placed on level of stable ground. Then, the slump cone was held tightly on the base plate and then filled with concrete mix. The cone was lifted vertically and allowed the concrete to flow freely.

The diameter of the concrete flow was measured in two directions using a measuring tape. The SCLFC must be in the range of 550–800 mm (BS EN 206-9:2010).

The mechanical properties testing for these hardened samples were a compression test. The compressive strength of the self-consolidating lightweight foamed concrete (SCLFC) specimens of size (100 mm x 100 mm x 100 mm) was tested at age of 3, 7, and 28 days. The dimensions of the specimens had been recorded before the testing. The specimens were dried before being tested. The load was applied on the smooth sides without shock and increased continuously until the failure of the specimen. This compressive strength tested was as follows to BS EN 12390-3:2009.

One of the hardened concrete properties tests that has been conducted is Ultrasonic Pulse Velocity (UPV) test. After the curing process, the specimens undergo a UPV test by using an Ultrasonic Pulse Velocity machine.

The durability of concrete includes the total volume of void in the concrete. Therefore, a porosity test has been conducted as in the RILEM porosity test of concrete. After the curing days, the samples were dried in the oven for 24 h. The weight of concrete as oven dried was recorded. The samples then were immersed in water to be weighed and the reading was recorded. Moisture on the surface of the concrete then was removed to take the weight of the Saturated Surface Dried (SSD) of the concrete.

After the curing process had reached days 3, 7, and 28, the specimens were undergoing a water absorption test as in the RILEM water absorption test. The specimens were placed in the oven at a temperature of 105 °C for 24 h. Then, it was taken out to cool down to room temperature and weight. After the weight of the specimens had been taken, it was placed in the vacuum desiccator for an hour to make sure the specimen was totally dry. The next step was pouring water into the vacuum desiccator until the specimens were submerged. Left it for an hour with a vacuum pressure of 30–40 kPa. Turn off the vacuum after one hour and left the specimens submerged in the water for 24 h. Lastly, the specimen was surface-dried and weighted.

3 Result and Discussion

3.1 Slump Flow Test

In order to be categorized as a self-consolidating lightweight foamed concrete (SCLFC), the concrete has to undergo the slump flow test required by EFNARC. Therefore, for this project, a slump flow test is required to be conducted to test the workability of the SCLFC containing the addition of WPSA. This test has been conducted for every batch of specimens. The results of the slump flow test are shown in Fig. 1.

Figure 1 presents the results of workability for (SCLFC), conducted to achieve self-consolidating concrete. The slump flow diameter must be in the range of 500

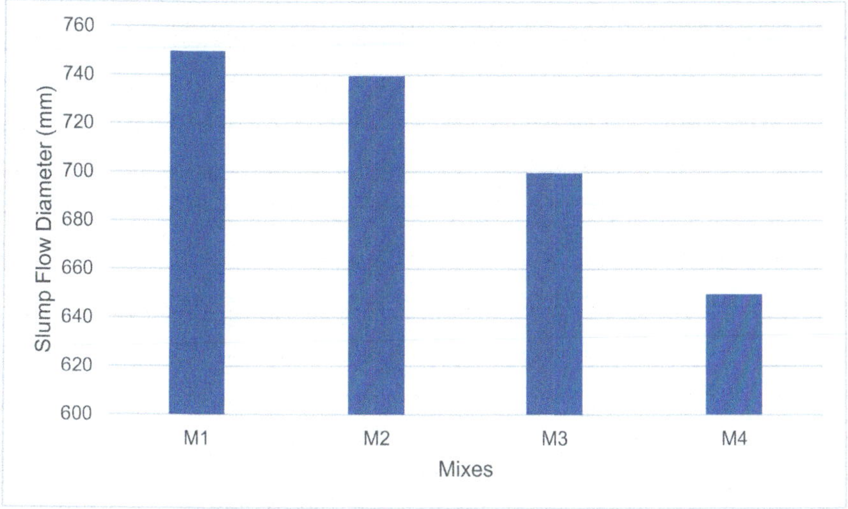

Fig. 1 Slump flow diameter of SCLFC with the addition of WPSA

to 700 mm for self-consolidating concrete (Sharipudin & Ridzuan, 2013). From the graph, it shows the average slump flow diameter for normal SCLFC was 700 mm, for 5% addition of WPSA was 690 mm, for 10% addition of WPSA was 650 mm, and for 15% addition of WPSA was 600 mm. The graph showed the highest slump was observed for the control concrete sample. In addition, the graph showed that the average slump flow diameters for all samples were in the range of 500 to 700 mm and thus all the samples were self-consolidating concrete. The graph showed the average slump flow diameter of SCLFC was decreased when the amount of WPSA was increased. This trend was similar to the other authors who found that the workability gradually decreased as the amount of WPSA increased in conventional concrete (Hamzah et al., 2015). Furthermore, the particles of waste paper sludge ash absorbed more water as compared to cement and thus the workability of the concrete mix was decreased (Sajjad Ahmad et al., 2013). In addition, the reduction of slump flow diameter of the samples has resulted from the lower density of waste paper sludge ash that gives a higher porosity and thus results in higher water demand (Bikila Meko Kejela, 2020).

With all the results obtained from the slump flow test, the concrete flow diameter has ranged from 600 to 750 mm. This value is still in accordance with the EFNARC guideline. The requirement from the EFNARC is as stated in Table 2.

Based on the standard from the EFNARC, all the SCLFC products are categorized in SF1 and SF2. The specimens passed the workability test required by the EFNARC to be classified as SCC. Thus, this concrete product is suitable to be used by the industry. The presence of superplasticizers to enhance the workability of the specimens without reducing the water content helps the specimens maintain their workability. From this test, it can be concluded that the workability of this SCLFC is

Table 2 Classification of self-consolidating concrete

Category	Range (mm)	Applications
SF1	550–650	Suitable for unreinforced or slightly reinforced concrete structures that are cast from the top with free displacement from the delivery point. Casting by pump injection system (e.g., tunnel linings) and for sections that are small enough to prevent long horizontal flow (e.g., piles and some deep foundations)
SF2	660–750	Suitable for many normal applications (e.g., walls and columns)
SF3	760–850	Typically produced with a small maximum size of aggregates and used for vertical applications in very congested structures, structures with complex shapes, or for filling under formwork

not affected too much with the addition of WPSA by the presence of superplasticizers and is still considered an SCC. To be compared with the previous work, the usage of waste materials such as fly ash and rice husk ash reduces the fluidity of the SCC but still maintains satisfactory workability.

3.2 *Compressive Strength*

Hardened concrete in this study was tested for compression test to determine the compressive strength of the SCLFC. The compressive strength with respect to different levels of WPSA contents (0, 5, 10, and 15%) had been compared and discussed. The compression test was performed on all specimens and the results had been plotted.

The graph of the strength development for this study was plotted in Fig. 2. The compressive strength of concrete was one of the most important properties in evaluating the quality of the type of concrete. The graph of strength development showed the trend of the strength of the concrete over a period of time.

Figure 2 shows the graph of strength development on the compression strength for hardened SCLFC in 3, 7, and 28 days. From the graph, curvature showed the strength development for the control SCLFC and SCLFC with the addition of WPSA was increased until 28 days of curing and showed a significantly different trend for each sample of SCLFC. The graph with a blue-coloured line showed the strength development for 0% addition of WPSA (M1) in SCLFC was increasing from the 3rd to 28th day of curing. The concrete strength after curing on day 3 was 8.743 kPa, on day 7 was 10.297 kPa, and on day 28 was 11.403 kPa.

In addition, the strength development for 5% addition of WPSA (M2) showed in orange coloured line was increased from the 3rd to 28th day of curing and the graph curvature was higher than SCLFC with 0% addition of WPSA. The concrete strength after curing on day 3 was 10.635 kPa, on day 7 was 11.38 kPa, and on day 28 was 12.37 kPa.

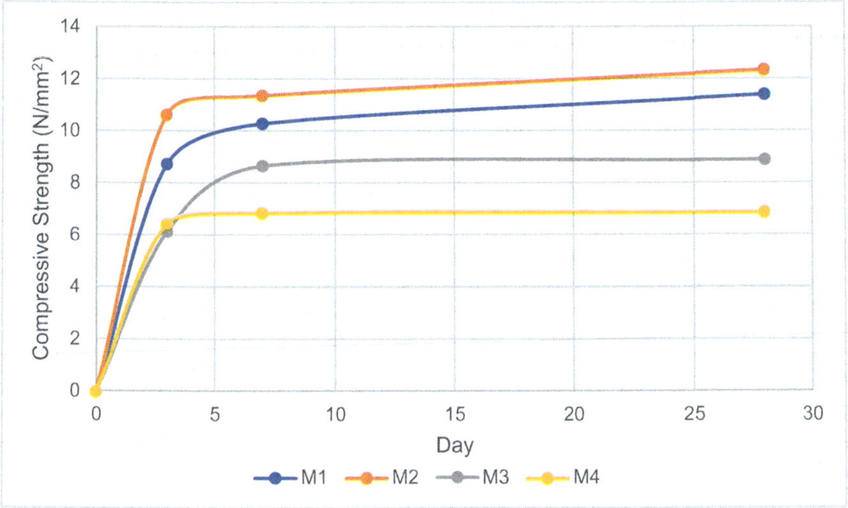

Fig. 2 Strength development of SCLFC containing WPSA

Furthermore, the strength development for the 10% addition of WPSA (M3) showed in grey coloured line was aggressively increased from the 3rd to the 7th day of curing and slowly increased from the 7th to the 28th day of curing. The graph curvature for the 10% addition of WPSA was lower than the 0% addition of WPSA from the 3rd to 28th day of curing. The SCLFC strength after curing on day 3 was 6.115 kPa, on day 7 was 8.645 kPa, and on day 28 was 8.875 kPa.

Meanwhile, the 15% addition of WPSA (M4) shown in the yellow-coloured line was slowly increased for strength development from the 3rd to 28th day of curing. The concrete strength after curing on day 3 was 6.44 kPa, on day 7 was 6.885 kPa, and on day 28 was 6. 915 kPa. The addition of 15% of WPSA (M4) in SCLFC was resulting in the lowest graph curvature for strength development.

On the other hand, when comparing all the data from the graph, the ones with higher strength development than the normal SCLFC were the SCLFC containing a 5% addition of WPSA. The higher compressive strength of the SCLFC with a 5% addition of WPSA may be attributed to the high value of pozzolanic reactivity of WPSA. In addition, the graph showed the strength increased with increasing age which indicated the presence of a pozzolanic reaction (Chowdhury et al., 2014). Furthermore, the SCLFC with a 15% addition of WPSA showed the lowest strength development of the normal SCLFC. This pattern was similar to other authors who found that the compressive strength gradually decreases as the amount of WPSA increases in conventional concrete (Sajjad Ahmad et al., 2013).

3.3 Ultrasonic Pulse Velocity Test

For hardened properties of SCLFC with the addition of WPSA, a non-destructive test has been conducted. The non-destructive test method applied in this study is the ultrasonic pulse velocity (UPV) test. The purpose of this test conducted was to determine the physical properties of the SCLFC in terms of quality. This test was to check the quality of the specimens. The performance of the concrete was observed by the ultrasonic pulse velocity that passes through the concrete. The results recorded from the test conducted are shown in Fig. 3.

Based on the graph in Fig. 3, the results from the UPV test conducted in the laboratory show a pattern of increasing with the day of curing. All specimens with the different percentage addition of WPSA which are 0% (M1), 5% (M2), 10% (M3), and 15% (M4) increased throughout the days. The results indicate that the control sample which is without any addition of WPSA has the lowest value of UPV while the specimens with a 5% addition of WPSA (M2) are better than the control sample. These results showed that the durability of SCLFC can be increased with the addition of WPSA. The UPV value for SCLFC with a percentage addition of 15% of WPSA (M4) is slightly decreased between the 10% addition of WPSA (M3). The UPV value is still increasing throughout the curing days, but it shows a lower value than the 10% addition specimens. All specimens have a high UPV value at 28 days. The highest value of UPV is at 28 days of 10% addition of WPSA. It has a value of 3820 m/s compared to 0%, 5% and 15% specimens which are 3680 m/s, 3740 m/s and 3700 m/s respectively. The grade of the specimens can be classified as a medium after 3 days of curing and improve to good quality concrete after 7 days of curing except for the control sample which is in medium class with a value of 3470 m/s.

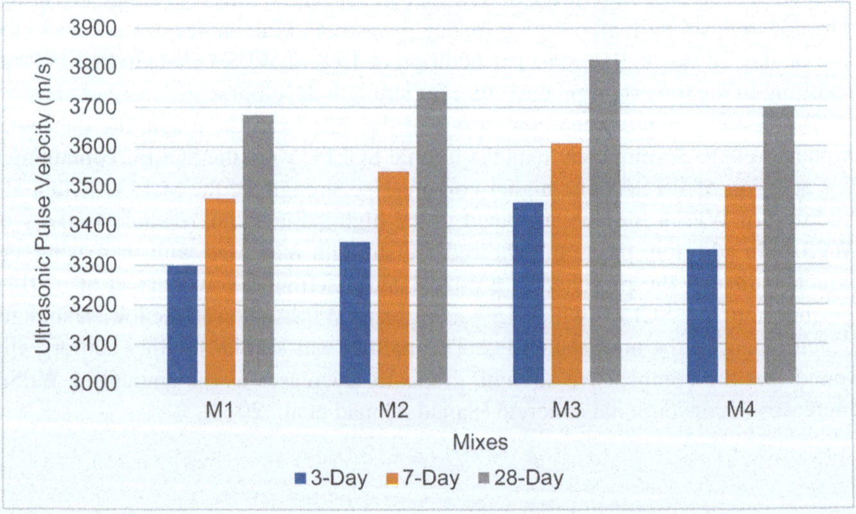

Fig. 3 Result of UPV test with different percentage addition of WPSA

Table 3 Classification of concrete

Ultrasonic pulse velocity (km/s)	Quality of concrete
Above 4.5	Excellent
3.5 to 4.5	Good
3.0 to 3.5	Medium
Below 3.0	Poor

At 28 days of curing, all of the specimens tested are classified in the good category. The classification of the concrete quality can be referred to the Table 3.

From the results obtained, we can conclude that the SCLFC with the percentage of 10% addition of WPSA (M3) has the highest value of pulse velocity. In line with the objectives of this study to compare the durability properties of SCLFC with the best mix proportion of SCLFC with the addition of WPSA, a 10% addition of WPSA is the optimum value to be added in the SCLFC to get a good quality of concrete. With all the results gathered from this test, SCLFC without any addition of WPSA has lower quality compared to the SCLFC with the addition of WPSA. This proved that WPSA can improve the durability of the concrete. All in all, the durability of the concrete can be improved with the addition of WPSA with an optimum addition of 10% of WPSA from cement content. Previous research by Heniegal (2009) indicates that SCC containing waste materials gives good quality concrete. To be compared with the previous study, WPSA is also acceptable to be used to enhance the physical properties of SCLFC.

3.4 *Water Absorption Test*

In this research study, water absorption has been carried out. This durability test was conducted to determine how much water can penetrate through the SCLFC. Figure 4 shows the graph plotted from the water absorption test of the SCLFC.

From the expression of the graph, the rate of water absorption of every batch is slightly decreasing with the age of curing. SCLFC with an addition of 5% and 10% of WPSA has a higher water absorption rate on day 3 of curing but both had a lower value after it reached day 28 of curing. This shows that the durability of the SCLFC is increasing by the day. The specimens with a 10% addition of WPSA had the lowest value of water absorption among the other specimens. The durability properties of the SCLFC in the rate of water absorption are at 7.2% at day 28 compared to 0%, 5%, and 15% which are 10.8, 107, and 10.1 respectively. The pattern of decrease in the rate of water absorption between day 3 and 7 is significant, while between day 7 and day 28, the decrease pattern is slow. This shows that the SCLFC becomes more durable in the range of 3 to 7 days until it reaches maximum durability on day 28.

From the graph plotted in Fig. 4, we can conclude that SCLFC with an additional WPSA is more durable than the control sample with the optimum addition of WPSA at 10% of the cement content. As to previous work by Bin Mohd Sani et al. (2011),

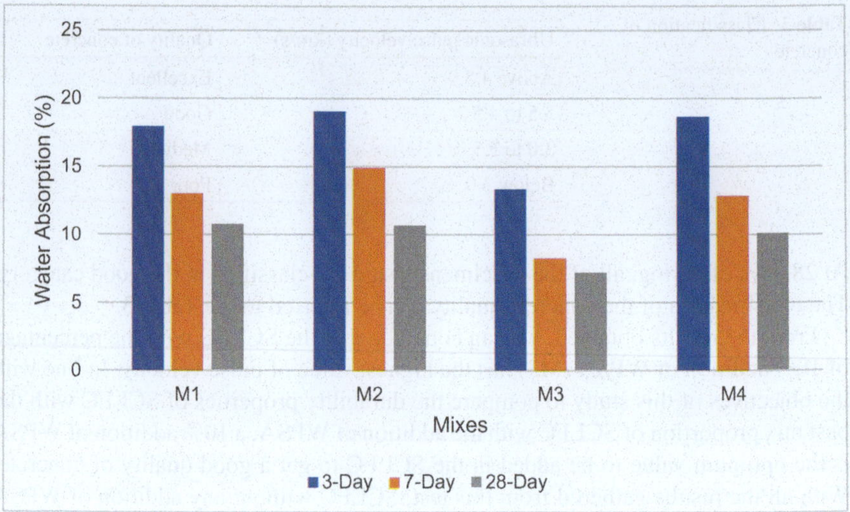

Fig. 4 Result of water absorption test with different percentage addition of WPSA

the replacement of cement with WPSA showed a low water absorption rate compared to the normal concrete. From the previous work, it can be supported that the addition of WPSA in SCLFC also showed a lower water absorption value than the normal SCLFC.

3.5 *Porosity Test*

The porosity test has been conducted for this study. This test was carried out to determine the total porosity of the SCLFC. The durability properties of the SCLFC can be determined by getting the total porosity of the concrete. In order to get the porosity of the SCLFC, all the specimens were weighed in the water.

From Fig. 5, the graph of the porosity test of the SCLFC has been plotted. At the earlier stage, all of the specimens are very high in permeability with higher than 20% total porosity. The total porosity for the SCLFC with the addition of 0%, 5%, 10%, and 15% of WPSA is 29.7%, 28%, 21.2%, and 28.5% respectively. However, the total porosity in concrete significantly decreases on day 7 of curing. This pattern of decreasing in value of total porosity keeps continuing to day 28 with a reading recorded 20.1%, 20.0%, 15.3%, and 18.7% for 0%, 5%, 10%, and 15% addition of WPSA respectively. To be compared, a control sample which is SCLFC with 0% addition of WPSA has a higher total porosity than SCLFC with an additional of WPSA. The graph shown in Fig. 5 indicates that the total porosity of the SCLFC is decreasing with the increasing percentage of WPSA in the specimens. Despite that,

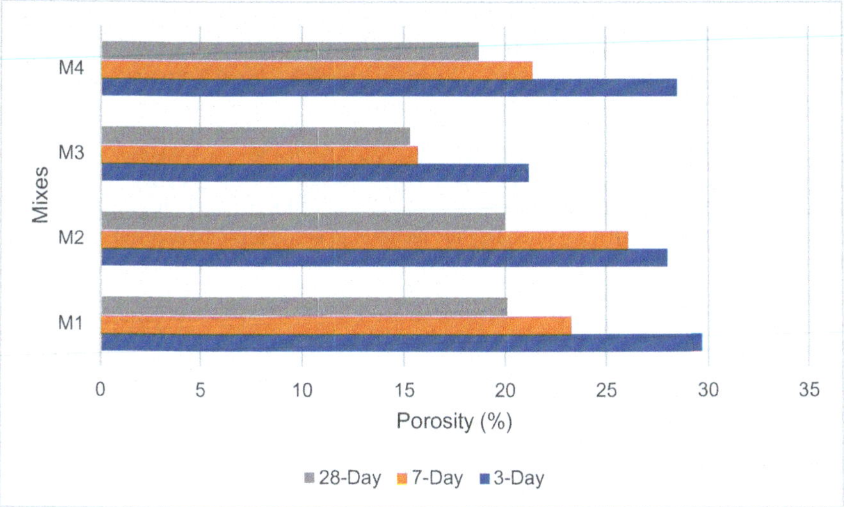

Fig. 5 Result of porosity test with different percentage addition of WPSA

the total porosity of the concrete is increasing at the 15% addition of WPSA but still lower value than controlled samples and 5% addition of WPSA.

This can be concluded that the optimum addition of WPSA in SCLFC is at 10%. The addition of WPSA in SCLFC can be validated by reducing the permeability of the concrete. Thus, the usage of WPSA help to produce more durable SCLFC in term of the porosity of the concrete.

4 Conclusion

The following conclusions can be drawn from the data recorded in the experimental works carried out for this study.

1. The workability properties of SCLFC containing WPSA achieved the standard of SCC by EFNARC. The average flow diameter of SCLFC with the addition of WPSA for all different percentages which is 5%, 10%, and 15% is proportionate to normal SCLFC.
2. The compressive strength of SCLFC containing WPSA increased as the percentage of WPSA addition increased. The compressive strength of SCLFC containing 5%, 10%, and 15% addition of WPSA had been comparable to normal SCLFC. The optimum addition of WPSA in SCLFC taken at 28 days strength was 5%.
3. The SCLFC containing WPSA viewing a high value of UPV compared to the normal SCLFC. The addition of WPSA produces a better quality of SCLFC with optimum addition at 10%.

4. The water absorption and porosity of SCLFC containing WPSA slightly decrease with the age of curing. SCLFC with an additional of WPSA is more durable than the control sample with optimum addition of WPSA at 10% of the cement content.

Acknowledgements The author would like to acknowledge the Research Management Institute for providing the "Supervisory Initiative Grant (GIP)" (600-RMC/GIP 5/3 (169/2021)), Institute for Infrastructure Engineering and Sustainable Management (IIESM), and College of Engineering, Universiti Teknologi MARA for providing the facilities to accomplish the experimental work for this research.

References

Bin Mohd Sani, M. S. H., Bt Muftah, F., & Ab Rahman, M. (2011). Properties of Waste Paper Sludge Ash (WPSA) as cement replacement in mortar to support green technology material. In *3rd ISESEE 2011 - International Symposium and Exhibition in Sustainable Energy and Environment*, June, 94–99. https://doi.org/10.1109/ISESEE.2011.5977117

BS EN 1008:2002, Mixing water for concrete. Specification for sampling, testing and assessing the suitability of water, including water recovered from processes in the concrete industry, as mixing water for concrete. British Standard, pp 1–22

BS EN 12390-3:2009 Testing hardened concrete. compressive strength of test specimens. British Standard, pp. 1–22.

BS EN 206-9: 2010, Concrete. Additional Rules for Self Compacting Concrete (SCC), British Standard, pp 1–16

BS EN 933–1:2012, Tests for geometrical properties of aggregates. Determination Of Particle Size Distribution. Sieving Method, British Standard, pp. 1–22

BS 1881: Part 125:2013 Testing concrete. *Methods for mixing and sampling fresh concrete in the laboratory* (pp. 1–16). British Standard

Chowdhury, S., Maniar, A., & Suganya, O. M. (2014). Strength development in concrete with wood ash blended cement and use of soft computing models to predict strength parameters. *Journal of Advanced Materials, 6*(6), 907–913.

EFNARC, F. (2002). Specification and guidelines for self-compacting concrete. European federation of specialist construction chemicals and concrete systems.

Elrahman, M. A., El Madawy, M. E., Chung, S. Y., Sikora, P., & Stephan, D. (2019). Preparation and characterization of ultra-lightweight foamed concrete incorporating lightweight aggregates. *Applied Sciences (Switzerland), 9*(7). https://doi.org/10.3390/app9071447

Hamzah, A. F., Wan Ibrahim, M. H., Jamaluddin, N., Jaya, R. P., & Zainal Abidin, N. E. (2015). Fresh properties of self-compacting concrete integrating coal bottom ash as a replacement of fine aggregates. *Advanced Materials Research, 1125*(July), 370–376.

Heniegal, A. M. (2009). Investigation of cracks in self compacting concrete by using ultrasonic pulse velocity. *Journal of Engineering Sciences, 37*(1), 63–84.

Kejela, B. M. (2020). Waste paper ash as partial replacement of cement in concrete. *American Journal of Construction and Building Materials, 4*(1), 8–13.

Kushwaha, M., Akhtar, S., & Rajput, S. (2013). Development of the self compacting concrete by industrial waste (Red Mud). *International Journal of Engineering Research and Applications, 3*(4), 539–542.

Muthusamy, K., Budiea, A. M. A., Zaidan, A. L. F., Rasid, M. H., & Hazimmah, D. S. (2017). Properties of concrete containing foamed concrete block waste as fine aggregate replacement. *IOP Conference Series: Materials Science and Engineering, 271*(1). https://doi.org/10.1088/1757-899X/271/1/012084

Okamura, H., & Ouchi, M. (2003). *Self-Compacting Concrete., 1*(1), 5–15.

Sajjad Ahmad, M., Malik, I., Wani, M. B., & Ahmad, R. (2013). Study of concrete involving use of waste paper sludge ash as partial replacement of cement. *IOSR Journal of Engineering, 3*(11), 6–15.

Sharipudin, S. S., & Ridzuan, A. R. M. (2013). Influence of waste paper sludge ash (WPSA) and fine recycled concrete aggregate (FRCA) on the compressive strength characteristic of foamed concrete. *Advanced Materials Research, 626*, 376–380. https://doi.org/10.4028/www.scientific.net/AMR.626.376

Siti, D., Bt, Q., Adenan, A., Nainy, I. A., Zuhair, M., & Rasid, A. (2016). High early strength of self-consolidating concrete incorporating Qua-Si-Rha. *4*(1), 3–11.

The Effect of Tendon Directions to The Analysis and Design of Transfer Slab—A Case Study

Ahmad Suhaimi Abdul Mutalib, Anizahyati Alisibramulisi, Norliyati Mohd Amin, Ekarizan Shaffie, and Adiza Jamadin

Abstract The focus of this study is to analyze and design a post-tensioned transfer slab by using RAM Concept as a finite element analysis tool. The analysis and design of transfer slab were in accordance with BS8110:1997—Structural use of concrete Part 1: Code of Practice for Design and Construction. Two types of tendon directions in terms of banded and distribution direction were applied. Basically, the procedure in analyzing and designing this element requires numerous calculations in order to reach the most desired and economical design. RAM Concept has the capability of analyzing problems related to moment, stresses, and deflection in a short time. This makes design calculations and procedures easier. It can be concluded that tendon direction does affect the analysis and design of the transfer slab studied. From the stresses and deflection results of two types of tendon directions, Type 1 is fully recommended. Whereas, for punching shear of the column surface, not much difference between the two types of tendon directions; both cases complied with the standard shear limit.

Keywords Tendon direction · Transfer slab · RAM concept · Prestressed · Post-tensioned · BS8110

A. S. Abdul Mutalib
YLSE Technologies (M) Sdn. Bhd., Sec. U9, Lot 7389, Jalan Ciku, Jalan Bukit Cherakah, 40150 Shah Alam, Selangor, Malaysia

A. Alisibramulisi (✉) · E. Shaffie
Institute for Infrastructure Engineering and Sustainable Management (IIESM), Universiti Teknologi MARA (UiTM), 40450 Shah Alam, Selangor, Malaysia
e-mail: aniza659@uitm.edu.my

E. Shaffie
e-mail: eka@uitm.edu.my

A. Alisibramulisi · N. Mohd Amin · E. Shaffie · A. Jamadin
School of Civil Engineering, College of Engineering, Universiti Teknologi MARA (UiTM), 40450 Shah Alam, Selangor, Malaysia
e-mail: norli830@uitm.edu.my

A. Jamadin
e-mail: adiza@uitm.edu.my

U. R. Hashim et al. (eds.), *Green Infrastructure*,
https://doi.org/10.1007/978-981-99-7003-2_14

1 Introduction

A building is an enclosed structure that has walls, floors, a roof, and usually windows. A "tall building" is a multi-story structure in which most occupants depend on elevators to reach their destinations. The most prominent tall buildings are called "high-rise buildings" in most countries and "tower blocks" in Britain and some European countries mentioned by Craighead (2009). High-rise buildings nowadays are designed to have an open space at the ground floor level. One of the solutions to this requirement is by terminating the support of the upper floor on a slab that is called "Transfer Floor." The transfer floor can be designed as a transfer beam or transfer slab. The transfer slab is usually known as "Transfer Plate" in America. Transfer slab receives load from column and wall support of the superstructure by transferring the load to a limited number of widely spaced supports. Generally, the wide space support is mega columns or shear walls. The type of transfer slab may be conventional reinforcement slab and prestressed transfer slab. Prestressed utilize the full section to remain in the compression as a result of the maximum section is utilized to bear the developed stresses. Prestressed transfer slab is more popular due to the thinner slab that contributes to smaller dead load and is cost-effective as compared to conventional design.

Prestressed concrete construction has obtained wide popularity in current building construction. Prestressed members can be constructed by utilizing unbonded or bonded strands. For bonded prestressed members most are post-tension. Post-tensioning is a method of prestressed concrete where the prestressed tendons are pulled and anchored after the concrete achieves the adequate concrete transfer strength, fci. The bond between strand and concrete is essential for the success of prestressed system. The bond in prestressed members may be categorized as transfer bond and flexural bond.

For pre-tensioning, initial tensioning of the strands causes a reduction of the strand's diameter due to Poisson's effect. After concrete reaches sufficient strength, the strands are released from the abutments, and the stress in the strands at the free ends of the members returns to zero. For the bonded prestressed structural members, the force is transferred from strands to concrete through end anchors, together with the bond between strand and concrete. Hussien et al. (2012) studied on behavior of bonded and unbonded prestressed normal and high strength concrete and concluded beams partially prestressed concrete beams with bonded tendons provide better behavior than those of unbonded tendons such as increased ductility, initial stiffness, and the ultimate deflection up to 265%, 13%, and 199%, respectively.

In Malaysia, prestressed systems have developed over the years and various companies have patented their products. Some examples of prestressed contractors in Malaysia are Utracon (M) Sdn. Bhd., Freyssinet PSC (M) Sdn Bhd, VSL Engineers (M) Sdn Bhd, BBR Construction System (M) Sdn Bhd, Strong Force (M) Sdn. Bhd., YLSE Technologies Sdn Bhd, and SIS Integrated Sdn. Bhd. For strand manufacture, they are using strand products either from Southern PC or Silvery Dragon. The diameter of strand varies depending on the type of application, but for prestressing,

12.7 mm diameter (183.7 kN per strand) and 14.24 mm diameter (260.7 kN per strand) are widely used.

Transfer slab can be designed either by using conventional reinforced concrete or post-tensioned prestressed concrete. The design force and associated reinforcement can be determined straight forward from analysis of the structure. The development of post-tensioned transfer slab offers advantages in terms of economics and potential of superior performance. The analysis and design of post-tensioned transfer slab tendon by considering one type of banded tendon with distribution tendon direction only may cause significant differences to the analysis and design efficiency. Therefore, an alternative tendon direction for transfer slab analysis is to be proposed to compare both directions of tendon. Basically, the post-tensioned transfer slab design requires basic knowledge of reinforced concrete design. The thick slab can be designed by using a one-way or two-way slab system which is the banded tendon supporting the distribution tendon. The ultimate objective is to meet the serviceability and safety of the structure. However, the need for comparison of tendon direction is required to compare which is more efficient to the analysis and design.

2 Methodology

This case study was conducted at the Bangi Apartment transfer slab. The building was proposed based on the multiple thickness of transfer slab. The transfer slab was located at Level 9, supporting 20 storey of apartment. Figure 1 shows the layout of the selected building for modeling. The transfer slab consists of 8 longitudinal spans and 4 transverse spans including 2 cantilever spans. The horizontal length of 7.9, 7.1, 10.275, and 8.5 m and the vertical span is 4.05 and 9.85 m, and various cantilever lengths with the shortest length of 1.075 m and the longest length of 5.625 m. The transfer slab is divided into 4 consequence zones based on the most practical area for concreting purposes.

The thickness of transfer slab is varied from 1800 mm (including drop) until 500 mm after gridline 4, followed by 1650 mm (including drops) and other 1800 mm thickness at 2125 m before gridline 9. All drop areas were ignored during modeling as it can be constructed as non-structural clements. Since the drop varied from 25, 50, and 75 mm, the maximum 75 mm drop is considered as non-structural element, and thus would be simplified in the modeling. The dead load of drop is considered as a superimposed dead load. The thicknesses that were used in the modeling are 1725 mm and 1575 mm, respectively.

The thicknesses of 1725 and 1575 mm may be practical for design, but may not be practical for site construction since it will cost a huge amount of falsework. So, it is recommended to design as multiple layers. 1st layer is designed to retain some percentage of design load and dead load of 2nd layer. The rule of thumb suggested that 1st layer design thickness is taken from 1/3 to 1/2 of design thickness. The 1st layer for 1725 mm slab is taken as 900 mm thickness as half of 1800 mm. The 1st

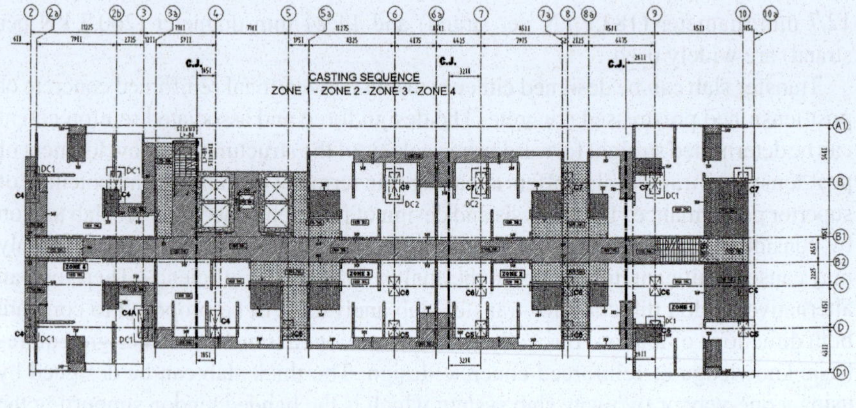

Fig. 1 Building's layout plan on Bangi apartment

layer for 1575 mm slab is taken as 750 mm, due to the same level with 900 mm thickness. The section is shown in Figs. 2 and 3.

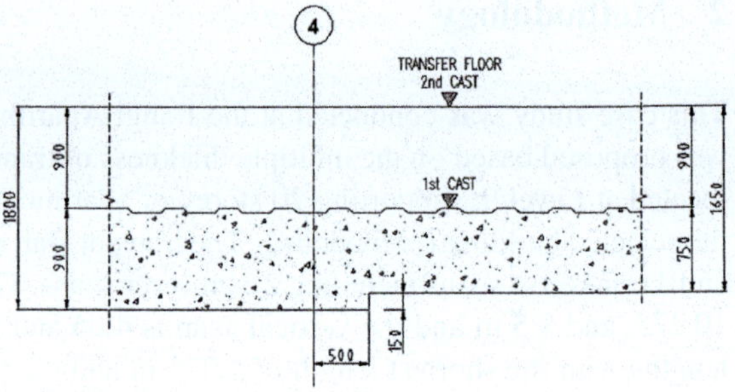

Fig. 2 Transfer slab section at Gridline 4 (all units in mm)

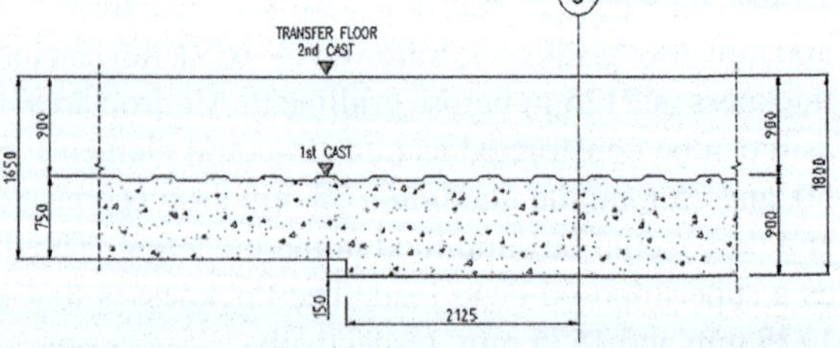

Fig. 3 Transfer slab section at Gridline 9 (all units in mm)

2.1 Modeling Process

The modeling process is shown in Fig. 4. RAM Concept was chosen because of the finite element analysis capabilities for a one-way or two-way post-tensioned slab system. Establishing finite element modeling that sufficiently well represents the real structural behavior in RAM Concept required several steps. The latitude and longitudinal modeling of transfer slab element by using RAM Concept will be discussed. All the design for the transfer slab is based on British standard and by other international recognized documents. The following British standard shall be used in the structural design whenever applicable: 1. (BS 8110) Structural Use of Concrete, Part 1: 1997 Code of Practice for Design and Construction. The model was analyzed, and the results of Top Stress, Bottom Stress, Deflection, and Reaction were obtained and compared.

2.2 Checking for Column Surface Punching

This study only considers the punching shear at the column surface. Since the relative economy of transfer slabs depends on their thickness which is governed by either deflection limits or punching resistance. So, it is important to check the punching around the column surface as shown in Fig. 5. For predetermination of slab thickness. In BS 8110-1:1997, clause 3.7.6.4, the maximum design shear stress should not exceed $0.8\sqrt{fcu}$ or 5 N/mm^2 if less. In the design, it is suggested the column to be constructed with a drop cap to reduce the punching effect. In this study, the models already consider the drop cap on some of the columns. So, the checking still needs to be carried out and determine the shear at column face. The formula for determining column surface punching shear is shown in Eq. 1.

$$v = R/A \tag{1}$$

v = Shear at column surface

R = Reaction at column.

A = Column Perimeter × d, effective depth of slab.

If there is any shear wall loading on top of column, the reaction can be reduced by minus the shear wall load. The excel spreadsheet was used to simplify the calculation.

3 Results and Discussions

There are two types of models for analysis and design on transfer slab which is:

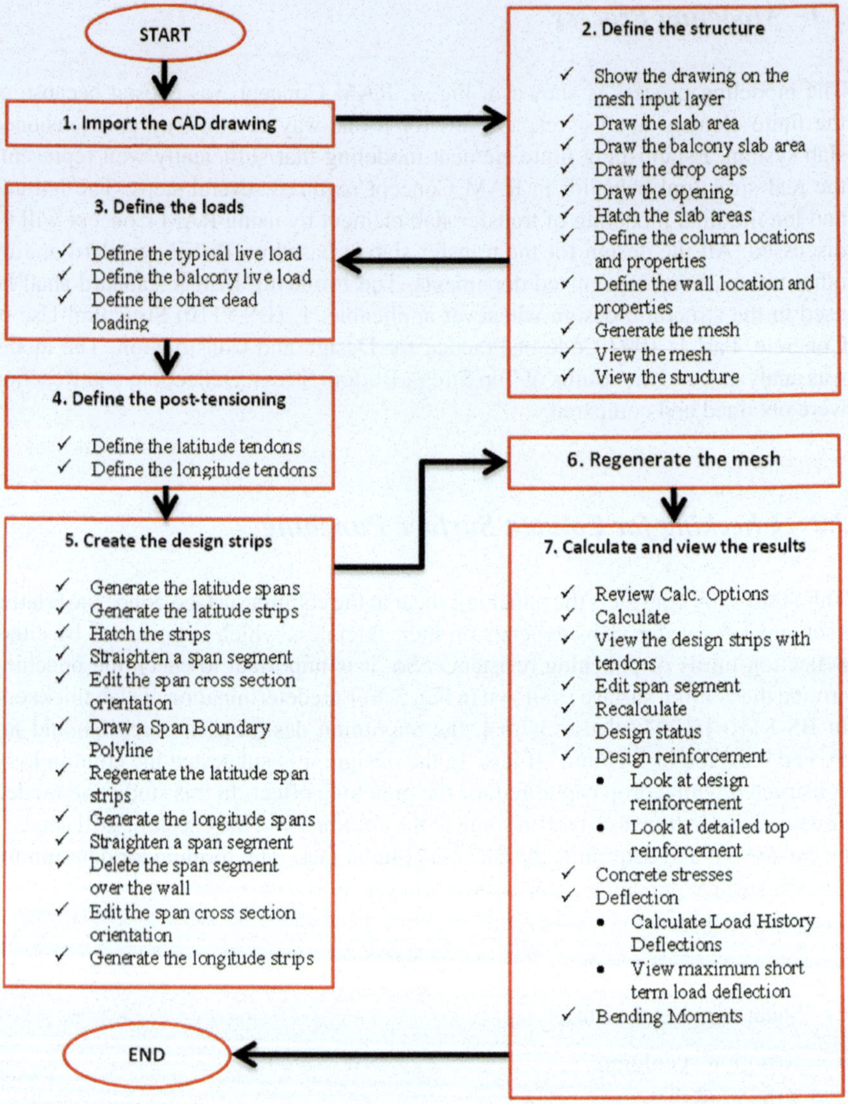

Fig. 4 Modeling flow chart

Type 1: Banded tendon in Horizontal direction, Distribution Tendon in Vertical direction.

Type 2: Banded tendon in Vertical direction, Distribution Tendon in Horizontal direction.

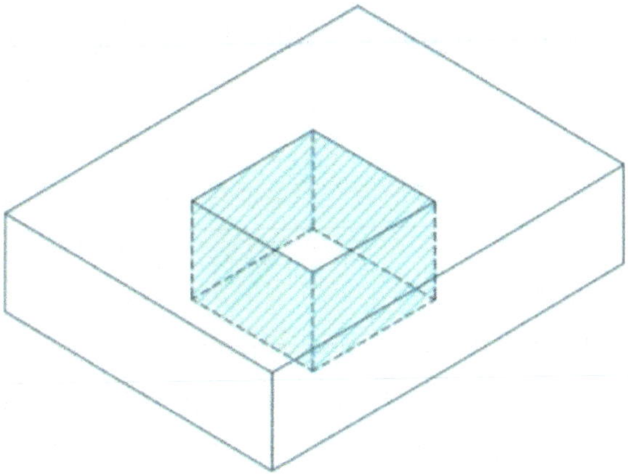

Fig. 5 Critical section for punching shear

Figures 6, 7, 8, 9, 10, 11, 12, 13, 14, 15, 16, 17, 18, and 19 shows the summary of result obtained from finite element programming, which is the RAM Concept (Tables 1 and 2).

(1) Model Type 1

Banded tendon in Horizontal direction, Distribution Tendon in Vertical direction.

(2) Model Type 2

Banded tendon in Vertical direction, Distribution Tendon in Horizontal direction.

(3) Stresses and Deflection Comparison

Results obtained from RAM Concept as Analysis and Design program have been used for comparison of stresses and deflection values. From finite element analysis, the result for Type 1 and Type 2 analysis and design is tabulated. The manual calculation has been done for maximum Top Stress and Bottom Stress. The summary of the results was tabulated in Tables 3 and 4.

According to Table 3, the values of top stress and bottom stress clearly showed the difference in percentage of top stress for Type 1 is 9.55% more than Type 2 at the same location. For bottom stress, Type 1 is 18.76% less than Type 2 at the same location, gridline GL B-C/9. This is due to more distribution tendon being provided for Type 1 compared to banded tendon provided for Type 2. The stresses for both Type 1 and Type 2 analysis still complied with BS8110:1997 clause 4.3.4 at serviceability limits state which limit to $0.33f_{cu}$ for compressive stress. RAM Concept also provided the top and bottom rebars to counter stresses.

Based on Table 4, the percentage difference in maximum deflection value for Type 1 compared to Type 2 is 4.57%. Type 1 may provide less banded tendon but quite a number of short distribution tendon. Type 2 tendon may offer better banded tendon direction due to longer tendon to sustain the load at this localized area.

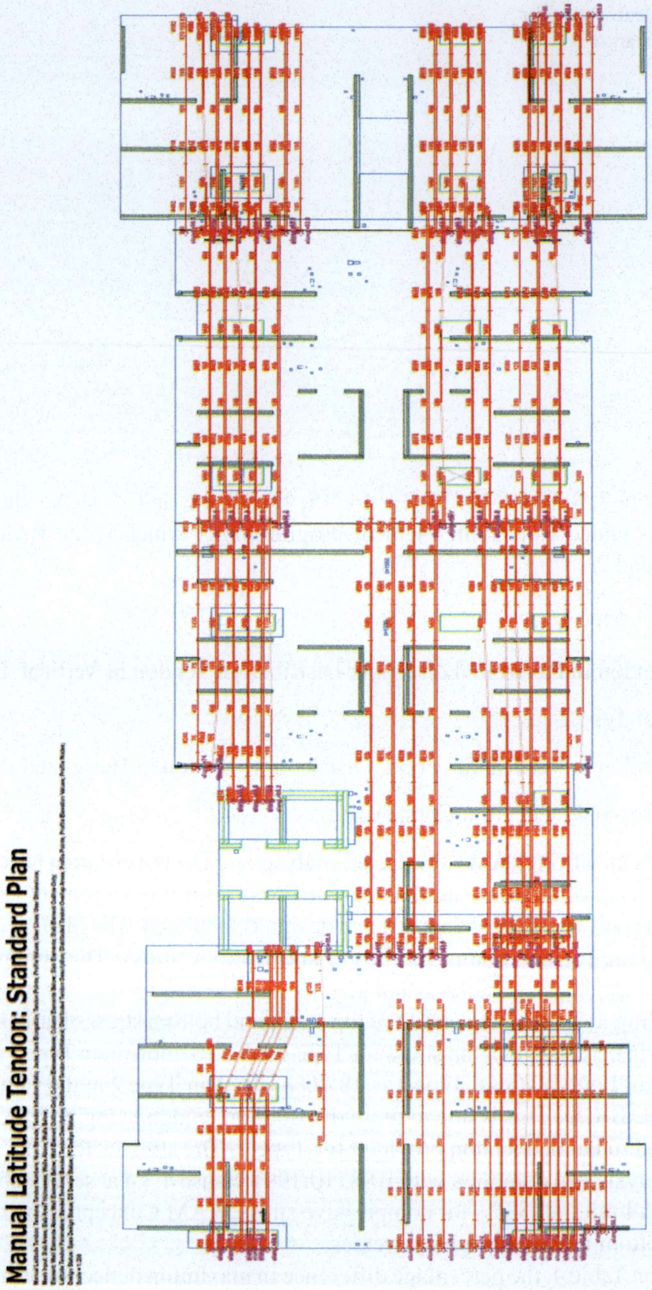

Fig. 6 Type 1 banded tendon in horizontal direction

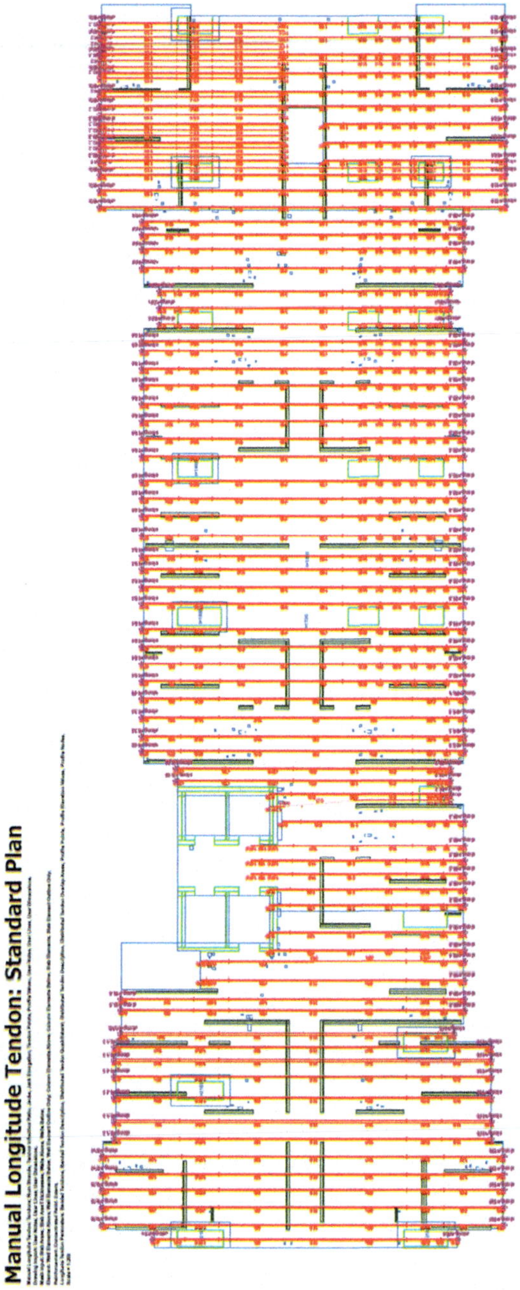

Fig. 7 Type 1 distribution tendon in vertical direction

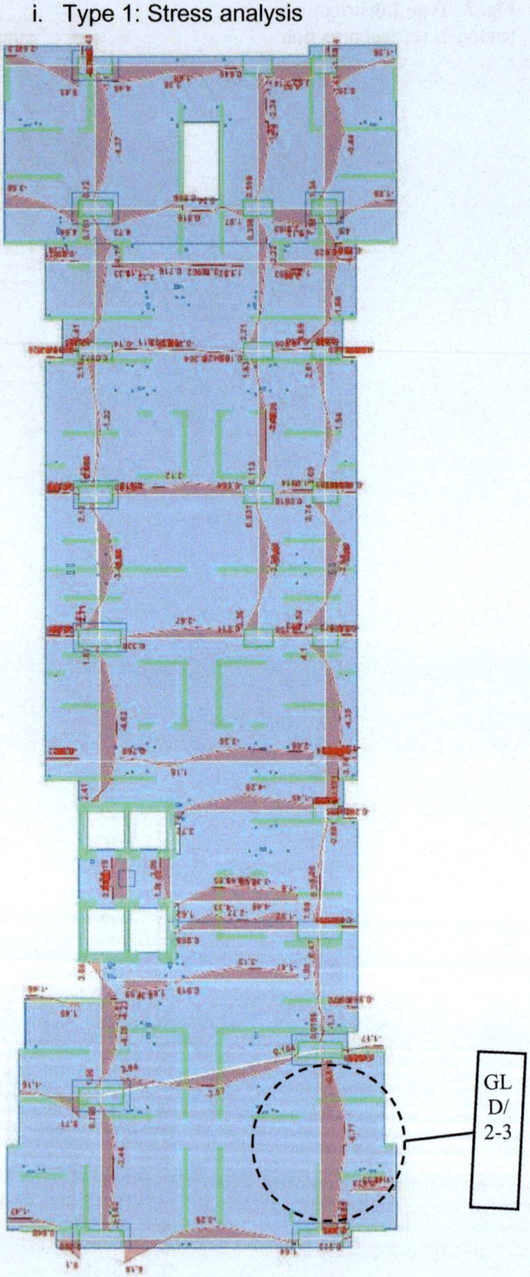

Fig. 8 Type 1 top stresses (top fiber of slab)

Fig. 9 Type 1 bottom stresses (bottom fibre of slab)

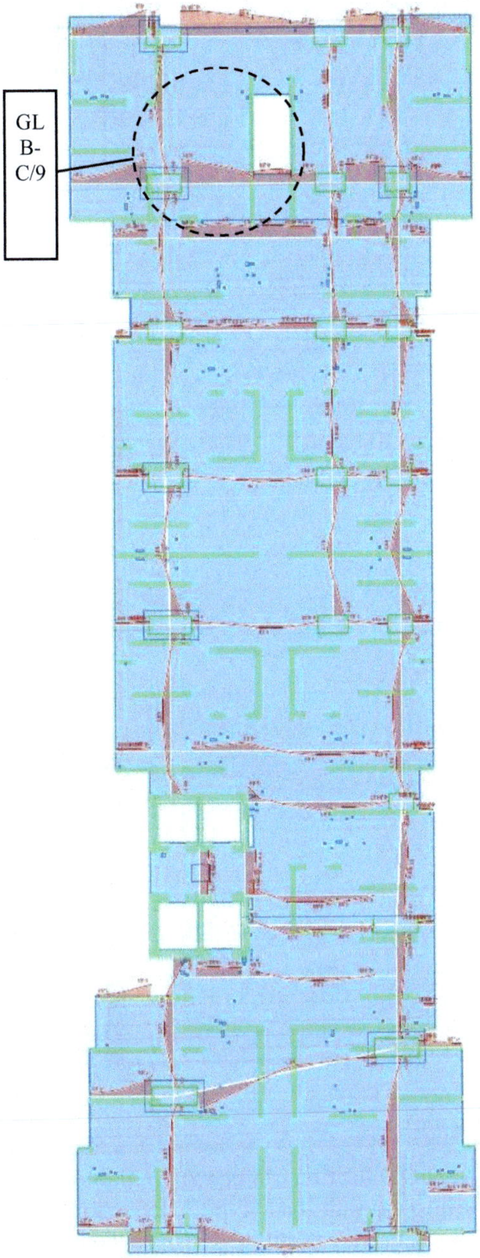

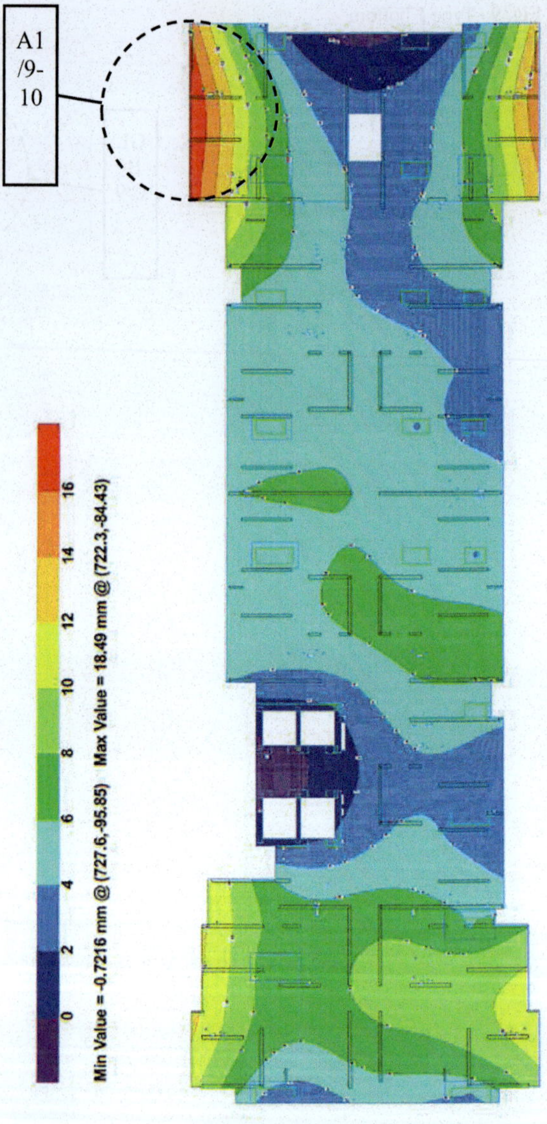

Fig. 10 Type 1 short-term service deflection

This deflection can be expected due to the span are 5.5 m cantilever and loaded with shear wall load. Type 1 deflection is more than the deflection for Type 2, but both deflection is still less than L/250 as stated in BS8110 (Clause 3.4.6.3).

(4) Punching Shear Check at Column Surface

Manual calculation has been done in Table 1 for Type 1 and Table 2 for Type 2. The summary of the results for punching shear was tabulated in Table 5.

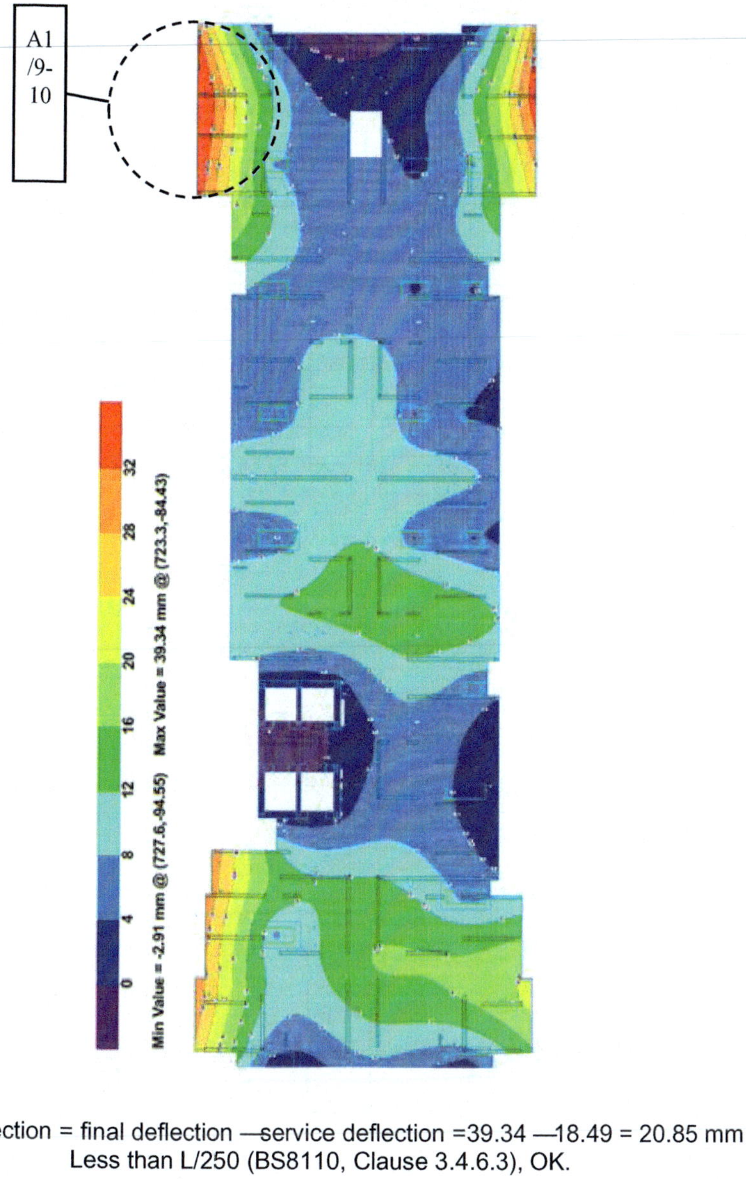

Fig. 11 Type 1 long-term deflection

Fig. 12 Type 1 column reaction

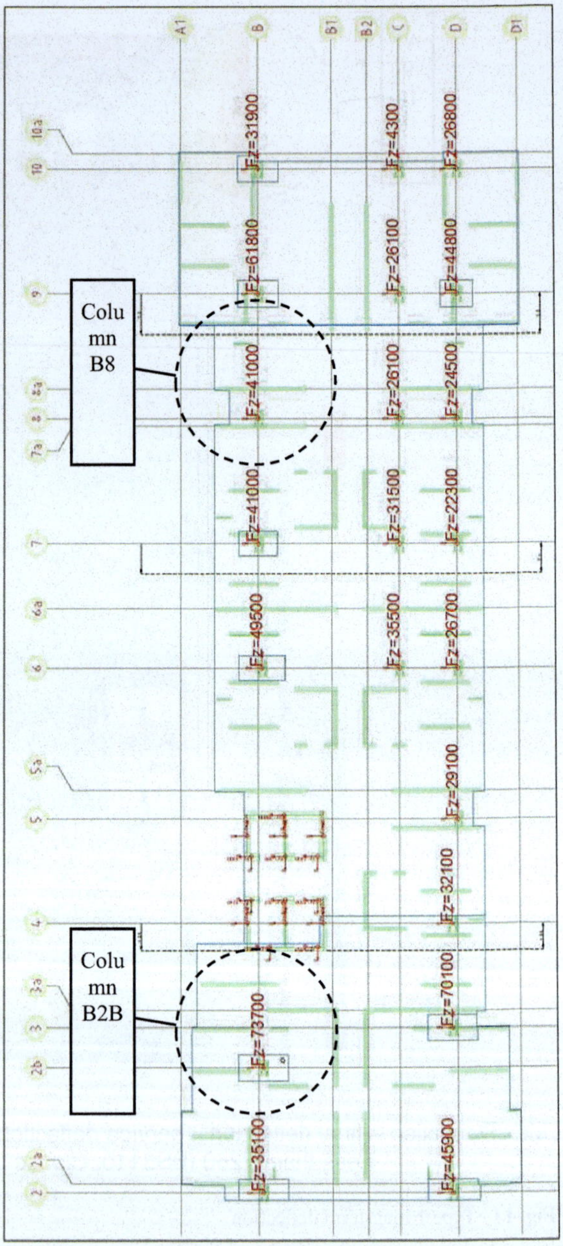

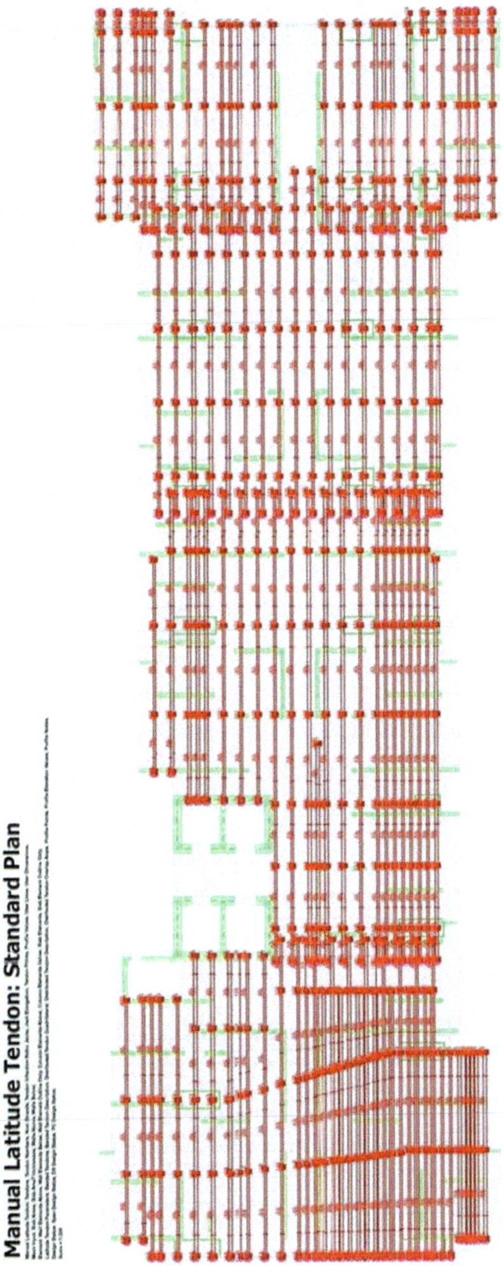

Fig. 13 Type 2 distribution tendon in horizontal direction

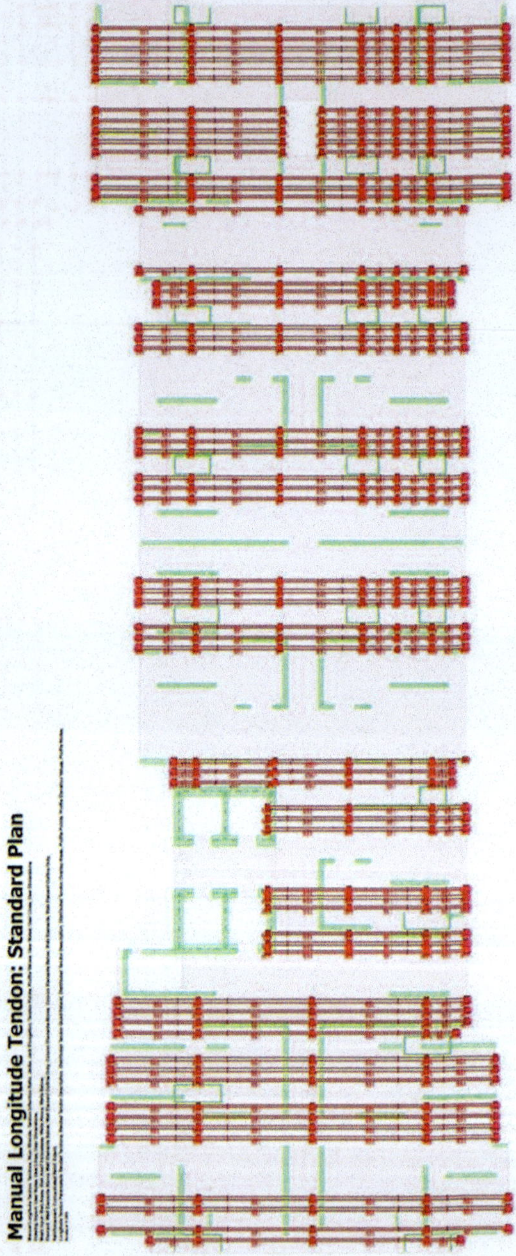

Fig. 14 Type 2 banded tendon in vertical direction

Fig. 15 Type 2 top stresses (top fiber of slab)

Fig. 16 Type 2 bottom stresses (bottom fiber of slab)

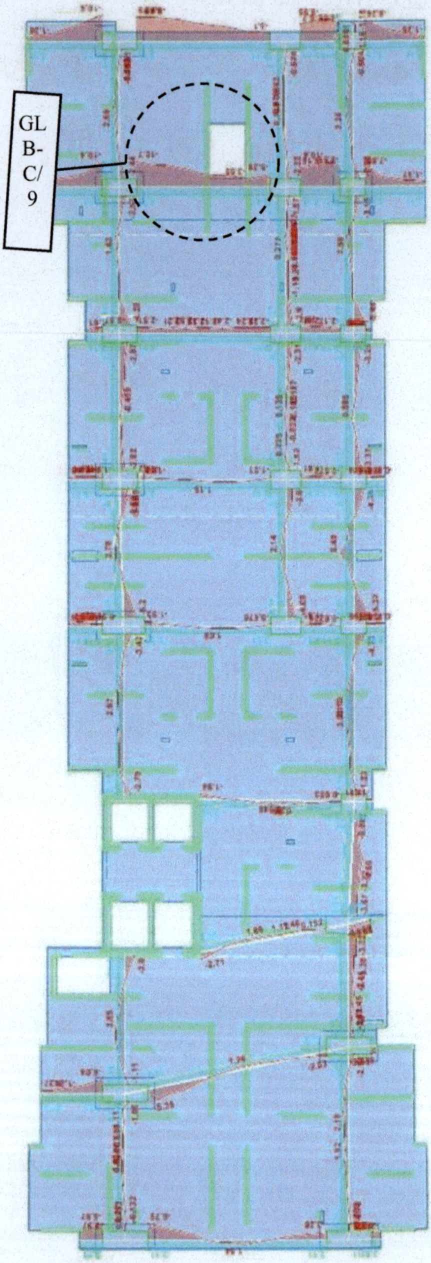

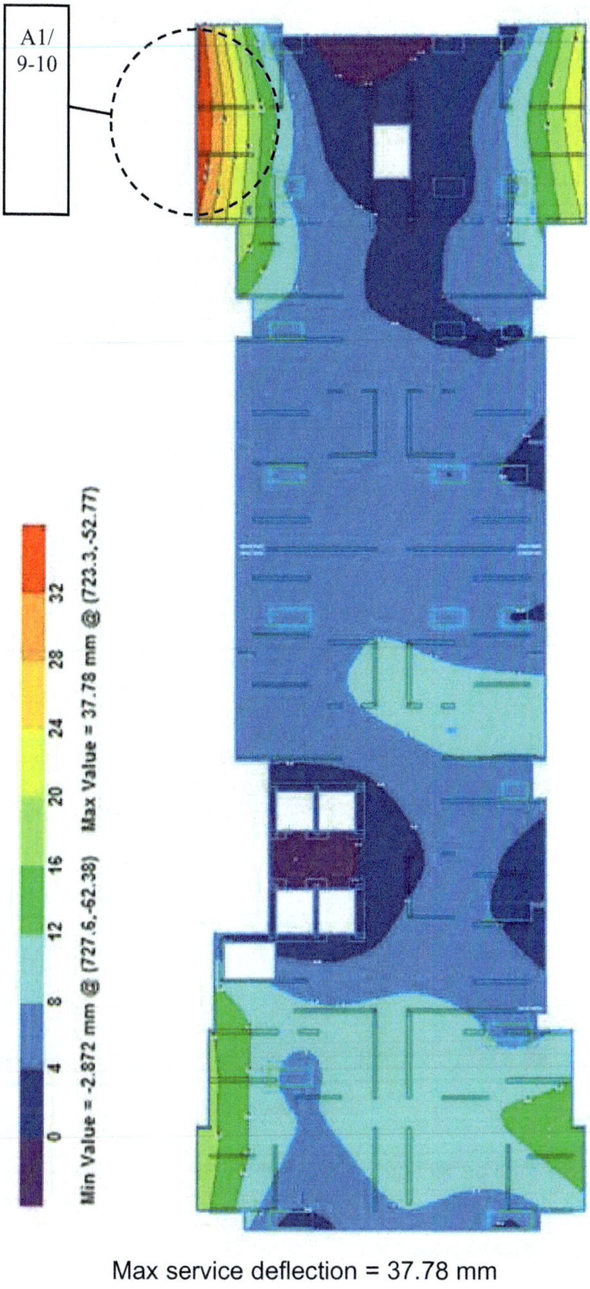

Fig. 17 Type 2 short-term service deflection

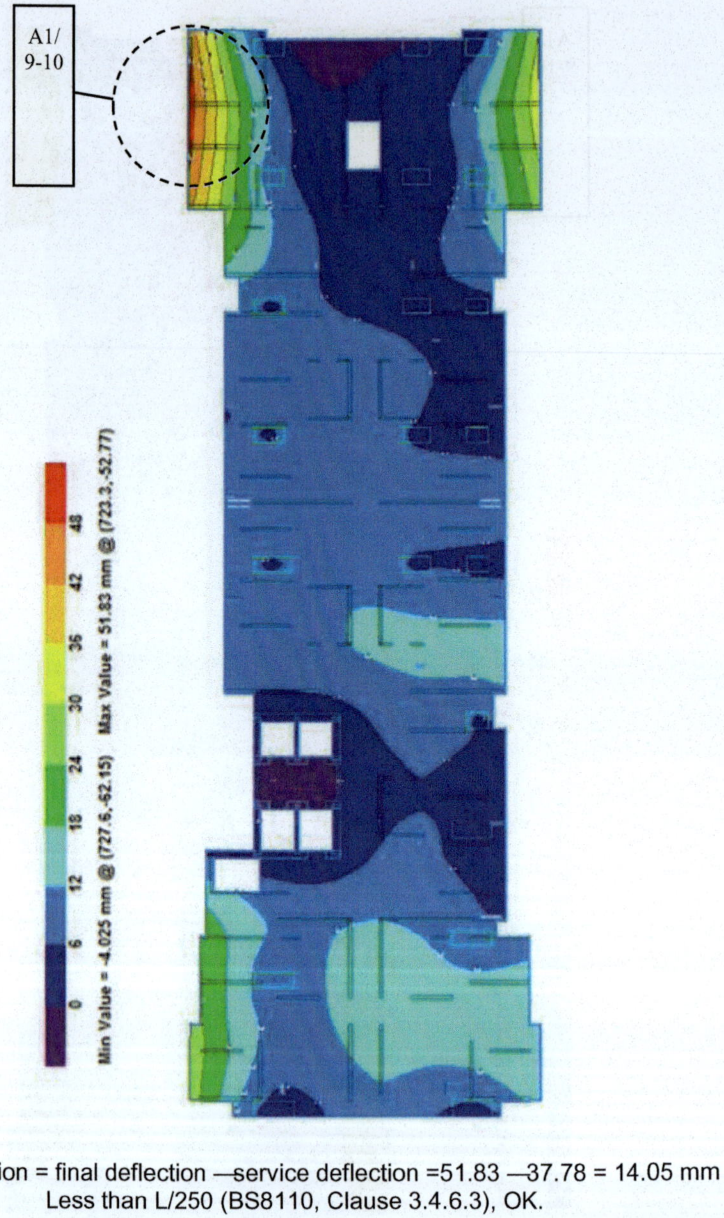

Fig. 18 Type 2 long-term deflection

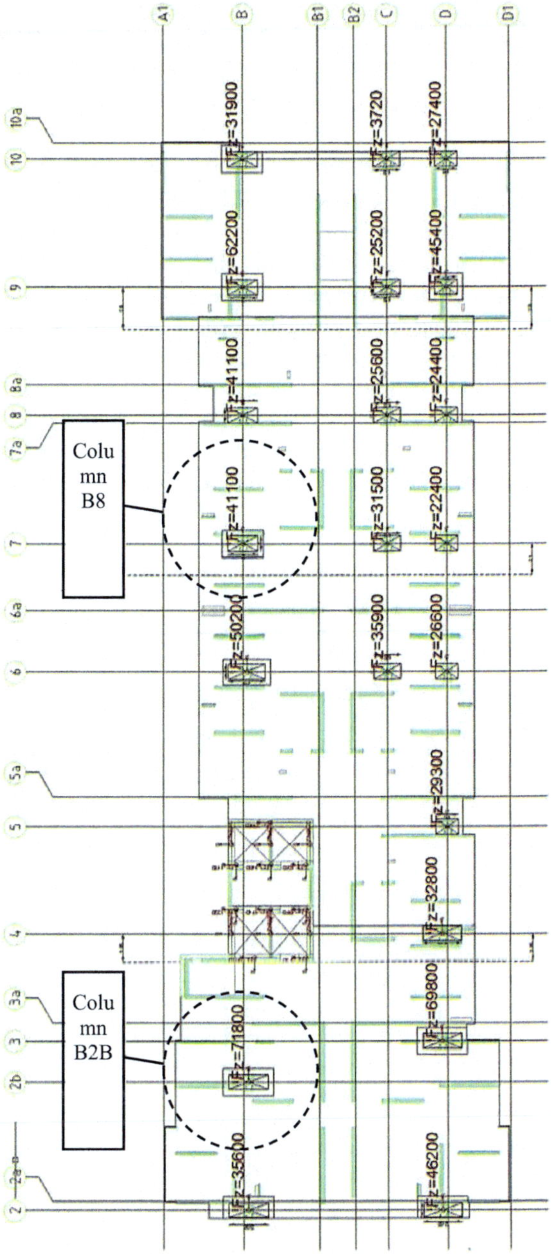

Fig. 19 Type 2 column reaction

At the location of maximum reaction, Column B2b, the difference between Type 1 and Type 2 shear is 2.65%. This is due to redistribution of load from different banded and distribution for Type 1 tendon is more on this column although the assigned load criterion is similar. By referring to the table for location of maximum shear at column B8, shear for Type 1 was slightly less than Type 2 is due to the column reaction different by 100 kN only. The Type 1 design shown banded tendon at longer span was acting as the concentrated support for distribution tendon at a shorter span. Both cases are complied with BS8110 Clause 4.3.8.2 which limits the shear to $0.8\sqrt{f_{cu}}$ or 5 N/mm^2. The distribution of loading is due to distribution tendon at longer span may cause the reaction to differ slightly.

Punching shear as mentioned by Aalami (2014) considered the area of strands that cross a predefined distance from the face of support. Other than that, the result is also influenced by the strength of concrete use for post-tension design. According to BS8110 Clause 4.1.8.1, the minimum recommended strength is 35 N/mm^2, based on the characteristic cube strength, f_{cu} determined at 28 days. In both cases the concrete strength at transfer, f_{ci} should be not less than 25 N/mm^2. In this study, concrete at

Table 1 Type 1 column surface punching shear

Column Marking	Column		Column Head		Column Head Depth	Rult	Edge				Column Face		Column Head		Load within μo	Column Axial	Net Punching Force, V	At Col Face	At Col Head
	B	0	B	0			1	2	3	4	μo (mm)	dl (mm)	μo (mm)	d2 (mm)					
	(mm)	(mm)	(mm)	(mm)	(mm)	(kI-I)	(mm)	(mm)	(mm)	(mm)					(kN)	(N/mm')	(kN)	(N/mm')	(N/mm')
B2	142S	3450	–	–	–	35100	3450	3450	142S	142S	9750	1670	–	–	0	7.14	35100	2..16	–
B2b	1800	3450	–	–	–	73700	3450	3450	1800	1800	10500	1670	–	–	1713	11.87	71'87	4.11	–
B6	1700	3200	–	–	–	49500	3200	3200	1700	1700	9800	IS20	–	–	0	9.10	49500	3.32	–
B7	1800	2800	–	–	–	41000	2800	2800	1800	1800	9200	1520	–	–	0	8.13	41000	2..93	–
B8	1000	2000	–	–	–	41000	2000	2000	1000	1000	6000	1520	–	–	1.645	20.SO	39355	4.32	–
B9	1800	2800	–	–	–	61800	2800	2800	1800	1800	9200	1670	–	–	2614	U.26	59186	3.85	–
B10	1800	2800	–	–	–	31900	2800	2800	1800	1800	9200	1670	–	–	2696	6.JI	29204	1.90	–
C6	1000	1800	–	–	–	35500	1800	1800	1000	1000	5600	1520	–	–	0	19.72	35500	4.17	–
C7	1000	1800	–	–	–	31500	1800	1800	1000	1000	5600	1520	–	–	0	17.SO	31500	3.70	–
C8	1000	1800	–	–	–	26100	1800	1800	1000	1000	5600	1520	–	–	1156	14.SO	24944	2..93	–
C9	1000	1800	–	–	–	26100	1800	1800	1000	1000	5600	1670	–	–	0	14.SO	26100	2.79	–
C10	1000	1800	–	–	–	4300	1800	1800	1000	1000	5600	1670	–	–	0	2..39	4300	0.46	–
D2	1425	3450	–	–	–	45900	3450	3450	1425	1425	9750	1670	–	–	0	9.34	45900	2..82	–
D3	1800	3375	–	–	–	70100	3375	3375	1800	1800	10350	1670	–	–	4101	11.54	65"9	3.82	–
D4	1000	2650	–	–	–	32100	2650	2650	1000	1000	7300	1520	–	–	UI	u.u	31'79	2..88	–
D5	1000	1500	–	–	–	29100	1500	1500	1000	1000	5000	1520	–	–	0	19.40	29100	3.83	–
D6	1000	1500	–	–	–	26700	1500	1500	1000	1000	5000	1520	–	–	0	17.80	26700	3.51	–
D7	1000	1500	–	–	–	22300	1500	1500	1000	1000	5000	1520	–	–	0	14.17	22300	2..93	–
D8	1000	1500	–	–	–	24500	1500	1500	1000	1.000	5000	1520	–	–	1301	16.33	23200	3.0S	–
D9	1800	2300	–	–	–	44800	2300	2300	1800	1800	8200	1670	–	–	2687	10.82	42113	3.08	–
D10	1000	1500	–	–	–	26800	1500	1500	1000	1000	5000	1670	–	–	2929	17.87	23871	2..86	–

Table 2 Type 2 column surface punching shear

Column Marking	Column		Column Head		Column Head Depth	R_{ult}	Edge				Column Face		Column Head		Load within μo	Column Axial	Net Punching Force,V	At Col Face	At Col Head
	B	D	B	D			1	2	3	4	μo (mm)	dl (mm)	μo (mm)	d2 (mm)					
	(mm)	(mm)	(mm)	(mm)	(mm)	(kN)	(mm)	(mm)	(mm)	(mm)					(kN)	(N/mm²)	(kN)	(N/mm²)	(N/mm²)
B2	1425	3450		–	–	35600	3450	3450	1425	1425	9750	1670	–	–	0	7.24	35600	2.19	–
B2b	1800	3450		–	–	71800	3450	3450	1800	1800	10500	1670	–		1713	11.56	70087	4.00	–
B6	1700	3200	–	–	–	50200	3200	3200	1700	1700	9800	1520	–	–	0	9.23	50200	3.37	–
B7	1800	2800	–	–	–	41100	2800	2800	1800	1800	9200	1520	–		0	8.15	41100	2.94	–
B8	1000	2000		–	–	41100	2000	2000	1000	1000	6000	1520	–	–	1645	20.55	39455	**4.33**	–
B9	1800	2800	–	–	–	62200	2800	2800	1800	1800	9200	1670	–	–	2614	12.34	59586	3.88	–
B10	1800	2800	–	–	–	31900	2800	2800	1800	1800	9200	1670	–	–	2696	6.33	29204	1.90	–
C6	1000	1800	–	–	–	35900	1800	1800	1000	1000	5600	1520	–	–	0	19.94	35900	4.22	–
C7	1000	1800	–	–	–	31500	1800	1800	1000	1000	5600	1520	–	–	0	17.50	31500	3.70	–
C8	1000	1800		–	–	25600	1800	1800	1000	1000	5600	1520	–	–	1156	14.22	24444	2.87	–
C9	1000	1800	–	–	–	25200	1800	1800	1000	1000	5600	1670	–		0	14.00	25200	2.69	–
C10	1000	1800	–	–	–	3720	1800	1800	1000	1000	5600	1670	–		0	2.07	3720	0.40	
D2	1425	3450	–	–	–	46200	3450	3450	1425	1425	9750	1670	–	–	0	9.40	46200	2.84	–
D3	1800	3375		–	–	69800	3375	3375	1800	1800	10350	1670	–	–	4101	11.49	65699	3.80	–
D4	1000	2650	–	–		32800	2650	2650	1000	1000	7300	1520		–	121	12.38	32679	2.95	–
D5	1000	1500	–	–	–	29300	1500	1500	1000	1000	5000	1520	–	–	0	19.53	29300	3.86	–
D6	1000	1500	–	–	–	26600	1500	1500	1000	1000	5000	1520	–	–	0	17.73	26600	3.50	–
D7	1000	1500	–	–	–	22400	1500	1500	1000	1000	5000	1520	–	–	0	14.93	22400	2.95	–
D8	1000	1500		–	–	24400	1500	1500	1000	1000	5000	1520	–	–	1301	16.27	23100	3.04	–
D9	1800	2300	–	–	–	45400	2300	2300	1800	1800	8200	1670	–		2687	10.97	42713	3.12	–
D10	1000	1500	–	–	–	27400	1500	1500	1000	1000	5000	1670	–		2929	18.27	24471	2.93	

Table 3 Stresses comparison

	Allowable stresses (N/mm^2)	Type of model		Percent of differences (%)
		Type 1	Type 2	
Top stresses (N/mm^2)	$0.33f_{cu} = -14.85$	GL D/2-3 = −6.77	GL D/2-3 = −6.18	9.55
Bottom stresses (N/mm^2)	$0.33f_{cu} = -14.85$ (Cantilever)	GL B-C/9 = −9.01	GL B-C/9 = −10.7	18.76

Table 4 Deflection comparison

	Deflection limit (mm)	Type of model		Percentage difference (%)
		Type 1	Type 2	
Maximum deflection (mm)	L/250 8500/250 = 34 mm	GL A1/9-10 = −20.85	GL A1/9-10 = −14.05	4.57

Table 5 Punching shear at column surface

	Limiting shear (N/mm^2)	Type of model		Percentage of difference (%)
		Type 1	Type 2	
Location at maximum reaction (kN)	5.0	$v = 4.11$ N/mm^2 (R = 73,700 kN) Column B2b	$v = 4.0$ N/mm^2 (R = 71,800 kN) Column B2b	2.65%
Location at maximum shear (N/mm^2)	5.0	$v = 4.32$ N/mm^2 (R = 41,000 kN) Column B8	$v = 4.33$ N/mm^2 (R = 41,100 kN) Column B8	0.23%

ultimate f_{cu} was equal to 45 N/mm^2 as mentioned in Table 4.1, and f_{ci} was equal to 35 N/mm^2 which is more than 75% of f_{cu}.

4 Conclusion

Based on this research study, the following conclusions can be drawn:

(a) Stresses

Type 1 is more practical in terms of maximum stresses at GL D/2–3 (Top −6.77 N/mm^2 at, Bottom −9.01 N/mm^2) compared to Type 2 maximum stresses at GL B-C/9 (Top −6.18 N/mm^2, Bottom −10.7 N/mm^2).

(b) Deflection

Type 2 is more practical in terms of deflection at GL A1/9–10, with 14.05 mm compared to Type 1 deflection at GL A1/9–10 that is 20.85 mm.

(c) Punching Shear Check
 i Type 1 maximum column reaction at Column B2B is 73700 kN that is more than Type 2 column reaction at Column B2B that is only 71,800 kN due to redistribution of loading from tendon arrangement that contributes to the value of column punching shear.
 ii The column size and slab thickness are important factors affecting punching shear. Adoption of drop cap at the column provides better punching shear resistance at column area.

Deflection can be controlled by additional tendon, increase the thickness of slab, additional reinforcement, and pre-camber of the slab before concreting. With additional tendon, more uplift force can be utilized, which improves the cracking and deflection behavior, in return increases the stiffness. In case of additional tendon increase P/A of the section, the slab thickness of the structure can be increased up to practical thickness. Other than that, pre-camber can be carried out by referring to BS8110 Clause 6.2.8.4.

For the effect of tendon direction, by comparing Type 1 and Type 2 deflection and stresses, the Type 1 deflection is 20.85 mm is more than Type 2. The Type bottom stress is -9.01 N/mm^2, less than bottom stress in Type 2. Since deflection can be controlled in many ways, Type 1 is fully recommended for analysis and design of post-tensioned transfer slab for this case study.

The different result in column surface punching shear at column B8 for Type 1 that is 4.32 N/mm^2 and Type 2 which is 4.33 N/mm^2 is about 0.23% only which is very minimal. Both cases complied with BS8110 Clause 4.3.8.2 which limiting the shear to $0.8\sqrt{fcu}$ or 5 N/mm^2. The distribution of loading is due to distribution of tendon at longer span and short span in both cases may cause the reaction is slightly different. In any case the punching shear is more than the allowable shear, it is suggested the column to be constructed with drop cap to reduce the punching shear effect as recommended in BS8110 Clause 3.7.1.5.

During the construction stage on site, the structure is considered prestressed structure after stressing work is carried out. Before the stressing works, the structure behaves as normal reinforced concrete. Due to this behavior, minimum reinforcement is required. Minimum areas of steel as recommended in BS8110 Clause 3.12.5.3 are needed to be provided at bottom and top slab for temporary load support.

The scope of study can be increased with the following recommendations and extended works:

1. There are a lot of codes of practice used around the world. In this study, only BS8110:1997 normally used in Malaysia was considered. It can be extended in comparison with other codes of practice.
2. This study only considers transfer slab. In future research, it may be extended to a car park or foundation structure.
3. The structure layout has an irregular shape, dual thickness and a lot of openings, in which for further research use symmetrical layout to study the effect of tendon direction in transfer slab.

4. For further research, it is possible to use different Finite Element Programming for structural analysis since there is a lot of computer programming to analyze post-tensioned structure.
5. Other than that, there are several methods in analyzing transfer slab element. Comparisons can be made.

Acknowledgements The authors would like to acknowledge the software support for this study from YLSE Technologies (M) Sdn. Bhd and other supports from Institute for Infrastructure Engineering and Sustainable Management (IIESM) and School of Civil Engineering, Universiti Teknologi MARA (UiTM) Shah Alam, Selangor, Malaysia. This support is gratefully acknowledged.

Referencess

Aalami, B. O. (2014). Post-tensioned building : Design and construction. *ACI Structural Journal, 2*, 11–12.

BS 8110-1:1997. (1997). Structural use of concrete. Part 1: Code of practice for design and construction, no. March, p. 160.

Craighead, G. (2009). High-rise building definition, development, and use. In *High-rise secur. fire life saf.* 3rd edn, no. August 2005 (pp. 1–26).

Hussien, O. F., Elafandy, T. H. K., Abdelrahman, A. A., Abdel Baky, S. A., & Nasr, E. A. (2012). Behavior of bonded and unbonded prestressed normal and high strength concrete beams. *HBRC Journal, 8*(3), 239–251.

Perception on Impact Land Reclamation from Pan Borneo Highway Project—Pilot Study

Mohamad Shakri Mohmad Shariff, Nazaruddin Abdul Taha, Mohd Azizul Ladin, Azwa Safiqah Darawati, Akhtarul Faiza Che Nen, and Che Maznah Mat Isa

Abstract "Land reclamation" is defined as creating new land from the sea, lake, or riverbed. Land reclamation occurs to construct residential areas, commercial or industrial estates, major roads, or other public purposes. In the context of road and highway construction, the effects of land reclamation have always been overlooked. Therefore, in this research, the perceptions of industry players were obtained to determine the perspective on the land reclamation project for the Pan Borneo Highway project. This paper presents some of the effects of land reclamation resulting from the Pan Borneo Highway Project. The Statistical Package for Social Science (SPSS) software was used to analyse the questionnaire survey data. The results showed that for socioeconomic impact, two variables were significantly associated with the profession: resettlement for locals who had their homes demolished and an increase in land value. In terms of environmental impact, five variables were significantly associated with the profession: disturbance of flora and fauna habitats due to deforestation, soil erosion, floods, landslides, and noise pollution.

Keywords Land reclamation · Pan Borneo Highway · Socioeconomic impacts · Environmental impact

M. S. Mohmad Shariff
Centre for Geotechnics and Transportation, Faculty of Engineering and Quantity Surveying, INTI International University, Nilai, Negeri Sembilan, Malaysia

N. Abdul Taha · M. A. Ladin · A. S. Darawati
Faculty of Engineering, Universiti Malaysia Sabah, 88400 Kota Kinabalu, Sabah, Malaysia

A. F. Che Nen
School of Civil Engineering, College of Engineering, Universiti Teknologi MARA, 40450 Shah Alam, Selangor, Malaysia

C. M. Mat Isa (✉)
Center of Civil Engineering Studies, Universiti Teknologi MARA Pulau Pinang Branch, Permatang Pauh Campus, 13500 Pulau, Pinang, Malaysia
e-mail: chema982@uitm.edu.my

U. R. Hashim et al. (eds.), *Green Infrastructure*,
https://doi.org/10.1007/978-981-99-7003-2_15

1 Introduction

Economic development is practically synonymous with urbanisation, and with limited land to accommodate the growing population, land reclamation has been a solution to this problem. Land reclamation is one of the main areas of interest for the dredging industry, including dredging massive amounts of sea sand and transporting it to create new land for industrial or infrastructure purposes (Mostafa, 2012). In the last decade (2002–2013), Malaysia has constructed 133,000 km of official roads, and under the 11th Malaysia Plan (2016–2020), there are ambitious proposals for further expansion of road networks with an emphasis on rural connectivity (Alamgir et al., 2017). Within Sabah, land reclamation usually occurs within the river, sea, foreshore area, or wetland forest to build residential areas, commercial or industrial estates, construction of major roads, or other public purposes (Alam Sekitar, 2012).

Considering the contribution that good roads will make to economic development and rural connectivity, expanding the country's road network is entirely justifiable. Alamgir et al. (2018) also state that problems may arise, however, when road proponents are arguably unaware of the full extent of the risks that bad roads can pose to the economy, society, and environment. Even though land reclamation boosts rapid economic development, it also has serious cumulative effects on the ecological environment. Dredged sediments that are not treated before being used for land reclamation will lead to long-term environmental impacts, including effects on the marine environment and fishing industries. This study aims to research the perception of the effect of land reclamation on the Pan Borneo Project.

2 Literature Review

Land reclamation is the process of creating new land from the oceans, seas, riverbeds, or lake beds. Basically, land reclamation is the improvement of lands to make them suitable for intensive use. The initial plan is to upgrade the current single carriageway to a four-lane or dual carriageway with the JKR R5 Standard. As has been highlighted by Nadzir et al. (2014), land reclamation is the process of creating new land from the sea. Naimi et al. (2018) also stated that land reclamation is the gaining of land from water bodies, such as the sea, coastal shores, or wetlands. For this study, land reclamation is needed to build a road for the Pan Borneo Highway project. This is likely to happen in areas near rivers or wetlands.

The purpose of land reclamation is to accommodate the rapid growth and development of the human population due to the shortage of land. In this research, developing nations' growing populations and aspirations have created a powerful impetus for road development. However, the tropics also hold most of the planet's biodiversity and other environmental services for humanity and nature. Roads can positively impact economic growth and social development and provide local producers with easy access to larger urban markets if they are strategically located and built properly

(Alamgir et al., 2018). Several countries that have adopted the reclamation project in Asia are Malaysia, Indonesia, Singapore, China, Japan, Bahrain, and Hong Kong. Poorly planned or executed roads will create numerous environmental, economic, and socio-political problems.

The social aspects of the community have changed as a result of land reclamation. Previous studies in Makassar coastal areas found that the community that lives in that area has experienced the changes (Saleh et al., 2016). Reclamation has forced the local people to change their occupations. Most of the people in the Makassar coastal area have jobs related to the sea, which are fishermen, boat makers, fishmongers, or searching for mussels. Saleh et al. (2016) stated that 45 (forty-five) families living in the coastal areas lost their houses and 456 fishermen in Panambungan lost their jobs. Sandirasegaran and Manap (2016) conducted research which showed that as land reclamation occurs, the income of local people may be reduced, and the businesses of local people may be affected. In this research, which was done in Kampung Teluk Empang and Kampung Langkah Baik, most of the businesses had to move to other places, which led to local people having big losses, such as on fixed assets and temporary assets. This shows that before land reclamation projects are to be done in an area, stakeholders need to gather information on socioeconomic so that society will not be affected much.

The environmental impacts due to reclamation works are the loss of shores, seabed, flora, and fauna within the reclaimed areas. Due to reclamation work, natural resources such as forests and fisheries are being put under pressure. Besides that, poorly planned reclamation work will increase the water table, cause intrusion of seawater, and increase salinity in soils and groundwater. Research on the economic impacts of reclamation in coastal areas claimed that it would disturb the businesses of local people. The social impacts of land reclamation are that local people who live in coastal areas need to change their occupation since their main income comes from the sea, which has been reclaimed.

Research has been done by Alamgir et al. (2018) to provide three risks which are associated with road development, namely environmental risk, economic risk, and socio-political risk. In terms of environmental risks, road development poses threats to biodiversity, increased surface erosion, landslides in steep terrain, deforestation, and increasing effects on aquatic biodiversity and downstream community income, such as pollution of fishing areas. For economic risks, increased construction costs due to the steep terrain of tropical regions, more cut and filled tunnels, and landslide alleviation are required. Another contributing factor to economic risk is that roads constructed on flatter or swampy ground need culverts, bridges, and drainage. This research paper was intended to determine the respondent's perception of land reclamation for the Pan Borneo Highway Project. Perception is the subjective process of acquiring, interpreting, and organising sensory information (Mcleod, 2018).

3 Research Methodology

Several stages of the methodology are literature and desk studies, interviews, questionnaires, and finally, data collection and analysis. To indicate the actual reflection on perception results, research samples should be needed of approximately 384 respondents for a confidence level of 95%, with the population size of Kota Kinabalu at around 500,000. As an initial investigation, a pilot study was done with industrial players being chosen as the respondents. 62 respondents from different professions have given their feedback.

The questionnaire consists of 15 questions divided into 6 main sections: personal details, awareness of land reclamation, socioeconomic impacts, environmental impacts, and factors that affect the project cost of Pan Borneo Highway Construction. The questionnaire survey data were collected and entered into the database and analysed using descriptive analysis, correlation, and the Chi-Square Test. The data analysis was done with IBM SPSS because it can do both parametric and non-parametric comparison analysis (Azman Ong & Puteh, 2017).

4 Results and Discussion

The first part of the questionnaire is regarding the respondent's details. Figure 1 shows the data obtained from 62 respondents who have answered the questionnaire. From the 62 respondents, the data shows that most of the respondents are degree holders with 87.1% (54 responders), diploma holders with 8.1% (5 responders), Ph.D. holders with 3.2% (2 responders), and lastly, Master's holders with 1.6% (1 responder). This shows that the level of education exposed to the respondents was different.

For this study, the higher percentage of respondents were civil engineers, with 50% (31 respondents), half of the total number of respondents. The second most common occupation, with an 8.1% (5 responders) share, is quantity surveyor, followed by consultants, environmental controller officers, and government officers (EPD), each with a 6.5% (4 responders) share. Next, environmental engineers, with 4.8% (3 responders), followed by other professions with 1.6% (1 responder). There are so many civil engineers with the highest percentage because researchers gave out most of the survey questionnaires to civil engineers working in Jabatan Kerja Raya (Cawangan Pan Borneo) and the Department of Irrigation and Drainage Sabah. Mostly, there is only one environmental engineer or environmental control officer who works in the Environment Protection Department.

In Fig. 2, most respondents had less than 5 years of experience, at 41.9% (26 respondents). Next, 12 respondents (19.4%) had less than 15 years of experience, followed by 10 respondents (16.1%) who had less than 20 years of experience. Lastly, 7 respondents (11.3%) had less than 10 years and more than 21 years of experience.

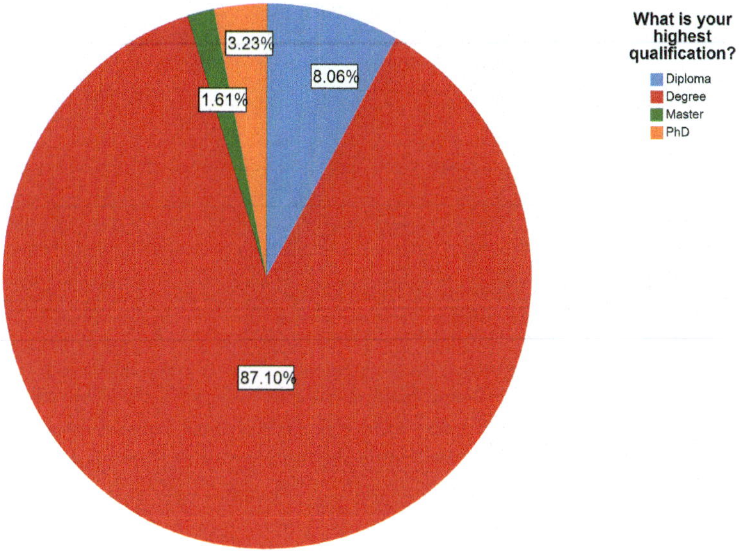

Fig. 1 Highest qualification of the respondent

From this experience level, different respondents might have more knowledge about land reclamation since they have been in the field for so long.

In Fig. 3, most of the respondents who contributed to the feedback were from the private sector, with 62.9%, and only 37.1% of the respondents came from the government sector. This is because most of the respondents working in the private sector had mostly distributed the online survey form to many respondents. In addition to that, since more than half of the respondents were from the private sector, the

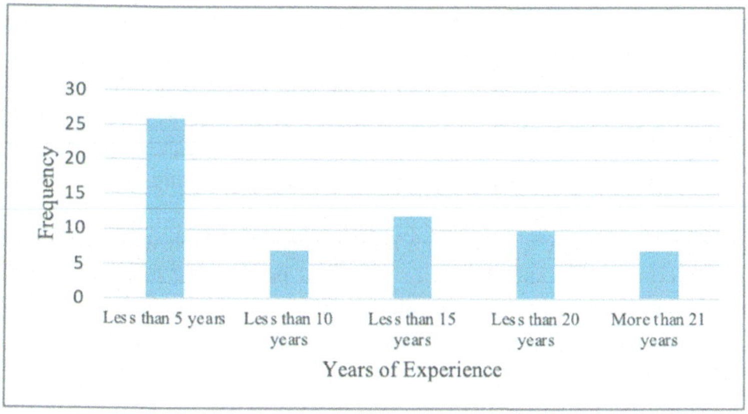

Fig. 2 Years of experience of respondents

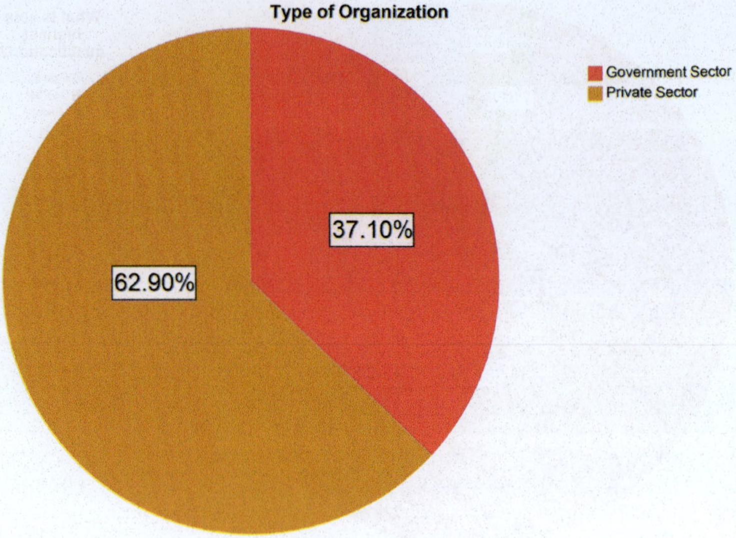

Fig. 3 Type of organisation where the respondent works

different professions also varied. Section B is the respondents' awareness towards land reclamation.

From Fig. 4, most respondents have heard the term "land reclamation" by about 91.94%, while 8.06% have never heard of the term "land reclamation." From the data, most of the respondents know about land reclamation because of their profession and the respondents' exposure during their working experience.

Figure 5 shows the respondents' level of knowledge about land reclamation with a rate of 1 as the minimal amount of knowledge and 5 as the maximum amount of knowledge. The graph shows that the histogram is skewed to the left; most of the respondents had more knowledge about land reclamation. Section C displays the socioeconomic impacts of land reclamation.

Table 1 presents the findings of the perception of the respondents towards the socioeconomic impacts of land reclamation due to the Pan Borneo Highway Project. The investigated socioeconomic impacts consist of; an increase of migrant workers/colonists, which affect the inhabitants, the increase in land value; residential building adjacent to a highway that is being constructed will decrease in demand because it disturbs the aesthetics; new or change in access roads which will increase journey time; traffic congestion; and relocation for local people that have their houses for demolition. The highest mean was relocation for local people whose houses are being demolished ($M = 2.89$), followed by increases in land value ($M = 2.68$), traffic congestion ($M = 2.27$), an increase in migrant workers/colonists affecting local residents ($M = 2.23$), new/changed access roads that increase journey time ($M = 2.11$), and residential building adjacent to a highway that is being built will decrease in demand because it disturbs the aesthetics ($M = 1.81$). The results obtained show

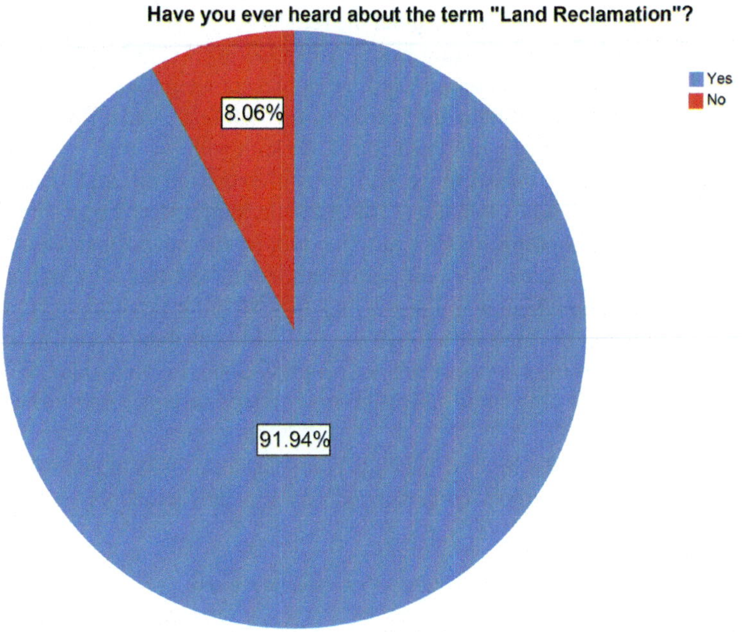

Fig. 4 Respondents' knowledge about the term "Land Reclamation"

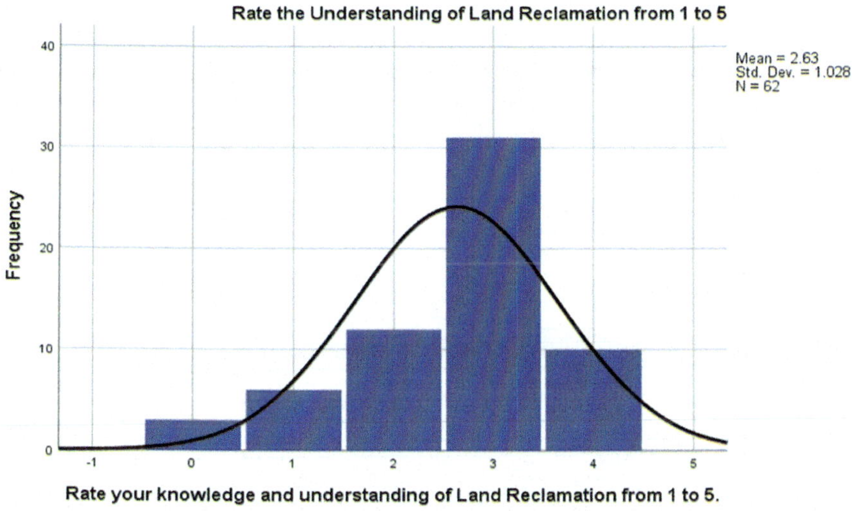

Fig. 5 Rate of understanding of respondents' knowledge of land reclamation

that the respondents' main concern about the relocation of settlement local people is the huge impact of land reclamation from the Pan Borneo Project.

Table 2 shows the results of the perceptions from respondents about environmental impacts. The disturbance of flora and fauna caused by forest clearing had the highest mean ($M = 2.95$), followed by soil and coastal erosion with ($M = 2.77$), particularly in coastal areas where some of the areas have suffered from severe erosion due to previous projects on land reclamation. Thirdly is flooding ($M = 2.65$). The respondents presumed that flooding could occur if the project was not properly handled and organised by the contractor. Thus, air pollution ($M = 2.48$) and noise pollution ($M = 2.47$) could happen while the project is ongoing. Then for landslides ($M = 2.44$), Sabah state, which is covered with hilly terrain, could contribute to landslide events if the design and construction work are not properly done. Finally, material leakage with $M = 1.922$ is possible while construction is underway, particularly when heavy machinery is used.

Table 3 shows that the highest factor is compensation for land acquisition. The Pan Borneo Highway project cannot avoid running into residents' houses when

Table 1 Mean and standard deviation of the socioeconomic impacts

No.	Variables	*N*	Min	Max	Mean	Std. dev.
1.	Relocation for local people who have their houses for demolition	62	1	4	2.89	1.026
2.	Traffic congestion	62	0	4	2.27	1.074
3.	New/change in access road which increase journey time	62	0	4	2.11	0.994
4.	Residential building adjacent to the highway that being constructed will decrease in demand because it disturbs the aesthetic view	62	0	4	1.81	1.099
5.	Increase in land value	62	0	4	2.68	1.113
6.	Increase of migrant workers/colonists which affect the local inhabitants	62	0	4	2.23	1.137

Table 2 Mean and standard deviation of environmental impacts

No.	Variables	*N*	Min	Max	Mean	Std. dev.
1.	Disturb the flora and fauna habitat due to forest clearing	62	0	4	2.95	1.165
2.	Soil and coastal erosion	62	0	4	2.77	1.078
3.	Flooding	62	0	4	2.65	1.088
4.	Landslides	62	0	4	2.44	1.002
5.	Air Pollution	62	0	4	2.48	1.141
6.	Noise pollution	62	0	4	2.47	1.051
7.	Leakage of dangerous material on road surface	62	0	4	1.92	1.076

Table 3 Mean and standard deviation of factors affecting the project cost

No.	Variables	*N* valid	Min	Max	Mean	Std. dev.
1.	Compensation for land acquisition	62	0	4	3.03	1.024
2.	Maintenance cost for roads	62	1	4	2.69	0.822
3.	Constructing roads in steep, dissected, or swampy terrain	62	0	4	2.90	0.882
4.	Restoration cost for other sector of the economy such as fisheries and agriculture that is affected by road construction	62	0	4	2.32	1.004
5.	Increase in labourers due to large project size and time constraint	62	0	4	2.52	0.936
6.	Fluctuation in the price of materials	62	1	4	2.27	0.872
7.	Delay in materials procurement	62	0	4	2.24	1.019
8.	Equipment availability and failure	62	0	4	2.24	1.112

constructing the road. Thus, according to Sect. 3 of the Sabah Land Acquisition Ordinance, the land acquisition process allows the government to acquire land compulsorily for public purposes (Ordinance & Amendments, 2012). Two variables correlate significantly with years of experience: compensation for land acquisition and the construction of roads in steep, dissected, or swampy terrain. According to research, roads on steep, dissected, or swampy terrain are costly to build. These higher costs stem from a need for more bridges, tunnels, culverts, and cut-and-fill operations, along with the complexity of building techniques (Alamgir et al., 2017). According to one of the respondents, even though it is costly to build, constructing a viaduct to replace the conventional method of reclaiming sea, river, or swamp areas is another method to reduce the effect of land reclamation. Based on what people said, the best way to reduce the negative effects of land reclamation is to do a lot of planning before starting the project.

5 Conclusion

This study examined the perceptions of 62 industrial players towards land reclamation projects for the Pan Borneo Highway based on their professional views due to years of working experience. The findings show that relocation for local people with demolished houses was significantly affected due to socioeconomic impacts, and the disturbance to flora and fauna due to forest clearing was the most severely affected. It was evident that the significant factors contributing to the project cost of the Pan Borneo Highway were the compensation for land value, construction of roads in steep, dissected, or swampy terrain.

Acknowledgements This publication resulted from research supported by the INTI International University for the research grant, INTI-FEQS-01-14-2023. The content is solely the responsibility of the authors.

References

Alamgir, M., Campbell, M. J., Sloan, S., Goosem, M., Reuben Clements, G., Mahmoud, M. I., & Laurance, W. F. (2017). Current biology review economic, socio-political and environmental risks of road development in the tropics. *Current Biology, 27*, R1130–R1140. https://doi.org/10.1016/j.cub.2017.08.067

Alamgir, M., Campbell, M. J., Sloan, S., Phin, W. E., & Laurance, W. F. (2018). Road risks & environmental impact assessments in Malaysian road infrastructure projects. *Jurutera,* (February), 13–16.

Azman Ong, M. H., & Puteh, F. (2017). Quantitative data analysis: Choosing between SPSS, AMOS in social science research. *International Interdisciplinary Journal of Scientific Research, 3*(1).

Alam Sekitar, J. P. (2012). *EIA guidelines for land reclamation activities*. Environment Protection Department, ISBN: 978-967-11712-0-2.

Mcleod, S. (2018). Visual perception simply psychology gregory (1970) and top down processing theory, 1–10.

Mostafa, Y. E. S. (2012). Environmental impacts of dredging and land reclamation at Abu Qir Bay, Egypt. *Ain Shams Engineering Journal, 3*(1), 1–15. https://doi.org/10.1016/j.asej.2011.12.004

Nadzir, N. M., Ibrahim, M., & Mansor, M. (2014). Impacts of coastal reclamation to the quality of life: Tanjung tokong community Penang. *Procedia - Social and Behavioral Sciences, 153*, 159–168. https://doi.org/10.1016/j.sbspro.2014.10.050

Naimi, A. Al, Karani, G., & Littlewood, J. (2018). Stakeholder views on land reclamation and marine environment in Doha, Qatar. *Journal of Agriculture and Environmental Sciences,* 6(1), 32–39. https://doi.org/10.15640/jaes.v7n1a4.

Ordinance, L. A., & Amendments, L. O. F. (2012). State of Sabah. 7(May), 1–22.

Saleh, M. S., Hakim, A., & Tamsil, A. (2016). Public Perception towards reclamation of coastal areas in Makassar. *Resources and Environment, 6*(3), 47–52. https://doi.org/10.5923/j.re.20160603.01

Sandirasegaran, K., & Manap, N. (2016). Impacts of dredging and reclamation projects. *Jurnal Teknologi,* 78(7), 139–143. https://doi.org/10.11113/jt.v78.9506

Challenges, Characteristics, and Success Factors in Implementing Green Highway Using SEM-PLS Model

Fatin Najwa Mohd Nusa, Che Maznah Mat Isa, Siti Zaharah Ishak, and Intan Rohani Endut

Abstract Green highway differs from conventional due to their concept, fundamentals, characteristics, technologies, and innovations. Although the Malaysian Highway Authority (MHA) has been introducing guidelines for green highways, the implementation in Malaysia is still far behind because of the challenges faced by the highway concessions in complementing the guideline requirement. This study comprises four objectives: (1) to investigate the green highway challenges faced by highway concessions in Malaysia; (2) to determine the green highway characteristics in the Malaysian context; (3) to identify the green highway success factors implementation in Malaysia; and finally (4) to develop a relationship model between characteristics and challenges towards the success factors impacting green highway. This study used triangulation research to obtain the primary data using unstructured interviews and questionnaire surveys. The data were analysed using content analysis, descriptive analysis, Criticality Index assessment, and Structural Equation Modelling-Partial Least Squares (SEM-PLS). 12 unstructured interviews and 63 usable postal questionnaires were received to establish the model. Results show that (1) legal and regulatory; technical and physical; financial and cultural are the green highway challenges; (2) planning, design, construction; operation, and maintenance stages are characteristics of implementing green highways; (3) innovation; construction activities; material and resources are success factors in implementing green highway and proposes; and (4) a relationship model shows that the green highway

F. N. Mohd Nusa · S. Z. Ishak
Malaysia Institute of Transport (MITRANS), UiTM Shah Alam, Selangor, Malaysia
e-mail: fatinnajwa@uitm.edu.my

S. Z. Ishak
e-mail: sitiza406@uitm.edu.my

F. N. Mohd Nusa · S. Z. Ishak · I. R. Endut
School of Civil Engineering, College of Engineering, UiTM Shah Alam, Selangor, Malaysia

C. M. Mat Isa (✉)
Center of Civil Engineering Studies, Universiti Teknologi MARA Pulau Pinang Branch, Permatang Pauh Campus, 13500 Pulau Pinang, Malaysia
e-mail: chema982@uitm.edu.my

U. R. Hashim et al. (eds.), *Green Infrastructure*,
https://doi.org/10.1007/978-981-99-7003-2_16

characteristics and challenges have a significant and positive effect in implementing green highway.

Keywords Sustainable highway · Best practices · Highway applications · Acceptance · Model

1 Introduction

People need highways to carry out activities in their daily lives. As the essential infrastructure, the highway plays a role as a "backbone" for the nation's social and economic development (MHA, 2015). But in the early 1960s, highway projects had adversely affected the environment, resources, raw materials, and people, which later were obliged to enforce the Environmental Impact Assessment (EIA) through legislation for sustainable development (Lim, 2009; Ogola, 2007). Sustainable development goal elements have been extensively considered in the construction industry. However, the majority of highway projects give rise to problems such as the greenhouse effect and ozone depletion; significant historical and archaeological resources diminution; substantial infrastructure impairment such as public parklands, public facilities, and prime agricultural lands; chemical contamination; and hazardous materials exposure (Wright et al., 2004). According to Clevenger et al. (2013), Saqr and Musa (2011), and Clark et al. (2009), highway projects contribute to a severe environmental problem, especially concerning carbon dioxide emission from the transportation sector.

The green highway is a concept and a tool for roadway design that can help to reduce the negative impacts on the environment. Various initiatives have been developed to enhance green highways (Rafidah & Zaimi, 2014; Mukta et al, 2020; Zeiada et al., 2022). Malaysia Highway Authority (MHA) has introduced a green highway assessment tool, namely the Malaysia Green Highway Index (MyGHi), to promote green highways (Zakaria et al., 2013). MyGHi is a tool to confirm the green credentials of highway construction aspects for every phase of highway construction. MyGHi aims to reduce environmental problems and excessive resource consumption in highway projects. It also promotes green highway applications in the current highway construction practice, which covers the use of recycled materials, ecosystem management, and energy reduction program that can increase the water quality of stormwater runoff and maximise overall societal benefits (MHA, 2015; Attahiru et al., 2019; Khoshnava et al., 2020; Shang et al., 2022).

Although Malaysia has officially launched the MyGHI manual in the past seven years, this concept faced challenges in being implemented in Malaysia. The challenges include deteriorating market condition, traffic congestion, low green energy supply and shortfall of maintenance funding, and capacity expansion to meet the increasing demand (MHA & UTM, 2015; Nusa et al., 2015; Rafidah & Zaimi, 2014). Until now, no enforcement of using MyGHI to assess highways in Malaysia, and most of the research and development is focusing on building projects using the Green Building Index (GBI) (Huang & Yeh, 2008).

Thus, little attention has been paid to examining factors impacting green highway implementation in an integrated model, which can show the relationship between green highway challenges and green highway characteristics that influence the success factors of green highway implementation. Therefore, a green highway model has been developed in this study to identify the success factors and best long-lasting alternative measures for green highway guidelines.

2 Methodology

This study used a questionnaire survey to obtain primary data and analyse quantitative data using descriptive analysis (SPSS version 22.0, Microsoft Excel, and Criticality Index assessment) and the green highway implementation model development using Partial Least Squares-Structural Equation Modelling (PLS-SEM).

2.1 Literature Review

The first stage of the research study involves a thorough review of literature in the area of green highway construction to identify the similarities and differences between the key study concepts, variables, and factors, and to find gaps in the existing literature. This research dealt with the available data from various sources such as books, journals, and conference proceedings regarding green highway challenges, characteristics, success factors to implement green highway, and the current green highway practice. The Malaysia Green Highway Index (MyGHI) manual was used as the guidance to design questionnaire survey.

2.2 Unstructured Interview

The unstructured interview was addresses each of the research areas highlighted: current practice in highway construction industry and problems related to the implementation of green highway practice in Malaysia. The unstructured interviews were conducted with 12 expert respondents. Findings from the unstructured interview were recorded and transcribed verbatim using content analysis and then documented accordingly into categories of challenges, characteristics, and success factors to develop the green highway model from the perspectives and opinions of the respondents. The findings from the respondents were taken into account for the development of the questionnaire survey.

2.3 Design of Questionnaires

The questions consisted of 3 sections. Part A was the respondent's information. Part B was a set of multiple-choice questions about green highway challenges for Malaysia. Part C of the survey questionnaire was about success factors for green highway implementation in Malaysia. The last part D of the survey questionnaire was about green highway characteristics for Malaysian practice.

2.4 Pilot Test

Before administering the online questionnaire survey, a pilot study was carried out among 12 respondents from government and private agencies to validate the questions in the survey. The population for this study comprised the personnel and representatives from the twenty-seven (27) highway concessions that manage tolled highways. A stratified random sampling technique was used to determine the sample size in this study from the overall population. The list of respondents in this study was obtained from "Persatuan Syarikat-syarikat Konsesi Lebuhraya" (PSKLM) within the purview of the Malaysian Highway Authority. This study employed the mixed mode survey as suggested by Converse et al. (2008) and Dillman et al. (2008). Postal and web-based questionnaires using Google Forms are fast, simple, and easy to administer and thus suitable for data collection of quantitative data (Lee & Egbu, 2006).

2.5 Criticality Index Assessment

The criticality index is a reliable statistical analysis method to identify and rank the main factors for each category generated (Che Nen et al., 2020; Shehu & Akintoye, 2009). Therefore, to identify and rank the main factors of challenges, characteristics, and success factors in implementing green highways in this study, the criticality index was used because of its popularity, reliability, and, effectiveness. The weights used for the importance index are 0, 0.25, 0.50, 0.75, and 1 (Cheng, 2002). Using this method, the weighting is substituted into the Likert scale and the maximum value of the criticality index is indicated as 1. Hence, any factors that show the highest values between 0 and 1 are considered critical factors. Equation 1 shows the relative criticality index.

$$C_m = \frac{\sum_{i-1}^{5} W_i X_i}{\sum_{i-1}^{5} X_i} \tag{1}$$

where:

Cm = *Criticality Index*

I = *Responses Category Index* = *1,2,3,4, and 5 (Position on the Likert Scale).*

Wi = *Weight Assigned to Ith Response* = *0, 0.25, 0.5, 0.75, and 1, Respectively*

Xi = *Frequency of the Ith Response Given as a Percentage of the Total Responses*

Using Eq. 1, data collected from the questionnaire survey was evaluated to calculate the respective criticality index. Higher mean scores of the Likert scale ranging from 1 (very weak) to 5 (very strong) will indicate higher criticality for the main factors in this study (Shehu & Akintoye, 2009).

2.6 Structural Equation Modelling (SEM)

SEM is one of the most convenient statistical analyses and a well-known second-generation technique for social and business sciences (Hair et al., 2014; Henseler et al., 2014; Sarstedt et al., 2011). It is a powerful tool that can show simultaneous relationships between variables with latent variables and simultaneous relationships among the latent variables using multivariate techniques (Yue, 2004). SEM is considered confirmatory when dealing with hypotheses of existing theories and concepts and classified as exploratory when searching for latent patterns in the data collection (Hair et al., 2014). This study selected the PLS-SEM (Partial Least Squares) based on the aim of the study; to explore the respondents' opinions and to predict the key target of the constructs. SEM applies the model structure control by allowing the investigation of latent variables. This is crucial for developing a relationship between characteristics and challenges constructs towards the success of the green highway implementation because the model structure does not directly measure the latent variables.

2.7 Partial Least Squares (PLS)

The purpose of using partial least squares is to make possible or maximise a prediction rather than to fit a prediction suitable to theory development (Hair et al., 2014). The model for hypothesis testing in this study was developed using the SmartPLS 3.0 software package (Chin, 2010). There are two main objectives of using PLS: (i) to maximise the explained variance in the dependent constructs; and (ii) to evaluate the quality of the data based on the characteristics of the measurement (Hair et al., 2014). Therefore, the PLS procedure is similar to the procedure of using multiple regression analysis. It is a useful technique for developing a model of latent constructs under the conditions of non-normality for small to medium sample sizes (Degross et al., 1996).

2.8 Partial Least Squares—Structural Equation Modelling (PLS-SEM)

PLS-SEM estimates latent variable scores by assessing their exact linear combinations of each associated manifest variable (Fornell & Bookstein, 1982). PLS-SEM was used in this study to state the effect relationship of one parameter to another parameter, to clarify how influential each parameter is to the others, and to predict a theory of a certain topic (Henseler et al., 2014). The formative construct will be part of the structural model in this study. Based on the non-normal data condition in this study and the small sample size, latent variable scores will be used in the subsequent analysis of the model. Data characteristics that suit the usage of PLS-SEM in this study are small sample sizes (N = 63), involving the formative measurement of latent variables. The flowchart of research methodology for this research is explained in Fig. 1 as follows:

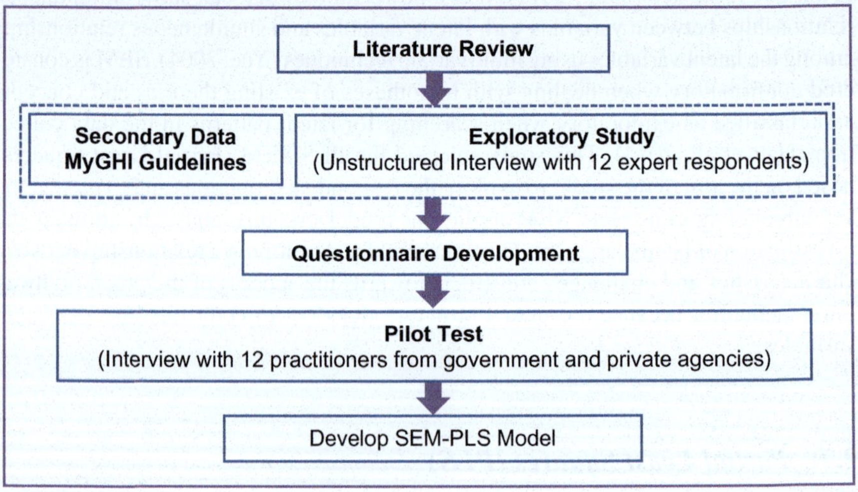

Fig. 1 Process of SEM-PLS model development

3 Results and discussion

3.1 Demographic Profiles of Respondent

Table 1 shows the respondent's gender and their sector in highway projects. 41 male and 22 female respondents were involved in this study. Contractors, highway concessions, government professionals, consultants, private professionals, and academicians were the study's sectors.

The table indicates that the majority of respondents involved in this study were in the highway concessions sector with 33.3%, while the minority of respondents were in the academic sector with 3.2%. These groups of professional respondents were directly involved with the planning, design, construction, operation, and maintenance of highway projects for tolled highways in Malaysia.

Table 2 summarises the area of responsibility and working experience. Based on the responses, 44.4% of the respondents refused to mention their current position. Most respondents, which accounted for 17.5%, were project engineers in highway construction. Furthermore, respondents among the project manager and site engineer showed an equal value of participation which was 12.7%, and the rest of the respondents were among the design engineer and site supervisor with 9.5% and 3.2%, respectively. Meanwhile, the majority of respondents, or 30.2%, had one year of working experience, followed by 27% of respondents with 1–3 years of working experience in the highway construction industry involved in this study. However, this study also received 19% responses from professionals, particularly those who had working experience of more than ten years. In addition, there were responses from respondents with working experience of 3–5 years (12.7%) and 5–10 years (11.1%).

Table 1 Demographic information (N = 63)

Variable	Demographic information	
	Frequency	Percentage
A1: gender		
Male	41	65.1
Female	22	34.9
Total	63	100
A2: sector		
Highway concessions	21	33.3
Contractors	18	28.6
Government professionals	14	22.2
Consultants	4	6.3
Private professionals	4	6.3
Academician	2	3.2
Total	63	100

Table 2 Area of responsibility and working experience (N = 63)

Variable	Area of responsibility and working experience	
	Frequency	Percentage
A3: current position		
Project engineer	11	17.5
Project manager	8	12.7
Site engineer	8	12.7
Design engineer	6	9.5
Site supervisor	2	3.2
Others	28	44.4
Total	63	100
A4: working experience		
1 year	19	30.2
1–3 years	17	27
3–5 years	8	12.7
5–10 years	7	11.1
>10 years	12	19
Total	63	100

3.2 Main Challenges of Green Highway

Table 3 shows the criticality index analysis to rank sub-categories based on four categories; (i) legal and regulatory, (ii) financial, (iii) technical and physical, and (iv) cultural. Responses show that the sub-category of "lack of cooperation between agencies and communities" (cultural) and "perceived high costs; short-term and/or long-term" (financial) scored the high criticality index value of 0.766 and 0.758, respectively. This result shows that respondents believed the cultural and financial were the main challenges to implementing green highways in Malaysia. Thus, cultural and financial could be seen as the most crucial parts to be focused on by the highway project stakeholders to realise green highway implementation in the future.

Meanwhile, from the legal and regulatory category, respondents agreed that "local laws are unclear" as the main reason for the challenge of green highway implementation with a criticality index of 0.746, the highest ranking in this category. In contrast, the lowest ranking in this category was "no long-term government structure for maintenance and ownership," such as policies, guidelines, or standards which burdensome to implement green highways in Malaysia. This result was not surprising because there are no available policies and local rules for green highway in MyGHI."

The main challenge that affected the technical and physical category was the "research has not proven benefits yet" towards green highways in Malaysia with a criticality index value of 0.745. This statement is supported by the interviewee statement that usually, stakeholders in highway projects will face challenges related to

Table 3 Criticality index and overall ranking for main challenges

Assessment items	Results			
	Variables	Criticality index	Rank	Overall rank
Green highway challenges items				
Legal and regulatory				
Local laws are unclear	B3	0.746	1	4
State policies are unclear	B5	0.738	2	7
Local rules are lacking or restrictive	B2	0.734	3	9
State policies are lacking or restrictive	B4	0.726	4	11
No long-term government structure for maintenance and ownership	B1	0.663	5	14
Financial				
Perceived high costs	B7	0.758	1	2
Not enough incentives	B6	0.738	2	8
Not enough information about costs and benefits	B10	0.734	3	10
Lack of funding for ongoing maintenance	B9	0.718	4	12
Lack of funding for design and implementation	B8	0.691	5	13
Technical and physical				
Research hasn't proven its benefits yet	B11	0.745	1	5
Limited appropriate sites	B12	0.743	2	6
Limited access to necessary materials	B13	0.639	3	17
Lack of design standards and best management practices	B14	0.580	4	18
Not enough technical knowledge and experience in development community	B15	0.579	5	19
Cultural				
Lack of cooperation between agencies and communities	B18	0.766	1	1
Not enough information	B19	0.754	2	3
Community is not convinced about its effectiveness	B17	0.659	3	15
Green highway is undervalued by the development community	B16	0.643	4	16
Green highway is aesthetically less pleasing	B20	0.484	5	20

the (i) limited appropriate sites for scheduled waste management and applied strategic detention pond, (ii) limited access to necessary materials, for example porous asphalt as green material to construct highway, (iii) lack of green highway design standard and best management practices, and (iv) lack of technical knowledge and experience in the green highway (Nusa et al., 2015). Thus, this analysis shows that there is a need to prove the benefits of the green highway to the community and stakeholders in highway projects, especially the government and private organisations, to enforce the implementation of the green highway concept in the highway project. Thus, based on the results obtained from the criticality index assessment results, the study question regarding the main challenges of green highways in Malaysia was addressed and answered.

3.3 Main Characteristics of Green Highway

Table 4 shows the criticality index and the overall ranking value for each sub-category related to green highway characteristics. There are four stages; (i) green highway at the planning stage, (ii) green highway at the design stage, (iii) green highway at the construction stage, and (iv) green highway at operation and maintenance stages. Most respondents agreed that the design stage is crucial in determining the green highway characteristics, showing the highest criticality index value of 0.833 for the sub-category of "pavement design life is important for the green highway." In addition, respondents recommended to the stakeholders in highway projects to focus on sustainable pavement design, which can extend the pavement lifetime for road user safety and comfort.

This study hypothesised that H_{2b} = Green highway characteristics would significantly influence the success of implementing green highways. Respondents highly recommended that stakeholders should do early planning at the design stage to ensure the successful implementation of the green highway. This statement is supported by the high value of the criticality index for these categories: (i) Light Emitting Diode (LED) with solar power shall be installed at the highway, parking areas, toll plaza office, rest and service area, operation, and maintenance building, supervision building, lay-by, integrated station building, and regional office because it is very cost-effective, no standing charge or ongoing cost once installed and environmentally friendly (Crd = 0.822); (ii) Designer shall consider using solar energy to generate electricity for street lighting, and other facilities at RSA, toll plaza, and lay-by for green highway purpose (Crd = 0.809); and (iii) Rainwater harvesting system shall be installed at RSA, toll plaza office, supervision building, and regional office for general cleaning purpose, toilet usage, landscaping, and to reduce the use of treated water for green highway (Crd = 0.802). Ashoka Highway Research Centre (AHRC) (2002) also supported that the stakeholders should customise a green highway project to follow the minimum utilisation of natural resources, materials, and products at the design stage. Respondents also clarified that green highway characteristics were not final and may continuously change depending on the innovative practices, research, and development of various aspects of the green highway concept. Thus, based on the results obtained from the criticality index assessment, the study question regarding the main characteristics of green highways in Malaysia was addressed and answered.

Table 4 Criticality index and overall ranking for main characteristics

Assessment items	Results			
	Variables	Criticality index	Rank	Overall rank
Green highway characteristics items				
Green highway at the planning stage				
Design should consider comments and suggestions from all stakeholders and beneficial to both project owner and public	D4	0.782	1	9
Design should linked to the public transport such as bus station or railway station	D5	0.762	2	13
Design should given a high priority to wildlife crossing structures	D6	0.758	3	14
Mitigation measures to give minimal environmental impact, such as tunnel, viaduct, bridge, etc., should be introduced	D3	0.754	4	21
The protected wild plant shall be identified and moved to save places such as national parks during EIA at the reserved forest project area for the future	D7	0.746	5	25
The number of access and interchange shall be minimised to avoid traffic weaving and braking	D1	0.659	6	53
In the hilly and mountainous area, the proposed design speed is within 80 km/h to 90 km/h	D2	0.575	7	55
Green highway at design stage				
Pavement design life is essential for green highways	D9	0.833	1	1
Light emitting diode (LED) with solar power shall be installed on the highway, parking areas, toll plaza offices, rest and service areas, operation and maintenance buildings, supervision buildings, lay-by, integrated station buildings and regional offices	D23	0.822	2	2
Usage of solar energy to generate electricity for street lighting and other facilities at RSA, toll plaza, and lay-by	D22	0.809	3	3

(continued)

Table 4 (continued)

Assessment items	Results			
	Variables	Criticality index	Rank	Overall rank
Rainwater harvesting system shall be installed at RSA, toll plaza office, supervision building, and regional office	D24	0.802	4	4
Retention and enhancement of view can be added to the aesthetic pleasure of the RSA	D28	0.794	5	7
The structure's surface that allows creepers to grow but still maintains the structure's integrity	D12	0.786	6	8
Detention pond must be strategically designed to store surface water naturally	D14	0.774	7	10
Sight lines and visibility are vital at junctions and must frequently maintain by using either low planting of taller shrubs or trees with clear lower limbs	D31	0.774	8	11
All buildings in highway reserves, such as toll plazas, operation offices, depots, and storerooms, shall fulfill the green building assessment criteria requirement	D8	0.758	9	15
Planting in and around the RSA can reduce noise from traffic, buildings, and car parking. for safety, some gaps should be left in the vegetation along the road boundary to allow passive observation for passing motorists	D25	0.758	10	16
Colourful and textured planting adds to the pleasurable experience of RSA instead of providing shelter, shade, and visual enhancement	D29	0.758	11	17
Vegetation within medians helps the visual separation of the roadways	D34	0.758	12	18
Foundry sand, the spent sand used in metal casting, can be recycled for road embankments, manufactured soils, flowable and structural fill, and road-base and sub-base	D17	0.754	13	22
Trees and other vegetation if are high enough, wide enough, and dense enough, can be planted as noise barriers for the green highway	D20	0.754	14	23

(continued)

Table 4 (continued)

Assessment items	Results			
	Variables	Criticality index	Rank	Overall rank
Construction and demolition debris such as concrete and reclaimed asphalt pavement (RAP); gypsum and drywall; brick, and block can be recycled	D16	0.750	15	24
Large junctions, including interchanges, should be well integrated with extensive planting	D30	0.746	16	26
Medians wider than 1 m can provide enough space for plants to grow and allow maintenance access	D33	0.746	17	27
Grass areas within medians should be replaced with less long-term maintenance plants	D38	0.746	18	28
Depressed medians for drainage function shall be planted with wetland species	D36	0.742	19	29
Medians and roundabouts associated with junctions should be planted or paved	D32	0.742	20	30
Wind energy can be used to generate electricity for street lighting and a variable message System (VMS)	D21	0.742	21	31
Vegetation reduces the effect of oncoming headlight glare by using low-growing shrubs for clear visibility	D35	0.738	22	32
Using the locally sourced native plant species can emphasise the indigenous character of an area while reinforcing the ecological value of RSA planting	D27	0.738	23	33
Coal combustion materials such as fly ash, boiler slag, flue gas desulfurisation, and bottom ash can be recycled to produce Portland cement and concrete, wallboard panel products, and foldable and structural fill	D15	0.734	24	35
Consumer by-products such as crushed glass and shredded tires can be used for the base; sub-base, fill and filter media; road material mixes, and energy reduction	D19	0.734	25	36
Porous asphalt pavement is an initiative for the most natural process and is sustainable for the green highway	D11	0.730	26	37

(continued)

Table 4 (continued)

Assessment items	Results			
	Variables	Criticality index	Rank	Overall rank
Bio-retention swale consists of a soil bed planted with suitable vegetation that shall be constructed for stormwater runoff infiltration of the green highway	D13	0.730	27	38
Soft frangible plants can enhance safety, acting as buffer zones between the highway and the RSA area	D26	0.730	28	39
Planting can reduce ongoing maintenance costs of mowing grassed medians	D37	0.719	29	41
Steel production materials such as electric arc and blast furnace slag can be recycled and added to bio-retention or infiltration beds to remove phosphorous while added to the concrete as aggregate for road-base and sub-base	D18	0.718	30	42
Grassed areas should retain limited room for vehicle recovery	D40	0.718	31	43
Regionally appropriate ground cover, grass, and shrub species should be considered for medians where growth patterns will not extend beyond the median	D41	0.718	32	44
Concrete pavement is an initiative for the most economical method in terms of life cycle costs for the green highway	D10	0.690	33	48
Frangible plants such as small shrubs and ground cover must be taken care of if safety barriers do not shield them	D39	0.683	34	50
Green highway at construction stage				
RSA requires well-planned scheduled waste management that complies with the environmental quality Act 1974	D46	0.802	1	5
Construction should fulfill the requirement as set by manual Saliran Mesra Alam (MASMA) and by the department of irrigation and drainage (DID)	D43	0.801	2	6
Construction aims to minimise the expansion of existing roads due to limited space in the urban area and to avoid the high cost of land acquisition to build elevated urban highways on the existing road alignment	D45	0.757	3	20
Installation of emission reduction strategies that complies with the factory and machineries Act 1967	D44	0.670	4	51
Green highways shall follow the minimum standards stated by environmental laws and regulations	D42	0.663	5	52
Signboard to inform vehicle drivers to switch off their engines as an initiative to reduce pollution at RSA	D52	0.773	1	12
Electric vehicle charging station shall be provided at RSA for the green highway	D55	0.758	2	19

(continued)

Table 4 (continued)

Assessment items	Results			
	Variables	Criticality index	Rank	Overall rank
All tenants must use a paper bag or biodegradable materials at RSA	D47	0.738	3	34
Use recycled pavement for highway rehabilitation and maintenance, such as hot in place recycling (HIPR) and cold in place recycling (CIPR)	D48	0.730	4	40
New pavement construction should use the crushed and recycled concrete pavement as aggregates, granular fill, sub-base material, or base course for new pavement	D50	0.710	5	45
Applications such as embankment fill, gravel roads, and road ballast to reduce carbon dioxide (CO_2) emissions and global warming must use recycled materials	D49	0.706	6	46
The bio-oxygen demand (BOD), carbon oxygen demand (COD), and other readings from the water treatment plant shall follow the EIA requirement before discharge from RSA	D51	0.702	7	47
Electronic toll collection (ETC) system at all toll plazas shall be provided	D54	0.687	8	49
All patrol and maintenance vehicles shall use biofuel or hybrid engines	D53	0.655	9	54

Note Interpretation of criticality index: 0.00–0.25 (weak); 0.26–0.50 (Moderate); 0.56–0.75 (High); and 0.76–1.00 (Very high)

3.4 Main Success Factors to Implement Green Highway

Table 5 shows the criticality index and the overall ranking value for green highway success factors sub-categories. There are three categories namely; (i) innovation, (ii) construction activities, and (iii) material and resources. Overall, respondents prioritised that the success factors under the construction activities category were essential in implementing green highways. Under these categories, the sub-category of "apply schedule waste management at the site" obtained the highest criticality index value of 0.841. The second highest priority of success factors indicated by the respondents was the "encourage recycle and reuse of highway materials" factor under the innovation category, which obtained a criticality index value of 0.826. "Encourage biodegradable material usage" and "encourage wildlife corridor and passage" were recorded as the fourth and fifth main priority of success factors which scored Crd = 0.813 and Crd = 0.794, respectively, under the innovation category. Respondents from the previous interview session also agreed on the similar sub-category of "encourage recycle and reuse of highway materials", "encourage recycle pavement", and "encourage wildlife corridor and passage" under the innovation category. The interviewees added that the responsible authorities, especially the government and policymakers, should address and promote the advantages of the green highway to the stakeholders in highway projects by providing seminars, technical conferences, engineering forums, or technical workshops to develop highway engineering skills among the highway practitioners. The third main priority for the success factors category indicated by the respondents is "encourage minimisation of waste" under the material and resources category.

Despite that, sub-categories such as "use natural method for runoff flow" under the material resources category, "apply schedule waste management at site", and "install a solar panel to generate electricity" under the construction activities category were also agreed by the respondents during the interview. The respondents mentioned that if the green contracting provision is enforced in Malaysia, there will be no problems such as the lack of exposure to the green contracting provisions in highway projects among the highway practitioners. Thus, based on the results obtained from the criticality index assessment, the study question regarding the main success factors of green highways in Malaysia was addressed and answered.

3.5 Model Development

The green highway implementation model involved three latent variables; Green Highway Challenges (CHALL), Green Highway Characteristics (CHAR), and Success Factors in implementing Green Highway (IMPLMT). The measurement model and the structural model were analysed to validate the green highway implementation model using indicator reliability analysis, internal consistency analysis, and validity (average variance extracted) analysis.

Table 5 Criticality index and overall ranking for main success factors

Assessment items	Results			
	Variables	Criticality index	Rank	Overall rank
Success factor for green highway implementation				
Innovation				
Encourage recycling and reuse of highway materials	C1	0.826	1	2
Encourage biodegradable material usage	C3	0.813	2	4
Encourage wildlife corridor and passage	C2	0.794	3	5
Encourage electronic toll collection (ETC) system usage	C4	0.766	4	11
Encourage diesel retrofit device usage	C5	0.73	5	14
Construction activities				
Apply schedule waste management at the site	C6	0.841	1	1
Encourage landscaping work	C10	0.79	2	6
install strategic detention pond	C9	0.782	3	8
Install solar panels to generate electricity	C7	0.754	4	12
Install bio-retention system	C8	0.746	5	13
Material and resources				
Encourage minimisation of waste	C13	0.814	1	3
Use natural methods for runoff flow	C12	0.79	2	7
Encourage the application of durable materials	C15	0.782	3	9
Encourage recycle of pavement method	C11	0.778	4	10
Encourage minimisation of excavation and earthmoving	C14	0.694	5	15

3.5.1 Indicator Reliability Analysis

The measurement models and structural models need to go through two phases of separate assessment in PLS-SEM. The first phase examines whether the reliability and validity measures are in accordance with specific criteria related to the formative and reflective measurement model specifications.

The reliability of the individual items to their respective constructs. However, Hulland also stated that in the case of exploratory research, a value of 0.4 or higher for all individual items in the reflective outer model is acceptable. Since this research is exploratory research, this section refers to the second opinion from Hulland in measuring the composite reliability value of 0.4 or higher. If the loadings are less than 0.4, the indicator in the reflective model must be eliminated from the measurement model. Table 6 summarises the results of the reflective outer model. By referring to Table 6 (results summary for reflective outer model), the calculation of the PLS

Table 6 Results summary for reflective outer model

Latent variable	Indicators	Loadings	Indicator reliability (i.e., loadings^2)	Composite reliability	AVE
CHALL	CULTURAL	0.707	0.500	0.798	0.505
	FINANCIAL	0.769	0.591		
	LEG_RGL	0.824	0.679		
	TECH_PHY	0.499	0.249		
CHAR	CONSTRCTN	0.802	0.643	0.869	0.683
	DESIGN	0.911	0.830		
	MAINTENCE	0.836	0.699		
	PLANNING	0.749	0.561		
IMPLMT	CONST_AC...	0.970	0.941	0.950	0.863
	INNOV	0.890	0.792		
	MAT_RSOR...	0.925	0.856		

algorithm showed that the loadings for all indicators in the proposed model were higher than 0.4. Therefore, all items in the measurement model remained the same, and no items were eliminated.

3.5.2 Internal Consistent Reliability Analysis

The Composite Reliability assessment was used to examine latent variables in this study. The bootstrapping step in PLS-SEM will calculate the standard error for CR and increase a confident interval to evaluate reliability accuracy in the range of plausible values of its population.

Table 7 shows the composite reliability results. All of the constructs for CHALL, CHAR, and IMPLMT showed values of 0.798, 0.896, and 0.950 for composite reliability, respectively. Results showed that all values were higher than 0.6. Therefore, composite reliability was stipulated and acceptable for the model.

Table 7 Composite reliability (CR) results

Latent variable	Composite reliability (CR)
CHALL	0.798
CHAR	0.896
IMPLMT	0.950

3.5.3 Validity (Average Variance Extracted) Analysis

There are two constraints in reflective measurement model validity assessment. The first constraint is the convergent validity and the second is the discriminant validity (Hair et al., 2014). Below are the descriptions to assess each constraint in reflective measurement model validity. For convergent validity in measurement model validity, the value of Average Variance Extracted (AVE) is required. If the value of AVE obtained from the results is 0.5 or higher, the measurement model indicates a sufficient degree of convergent validity. In other words, each of the latent variables in the model explains more than half of its indicators. In SmartPLS 3.0, the AVE value can be obtained under the PLS algorithm calculation (Wong, 2013). Results shown in Table 8 indicated that AVE values were higher than 0.5, and all factors and items loaded significantly on the designated constructs. This showed that there was no evidence of any cross-loading happened between the constructs.

Discriminant validity assessment has become dominant in analysing relationships between latent variables. Discriminant validity will determine whether a constructed measure is empirically unique and represents phenomena of interest that other measures in a structural equation do not capture to improve results with better validity and reliability (Hair et al., 2014). The Heterotrait-Monotrait ratio of correlations (HTMT) is a new criterion for assessing discriminant validity in variance-based structural model equation modelling. For this study, the HTMT criterion was computed using SmartPLS 3.0. HTMT measures two (2) advantages in dealing with disattenuated construct score correlation where HTMT does not require a factor analysis and calculation of construct scores to obtain factor loadings (Henseler et al., 2014). Table 9 shows the HTMT results for this study. HTMT-based approaches are $HTMT_{0.85}$, $HTMT_{0.90}$, and $HTMT_{inference}$ to set up the most conservative benchmark and critical criteria value. However, the strictest standard must be followed using the guideline for discriminant validity assessment in variance-based SEM, where $HTMT_{0.85}$ is required to assess for discriminant validity (Henseler et al., 2014). The HTMT approach set up for this study was $HTMT_{0.90}$ (which means all HTMT values must be below 0.90). From Table 10, the results marked in bold indicated discriminant validity problems according to the $HTMT_{0.85}$ criterion. At the same time, there were no discriminant validity problems according to the $HTMT_{0.90}$ criterion and $HTMT_{inference}$ in the dataset. This result concluded that the discriminant validity was established accordingly, with acceptable high sensitivity rates conforming to the HTMT standard.

Confidence Intervals (CI) and bootstrapping procedure were computed under no sign changes; two-tailed test type, significance level of 00.5, and Confidence Interval

Table 8 Average variance extracted (AVE)

Latent variable	AVE
CHALL	0.505
CHAR	0.683
IMPLMT	0.863

Table 9 HTMT results

	CHALL	CHAR	IMPLMT
CHALL			
CHAR	0.626		
IMPLMT	0.734	**0.852**	

Table 10 Confidence intervals (CI) results

Confidence interval (Lower)	Confidence interval (Upper)
0.007	0.413
0.294	0.725
0.501	0.859

Method of Bias-Corrected and Accelerated (BCa) Boot. Table 5.15 shows the results for the Confidence Interval. Suppose the correlation between the two constructs is 1.0. In that case, one expectation of discriminant validity problems will be detected with a frequency close to 100% regarding all criteria in the total analysed constellations. Table 10 shows that the upper confident interval limits were below 1.0, which indicated that all HTMT inference criteria values were significantly different from 1.0. Therefore, discriminant validity was well-established and fulfilled the requirement for this study. Consequently, overall results for the validity and reliability test analysis for the green highway implementation model in this study were verified and adequate to proceed to the next stage of evaluation without eliminating any of the eleven (11) observed assessment items of indicators.

3.5.4 Green Highway Implementation Model and Hypotheses Test

The list of hypotheses tested for the green highway implementation model is shown in Table 11, illustrating the reciprocal relationship between green highway challenges, characteristics, and success factors.

Furthermore, Fig. 2 shows the green highway implementation model with eleven (11) observed assessment items of indicators remaining in the model. Research

Table 11 Hypotheses

Hypotheses		Definition
H_1	a	Green highway challenges will significantly influence the success of implementing green highway
H_2	a	Green highway characteristics will significantly influence green highway challenges
	b	Green highway characteristics will significantly influence the success of implementing green highway

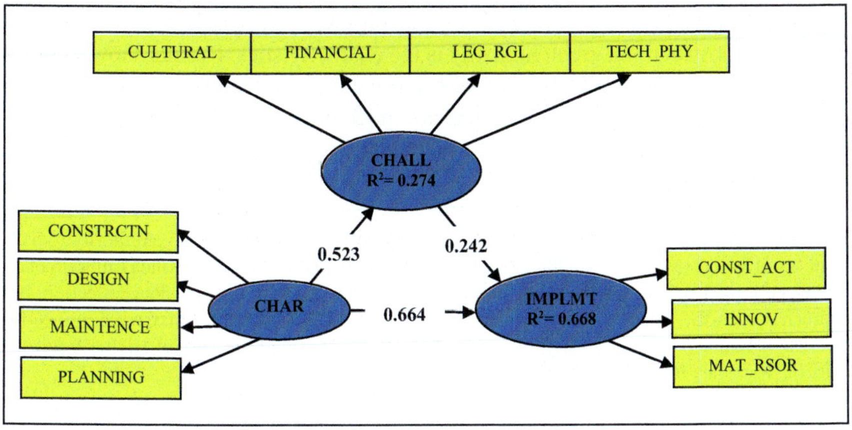

Fig. 2 Green highway implementation model

hypotheses that illustrate the reciprocal relationship between latent constructs of green highway challenges, green highway characteristics, and success factors in implementing green highways were also labelled as follows:

4 Conclusion

The overall goal of this study was to develop a model that can relate the driving factors of green highway characteristics and green highway challenges to the success factors of implementing green highways in the Malaysian context. This study discovered that the main green highway challenges are local laws are unclear (legal and regulatory) and perceived high costs for short-term and/or long-term green highway projects (financial). Therefore, it is observed that there is a need for a comprehensive policy regarding green highways for practitioners in highway projects to have a standard of procedure for implementing green highway practices.

Furthermore, this study found that the green highway practice at the design stage is the most crucial stage in a project life cycle. Respondents suggest that the main green highway characteristics should be emphasised, such as the pavement design life, designer shall consider using solar energy to generate electricity for street lighting and other facilities at RSA, toll plaza, and lay-by for green highway purpose; and rainwater harvesting system shall be installed at RSA, toll plaza office, supervision building, and regional office. This is to ensure the stakeholders can control the cost, time, and quality of green highway projects from time to time.

Moreover, this study also discovered that in-depth focus should be given to innovation and construction activities as important initiatives such as (i) applying scheduled waste management at the site (construction activities), (ii) encouraging recycling and reuse of highway material (innovation), and (iii) finally encouraging minimisation of waste (material and resources). The green highway implementation model shows that a key construct of a green highway is the characteristics. Results show a significant relationship and positive effect between characteristics and challenges in implementing green highways. It is essential to understand that by utilising the body of knowledge in this model in highway projects can help the successful implementation of the green highway.

Acknowledgements The author would like to thank the Malaysian Highway Authority (MHA), Ministry of Works (MoW), Road Planning Division, Jalan Sultan Salahuddin, Kuala Lumpur, PLUS Malaysia Berhad, and all highway concessionaires for supporting the primary data collection. Best appreciation to the Malaysia Institute of Transport (MITRANS), School of Civil Engineering, College of Engineering, University Teknologi MARA (UiTM), and Ministry of Higher Education (MoHE) Malaysia for the opportunity to conduct this research.

Referencess

Ashoka Highway Research Centre (AHRC). (2002). *Green highways.* Retrieved January 31, 2014, from http://www.ahrc.in/Green_Highways.pdf

Attahiru, Y. B., Aziz, M. M. A., Kassim, K. A., et al. (2019). A review on green economy and development of green roads and highways using carbon neutral materials. *Renewable and Sustainable Energy Reviews, 101*, 600–613.

Cheng, J. (2002). Discussion of importance index in technology foresight. National centre fro science and technology for development ministry of science and technology of the people republic of China. *International Journal of Foresight and Innovation Policy, 1*, 1–2. https://doi.org/10.1504/IJFIP.2004.004620

Chin, W. W. (2010). *Handbook of partial least squares.* In V. Esposito Vinzi, W. W. Chin, J. Henseler, & H. Wang (Eds.). https://doi.org/10.1007/978-3-540-32827-8

Clark, M., Paulli, C., Tetreault, Z., & Thomas, J. (2009). *Green guide for roads rating system.*

Clevenger, C. M., Ozbek, M. E., & Simpson, S. (2013). Review of sustainability rating systems used for infrastructure projects. In *49th Associated Schools of Construction (ASC) Annual International Conference Proceedings.* Associated Schools of Construction.

Converse, P. D., Wolfe, E. W., Huang, X., et al. (2008). Response rates for mixed-mode surveys using mail and E-mail/Web. *American Journal of Evaluation, 29*, 99–107. https://doi.org/10.1177/1098214007313228

Degross, J. I., Chin, W. W., & Marcolin, B. L. (1996). Proceedings of the seventh a partial least squares latent variable modelling approach for measuring interaction effects: Results from a Monte Carlo simulation study and voice mail emotion/adoption study.

Dillman, D.A., Smyth, J. D., & Christian, L. M. (2008). Internet, mail, and mixed-mode surveys: The tailored design method. In *Internet Mail and MixedMode Surveys the Tailored Design Method.* https://doi.org/10.2307/41061275

Fornell, C., & Bookstein, F. L. (1982). Two structural equation models: LISREL and PLS applied to consumer exit-voice theory. *Journal of Marketing Research, 19*(4), 440–452.

Hair Joseph, F., Hult, G. T. M., Ringle, C. M., et al. (2014). A primer on partial least squares structural equation modelling (PLS-SEM). In V. Knight, K. Koscielak, L. Barrett, & G. Dickens (Eds.). Sage Publications.

Henseler, J., Ringle, C. M., & Sarstedt, M. (2014). A new criterion for assessing discriminant validity in variance-based structural equation modelling. *Journal of the Academy of Marketing Science*, 115–135. https://doi.org/10.1007/s11747-014-0403-8

Huang, R., & Yeh, C. (2008). Development of an assessment framework for green highway construction. *Journal of the Chinese Institute of Engineers, 31*(4), 573–585.

Khoshnava, S. M., Rostami, R., Zin, R. M., et al. (2020). Green efforts to link the economy and infrastructure strategies in the context of sustainable development. *Energy, 193*, 116759.

Lee, C. C., & Egbu, C. O. (2006). The development of a methodology to match the client's project requirements with the knowledge of the project team in refurbishment projects. In E. Sivyer (Ed.), *Proceedings of the annual research conference of the royal institution of chartered surveyors* (pp. 7–8). ISBN:978-1-84219-307-4.

Lim, S. K. (2009). *Framework and process for enhancing sustainability deliverables in Australian road infrastructure projects*. Queensland University of Technology.

MHA, & UTM. (2015). *Malaysia green highway index manual V1.0 (MyGHI)* (First Edit; UTM, Ed.). Perpustakaan Negara Malaysia.

Mukta, M. Y., Rahman, M. A., Asyhari, A. T., & Bhuiyan, M. Z. A. (2020). IoT for energy efficient green highway lighting systems: Challenges and issues. *Journal of Network and Computer Applications, 158*, 102575.

Nen, A. N. C., Isa, C. M. M., Nusa, F. N. M., et al. (2020). Complexity factors during pre-construction stage affecting quality performance in infrastructure projects. In *2020 11th IEEE Control and System Graduate Research Colloquium, ICSGRC 2020-Proceedings* (pp. 253–258). https://doi.org/10.1109/ICSGRC49013.2020.9232566

Nusa, F. N. M., Endut, I. R., & Ishak, S. Z. (2015). Challenges of green highway concept towards implementation of green highway. *747*, 3–6. https://doi.org/10.4028/www.scientific.net/AMM.747.3

Ogola, P. F. A. (2007). Environmental impact assessment. In *Short course II on surface exploration for geothermal resources* (pp. 1–16). United Nations University Geothermal Training Programme (UNU-GTP) and Kenya Electricity Generating Co., Limited (KenGen).

Rafidah, R., & Zaimi, M. (2014). Sustainable design and construction assessment tool for green highway in Malaysia. *Journal of Applied Science and Agriculture, 9*(21), 35–40.

Saqr, K. M., & Musa, M. N. (2011). A perspective of the Malaysian highway energy consumption and future power supply. *Energy Policy, 39*(6), 3873–3877. https://doi.org/10.1016/j.enpol.2011.03.034

Sarstedt, M., Henseler, J., & Ringle, C. M. (2011). Advances in international marketing emerald book chapter: Multigroup analysis in partial least squares (PLS) path modeling: Alternative methods and empirical results. https://doi.org/10.1108/S1474-7979(2011)0000022012.

Shang, T., Ye, K., Wu, L., & Shang, J. (2022). Opportunism or symbiosis? A case study on contractors' unsustainable highway construction in China. *Resources, Conservation & Recycling Advances, 14*, 200082.

Shehu, Z., & Akintoye, A. (2009). The critical successs factors for effective programme management: A pragmatic approach. *2*, 1–24

Wong, K. K. (2013). Partial least squares structural equation modeling (PLS-SEM) techniques using smartPLS. *Marketing Bulletin, 24* (Technical Note 1).

Wright, P. H., Dixon, K. K., & Meyer, M. (2004). *Highway engineering* (7th ed.). John Wiley & Sons, Incorporated.

Yue, W. (2004). *Predicting the citation impact of clinical neurology journals using structural equation modeling with partial least squares*. University of New South Wales.

Zakaria, R., Seng, F. K., Majid, M. Z. A., et al. (2013). Energy efficiency criteria for green highways in Malaysia. *Journal of Technology, 3*, 91–95.

Zeiada, W., Hamad, K., Khalil, M. A., et al. (2022). Development of green pavement rating system in the UAE. *Innovative Infrastructure Solutions, 7*(1), 1–16.

Proposed Development of an Integrated Framework for Public–Private Partnership and Value for Money Evaluation System of Urban Rail Transit in China

Gao Ying, Che Maznah Mat Isa, Nur Izzati Ab Rani, and Nur Kamaliah Mustaffa

Abstract With the advancement of urbanization in China, the construction of rail transit has become an important choice for urban transportation infrastructure. The continuous and rapid construction of rail transportation is becoming the epitome of China's promotion of the urbanization process. However, considering the macroeconomic background, China's economic growth rate has gradually changed from rapid to steady growth due to upgrading its industrial structure, which has led to a decline in tax revenue growth and increased pressure on local government debt. At the same time, the central government continues to emphasize the importance of "reducing leverage" and pressure on local government financing, so there is an urgent need for social capital to participate in rail transportation construction through public–private partnership (PPP) projects. Currently, China's value for money (VFM) qualitative evaluation is based on the Ministry of Finance documents, with many shortcomings in the actual operation. Various issues are related to the content of the indicator system that involves primarily economic benefit-oriented evaluation, with less concern about social and environmental problems caused by the PPP, and less consideration is given to technical, legal, and policy factors. Thus, this paper proposed a study toward developing a framework integrating the PPP model and the VFM evaluation system for urban rail transit projects in China.

G. Ying · N. I. Ab Rani · N. K. Mustaffa
School of Civil Engineering, College of Engineering, Universiti Teknologi MARA, Shah Alam, Selangor, Malaysia
e-mail: gaoying@Ixledus.cn

N. I. Ab Rani
e-mail: izzati.rani@uitm.edu.my

N. K. Mustaffa
e-mail: nurkamaliah@uitm.edu.my

C. M. Mat Isa (✉)
Center of Civil Engineering Studies, Universiti Teknologi MARA Pulau Pinang Branch, Permatang Pauh Campus, 13500 Pulau Pinang, Malaysia
e-mail: chema982@uitm.edu.my

U. R. Hashim et al. (eds.), *Green Infrastructure*,
https://doi.org/10.1007/978-981-99-7003-2_17

Keywords Public–private-partnership · Urban rail transit · Value for money

1 Introduction

With the rapid development of China's economy and the rapid increase of urbanization rate, the number of motor vehicles is increasing year by year and the problem of traffic congestion is becoming more and more prominent, relying closely on the new construction and expansion of urban roads can no longer meet people's demand for transportation. Urban rail transit, as an energy-saving and green transportation system, can effectively alleviate urban congestion and is valued and favored by many cities. By the end of 2021, a total of 50 cities in mainland China had opened 283 urban rail transit operating lines, with an operating line total of 9,206.8 km, and a total construction investment of 585.98 billion yuan completed in the whole year of 2021(China Association of Metros, 2022).

The rapid development of urban rail transit in China also poses a challenge to the financing needs of cities. In order to solve this series of problems, PPP mode is being widely introduced into the investment and operation of urban rail transit infrastructure. In December 2014, China Public–Private Partnership Center was established. On December 4, 2014, Ministry of Finance of the People's Republic of China issued the Operational Guidelines for Government and Social Capital Cooperation Model (for Trial Implementation) (Ministry of Finance of the Reople's Republic of China, 2014), which put forward the identification of infrastructure PPP projects, project preparation, project procurement, project execution, and project transfer operational guidelines. In the same period, the National Development and Reform Commission published the guidance on the development of government-social capital cooperation (NDRC Investment [2014] NO. 2724) (NDRC, 2014), which provided guidance on accurately grasping the main principles of government-social capital cooperation, project scope and project mode, working mechanism, and project supervision. After that, The Ministry of Finance and the Development and Reform Commission issued a series of policies to promote the investment and financing direction of urban rail transit projects more effectively under the PPP model. In 2004, China started the exploration of rail transit PPP model in Beijing Metro Line 4 project, followed by Shenzhen, Shanghai, Hangzhou, and other places to carry out pilot projects. By the end of 2021, more than 90 rail transportation projects nationwide have implemented the PPP model (Bin, 2021).

Although some of the urban rail transit PPP projects in China have achieved certain success, with the passage of time, the financial support requirements of urban rail transit projects for local governments are gradually increasing, and the operational requirements of PPP projects are also increasing, and the domestic urban rail transit PPP projects have focused on revealing some outstanding problems. As rail transportation is a public welfare product, most participants in China's rail transport PPP projects are large state-owned enterprises due to China's national conditions, and only a few private capitals participate, which does not effectively stimulate the participation of all social capitals (Xin, 2021). From the type of enterprises, most enterprises are construction and equipment manufacturers, and there are fewer operational enterprises, which can only rely on government subsidies, which is not in line with the requirements of the Guidelines that do not rely on third-party subsidies (Hongyan, 2022).

Meanwhile, China's VFM qualitative evaluation system is not perfect. At present, China's VFM qualitative evaluation is based on the Ministry of Finance documents (Ministry of Finance of the People's Republic of China, 2014), and there are many shortcomings in the actual operation. The content of the indicator system is not perfect. It is mainly an economic benefit-oriented evaluation system, but less concern about social and environmental problems caused by PPP and lacks external indicators such as social and environmental benefits, and considers less technical, legal, and policy factors. There is a lack of comprehensiveness of indicators and the division between basic and supplementary indicators is rough (Jing & Hongyu, 2016), and the specificity of indicators is insufficient (Wei et al. 2014). The qualitative evaluation of VFM in China lacks the design of scoring criteria, which leads to the inconsistency of scoring criteria and the existence of some arbitrariness. Due to the lack of basis in the scoring criteria, the evaluation results can be adjusted to a certain extent by the implementor, which ultimately leads to the low accuracy of the qualitative evaluation of VFM (Huishan & Shou, 2017). Each consulting organization has different perceptions of the weight scale of qualitative evaluation indexes, and the subjectivity of index weight setting can be used as the "reasonableness" of their adjusted scores, which often results in large deviations in the final scores (Kun et al., 2018; Zhijun, 2020).

The calculation model of VFM quantitative evaluation needs to be improved (Lingxia et al., 2018). The current quantitative evaluation of value for money in China adopts the Public Sector Comparator (PSC) method, which compares the PSC value with the PPP value by discounting to obtain the VFM value and give the conclusion of quantitative evaluation of VFM. Both PSC value and PPP value are the net present value of Life Cycle Cost (LCC) of PPP projects. Whether the discount rate can be determined scientifically and reasonably will ultimately affect the accuracy and authority of the quantitative evaluation results of VFM. The discount rate of most PPP projects is reasonably determined with reference to the local government bond income in the same period. The discount rate selected by this method is essentially a risk-free rate of return, even if the bank's loan interest rate is considered, but it fails to effectively reflect the return on investment and inflation rate, not to mention the principle of risk–benefit equivalence, and there is a strong subjective consciousness and a strong subjective color, which makes it difficult to grasp the scientific of the discount rate. Insufficient accuracy of risk cost quantification. In the quantitative evaluation, the calculation of PSC value requires the calculation of all the risk costs borne by the government for PPP projects; similarly, the calculation of PPP value also requires the calculation of the government's own risk costs under the PPP mode. Therefore, in order to arrive at accurate PSC and PPP values, it is necessary to accurately account for the possible costs associated with the occurrence of risks. In practice, due to the lack of relevant evaluation knowledge and experience, the government finance department usually entrusts the preparation of the quantitative VFM evaluation report to a third-party consultant and is responsible for the review of the evaluation results (Tong, 2016). In the process of determining the risk-bearing cost, the third-party consultant often adopts the proportional method due to the lack of accumulated risk data from past projects. The advantage of using this method is that it is convenient, quick, and hassle-free.

In China, quantitative evaluation is still in its infancy, and the existing methods only consider the economics of the project, without considering the social benefits of the project. As an infrastructure project, rail transit can generate good social and environmental benefits, and its social benefits are usually greater than its own economic benefits. The lack of consideration of social benefits in the existing VFM quantification system will exclude the prediction of potential social benefits during the construction and operation periods, which reduces the social sustainability of the project. Meanwhile, there is still no unified opinion on the calculation of government subsidy model, operating cost and revenue model in rail transit projects, and the whole life cycle VFM quantitative evaluation model needs to be improved (Yanhong et al., 2019; Hanwen et al., 2021).

Based on the literature review, the research questions for this study are as follows:

(1) What are the characteristics and problems of China's rail transit PPP model?
(2) How to conduct qualitative evaluation and parameter determination of rail transit VFM?
(3) How to establish the quantitative calculation model of whole life cycle VFM?

In order to answer the research questions, the specific objectives are as follows:

1. To determine the characteristics and problems of China's rail transit PPP model through document review;
2. To evaluate the implementation of the rail transit VFM by conducting qualitative evaluation and parameter determination to establish the quantitative calculation model of whole life cycle VFM; and
3. To develop a comprehensive and validated framework for PPP-VFM system for urban rail transit in China.

The following section presents a literature review to evaluate the various PPP modes of urban transportation by focusing on the current situation of rail transit PPP model in China, followed by an evaluation on value for money (VFM) qualitative and quantitative studies.

2 Literature Review

Since its introduction, the PPP model has been highly valued and widely used in the international arena. In developed countries, the PPP model has been widely used in public transportation, energy, and power, health care, defense and security, justice and security, water industry, etc. Australia's share of PPP projects has reached 10%

to 15% of the annual infrastructure investment. Based on the prominent contradiction between infrastructure shortage and government financial pressure, developing countries are also accelerating the implementation and application of PPP model to encourage private funds to enter the infrastructure construction field.

From the existing literature, scholars at home and abroad have conducted a lot of theoretical analysis and empirical research on the operation mode of rail transit PPP projects and related contents, and several excellent results have been formed. A very reasonable analysis has been made on the development of rail transportation, mode selection of rail transportation PPP projects, and VFM evaluation methods. It has laid a solid foundation for the subsequent research scholars.

2.1 Research on PPP Mode of Urban Rail Transportation

The PPP model has been studied earlier in developed countries and today it is widely used all over the world. Wilmoth (1990) has analyzed the financing problems of urban infrastructure in Australia, especially rail transit, and the PPP pilot in the reform direction proposed by the Australian government. Carpintero and Siemiatycki (2016) have illustrated the influence of political factors on transport projects through a case study of a light rail project in Spain and suggested that urban rail projects built using PPP models should reduce the influence of political factors on project-level decisions. Sock-Yong (2007) has summarized the rail transit PPP project models of seven international cities into four different operating modes and made a detailed analysis of the risk-sharing strategies of each model. Anderson (2014) concluded that even though rail transit in Los Angeles accounts for only 1% of the population miles traveled by Los Angeles residents, the benefits of social capital participation in PPP projects are much greater than expected. Wu et al. (2012) improved the government investment decision conditions of the concession period BOT-CCM decision model based on the current situation of rail transit PPP projects in many countries, so that the decision model can be adapted to the current situation of multi-regional development. Saskia and Proost (2012) studied the current situation of rail transportation in developing countries and used a game model from the perspective of central government and local government to explain that local government may give more profit margin to social capital for their own performance in the context of incomplete information, which makes social capital more willing to participate in rail transportation construction with PPP projects.

Most of the studies on urban rail transit PPP model in China focus on a specific aspect, such as specializing in performance assessment and compensation model, construction and operation risk or investment and financing path of PPP projects, the positioning of government and system construction under PPP model, etc. By introducing the current situation of social capital cooperation in Taiwan's rail transit, Cheng (2010) argued that in the case of a certain PPP project cycle, a long concession period would reduce the period of use after the handover, and social capital would not invest too many resources in maintaining the rail transit PPP project due to interest considerations. Li Moxuan and Zhang Jiayi (2010) analyzed the advantages of Beijing Metro Line 4 PPP project in relieving the pressure of financing and promoting the expansion of investment and financing from the perspective of property rights relationship; and pointed out that the PPP project of introducing Hong Kong capital to Beijing Metro Line 4 gave too much support to social capital in cooperation. Yan Ling and Cui Jian (2011) compared Beijing Metro Line 4, Beijing Metro Olympic Branch Line, and Shenzhen Metro Line 5, which have completed PPP projects, in terms of investment and financing risks, project survey and design, and project construction. The concept of strong and weak PPP projects was proposed, and the study concluded that the adoption of strong/weak PPP projects is mainly determined by the financial pressure faced by the sponsors, and projects with high financial pressure adopt weak PPP projects to carry out; projects with low financial pressure should adopt strong PPP projects to cooperate. Wu Shuli (2014) analyzed the PPP project of Beijing Metro Line 4 as a case study and suggested that the government should build a perfect financing platform to help the project financing when the PPP project is adopted for rail transportation. He also proposed to learn from the Hong Kong model to explore the internal business opportunities of rail transit, and to take advantage of the high traffic flow of rail transit to recover the investment as soon as possible.

2.2 Current Situation of Rail Transit PPP Model in China

Gu Xin (2009) derived from the successful experience of Hong Kong rail transit PPP project to the problems of Shenzhen metro PPP project, and concluded that the existing government subsidy model for rail transit PPP project could not attract social capital to cooperate, and the way to break the dilemma was to transfer the operation right of the above-ground project, which was contrary to the existing national laws and regulations at that time, so the legislative level should be changed as soon as possible. Therefore, the legislation should be revised as soon as possible. Wei et al. (2012) took the current situation of China's transportation PPP project operation as the research object, under the assumption that the traffic volume obeys the geometric Brownian motion, took the maximization of project value of government and social capital as the decision objective, combined with option pricing and optimal stopping time theory to study the optimal investment timing of both parties under the uncertain environment, and then decided the feasible interval of the concession period, and then used the option game theory to construct a full information dynamic game theory. Then, a full information dynamic game model is constructed by using option game theory. Youyong and Dong (2017) analyzed that the current situation of PPP project operation in China still exists in the lack of reasonable protection mechanism for social capital and the discrimination status of social capital interest demand. And they argued that the sustainable development of rail transit PPP projects in China must respect the investment profit demand of social capital.

2.3 Value for Money (VFM) Qualitative Evaluation Study

Since the PPP model was proposed by the British government in 1982, a more complete project database has been developed over a long period of time. In order to better conduct VFM evaluation in the pre-decision of PPP projects, many countries have formulated relevant documents, and some of them have also constructed a special VFM qualitative evaluation system for VFM qualitative evaluation. The European Commission focused on analyzing the VFM evaluation decision process and evaluating the rationality of the scheme for PPP projects in which the European Investment Bank participated in the investment in the "Guidelines for Successful Public–Private Partnerships", and also discussed the value-added and hidden cost identification (Directorate-General-Region-Policy, 2003). The UK Treasury in the Value for Money Assessment Guide stipulates that qualitative evaluation should be included in the VFM evaluation of the PFI project cluster, project, and procurement phases, generally in terms of whether the project is feasible, beneficial, and achievable (HM Treasury, 2006). The Australian Infrastructure Centre proposes that quantitative and qualitative analysis together form a complete VFM evaluation, and that qualitative evaluation is particularly important when the comparative values in quantitative evaluation are very close or difficult to quantify (Australian Government, 2009). The Canadian PPP Center proposes that the development of evaluation guidelines and scoring methods, assessment and comparison of procurement models, sensitivity analysis of findings, and explanatory notes are the basic steps that a complete qualitative VFM evaluation should contain (PPP-Canada, 2014). The qualitative evaluation of VFM in developed countries is more complete and has formed a more systematic and standardized method and process.

With the vigorous promotion of PPP mode in China, VFM qualitative evaluation has become an important support to assist PPP project establishment decision. China's VFM qualitative evaluation system is based on reference and learning from the experience of other countries, and there are still some gaps in the selection of indicators and process setting. Tao et al. (2017) constructed an index system for value for money qualitative evaluation of urban water ecology PPP projects and used uncertain hierarchical analysis to determine the index weights and the index scores through the different weighting expert scoring method, and finally, the example analysis showed that this evaluation index system is feasible. Jing and Hongyu (2016) set up VFM qualitative indicators, weights, and scoring criteria with the characteristics of rail transit PPP projects, established a VFM qualitative evaluation system, and conducted a case study with Beijing Metro Line 4. Wen Junxia (2016) discussed the evaluation method selection, evaluation idea, and evaluation index system construction of VFM qualitative evaluation of PPP projects from the perspective of transaction cost. Hongyue et al. (2017) compared the qualitative evaluation of value for money of international PPP projects and put forward suggestions to improve the qualitative evaluation of VFM in China.

2.4 Value for Money (VFM) Quantitative Evaluation Study

The quantitative value for money (VFM) evaluation is to determine the VFM value of a project by selecting a suitable calculation method to quantify the project cost and risk. The choice of calculation method is the most critical part of VFM quantitative evaluation, which mainly includes Cost–Benefit Analysis (B/C) evaluation method, competitive bidding method, and Public Sector Comparator (PSC) evaluation method. Among them, the PSC evaluation method is the VFM quantitative evaluation calculation method adopted by most countries internationally, and this method is also used in China at present.

Ball et al. (2007) believed that the PSC system for PPP /PFI projects has been established in the UK, Germany, France, Australia, etc. For the establishment of PSC evaluation system, foreign governments have many guiding documents and set up special departments to collect and organize PSC data and make constant additions and updates according to practical applications. The UK Treasury proposed in 2004 that VFM evaluation should be carried out in the early stage of decision-making, and since the PSC evaluation method is flexible, projects can develop PSC with different accuracy according to the different stages of early decision-making. Shugart (2008) argues that the French Ministry of Health has worked out eight steps for PSC establishment. Katz (2006) elaborates that the Australian Victorian government in Partnerships Victoria requirements, the Victorian government of Australia has a separate list of Frequently Asked Questions for the development of PSC to address the issues that often arise during the establishment of PSC. Coulson (2008) argues that the current model of PSC calculation is based on too many ideal assumptions and its parameters are too complicated, which is easy to make mistakes in the calculation.

Most domestic scholars mainly focus on the application of PSC evaluation method and the accuracy of PSC value as the research objectives. Shen Yuyu and Du Jing (2008) initially constructed the calculation method of VFM after referring to the implementation rules of foreign VFM evaluation, and also introduced the components of PSC and whole life cost in detail; Guangsheng and Tong (2013) improved the three aspects of PSC value determination, discount rate and risk adjustment in VFM quantitative analysis, and suggested to combine VFM quantitative analysis with financial analysis in project feasibility analysis to improve the VFM quantitative evaluation method. Jiarong and Shouqing (2011) analyzed the components of the PSC system, pointing out that the assets will be transferred to the government at the end of the concession period without compensation, so the initial PSC does not consider the residual value of assets and other adjustments based on China's national conditions. Qing et al. (2021) combed through the measurement methods of shadow fare subsidy model and operating car-kilometer service fee subsidy model commonly used in urban rail transit PPP projects, analyzed the stability and adaptability of the two models with relevant cases, and proposed the design points and development suggestions of the return mechanism. Yingying et al. (2015) analyzed and established a practical VFM evaluation system, and proposed the calculation formulae of PSC and LCC, the core elements of VFM evaluation, in the light of project practice. Yanhong,

Ting, and Shuyao (2019) constructed a quantitative VFM evaluation model for rail transit PPP projects from the perspective of the whole life cycle and validated the constructed model through case studies.

In summary, the Chinese rail transit PPP model is in the development stage, and its theory is not fully mature, and it follows the practice of developed countries. Chinese scholars have conducted some studies on Chinese characteristics, but they have not systematically sorted out the characteristics of Chinese rail transportation and analyzed only for a certain project. In this paper, we will organize the generated Chinese rail transit PPP projects and analyze the characteristics of Chinese rail transit PPP projects, to determine the qualitative and quantitative evaluation system of VFM suitable for our reality.

3 Research Methodology

This study uses a mixed analytical approach of quantitative and qualitative analysis, which can well solve the problem of this study and achieve the research objectives. The hybrid research method is used by several scholars in the study of qualitative and quantitative analysis of PPP model VFM (Wei et al. 2014).

Wenxue (2019) distilled the factors affecting the value for money evaluation in China through literature research, set up a questionnaire to investigate the importance of the influencing factors, and organized the evaluation indexes Peiling (2017) adopted an expert consultation method to finalize the parameters of each index. For the weight of each parameter, the weight of each indicator is determined based on AHP and entropy value method, and Yasan and Yongpu (2020) calculate the comprehensive weight of evaluation indicators based on AHP and entropy value method to build a more complete PPP project pricing evaluation model. This study collates the national rail transit PPP projects through data collation, conducts research on Chinese rail projects, and summarizes the VFM evaluation indexes and weight analysis that have been carried out for rail transit at present. It also combines various literature studies to refine the value for money evaluation factors of rail transportation in China.

Jing and Hongyu (2016) conducted a questionnaire survey on research scholars, experts, and economists, etc. The experts who participated in the survey studied PPP and project management the most, 36% and 33%, respectively; 11% and 8% for engineering technology and economic research, respectively; 6% for both rail transportation and law. This study collected more than 90 rail transit PPP projects. The questionnaire of this study is dominated by project managers, PPP participants, and government units who have participated in rail transportation PPP projects and completed them nationwide, accounting for 90%, and the other 10% are other experts, which can effectively compensate for the problem of lack of experience.

Each evaluation factor is analyzed according to the questionnaire and finalized by organizing expert review, and the entropy value method is applied to calculate the comprehensive weight of evaluation indexes. Through the study of existing rail

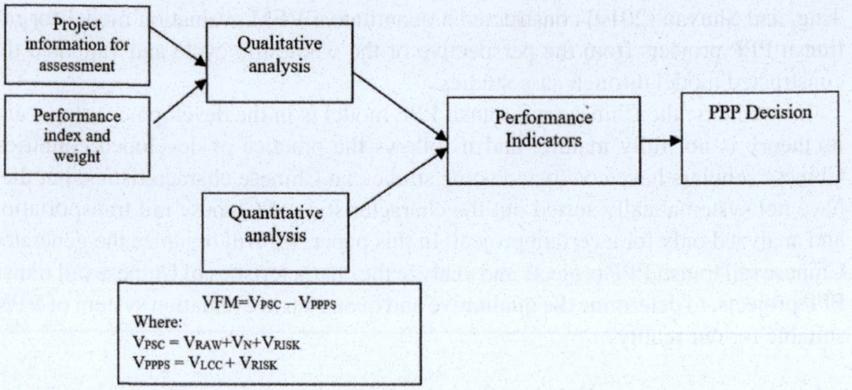

Fig. 1 Framework for PPP-VFM system for urban rail transit in China

transportation, the PSC calculation guidelines are clarified, and the rail transportation VFM evaluation system is finally determined.

Figure 1 shows the proposed framework for this study based on the PPP-VFM system for urban rail transit in China.

As shown in Fig. 1, the PPP Decision is the dependent variable supported by the performance indicators. The PIs are determined based on the qualitative analysis using project information for assessment and performance index and weightage, while for the quantitative analysis, a formula is established as shown in Fig. 1.

4 Conclusion

This paper proposes a study toward developing a framework integrating the PPP model and the VFM evaluation system for urban rail transit projects in China. The reviews indicate that the Chinese rail transit PPP model is still in the infancy stage and is not fully mature while following the practice of developed countries. Various studies of Chinese characteristics were carried out but unsystematically sorted out and analyzed only on certain projects. Various issues are related to the content of the indicator system that involves primarily economic benefit-oriented evaluation, with less concern about social and environmental problems caused by the PPP, and less consideration is given to technical, legal, and policy factors. It organizes the generated Chinese rail transit PPP projects and analyze the characteristics of Chinese rail transit PPP projects, to determine the qualitative and quantitative evaluation system of VFM suitable for China current situation. The study adopts a mixed-method analytical approach of quantitative and qualitative analysis to solve the problem and achieve the research objectives. It starts from the current situation of China's rail transit PPP model using qualitative analysis on the Ministry of Finance documents, with many shortcomings in the actual operation followed by an analysis of the characteristics of the participants of China's rail transit PPP model, and finally proposed a rail transit PPP model VFM evaluation system suitable for China's national conditions known as PPP-VFM framework. It is hoped that the proposed PPP-VFM framework helps to improve the systems for urban rail transit projects in China.

References

Anderson, M. L. (2014). Subways, strikes, and slowdowns: The impacts of public transit on traffic congestion. *American Economic Review, 104*(9), 2763–2796. https://doi.org/10.1257/AER.104.9.2763

Australian Government. (2009). *The national public private partnerships (PPP) policy framework and national PPP guidelines*. https://www.docin.com/p-1384326404.html

Ball, R., Heafey, M., & King, D. (2007). The private finance initiative in the UK. *Public Management Review*. https://doi.org/10.1080/14719030701340507

Carpintero, S., Siemiatycki, M. (2016). The politics of delivering light rail transit projects through public-private partnerships in Spain: A case study approach. *Transport Policy 49*, 159–167. https://doi.org/10.1016/j.tranpol.2016.05.001

Chen, W. (2019). *Design and implementation of the optimal scheme for the VFM evaluation system of PPP project*. Shandong University

Cheng, H.-K. (2010). Lessons learned from BOT infrastructure development in Taiwan: A case study of the Taiwan high speed railway (THSR) project. *Public Administration Quarterly, 34*(1), 51–75.

China association of metros. (2022). *China Urban rail transit association information*. https://oss.camet.org.cn/u/cms/www/202204/22154011ehvh.pdf

Coulson, A. (2008). Value for money in PFI proposals: A commentary on the UK treasury guidelines for public sector comparators. *Public Administration, 86*(2), 483–498. https://doi.org/10.1111/J.1467-9299.2008.00729.X

Directorate-General-Region-Policy. (2003). *Guidelines for successful public-private partnerships*. European Commission. https://ppp.worldbank.org/public-private-partnership/library/guidelines-successful-public-private-partnerships

Du, J., & Wu, H. (2016a). Analysis of the VFM evaluation of the Urban rail transit PPP project. *Journal of Engineering Management, 30*, 66–71. https://doi.org/10.13991/j.cnki.jem.2016.03.011

Guangsheng, L., & Tong, W. (2013). Exploring the improvement of PSC evaluation method for PPP projects' capital value Liu Guangsheng. *Journal of Industrial Technological Economics, 10*, 17–22.

Hongyan, Y. (2022). Analysis of current situation of Urban rail transit PPP projects. *Chinese Engineering Consultants, 4*, 100–106.

Hongyue, W., Jingfeng, Y., & Jing, D. (2017). Qualitative evaluation of value for money of international PPP project and enlightenment to China. *Construction Economy, 38*, 38–42.

Huishan, L., & Shou, W. (2017). Analysis of factors influencing VFM evaluation of infrastructure PPP projects. *Project Management Technology, 15*, 47–52.

Jiarong, L., & Shouqing, W. (2011). Improvement and perfection of the pre-decision system of PPP projects in China. *Project Management Technology, 9*, 17–22.

Katz, D. (2006). *Financing infrastructure projects: Public private partnerships (PPPs)*. https://www.researchgate.net/publication/237452854_Financing_Infrastructure_Projects_Public_Private_Partnerships_PPPs

Kun, L., Luxia, L., Zhen, Y., & Hao, W. (2018). Research and improvement on qualitative evaluation of value for money in rail transit PPP projects. *Project Management Technology, 16*, 40–43.

Li, P. (2017). The structure of qualitative evaluation system of value for money of PPP model. Guangdong University of Technology.

Liang, L., Han, F., Zhou, F., Zhang, J. (2018). Comparison at home and abroad on value for money assessment of PPP projects. *Journal of Civil Engineering and Management*, *35*, 182–88. https://doi.org/10.13579/j.cnki.2095-0985.2018.04.029.

Ling, Y., & Jian, C. (2011). Multi-case study on the selection of PPP mode transactions for Urban rail transit projects. *Science & Technology Progress and Policy, 28*, 110–115.

Liu, Z. (2020). Comparative study on qualitative value-for-money evaluation of PPP projects. *Accounting and Finance*, 83–88.

Ministry of Finance of the People's Republic of China. (2014). *The operational guidelines for government and social capital cooperation model (for Trial Implementation)*. http://jrs.mof.gov.cn/zhengcefabu/201412/t20141204_1162965.htm.

Moxuan, L., & Jiayi, Z. (2010). Analysis on the property relation and business. *Journal of Beijing Jiaotong University, 9*, 58–67.

NDRC. (2014). *The guidance on the development of government-social capital cooperation*. https://www.ndrc.gov.cn/xwdt/gdzt/cjnjtzzcwj/201606/t20160607_1033486.html?code=&state=123.

Peng, W., Chen, J., Cui, Q., & Cui, S. (2014). Comparative analysis of value for money assessment in public-private partnership projects. *Soft Science*, 28–5, 28–42. https://doi.org/10.13956/j.ss.2014.05.006.

PPP-Canada. (2014). *New building Canada fund: Procurement options analysis guide*. Macmillan Education UK. https://doi.org/10.1007/978-1-349-21159-3_4

Qing, Y., Shengbo, C., Yongping, L., & Min, C. (2021). Suggestions for the development of subsidy models and return mechanisms for Urban rail transit PPP projects. *Urban Mass Transit, 12*, 11–16.

Saskia, V. D. L., & Proost, S. (2012). *The European road pricing game: How to enforce optimal pricing in high-transit countries under asymmetric information* (pp. 1–33). Social Science Electronic Publishing. https://doi.org/10.2139/SSRN.1934136

Shugart, C. (2008). *Public private partnerships, the public sector comparator, and discount rates: Key issues for developing countries*. https://ideas.repec.org/p/qed/dpaper/4564.html

Shuli, W. (2014). Study on the introduction of PPP mode in Urban rail transit construction. *Expanding Horizons, 06*, 47–51.

Si, T. (2016). *Research on quantitative evaluation of value for money of PPP in China--PPP project evaluation by introducing real option*. Chinese Academy of Fiscal Sciences.

Sock-Yong, P. (2007). Urban rail transit PPPs: Survey and risk assessment of recent strategies. *Transport Policy, 14*(3), 214–231.

Tao, L., Xiaopeng, L., Lunyan, W., & Lei, G. (2017). Qualitative evaluation for the VFM research of Urban water ecological PPP project. *Yellow River, 39*, 87–91.

Treasury, H. M. (2006). *Value for money assessment guidance*. The Stationary Office. https://xueshu.baidu.com/usercenter/paper/show?paperid=5cf912d1749a295167e450822ee0f646

Wei, L., Junna, L., & Qing, Z. (2012). Concession period decision model for transportation BOT projects under revenue uncertainty. *Systems Engineering, 30*, 51–56.

Wen, J. (2016). *VFM evaluation system research of PPP project-based on the transaction cost*. Chang'an University

Wilmoth, D. (1990). Urban infrastructure finance issues in Australia: A review in the context of international experience. *Urban Policy and Research, 8*(4). 159–168. https://doi.org/10.1080/08111149008551441

Wu, M., Chau, K. W., Shen, Q., & Shen, L. Y. (2012). Net asset value-based concession duration model for BOT contracts. *Journal of Construction Engineering and Management, 138*(2), 304–308. https://doi.org/10.1061/(ASCE)CO.1943-7862.0000397

Xin, G. (2009). Enlightenment of rail transit planning and operation of Hongkong on Shenzhen. *City Planning Review, 08*, 87–91.

Xin, J. (2021). The development mode of PPP in Urban rail transit in China. *China Transportation Review, 43*, 48–54.

Yasan, Y., & Yongpu, Z. (2020). Ahp 7 Ppp Vfm. *Project Management Technology, 7*, 18–21.
Yingying, W., Ke, F., Jin, Y., & Shouqing, W. (2015). Construction and application of value-for-money evaluation model–take Urban rail transit PPP project as an example. *Project Management Technology, 13*, 21–27.
Youyong, Z., & Dong, Z. (2017). Imbalance of interests in Urban transportation PPP model and its rule of law construction. *HeNan Social Sciences, 25*(10), 89–98.
Yuyu, S., & Jing, D. (2008). The VFM evaluation on public project financing by means of PFI. *Consteuction Management Modernization, 03*, 53–55.
Zhang, Y., Zhao, T., & Shuyao, L. (2019). Construction and application of VFM quantitative evaluation model for the whole life cycle of rail transit PPP projects. *Construction Science and Technology*, 39–43. https://doi.org/10.16116/j.cnki.jskj.2019.21.008.
Zhu, H., Du, Y., Zhang, X., Yu, Y., & Zhang, M. (2021). Research on improvement of VFM evaluation method of rail transit PPP project based on sustainable development theory. *Construction Science and Technology, 432*, 29–33. https://doi.org/10.16116/j.cnki.jskj.2021.12.005

Pavement Maintenance in Malaysia: The Key to Pavement Sustainability

Ahmad Kamil Arshad, Ekarizan Shaffie, Mohd Izzat A. Kamal, Mat Zain Hussain, and Nuryantizpura Mohamad Rais

Abstract A road network needs to be maintained continuously to be able to continue providing an acceptable level of service to road users. Maintenance reduces road deterioration, lowers road user costs, and keeps the road open on a continuous basis. This paper presents the various pavement maintenance techniques used in Malaysia to preserve the condition of road pavements so that the road can be continuously operated without disruption due to major road rehabilitation activities. First, the evaluation and selection of the appropriate maintenance treatment for flexible pavement are discussed. Then, maintenance treatment methods are presented include crack sealing, chip seal, micro-surfacing, thin hot mix asphalt overlay, and hot in-place recycling. Finally, the treatment methods are evaluated to determine their cost effectiveness. The approach presented in this paper will assist those involved in road maintenance on the application of appropriate pavement maintenance treatment for flexible pavement.

Keywords Pavement Maintenance · Pavement treatment · Sustainable Pavements

---

A. K. Arshad (✉) · E. Shaffie · M. I. A. Kamal
Institute for Infrastructure Engineering and Sustainability Management (IIESM), Shah Alam, Selangor, Malaysia
e-mail: drahmadkamil@uitm.edu.my

M. Z. Hussain
Kumpulan Ikram Sdn. Bhd. Kajang, Kajang, Malaysia
e-mail: matzin@ikram.com

N. Mohamad Rais
Malaysia Institute of Transport (MITRANS), Universiti Teknologi MARA, Shah Alam,, Selangor, Malaysia
e-mail: yantiz@uitm.edu.my

U. R. Hashim et al. (eds.), *Green Infrastructure*,
https://doi.org/10.1007/978-981-99-7003-2_18

1 Introduction

The Government of Malaysia spends a substantial amount of money to maintain its road infrastructure. Maintenance of the road infrastructure is essential in reducing road deterioration, lowering of road user costs, and opening the road network on a continuous basis. The maintenance activities carried out on the roads in Malaysia can be divided into three scopes of work, namely routine maintenance, periodic maintenance, and emergency works (Sufian, 2009). Routine maintenance are works carried out on acyclic and needs basis to meet the minimum acceptable requirements of the road authorities. These activities include pavement maintenance, road shoulder maintenance, grass cutting, road furniture cleaning, bridges and culvert cleaning, and drainage cleaning.

Periodic maintenance refers to those works carried out as scheduled pavement rehabilitation works and unscheduled repair works. These also include preventive pavement maintenance works such as mill & pave, resurfacing with thin overlay, and other associated works. Pavement strengthening works are also in this category of work, which include thick pavement overlays, reconstruction works, and recycling. Other works in this category include repair and maintenance of road furniture, drainage structures and culverts, repairs to slope and embankment, minor repair to bridges, pedestrian crossings, and other similar structures.

Emergency works provide rapid response to emergency situations such as flood, landslide, embankment failures, road subsidence, bridge and culvert collapse, hazardous material spillage, and fallen trees by carrying out traffic management within one hour, managing temporary traffic diversion and clearing of debris and material to make the road passable within 24 h.

As for road pavements, the concept of preventive maintenance is being embraced by the road authorities in Malaysia such as JKR and LLM to preserve and prolong the road pavements in good condition. Preventive pavement maintenance treatments are applied to roads in relatively good condition before extensive amounts of deterioration are present. Figure 1 shows the relationship between pavement condition and categories of treatment that are suitable at different times and condition of the pavement.

Examples of preventive maintenance treatments include crack sealing, chip seals, micro-surfacing, thin hot mix overlays, and hot in-place recycling (Asphalt Institute, 1996). In addition, preventive pavement maintenance promotes environmental sustainability as it conserves energy, use of new materials, and lowers the production of greenhouse gases as the roads are being maintained properly (Tighe & Gransberg, 2013; Transportation Research Board, 2011). The purpose of this paper is to provide an overview of the pavement maintenance methods used in Malaysia to ensure that the road network is in a good condition and safe to be used throughout the design life of the pavement.

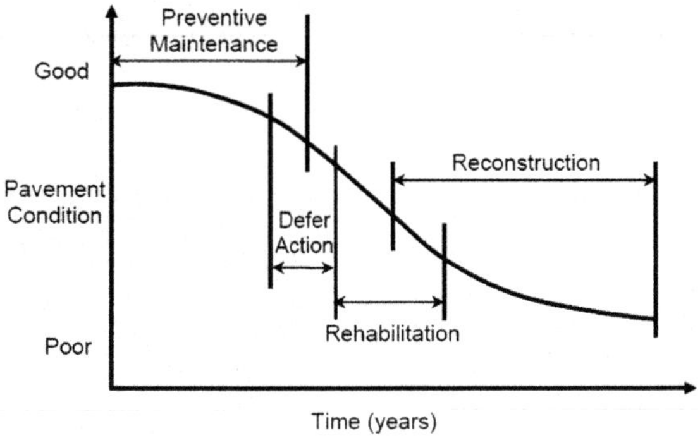

Fig. 1 Relationship between pavement condition and treatment categories (state of California department of transport, 2007)

2 Pavement Maintenance Techniques

2.1 Crack Sealing

Sealing cracks in flexible pavements is one of the most effective methods of maintaining flexible pavement. The use of a crack sealant material prevents moisture infiltration, retain material strength, and therefore decrease the potential for pavement distresses such as stripping and fatigue cracking. Crack sealing is most effective on pavements exhibiting little structural deterioration and when applied early in the life of the pavement before significant moisture can infiltrate the pavement through surface cracks.

Flexible pavements showing advanced stages of alligator cracking should not be treated by crack sealing. If the pavement has a structural deficiency, a structural solution such as a thick overlay is required. However, most types of longitudinal and transverse cracking (including block and non-load-related cracking) can be effectively sealed to prevent further deterioration. Cracks between 3 mm (0.10 in) and 25 mm (1 in) are recommended for crack sealing. Cracks greater than 25 mm (1 in) wide should be patched using standard patching materials and procedures (State of California Department of Transport, 2007).

Thermoplastic sealants are the most used sealants on flexible pavements. Among the more widely used hot-applied thermoplastic sealants are rubberized or polymer-modified asphalt. This type of sealant is produced by incorporating various types and amounts of polymers and melted rubber into asphalt. The resulting sealant possesses a greater resistance to high-temperature softening.

The crack sealing process includes crack refacing (reservoir cutting), crack cleaning, hot air blasting, sealant preparation, sealant application, project clean-up, and opening to traffic (Decker, 2016). During reservoir cutting, the contractor needs to ensure that the proper reservoir dimensions (width and depth) are being maintained. All dirt or debris should be blown from refaced cracks prior to sealant application. Hot air blasting with a hot compressed air lance is conducted on the cracks to minimize condensation formation and to maintain warm pavement temperatures

Fig. 2 Application of sealant on cracks

in preparation for the installation of a hot-applied sealant. The proper application of the sealant requires that the material be placed at a specified temperature, pressure, and rate that allows the sealant to penetrate and adhere to the area in and around the crack (Fig. 2). Traffic control procedures are required to avoid the use of roadway by vehicles during the curing time of the sealant.

2.2 *Thin Hot Mix Asphalt Surfacing*

Thin hot mix asphalt surfacing, also known as thin overlays, are typically surface mixes of 40 mm (1.5 inches) or less placed on a well-prepared flexible pavement surface (Asphalt Institute, 1996; NAPA, 2009). In Malaysia, thin overlays are surface mixes of 50 mm or less as the nominal maximum aggregate size (NMAS) is 14 mm for AC14 mix used for the overlays and the thickness must be at least three times the NMAS to ensure adequate compaction (Public Works Department of Malaysia, 2008). Thin overlay is suitable to correct pavement deficiencies such as raveling, minor alligator cracking, or rutting which are non-structural. In cases of minor rutting alligator cracking, the depth of rutting or cracking will dictate the type and extent of surface preparation for the thin overlay. Surface preparation depends on the level and

Fig. 3 Thin hot mix asphalt surfacing on existing road surface

depth of the distress present. Typical surface preparation for alligator crack include mill to crack depth, clean, and apply tack coat prior to the application of thin overlay. For rutting, surface preparation include mill to depth of rutting of existing surface layer, clean the surface, and apply tack coat before the application of thin overlay. A thin asphalt overlay may also be applied to correct functional problems such as skid resistance, ride quality, and noise generation. Figure 3 shows the application of thin hot mix asphalt surfacing on existing pavement.

2.3 *Chip Seal*

Chip Seal is an application of a binder in the form of an emulsion or bitumen, followed by an application of single-size aggregate as shown in Fig. 4 (Public Works Department of Malaysia, 2008). Chip seal is a three-stage process: after the existing surfacing has been prepared by patching, cracks are filled by spraying binder, followed by the application of single size aggregates (Fig. 5). The final operation is several passes of pneumatic tire roller.

Chip seal is used to restore skid resistance, provide water proofing layer, and retard binder hardening (Sufian et al., 1996). Chip seal is an economical surface treatment designed to protect and prolong pavement life. Chip seal on aged but structurally sound pavement surfacing will help prolong the pavement functional condition. Figure 6 shows the completed chip seal surface texture.

Fig. 4 Single-graded precoated aggregates used in the chip sealing work

Fig. 5 Spraying of polymer modified bitumen and spreading of precoated chippings

Fig. 6 Surface texture of the road after completed chip seal work

Chip seal should not be used for pavement experiencing structural deficiency /substandard structural strength. Flying aggregates causing windscreen breakages because of inadequate embedment of aggregates into the binder due to rolling after the breaking of emulsion.

2.4 *Micro Surfacing*

Micro-surfacing is a cost effective, long-lasting preventive pavement maintenance method that seals and protects the road surface from water, weathering, deterioration from ultraviolet rays, and mechanical wear thus prolonging the lifespan of the road pavement (Gujar et al., 2013). Micro-surfacing is a mixture of polymer-modified emulsion, crushed aggregate, water, and mineral filler that is pre-mixed and placed as a slurry onto a pavement surface. It was developed in Germany in the late 1960s to fill in wheel ruts in hot mix asphalt pavements on the autobahn. Pavement candidates may be flexible or concrete pavements in good structural condition with low-severity surface distresses, polished surfaces, raveling, and minor rutting.

The mix design for micro-surfacing is first carried out in the laboratory for the determination of suitable materials to be used, optimal curing time, and additives rates for a project. Before construction, any surface distresses are patched and crack filled,

Fig. 7 Application of micro-surfacing on existing pavement surface

localized pavement structural deficiencies are repaired, and the pavement surface is then swept and cleaned to remove loose dirt and contaminant. Micro-surfacing application requires a continuous flow mixing unit, a multiblade, double-shafted mixer, and a spreader box (Fig. 7). The polymer-modified emulsion, the aggregate, and mineral fillers are mixed in a continuous flow mixing unit and discharges the mixture on a continuous flow basis. The mixture is spread uniformly by a spreader box with a rear seal as the final strike off and is automatically feathered at the edges (Fig. 8). The quick setting polymer-modified emulsion allows traffic to return in about one hour. Figure 9 shows the completed micro-surfacing surface.

Micro-surfacing is typically available in two sizes, as specified in the ISSA specifications. Type II is the most common type used for all applications including arterial roads, residential areas, highways, and airport runways. Type III uses a coarser aggregate and is used on higher-traffic pavements, including expressways and high-speed roads, industrial applications, and runways.

2.5 Hot In-Place Recycling

In-place recycling is defined as the process of removing the top portion of an existing flexible pavement surface, reprocessing the material on site (such as combining the

Fig. 8 Finishing the micro-surfacing surface

removed material with a recycling agent and adding new aggregate/hot mix asphalt as necessary), and placing and compacting the material back on the pavement. There are two types of in-situ recycling used to maintain flexible pavement, the cold in-place recycling and hot in-place recycling. In Malaysia, the hot in-place recycling is normally used as a maintenance technique, while the cold in-place recycling is used for rehabilitation purposes (Arshad, 2004). This paper discusses the hot in-place recycling, which is used as a maintenance technique in Malaysia.

Hot in-place recycling is the process of heating the existing pavement surface to soften the surface layer, loosening the pavement surface material by scarifying, mixing the loosened material with a recycling agent, and relaying the recycled material in-situ and compacting the material back on the flexible pavement (Arshad, 2005). Hot in-place recycling is effective in correcting surface distresses that are limited to the top 25–50 mm of the surface layer. The surface distresses include rutting, corrugations, raveling, bleeding, lack of surface friction, and minor cracking. Hot in-placed recycling is applied only when the pavement is structurally sound, and the distress is confined to the top of the surfacing (up to a depth of 60 mm). The technique can be used to replace the conventional mill and pave method.

There are three types of HIPR process according to the American Recycling and Reclaiming Association (Asphalt Recycling & Reclaiming Association, 2015), namely:

Fig. 9 Completed micro-surfacing surface

- **Surface Recycling/Heater Scarification**–The process involves: (1) Heating the existing pavement surface, (2) Scarifying the softened surface, (3) Adding a recycling agent, (4) Mixing the loose recycled material, (5) Spreading and placing the recycled mix with a free-floating screed, (6) Compacting the mix with conventional rollers and procedures. Depths of 20 mm to 25 mm are common for this type of recycling method
- **Remixing**–The remixing process consists of the following steps: (1) Heating the pavement to a depth of 37.5 mm to 50 mm; (2) Scarification and collection of RAP in windrow; (3) Addition of virgin aggregate, recycling agent, or fresh HMA mix; (4) Mixing in a pugmill; and (5) Spreading recycled mix and compacting.
- **Repaving**–surface recycling is combined with a simultaneous overlay of new HMA to form a thermal bond between recycled and new layers. The recycled depth generally ranges from 25 to 50 mm.

In Malaysia, the most common method presently is the remixing method with a recycled depth of 50 mm as described above. Figure 10 shows the HIPR equipment train to carry out the remixing method. Figure 11 shows the completed HIPR surfacing under compaction, prior to the opening to the traffic.

Fig. 10 HIPR equipment train: preheater, remixer, tipper, rollers

3 Selection of the Preferred Maintenance Treatment

There are many factors to be considered before the selection of an appropriate maintenance treatment method required for a pavement. The following steps show the process for the selection of the preferred maintenance treatment (FHWA, 2004):

- Obtain available project information
- Establish existing condition of pavement
- Determine cause of distress
- Develop feasible alternatives
- Conduct engineering and economic analysis
- Select preferred alternative
- Design preferred alternative
- Follow up review of pavement performance

The process of pavement evaluation and establishing the cause of pavement distress requires a systematic process and is covered elsewhere (Arshad et al., 2022). The use of decision trees or decision matrices is helpful in sorting out the best short list of feasible alternatives (maintenance treatments) that should be considered for the project. A typical decision support matrix for flexible pavements is shown in Fig. 12 (Hicks et al., 2000).

Fig. 11 Completed HIPR surfacing under compaction

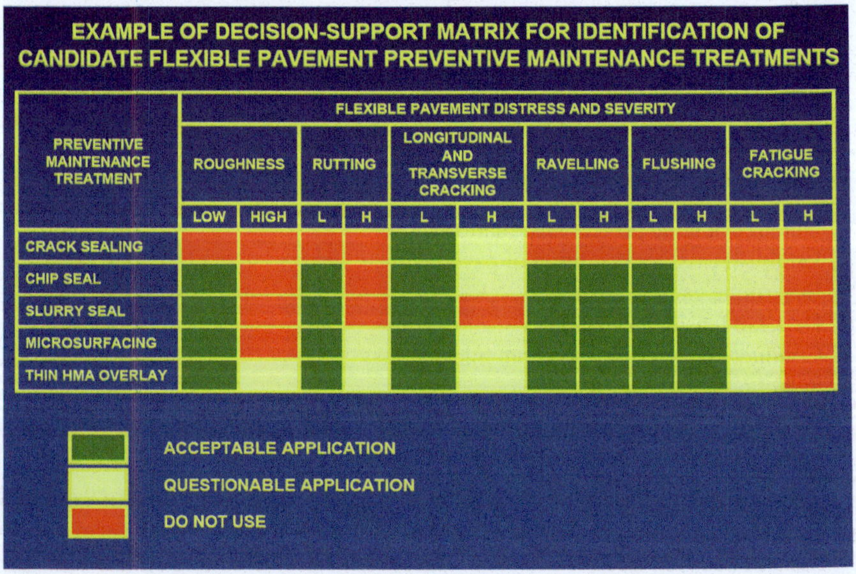

Fig. 12 Decision support matrix for maintenance treatment

In addition to the above, the use of life-cycle cost analysis (LCCA) for flexible pavement maintenance and rehabilitation strategies should also be carried out to select the most cost-effective strategy for a particular flexible pavement (FHWA, 2016; Offenbacker & Mehta, 2022).

4 Conclusion

Pavement maintenance techniques such as crack sealing, chip seal, micro-surfacing, and thin HMA overlay are currently being used for the maintenance of road pavement in Malaysia. Proper selection of the suitable maintenance technique is important in addressing the pavement distress requirements and thus extending the life of the pavement. Maintenance is essential for the continuous opening of the road network and the cost effectiveness of preserving the road pavement. Neglecting maintenance of the pavement will further increase the rate of deterioration of the road pavement and thus increase the future cost of rehabilitating the road pavement.

References

Arshad, A. K. (2004). Pavement recycling: An overview. *Jurutera, 2003*(2), 14–17. The Institution of Engineers.

Arshad, A. K. (2005). Hot in-place recycling: An introduction. *Jurutera, 2005*(2), 10–14. The Institution of Engineers.

Arshad, A. K., Shaffie, E., Kamal, M. I. A., Hussain, M. Z., & M. Rais, N. (2022). Application of pavement evaluation for road maintenance and rehabilitation. In *Green infrastructure: Materials and applications*. Springer Nature Singapore Pte Ltd.

Asphalt Institute. (1996). *Asphalt in pavement maintenance. Manual Series No. 16* (3rd ed.). Asphalt Institute.

Asphalt Recycling and Reclaiming Association. (2015). *Basic asphalt recycling manual* (2nd ed.). Federal Highway Administration.

Decker, D. S. (2016). Best practices for crack treatments in asphalt pavements. In *Proceedings of the 6th Eurasphalt & Eurobitume Congress, 1–3 June 2016, Prageu, Czech Republic*.

FHWA. (2004). *Pavement preservation: Design and construction of quality preventive maintenance treatments, Publication No. FHWA-NHI-04–171*. Federal Highway Administration.

FHWA. (2016). *Pavement life-cycle assessment framework, Report FHWA-HIF-16-014*. FHWA.

Gujar, R. S., Chauhan, K. A., & Dadhich, G. (2013), Microsurfacing–an eco-efficient tool for road safety and pavement maintenance. *International Journal of Sustainable Construction Engineering & Technology, 4*(2). Universiti Tun Hussein Onn.

Hicks, R. G., Seeds, S. B., & Peshkin, D. G. (2000). *Selecting a preventive maintenance treatment for flexible pavements. FHWA Report No.: FHWA–IF–00–027*. FHWA.

NAPA. (2009). *Thin asphalt overlays for pavement preservation*. Information Series 135, National Asphalt Pavement Association.

Offenbacker, D., & Mehta, Y. (2022). Assessing the life-cycle costs of pavement rehabilitation strategies used in long-term pavement performance program, ASCE. *Journal of Transportation Engineering, Part b: Pavements, 148*(1), 1–9.

Public Works Department of Malaysia. (2008). *Standard specification for road works (JKR/SPJ/2008-S4)*. Kuala Lumpur.

State of California Department of Transport. (2007). *Maintenance technical advisory guide volume 1-flexible pavement preservation* (2nd ed.). Caltrans.

Sufian, Z. (2009), Malaysia: Maintenance of federal roads–strategies and challenges. *Routes*, *343*, 66–77. PIARC.

Sufian, Z., Smith, H. R., & Ford, W. G. (1996). The benefits of using chip seals in Malaysia. In *Proceedings of the Second Malaysian Roads Conference, 10-13 June 1996, Kuala Lumpur* (pp. 445–453).

Tighe, S. L., & Gransberg, D.D. (2013). Demonstrating how sustainability can effectively be incorporated into pavement engineering and management. In *ASCE 2013 Airfield and Highway Pavements Conference, Los Angeles, California, June 9–12, 2013*.

Transportation Research Board. (2011). *Sustainable pavement maintenance practices*. Research Results Digest 365. TRB

Evaluation on Volumetric Properties of Stone Mastic Asphalt Mix Containing Steel Fibre Using Response Surface Method

Fionna Shiong, Ekarizan Shaffie, and Nuryantizpura Mohamad Rais

Abstract This paper presents the effects of different amounts of steel fibre on the volumetric properties of stone mastic asphalt (SMA) mixtures. Central composite design (CCD) method was used to design the experiments based on the response surface method (RSM) using Design Expert Software. Steel fibre (SF) content (0.3, 0.5 and 0.7%), degree of compaction (%) and stiffness (kg/mm) were selected as independent variables, while bulk specific gravity, Marshall stability, flow and air void of asphalt mixtures were chosen as dependent variables. In this research study, volumetric properties of SMA mixtures were measured by using Marshall Mix Design. The RSM analyses showed that all independent variables were significant factors for influencing the volumetric properties of the mixtures. In addition, analysis of the test results showed that the mixtures containing 0.3% steel fibre are the most optimum value to be used as a modifier in the SMA mixture. Furthermore, the developed models between the independent and dependent variables demonstrated acceptable levels of correlation. It was concluded that optimization using RSM is an effective approach for providing an appropriate empirical model for relating parameters and predicting the optimum performance of an asphaltic mixture.

Keywords Stone mastic asphalt · Volumetric · Steel fibre · RSM · CCD

F. Shiong · E. Shaffie (✉) · N. Mohamad Rais
School of Civil Engineering, College of Engineering, Universiti Teknologi MARA, 40450 Shah Alam, Selangor, Malaysia
e-mail: eka@uitm.edu.my

N. Mohamad Rais
e-mail: yantiz@uitm.edu.my

E. Shaffie
Institute for Infrastructure Engineering and Sustainability Management (IIESM), Universiti Teknologi MARA, 40450 Shah Alam, Selangor, Malaysia

N. Mohamad Rais
Malaysia Institute of Transport (MITRANS), Universiti Teknologi Mara, 40450 Shah Alam, Selangor, Malaysia

U. R. Hashim et al. (eds.), *Green Infrastructure*,
https://doi.org/10.1007/978-981-99-7003-2_19

1 Introduction

In order to withstand heavy loads from traffic and provide an excellent macro texture, stone mastic asphalt (SMA) mixtures were developed in Germany (Woodward et al., 2016). Due to its solid stone structure, excellent, bound asphalt mixture and stabilising agent, SMA is often utilised. To increase strength and ensure proper asphalt mixture stability, a stabilising ingredient, such as fibre, is added to the asphalt mixture (Shaffie et al., 2022). The asphalt mixture contains a stabilising substance (fibre) that stops the binder from draining. Numerous researchers have studied ways to change the SMA. To enhance SMA performance, a variety of additives have been employed.

In the industry, fibre-reinforced bituminous mixtures are no longer uncommon. It is now frequently utilised, particularly in SMA. A study by (Arabani & Shabani, 2019), ceramic fibre is used to enhance the qualities of asphalt binder. In addition, (Tanzadeh et al., 2019) improved the performance of an open-graded friction course using glass fibre. When the mixture is heated to a high temperature and the amount of binder is high, the binder drains down (Masri et al., 2016). Mineral and steel fibres may be added to the SMA to improve its function and stop the binder leak down from growing worse. Compared to other modified fibre, the cost of producing steel fibre-reinforced bituminous mixes may be quite high, but in the end, steel fibre may lower and minimise the maintenance costs (Jasni et al., 2020). Identifying the volumetric characteristics of SMA containing steel fibre is the objective of this research work. According to (Bindu, 2012), the inclusion of fibre, especially steel fibre, may have a considerable impact on the volumetric, mechanical and binder-draining characteristics of SMA mixtures. This study may provide suggestions for improving the performance of asphalt pavement's fibres.

The use of statistical modelling and optimization approaches is beneficial because they excel at handling a variety of restrictions and goals and articulating the interactions between dependent variables that influence a certain response (Kikuchi et al., 2012; Soltani et al., 2018). By using techniques like response surface methodology (RSM), a factorial design of experiments (DOE) is used to take into account several variables at once at various levels and provide an appropriate prediction model for the relationship between the different elements (Moghaddam et al., 2015). RSM techniques have been successfully used in a variety of fields, including biomass and clay sciences, material and mechanical engineering and concrete technology. RSM has recently gained more popularity as a tool for designing asphalt mixes (Bala et al., 2017).

It is obvious that adding steel fibre to the binder improves its performance qualities. The volumetric characteristics of steel fibre-modified SMA mixes are optimised and developed using response surface methodology (RSM) in the present work. The purpose of this research project is to define the properties and functionality of SMA containing steel fibre. This study may offer suggestions for improving the performance of asphalt pavement's steel fibre content.

2 Materials and Method

2.1 *Aggregate Properties*

The aggregate used in this research was obtained from Kajang Rocks Quarry, Selangor, Malaysia. The nominal maximum aggregate size was 20 mm. Tables 1 and 2 show the aggregate properties and aggregate gradation, respectively. The gradation of the combined coarse aggregate, fine aggregate and mineral filler was chosen to conform with the appropriate envelope in accordance to PWD Malaysia's Standard Specification for Road Works (JKR, 2008). Figure 1 shows the aggregate gradation curve of the SMA20 mixture. The aggregate properties have been tested and confirmed to Public Works Department of Malaysia's (PWD Malaysia) Specification for Road Works requirement which indicates good quality thus acceptable for use in road works.

Table 1 Aggregate properties

Aggregate test	Test result	PWD requirements (%)	Designation
Aggregate crushing value, ACV %	22.53	<30	BS 812: PART 110
Aggregate impact value, AIV %	22.4	<25	BS 812: PART 112:1990
Los Angeles abrasion value, LA %	21.38	<25	ASTM: C 131–96
Flakiness index, %	10.25	<25	MS30

Table 2 SMA aggregate gradation (JKR, 2008)

Sieve	Gradation		Percentage passing	Percentage retained	Mass passing	Mass retained
Size	Lower	Upper				
19.0	100	100	100	0.0	1200.0	0
12.5	85	95	90	10.0	1080.0	120
9.5	65	75	70	20.0	840.0	360
4.75	20	28	24	46.0	288.0	912
2.36	16	24	20	4.0	240.0	960
0.60	12	16	14	6.0	168.0	1032
0.30	12	15	13.5	0.5	162.0	1038
0.075	8	10	9	4.5	108.0	1092
Pan				9.0		108

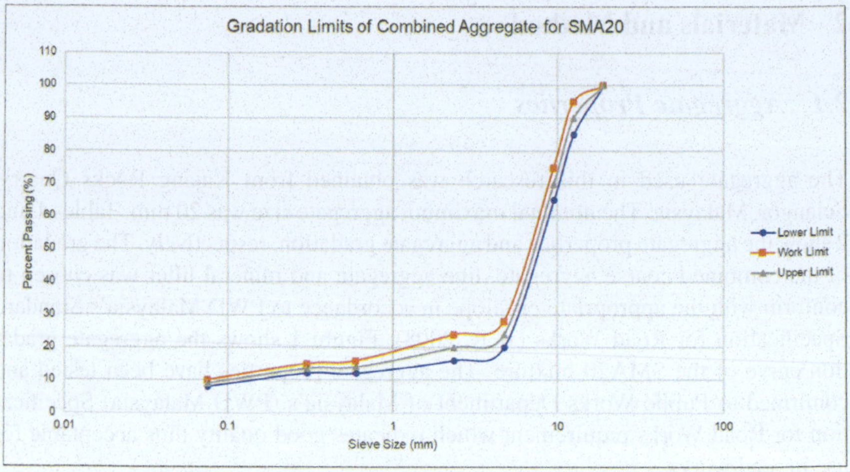

Fig. 1 Aggregate gradation curve for SMA20 mix

Table 3 Properties of asphalt binder used

Binder test	Test result	PWD requirements	Designation
Penetration (25 °C, 100Gm, 5 s), (1/10 mm)	60.2	60–70 mm	ASTM D-5
Softening point (Ring and Ball), °C	52	≥ 48 °C	ASTM D-36
Ductility, cm	144	≥100 cm	ASTM D-113

2.2 *Binder Properties*

Penetration Grade (PEN 60/70) asphalt binder was used in asphalt mixtures as binder. Binder for PEN 60/70 has been chosen based on the current practice for road in Malaysia. The properties of the binder are presented in Table 3.

2.3 *Steel Fibre*

Steel fibre in this research was sourced from Gardner Global Enterprise as shown in Fig. 2 to make a proportion as an addition to SMA with various percentage of 0%, 0.30%, 0.50%, and 0.70% by total weight of mixture. The properties of steel fibre are shown in Table 4. The properties of steel fibre are according to previous research (Shaffie et al., 2021, 2022) and it is for continuous study.

Fig. 2 Steel fibre

Table 4 Properties of steel fibres

Description	Length (mm)	Diameter (mm)	Density (Kg/m^3)	Tensile strength (Mpa)	Aspect ratio
Straight	13	0.2	7800	2600	65

2.4 *Marshall Mix Design*

Marshall tests were conducted to determine the volumetric properties of mixture. The volumetric properties determined included bulk specific gravity, air voids, stability and flow. The optimum binder contents of the control mixtures were determined using Marshall design method (ASTM D1559) with 50 blows on each side. The optimum binder contents is 6.2% for mixtures containing PEN 60/70.

As a method of dry blending method process, the steel fibre (0%, 0.3%, 0.5% and 0.7%) by the total weight of mixture was added directly to the mixture before the binder. The mixtures containing different amounts of steel fibre were made by the same optimum binder content, so that the amount of asphalt binder would not confound the analysis of the test results. There are significant effects on the Marshall properties with the inclusion of steel fibre in the asphalt mixes. The results of volumetric properties for SMA20 mixtures are shown in Table 5.

Table 5 Volumetric properties for SMA20 mixtures

Mixture type	SMA-control	SMA20-0.3SF	SMA20-0.5SF	SMA20-0.7SF
Bulk specific gravity	2.243	2.248	2.236	2.228
Stability (kN)	8.7	10.7	10.6	9.8
Flow (mm)	6.25	4.87	4.70	4.42
Air voids (%)	10.637	5.863	4.526	7.820

2.5 Central Composite Design Method (CCD)

In this study, volumetric properties of asphalt mixtures containing 0%, 0.3%, 0.5% and 0.7% steel fibre were measured by the Marshall mix design apparatus and the results were analysed using response surface method. For this purpose, central composite design (CCD) was used for designing the experiments. A fractional factorial experimental design known as the CCD may make a correlation between test and response variables at various factor levels (Nassar et al., 2016a). The test variables and response matrix for the experiment are shown in Table 6. Figure 3 depicts the study's organisational structure.

2.6 Response Surface Methodology

The Response Surface Method (RSM) is a set of statistical tools for creating mathematical models, planning experiments, analysing the effects of experiment parameters and streamlining the procedure (Golchin & Mansourian, 2017). Statistical methods were utilised to create mathematical models from the experimental test results. The F-test and lack of fit test were used to determine the model's significance and R-squared was used to determine how well the model matched the data. Analysis of variance was used to assess how the experiment parameters influenced the results (ANOVA). Equations (1), (2) and (3) show simple forms of linear, two-factor interaction (2FI) and quadratic models for two independent factors, respectively:

$$Y = C_0 + C_1X_1 + C_2X_2 \tag{1}$$

$$Y = C_0 + C_1X_1 + C_2X_2 + C_{12}X_1X_2 \tag{2}$$

$$Y = C_0 + C_1X_1 + C_2X_2 + C_{12}X_1X_2 + C_{11}X_1^2 + C_{22}X_2^2 \tag{3}$$

where Y is the response, X_1 and X_2 are the independent factors, and C_0, C_i, C_{ii} and C_{ij} are the intercept, linear, quadratic and interaction coefficients, respectively. Recently, RSM has been used in asphalt mixture studies (Haghshenas et al., 2015; Hamzah et al., 2015, 2017; Nassar et al., 2016b). Design-expert software is a popular statistic tool for designing of experiments and analysis based on RSM (Golchin & Mansourian, 2017). In this study, this software was used for designing the experiment, selecting appropriate mathematical models to fit data, analysis of variance (ANOVA) with 95% of confidence interval and plotting the graphs.

Table 6 Experiment matrix and obtained volumetric properties results

Test factors (Independent variables)			Test responses (Dependent variables)			
A: degree of compaction, %	B: stiffness, %	C: steel fibre, %	R1: bulk specific gravity, Gmb	R2: stability, kg	R3: flow, mm	R4: air void, %
92.418	185.673	0.3	2.242	1110.09	4.5	6.594
92.418	229.014	0.7	2.26	1016.31	3.96	5.815
92.418	185.673	0.5	2.243	1059.12	4.16	5.665
95.474	229.014	0.7	2.26	1016.31	3.96	5.815
92.418	229.014	0	2.255	822.63	4.7	10.43
89.362	229.014	0	2.218	822.63	4.7	11.402
92.418	142.331	0.5	2.243	1141.69	5.92	5.665
95.474	229.014	0.3	2.264	1323.14	5.02	5.473
95.474	142.331	0	2.255	952.09	7.25	10.43
95.474	185.673	0.7	2.26	1028.54	4.13	5.815
95.474	142.331	0.3	2.264	835.88	5.08	5.473
89.362	142.331	0.3	2.242	835.88	5.08	6.594
95.474	185.673	0.5	2.223	1059.12	4.16	3.743
95.474	185.673	0	2.255	893.99	6.8	10.43
92.418	229.014	0.5	2.243	1030.58	4.03	5.665
92.418	185.673	0.7	2.26	1028.54	4.13	5.815
95.474	142.331	0.5	2.223	1141.69	5.92	3.743
89.362	185.673	0.5	2.243	1059.12	4.16	5.665
89.362	229.014	0.7	2.2	1016.31	3.96	8.8
92.418	185.673	0	2.255	893.99	6.8	10.43
89.362	229.014	0.3	2.242	1323.14	5.02	6.594
95.474	229.014	0.5	2.223	1030.58	4.03	3.743
92.418	185.673	0.3	2.242	1110.09	4.5	6.594
95.474	229.014	0	2.255	822.63	4.7	10.43
92.418	142.331	0	2.255	952.09	7.25	10.43
95.474	185.673	0.3	2.264	1110.09	4.5	5.473
89.362	142.331	0.7	2.2	945.97	5.17	8.8
92.418	229.014	0.3	2.242	1323.14	5.02	6.594
92.418	185.673	0.7	2.26	1028.54	4.13	5.815
92.418	142.331	0.3	2.242	835.88	5.08	6.594
89.362	142.331	0	2.218	952.09	7.25	11.402
92.418	142.331	0.7	2.26	945.97	5.17	5.815
89.362	185.673	0.3	2.242	1110.09	4.5	6.594
92.418	185.673	0	2.255	893.99	6.8	10.43

(continued)

Table 6 (continued)

Test factors (Independent variables)			Test responses (Dependent variables)			
A: degree of compaction, %	B: stiffness, %	C: steel fibre, %	R1: bulk specific gravity, Gmb	R2: stability, kg	R3: flow, mm	R4: air void, %
89.362	185.673	0.7	2.2	1028.54	4.13	8.8
89.362	142.331	0.5	2.243	1141.69	5.92	5.665
89.362	229.014	0.5	2.243	1030.58	4.03	5.665
92.418	185.673	0.5	2.243	1059.12	4.16	5.665
95.474	142.331	0.7	2.26	945.97	5.17	5.815
89.362	185.673	0	2.218	893.99	6.8	11.402

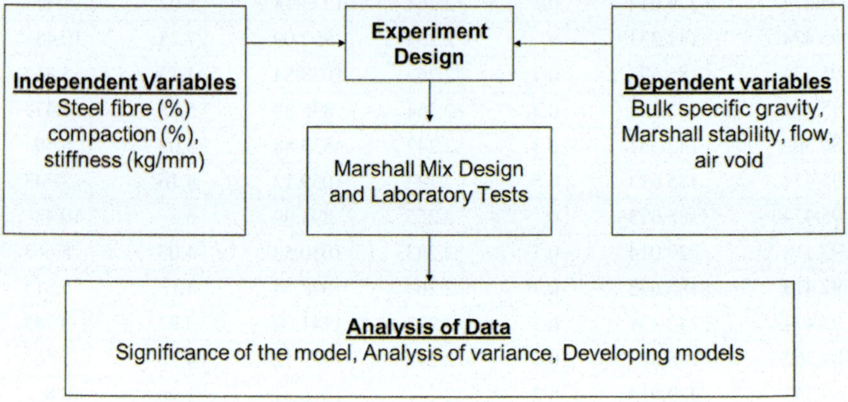

Fig. 3 Research framework

3 Results and Discussion

The impacts of test parameters (Bulk specific gravity, Marshall stability, flow and air void) on the volumetric properties are individually discussed in Sects. 3.1, 3.2, 3.3 and 3.4.

3.1 *Bulk Specific Gravity*, G_{mb}

To match the data for the prediction of bulk specific gravity, three regression models were examined: quadratic, linear and two-factor interaction models. As indicated in Table 7, data analysis revealed that the quadratic model was an appropriate regression model. This model showed a lower p-value in sequential F-test and acceptable R-squared value. Table 8 displays the ANOVA results. This table shows that steel fibre, degree of compaction and stiffness have significant effects on the bulk specific gravity of the mixture. Substantial independent variables and significant interactions between independent factors are shown by a "Prob > F" value less than 0.05. Significantly more important than steel fibre content and stiffness is the degree of compaction. A higher F-value indicates that the test factors are more significantly significant as in Table 8.

Equations (4), (5), (6), and (7), as given in Table 9, are suggested equations for predicting the bulk specific gravity of the mixture from a volumetric test. The bulk specific gravity of specimens rises as the degree of compaction increases, according to regression coefficients from these equations, however, the interaction between stiffness and degree of compaction causes the bulk specific gravity to decrease.

Figure 4 shows 3D-model of G_{mb} versus compaction and stiffness. From the figure, G_{mb} increases when stiffness increases. For instance, mixes with greater stiffness have the lowest value of this reaction. This is a result of the mixture of modifying ingredients. Additionally, increased Gmb levels are seen at higher compaction levels. For instance, this Gmb's highest value is 2.248. Figure 5 illustrates the predicted and actual values of bulk specific gravity that was used to determine the adequacy of the regression model between the experimental data and the model response. The regression line yielded R^2 value of 0.8338. Statistical R^2 values demonstrated an acceptable correlation between the predicted and actual experimental response values. It can be observed that both quadratic regression models fit realistically, thereby adequately expressing the experimental range studied.

Table 7 Models proposed for prediction of bulk specific gravity

Title	Sum of squares	DF	Mean square	F-Value	Prob > F	Model type
Regression	0.0108	14	0.0008	8.96	< 0.0001	Quadratic
Residual error	0.0022	25	0.0001			
R-squared	0.8338					
Lack of fit	0.0022	21	0.0001			

[a] DF is Degree of freedom

Table 8 Bulk specific gravity analysis of variance (ANOVA)

Factor	Sum of squares	DF	Mean square	F-value
A-compaction	0.0037	1	0.0037	42.58
B-stiffness	0.0000	1	0.0000	0.0000
C-steel fibre	0.0007	3	0.0002	2.67

Table 9 Bulk specific gravity equations obtained from the response surface method

Type	Steel fibre (%)	Equations	
G_{mb}	0	$y = -9.1706 + 0.2411A + 4.4394e - 16B - 3.6838e - 18AB - 0.0013A^2 - 1.7472e - 19B^2$	(4)
	0.3	$y = -8.9391 + 0.2387A + 4.0458e - 16B - 3.6838e - 18AB - 0.0013A^2 - 1.7472B^2$	(5)
	0.5	$y = 8.3155 + 0.2318A + 4.0476e - 16B - 3.6838e - 18AB - 0.0013A^2 - 1.7472B^2$	(6)
	0.7	$y = -9.5203 + 0.2449A + 4.0428e - 16B - 3.6838e - 18AB - 0.0013A^2 - 1.7472B^2$	(7)

[a] y is bulk specific gravity, A is degree of compaction, B is stiffness

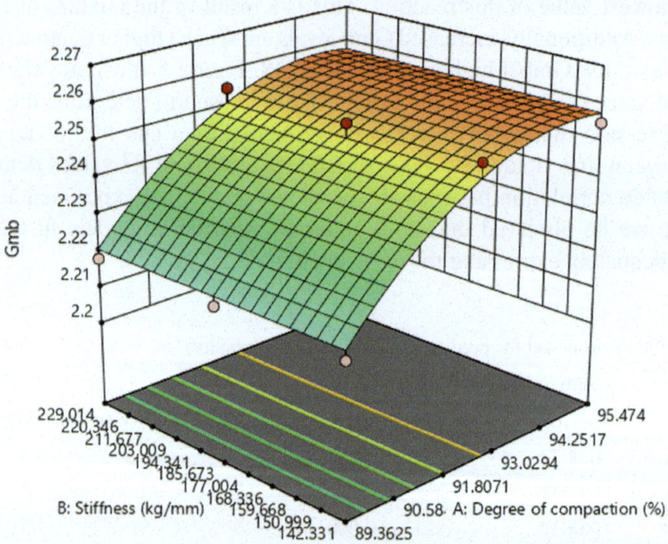

Fig. 4 3D-model of bulk specific gravity versus degree of compaction and stiffness

3.2 Marshall Stability

Marshall stability transformed values provide adequate regression parameters for formulating mathematical equations, according to regression analysis utilising RSM. A two-factor interaction regression model, as shown in Table 10, was used to measure the impact of steel fibre content, degree of compaction and stiffness on stability value. High R^2 and low Prob > F values are shown by this model. Table 11 displays the ANOVA results. According to the table, the amount of steel fibre, stiffness, degree

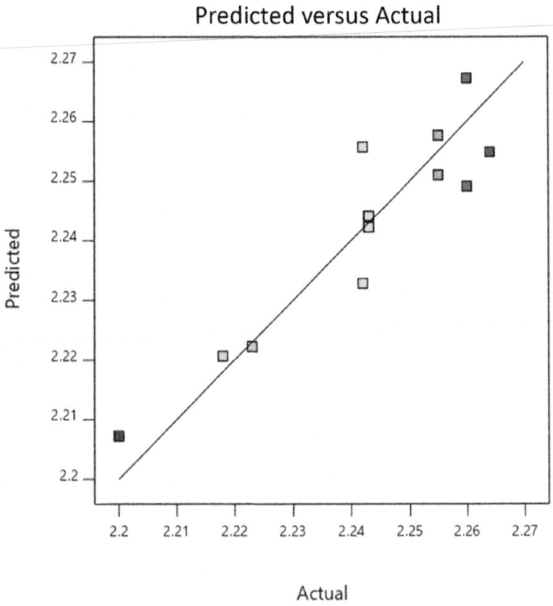

Fig. 5 Predictive and actual values of bulk specific gravity

of compaction and interaction between stiffness and degree of compaction all had a significant impact on the stability prediction value.

Equations (8), (9), (10) and (11) illustrate how test factors and stability relate to one another, as seen in Table 12. These regression equations show that the interaction between compaction and stiffness has a positive effect on stability while the degree of compaction and stiffness has negative effect on stability.

Table 10 Models proposed for prediction of Marshall stability

Title	Sum of squares	DF	Mean square	F-value	Prob > F	Model type
Regression	6.609E + 05	12	55,076.88	156.64	< 0.0001	2FI
Residual error	9493.59	27	351.61			
R-squared	0.9858					
Lack of fit	9493.59	27	412.76			

[a] DF is Degree of freedom

Table 11 Marshall stability analysis of variance (ANOVA)

Factor	Sum of squares	DF	Mean square	F-value
A-compaction	0.0000	1	0.0000	0.0000
B-stiffness	37,690.51	1	37,690.51	107.19
C-steel fibre	2.537E + 05	3	84,569.83	240.52

Table 12 Marshall stability Equations obtained from the response surface method

Type	Steel fibre (%)	Equations	
Marshall Stability	0	$y = 1167.3116 - 1.0715E-13A - 1.4935B + 1.1545E-16AB$	(8)
	0.3	$y = 48.0451 - 2.6686E-14A + 5.6212B + 1.1545E-16AB$	(9)
	0.5	$y = 1313.3234 - 3.9334E-14A - 1.2818B + 1.1545E-16AB$	(10)
	0.7	$y = 849.4337 - 1.3587E-14A - 0.8115B + 1.1545E-16AB$	(11)

[a] y is Marshall stability, A is degree of compaction, B is stiffness

Figure 6 shows a 3D-model of Marshall stability in relation to compaction and stiffness. As shown in the illustration, stability rises as stiffness falls. Increasing stiffness has no significant impact on stability at lower levels of compaction. On the other side, stability rises as stiffness falls with increasing levels of compaction. Therefore, the minimal value of this reaction occurs at high stiffness levels and low compaction levels. Additionally, Figure 7 displays the stability's actual and predicted values. The ability of the created regression models to predict the value of stability is shown by the proximity of the actual and predicted values to the 1:1 line. The regression line yielded R^2 value of 0.9858. Statistical R^2 values demonstrated a strong correlation between the predicted and actual experimental response values. It can be observed that both quadratic regression models fit realistically, thereby adequately expressing the experimental range studied.

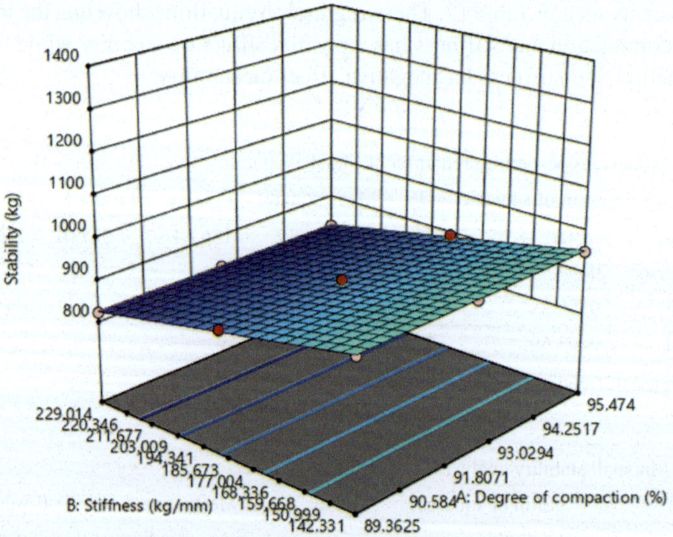

Fig. 6: 3D-model of Marshall stability versus degree of compaction and stiffness

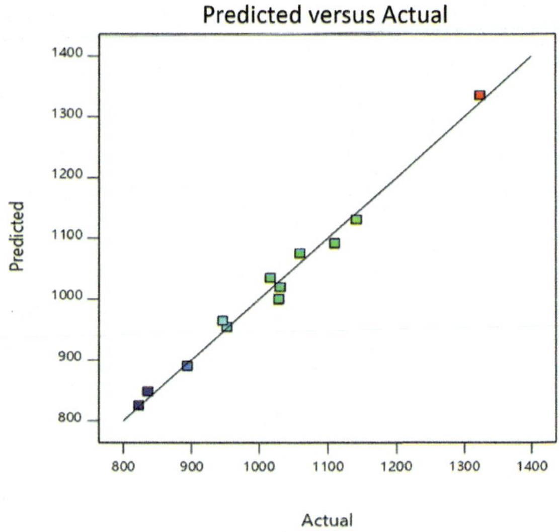

Fig. 7 Predictive and actual values of Marshall stability

3.3 *Flow*

A two-factor interaction regression model may be used for the prediction of the value of flow based on the variables listed in Table 13. The sequential F-test p-value for this model was lower, and the R-squared value was acceptable. According to the ANOVA findings in Table 14, stiffness and the degree of compaction have a significant impact on the value of flow.

The relationship between flow and test factors is expressed by Eqs. (12), (13), (14) and (15) as shown in Table 15. Figure 8 illustrates the impact of stiffness and

Table 13 Models proposed for prediction of fatigue properties

Title	Sum of squares	DF	Mean square	F-value	Prob > F	Model type
Regression	39.50	12	3.29	20.16	< 0.0001	2FI
Residual error	4.41	27	0.1633			
R-squared	0.8996					
Lack of fit	4.41	27	0.1916			

[a] DF is Degree of freedom

Table 14 Flow Analysis of variance (ANOVA)

Factor	Sum of squares	DF	Mean square	F-value
A-compaction	0.0000	1	0.0000	0.0000
B-stiffness	12.23	1	12.23	74.89
C-steel fibre	22.18	3	7.39	45.30

degree of compaction on the expected value of flow value in accordance with these equations. As shown in the picture, the flow value of specimens rises with increasing compaction or stiffness. It indicates that at all levels of degree of compaction, asphalt mixes having a large proportion of steel fibre will not exhibit a reduced level of flow value. Figure 9 compares the flow values for expected and actual values that were used to determine the adequacy of the regression model between the experimental data and the model response. The regression line yielded R^2 value of 0.8996. Statistical R^2 values demonstrated an acceptable correlation between the predicted and actual experimental response values. It can be observed that both quadratic regression models fit realistically, thereby adequately expressing the experimental range studied.

Table 15 Flow Equations obtained from the response surface method

Type	Steel fibre (%)	Equations	
Flow	0	$y = 11.7670 - 4.4521E - 15A - 0.0294B + 2.3427E - 17AB$	(12)
	0.3	$y = 4.9585 + 4.4808E - 15A - 0.0007B + 2.3427E - 17AB$	(13)
	0.5	$y = 8.6973 + 4.4539E - 15A - 0.0218B + 2.3427E - 17AB$	(14)
	0.7	$y = 6.9828 - 4.0072E - 15A - 0.0140B + 2.3427E - 17AB$	(15)

[a] y is Flow, A is degree of compaction, B is stiffness

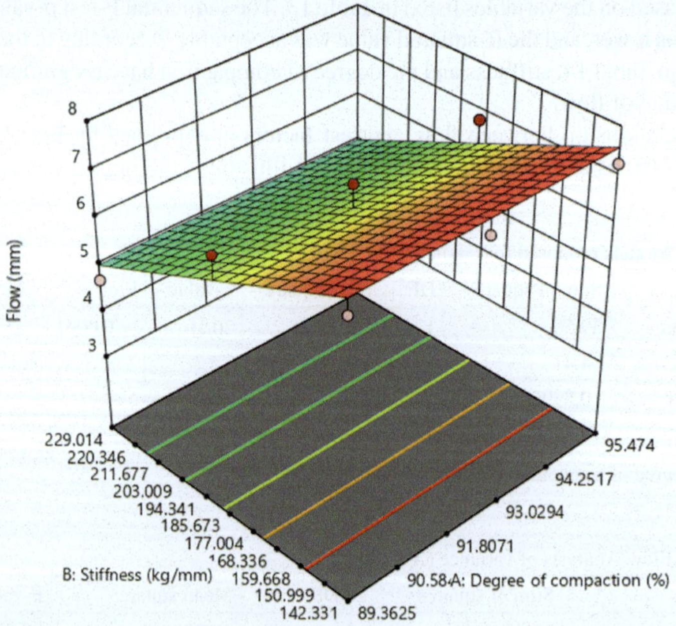

Fig. 8 3D-model of flow versus degree of compaction and stiffness

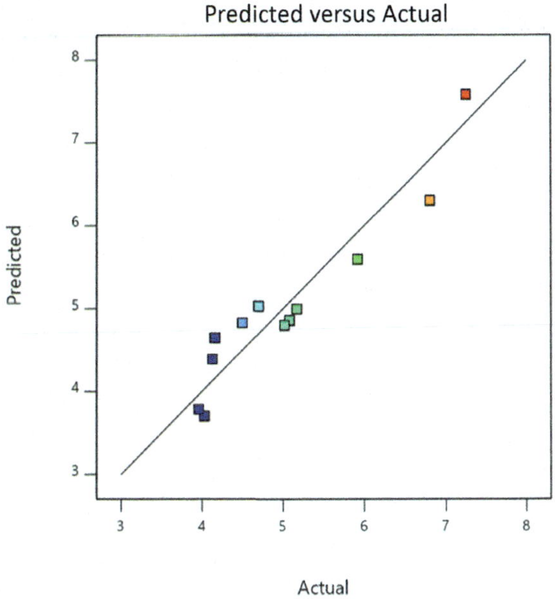

Fig. 9 Predictive and actual values of flow

3.4 *Air Voids, AV*

Transformed values of air voids in a regression study conducted using RSM demonstrated suitable regression parameters for formulating mathematical equations. A linear regression model was used to evaluate the impact of steel fibre content, degree of compaction and stiffness on air voids, as shown in Table 16. The R^2 value of this model is strong, while the Prob > F-value is low. Table 17 displays the ANOVA results. According to the table, the quantity of steel fibre, level of compaction and stiffness all significantly impacted the forecast value of air void.

Equations (16), (17), (18), and (19) depict how test variables and air voids interact as seen in Table 18. These regression equations show that the degree of compaction and stiffness has negative effect negative effect on air voids.

Figure 10 shows a 3D representation of an air void as a function of stiffness and compaction. The graphic shows that when compaction diminishes, air voids

Table 16 Models proposed for prediction of air voids

Title	Sum of squares	DF	Mean square	F-Value	Prob > F	Model type
Regression	198.25	5	39.65	106.01	<0.0001	Linear
Residual error	12.72	34	0.3740			
R-squared	0.9397					
Lack of fit	12.72	30	0.4239			

[a] DF is Degree of freedom

Table 17 Air voids Analysis of variance (ANOVA)

Factor	Sum of squares	DF	Mean square	F-value
A-compaction	18.38	1	18.38	49.13
B-stiffness	0.0000	1	0.0000	0.0000
C-steel fibre	179.88	3	59.96	160.31

Table 18 Air voids Equations obtained from the response surface method

Type	Steel fibre (%)	Equations	
AV	0	$y = 37.1851 - 0.2863A - 1.8552E - 17B$	(16)
	0.3	$y = 32.7212 - 0.2863A - 1.8552E - 17B$	(17)
	0.5	$y = 31.5519 - 0.2863A - 1.8552E - 17B$	(18)
	0.7	$y = 33.1740 - 0.2863A - 1.8552E - 17B$	(19)

[a] y is air void, A is degree of compaction, B is stiffness

grow. A considerable impact on air voids is shown when the degree of compaction is decreased, and the stiffness is increased. On the other side, air voids grow as stiffness diminishes at greater levels of compaction. Therefore, stiffness and greater levels of compaction provide the response's minimal value. Additionally, Figure 11 displays the predicted and actual air void values that were used to determine the adequacy of the regression model between the experimental data and the model response. The regression line yielded R^2 value of 0.9397. Statistical R^2 values demonstrated a strong correlation between the predicted and actual experimental response values. It can be observed that both quadratic regression models fit realistically, thereby adequately expressing the experimental range studied.

The model optimization process was conducted to determine the factor conditions that give desirable steel fibre content performance in asphalt pavement. All the factors and responses with their particular lower and upper limits were re-specified for the optimum state and best resistance to pavement failure based on the PWD specifications. Figure 12 illustrates the predicted optimum conditions and the response studied in this paper. The corresponding predicted steel fibre content was found to be 0.3%.

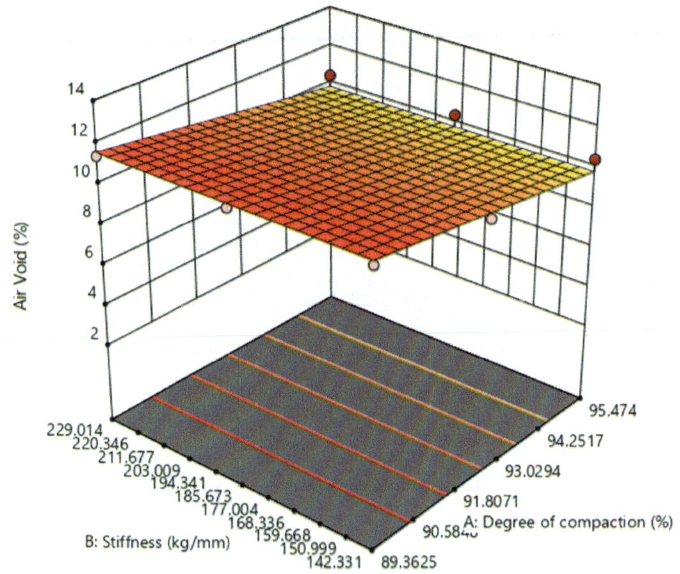

Fig. 10 3D-model of air void versus degree of compaction and stiffness

Fig. 11 Predictive and actual values of air void

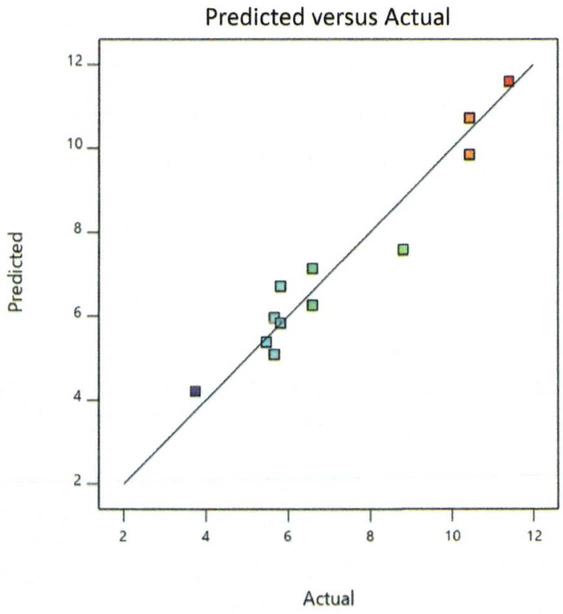

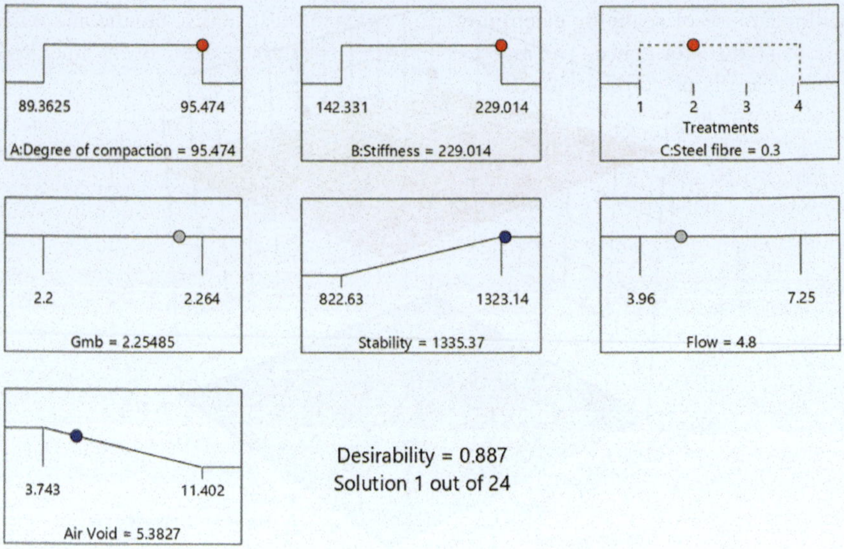

Fig. 12 Optimum conditions and responses

4 Conclusion

For standard asphalt binders in Malaysia, the volumetric characteristics of asphalt mixes including varying quantities of steel fibre were investigated at various variables and quantified by RSM in this research. RSM was utilised statistically for data analysis and experiment planning. Additionally, the effects of interactions between test factors were assessed concurrently. RSM analysis demonstrated a substantial relationship between steel fibre content, degree of compaction and stiffness and bulk specific gravity, stability, flow and air void. All response and test factor relationships were found to be simple linear, two-factor interactions (2FI), and quadratic regression models. The number of tests conducted for this study was limited. Based on these limitations, the obtained findings revealed that, except for air void, the values of bulk specific gravity, stability and flow increased when the degree of compaction was raised.

In volumetric testing, mixtures with a larger percentage of steel fibre shown increased bulk specific gravity and stability. Increasing the steel fibre content has no effect on the transformed stability and flow value at lower levels of degree of compaction. On the other side, when stiffness improves with increasing levels of compaction, air voids diminish. Consequently, the lowest value of this reaction occurs at greater levels of compaction and steel fibre content. Overall, specimens examined at optimization with 0.3% of steel fibre showed a better desirability outcome.

Overall research results indicate that steel fibre has high potential as a SMA modifier. However, to have a better knowledge of this component in the asphalt mixture, long-term inquiry has been proposed for future study. This investigation will make use of scanning electron microscope pictures, rutting, fatigue and other mechanical performance. The independent variables and responses were modelled, and the levels of correlation between the two were satisfactory. According to the research, the Response Surface Methodology (RSM) is a useful statistical technique for developing an adequate empirical model for correlating variables and forecasting the ideal performance of an asphaltic mixture to lower the likelihood of flexible pavement collapse. The procedure and laboratory tests conducted in this study can be targeted in future designs for asphalt mixtures incorporating steel fibre. Last but not least, this research has the potential for developing asphalt pavement mixes that are durable for use in highways and indirectly contribute to the Government's Strategic Reform Initiative by ensuring the nation's sustainability of growth.

Acknowledgements The Institute for Infrastructure Engineering and Sustainable Management (IIESM), University of Technology MARA (UiTM), Research Management Institute (RMI), and Ministry of Higher Education (MOHE) were greatly acknowledged for providing the fund 600-IRMI/FRGS 5/3 (379/2019) to enable this research to be carried out. The authors would also like to thank the people who participated directly or indirectly in this research.

References

Arabani, M., & Shabani, A. (2019). Evaluation of the ceramic fiber modified asphalt binder. *Construction and Building Materials, 205*, 377–386. https://doi.org/10.1016/j.conbuildmat.2019.02.037

Bala, N., Napiah, M., & Kamaruddin, I. (2017). Application of response surface methodology for mix design optimization of nanocomposite modified asphalt mixtures. *International Journal of GEOMATE, 13*(39), 237–244. https://doi.org/10.21660/2017.39.98554

Bindu, C. S. (2012). Influence of additives on the characteristics of stone matrix asphalt.

Golchin, B., & Mansourian, A. (2017). Evaluation of fatigue properties of asphalt mixtures containing reclaimed asphalt using response surface method. *International Journal of Transportation Engineering, 4*(4), 335–350.

Haghshenas, H. F., Khodaii, A., Khedmati, M., & Tapkin, S. (2015). A mathematical model for predicting stripping potential of hot mix asphalt. *Construction and Building Materials, 75*, 488–495. https://doi.org/10.1016/j.conbuildmat.2014.11.041

Hamzah, M. O., Gungat, L., & Golchin, B. (2017). Estimation of optimum binder content of recycled asphalt incorporating a wax warm additive using response surface method. *International Journal of Pavement Engineering, 18*(8), 682–692. https://doi.org/10.1080/10298436.2015.1121779

Hamzah, M. O., Omranian, S. R., Golchin, B., & Hainin, M. R. (2015). Evaluation of effects of extended short-term aging on the rheological properties of asphalt binders at intermediate temperatures using respond surface method. *Jurnal Teknologi, 73*(4), 133–139. https://doi.org/10.11113/jt.v73.4306

Jasni, N. E., Masri, K A., Ramadhansyah, P. J., Arshad, A. K., Shaffie, E., Ahmad, J., Norhidayah, A. H. (2020). Mechanical performance of stone mastic asphalt incorporating steel fiber. *IOP Conference Series: Materials Science and Engineering, 712*(1). https://doi.org/10.1088/1757-899X/712/1/012026

JKR. (2008). *Standard specification for road works-section 4: Flexible pavement (JKR/SPJ/2008-S4)*. Jabatan Kerja Raya Malaysia.

Kikuchi, S., Kronprasert, N., & Easa, S. M. (2012). Aggregate blending using fuzzy optimization. *Journal of Construction Engineering and Management, 138*(12), 1411–1420. https://doi.org/10.1061/(asce)co.1943-7862.0000557

Masri, K. A., Arshad, A. K., & Samsudin, M.S. (2016). Mechanical properties of porous asphalt with nanosilica modified binder. *Jurnal Teknologi, 78*(7–2), 139–146. https://doi.org/10.11113/jt.v78.9509

Moghaddam, T. B., Soltani, M., & Karim, M. R. (2015). Stiffness modulus of polyethylene terephthalate modified asphalt mixture: A statistical analysis of the laboratory testing results. *Materials and Design, 68*, 88–96. https://doi.org/10.1016/j.matdes.2014.11.044

Nassar, A. I., Thom, N., & Parry, T. (2016). Optimizing the mix design of cold bitumen emulsion mixtures using response surface methodology. *Construction and Building Materials, 104*(February), 216–229. https://doi.org/10.1016/j.conbuildmat.2015.12.073

Nassar, A. I., Thom, N., & Parry, T. (2016a). Optimizing the mix design of cold bitumen emulsion mixtures using response surface methodology. *Construction and Building Materials*, 104(October 2017), 216−229. https://doi.org/10.1016/j.conbuildmat.2015.12.073

Shaffie, E., Rashid, H. A., Shiong, F., Arshad, A. K., Ahmad, J., Hashim, W., Jaya, R. P., & Masri, A. (2021). Performance characterization of stone mastic asphalt using steel fiber. 2, 22−33

Shaffie, E., Ahmad, H. E., Arshad, A. K., Hashim, W., Yaacob, H., & Shiong, F. (2022). Performance evaluation of stone mastic asphalt containing steel fibre as additive. In *Green Infrastructure* (pp. 179–195). Springer Nature Singapore Pte Ltd.

Soltani, M., Moghaddam, T. B., Karim, M. R., Shamshirband, S., & Sudheer, C. (2018). Stiffness performance of polyethylene terephthalate modified asphalt mixtures estimation using support vector machine-firefly algorithm. *Measurement: Journal of the International Measurement Confederation, 121*, 360. https://doi.org/10.1016/j.measurement.2018.03.048

Tanzadeh, R., Tanzadeh, J., & Amid, S. (2019). Experimental study on the effect of basalt and glass fibers on behavior of open-graded friction course asphalt modified with nano-silica. *Construction and Building Materials, 212*, 467–475. https://doi.org/10.1016/j.conbuildmat.2019.04.010

Woodward, D., Millar, P., Lantieri, C., Sangiorgi, C., & Vignali, V. (2016). The wear of stone mastic asphalt due to slow speed high stress simulated laboratory trafficking. *Construction and Building Materials, 110*, 270–277. https://doi.org/10.1016/j.conbuildmat.2016.02.031

Envisaging the Potential Use of Resistance Micro Drilling on Wood Density Assessment: A Review

J. Joseph, R. D. Maripa, and M. H. Phua

Abstract Recurrent assessment of in-situ tree conditions through destructive methods has resorted to the critical need to adopt a more convenient and practical technique. Due to this, a Non-Destructive Testing (NDT) is indispensable to predict the wood density (WD) of trees without jeopardizing the tree itself. Resistance micro-drilling is a promising non-destructive tool, provided by its broad application in forestry to the extent of tree care industry assessment. Applying a Resistograph to assess the wood resistance is a breakthrough in sustainable forest management operations. This paper provides an insightful review of the research methodologies on Resistograph. It will discuss the use of a micro-drilling device measuring WD in standing trees, including the potential use of resistance micro-drilling to predict WD for the forest AGB and its possible use in the context of green infrastructure and sustainable management.

Keywords Resistograph · Sustainable management · Wood density · Forest aboveground biomass · Green infrastructure

1 Introduction

The density prediction of wood is a present-day research topic (Inakagi et al., 2014). The present non-destructive evaluation is essential to evaluate the wood's intrinsic properties, such as stiffness, abnormalities, strength in tension and compression and density. In addition, the density of wood species is critical for data conversion on wood volume to biomass. It has made WD an essential variable for accurately quantifying woody biomass and carbon stocks. Disregarding WD variations could lead to the poor

J. Joseph (✉)
Faculty of Plantation and Agrotechnology, Universiti Teknologi MARA UiTM Sabah Branch, Kota Kinabalu Campus, Locked Bag 71, 88997 Kota Kinabalu, Sabah, Malaysia
e-mail: jacqueline@uitm.edu.my

R. D. Maripa · M. H. Phua
Faculty of Tropical Forestry, Universiti Malaysia Sabah, Jalan UMS, 88400 Kota Kinabalu, Sabah, Malaysia

U. R. Hashim et al. (eds.), *Green Infrastructure*,
https://doi.org/10.1007/978-981-99-7003-2_20

overall prediction of the stand AGB as indicated by Baker et al. (2004). Therefore, it is crucial to acquire WD data that is feasible for extensive analysis and accomplished in conservation and protected areas. As WD data is needed to improve the allometry, various tools have been used and explored to obtain WD in standing trees without sacrificing excessive energy and costs.

To date, arborist has recognized the application of Resistograph as a tool in evaluating sustainable urban development that deals with managing natural resources, including tree populations. Resistance micro-drilling has been seen as a potential alternative to acquire WD's predicted value in standing trees. In addition to its application, arborists used it to measure trees' age and assess their health. The assessment includes:

- The diagnosis of urban and historic trees.
- Natural monuments for aesthetic preservation.
- Early precautions for public safety.

Arborists anticipated the results of each assessment to contribute remarkably to the decision-making process in sustainable conservation practices. Researchers are foreseeing the potential application of Resistograph in green infrastructure and sustainable management. Its importance is in evidence by a significant amount of research conducted to explore WD assessment using resistance micro-drilling for various applications such as timber structural analysis, tree health assessment, tree breeding and ages investigation (Kubus, 2009; Isik et al., 2019; Isik & Li, 2003; Bouffier et al., 2008; Oh et al., 2019). Several studies addressed the correlation between Resistograph readings (amplitude) and wood density (Couto et al., 2013; Downes et al., 2022; Joseph et al., 2022; Rinn, 2012; Silva et al., 2020). Nevertheless, studies still need to be conducted on micro-drilling applications to predict and estimate WD value on various tropical tree species for forest AGB and carbon stock assessment.

2 Non-destructive Testings for WD Assessment

2.1 Incremental Borer

Incremental borer, also known as Pressler borers or Swedish increment borers, has been proven the most reliable tool to assess the WD of living trees but is only suitable for softwoods and barely used in hardwood species (Fay & De Berker, 2018). Most of the research used incremental borer to extract the fresh wood cores across the tree trunk and determine the basic wood density using volumetric methods. The coring hole is slightly bigger and visible, which makes it an easy entry for pests (Fig. 9). Nevertheless, the healing and the hole-covering process will take place naturally. Even though its application demands upper body exertion and is not ergonomically practical, researchers used it as independent validation towards the WD assessment using the micro-drilling device. Joseph et al. (2022) have made a modified chuck to fit the borer with the impact wrench to facilitate the thrust speed into the tree trunk without using maximum body force for tree coring. Several studies also used incremental borers to work with the micro-drilling device to investigate the WD of Norway spruce (Vlad et al., 2022) and pendulate Oak (Tomczak et al., 2022).

2.2 *Pilodyn Wood Tester*

Pilodyn is a tool used to measure the WD of standing trees, and it is cheaper and faster for extensive data sampling. Despite its practicability usage, removing the tree bark is necessary before the insertion of Pilodyn. Meanwhile, there is no need for bark removal before using the micro-drilling device. However, one can opt to trim or discard the bark during the data processing using the DECOM™ software, developed explicitly for Resistograph data processing (Joseph et al., 2022). As Pilodyn could only penetrate the first few outer rings, the micro-drilling device could travel across the entire tree cross-section, which gives more detailed information on the variability of the tree's density. Shijun et al. (2010) observed that Pilodyn did not provide data accuracy on data measurement towards Eucalyptus clones but recommended Pilodyn as an effective tool for estimating wood properties.

3 Types of Resistance Micro Drilling, Resistograph

There are two types of resistance micro-drilling, namely; Rinntech Resistograph®, Heidelburg, Germany (Fig. 1) and IML Instrumenta Mechanik Labor System GmbH, Wiesloch, Germany (Fig. 2). Different models of resistance micro-drilling IML Resistograph (RESI) were highly opted in the number of studies conducted by researchers, as summarizes in Table 1. Both have a similar working mechanism where the torque (rotating force) will be electronically measured as a small needle is driven into a tree with constant energy. The power being consumed and absorbed during the drilling will reveal the WD variation. The drill resistance profiles will appear as a succession of peaks and valleys corresponding to the degree of wood penetrability by the resistance drill (Fig. 7).

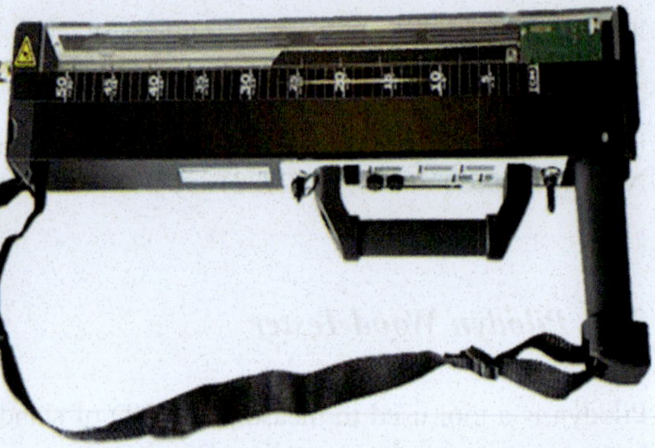

Fig. 1 Rinntech Resistograph®, Heidelburg, Germany (*Source* www.rinntech.info)

Fig. 2 IML instrumenta mechanik labor system GmbH, Wiesloch, Germany (*Source* www.alberosano.it)

4 Roles of Resistograph in Green Infrastructure and Sustainable Management

4.1 Tree Care Industry

Green areas are the essential components of the natural environment and the urban landscape. Hence, it is crucial to monitor dan conduct a necessary assessment of the tree health, including the tree monuments. Resistograph is widely used to test timber structures' structural integrity, such as bridge timbers and power holes. It gives solid decision-making for improving the green space in urban in terms of proper maintenance. Hence, the assessment will help to decide the possibility of restoration and the kind of support system needed.

Table 1 Types of resistance micro drilling model used on previous study related to wood density assessment

Researcher/authors	Resistance micro drilling model	Research ritle
Gwaze and Stevenson (2008)	IML RESI F400-S	Genetic variation of wood density and its relationship with drill resistance in shortleaf pine
Isik and Li (2003)	IML RESI B-400 Pro	Rapid assessment of wood density of live trees using the Resistograph for selection in tree improvement programmes
Downes et al. (2018)	IML RESI PD400	Application of the IML Resistograph to the infield assessment of basic density in plantation eucalypts
Oliveira et al., (2017)	IML RESI F300	Assessing specific gravity of young eucalyptus plantation trees using a resistance drilling technique
Rinn et al., (1996)	Rinntech Resistograph®	Resistograph and X-ray density charts of wood comparative evaluation of drill resistance profiles and X-ray density charts of different wood species
Icel and Guler (2016)	IML RESI F500-S	Nondestructive determination of spruce lumber wood density using drilling resistance (Resistograph) method
Morales-Condea et al., (2014)	IML RESI B1280	Predicting the density of structural timber members in service. The combined use of wood cores and drill resistance data
Bouffier et al., (2007)	IML Resi F300S	Can Wood Density Efficiently Selected at the Early Stage in Maritime Pine (*Pinus pinaster* Ait.)?
Silva et al. (2020)	Rinntech Resistograph®	Comparison between Resistograph analysis with physical properties of the wood Brazilian native tree species

4.2 Tree Stand Assessment in Forest Sustainable Management Operation

Resistograph is proven useful to evaluate the timber grading based on the wood mechanical properties for harvesting purposes. Strength grading represents the importance of WD in the timber industry. Resistograph is seen as a convenient technique to acquire the density of the timber without severely damaging the tree itself. The assessment derived from the Resistograph helps the stakeholders to sort and market timber to increase the revenue (Wang et al., 2007). Hovde et al. (2022) confirmed that there is a need to use micro-drilling device to assess the intrinsic wood qualities of dead standing tree prior to salvaging for the purposes of wood marketing.

5 Wood Density Assessment and Prediction Using Resistance Micro Drilling

Silva et al. (2020) identified strong and moderate positive correlations between the Resistograph amplitude (RA) with the basic density (BD) of five Brazilian tree species. Their study confirms that RA can validate the species BD density with the density data from the previous research findings and benchmark the correct classification of density classes of tree species. Nevertheless, it is crucial to investigate the correlations of RA with wood specific gravity (WSG) in the future as most compiled WD databases lacks SG values for tropical species (Wiemann & Williamson, 2013).

Fundova et al. (2018) recommended that Resistograph® is a more reliable tool than Pilodyn for indirect WD assessment. Similar findings reported by Couto et al. (2013), implied that Resistograph® provided greater accuracy in comparison to Pilodyn based on the fitter linear models. Meanwhile, WD data from SilviScan X-ray analysis is chosen as a benchmark on the non-destructive WD assessments of Scots pine (*Pinus sylvestris* L.). There is strong positive genetic correlations of the SilviScan density (DEN) with Resistograph mean unadjusted density, RES (0.89) and Resistograph mean detrended and debarked, RESTB (0.96). Meanwhile, Pilodyn with two traits, namely depth of Pilodyn's pin penetration with bark, PIL and without bark, PILB had recorded moderate positive genetic correlation genetic correlations with DEN, with the value of 0.59 and 0.71, respectively. From the results, the study underlines that adjustment on Resistograph drilling profiles are the key points to improving WD estimates' accuracy. This process can be accomplished before WD estimation analyzes by executing the profile detrending and, subsequently, profile debarking, as shown in Fig. 3.

Oliveira et al. (2017) revealed that internal friction could significantly affect the efficient use of Resistograph in specific gravity (SG) assessment for *Eucalyptus* trees, particularly when drilling hardwood species. Friction on the drill's shaft increases and causes the resistance amplitude (resAmp) gradually increases when penetration reaches the pith and moves beyond the pith midpoint. Figure 4 depicts the correlation that starts to decrease dramatically as the drilling depth exceeds 40 mm, due to accumulated friction but still manageably acquire a moderate positive relationship with density, using combined data (34 and 64 months old). In short, executing partial drilling (from bark to pith) can improve the weak correlation between average resAmp and SG rather than complete drilling to avoid outlier readings. Researchers should

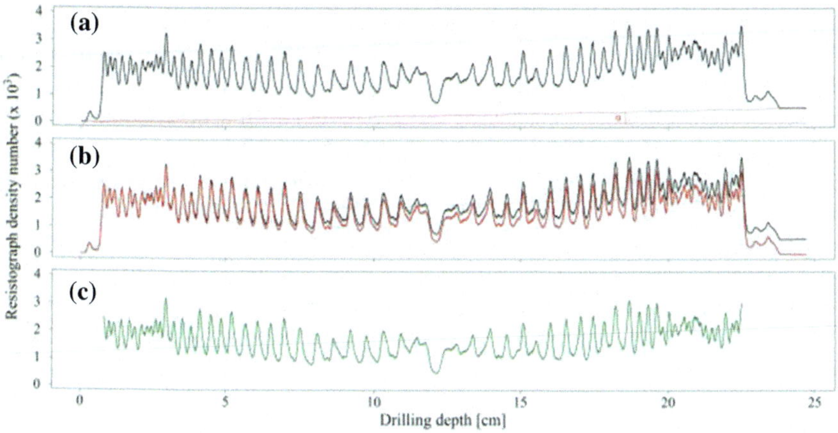

Fig. 3 **a**; unedited resistogram, **b**; detrended profiles, **c**; after detrended and debarked (Fundova et al., 2018)

adopt this approach by experimenting with different tropical trees and ages in future research.

Meanwhile, Gwaze and Stevenson (2008) observed that amplitude (%) is a reliable variable to predict the WD of shortleaf pine (*Pinus echinata* Mill.) based on the moderate positive linear relationships ($R^2 = 0.47$) of WD with amplitude at the family mean level. The result also coincides with the forest genetic studies done by Bouffier et al., (2008), Isik and Li (2003) and Rinn (1996). Their study disregarded the amplitude readings beyond the pith midpoint due to inconsistent readings triggered by the accumulated friction on the drill's shaft. As the Resistogram profiles are found for not being debarked and detrended in this study, it could be one of the factors that contributed to poor relationships between density and amplitude at individual mean level ($R^2 = 0.23$).

Fig. 4 Correlation between tree age and drilling depth (Oliveira et al., 2017)

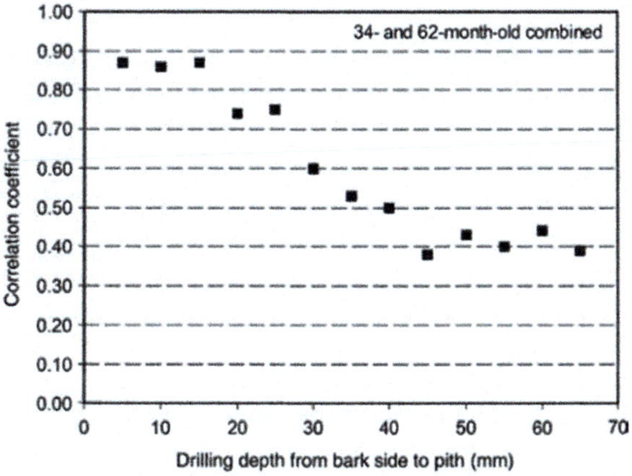

According to Conde, Linan and Machadol (2018), there was a good relationship, $R^2 = 0.73$, between the Resistographic measurement, taken at the first 20 mm exterior layer of wood compared to results obtained, $R^2 = 0.70$ from complete drilling. The results reveal that Resistograph data obtained at partial drilling depth could be used in in-situ prediction of stand's density. This study relates to Isik et al., (2016) methods as they only analyzed the first half of the resistogram profiles to avoid biassed readings due to shaft friction.

Karlinasari et al. (2017) reported that the density and harness evaluation on *Aquilaria malaccensis* (*Gaharu* or *agarwood*) using Resistograph has low coefficient determination, $R^2 = 0.25$ between the amplitude (%) and density (g cm^{-3}) (Fig. 5) and between amplitude and hardness, $R^2 = 0.24$ (Fig. 6). Judging by the unsatisfactory relationship, they pointed out that other parameters should be taken into account aside from WD and thus requires further investigation to improve the coefficient relationship. The presence of extractives in *A. malaccensis*, found in other tropical tree species such as from Sapotaceae, Euphorbiaceae, Apocynaceae and Moraceae, could potentially contribute to the poor relationship between the amplitude and WD and hardness (Chave, 2006). Hence, the presence of resin, sap, chemical extractives in tree species should be meticulously considered, particularly on WD assessment using micro-drilling resistance.

6 Research Challenges and Way Forwards to Predict WD in Living Tree Using Resistance Micro Drilling

The reliability of resistance micro-drilling as a tool to provide indirect predicted WD value as described on the following table (Table 2).

Icel and Guler (2016) eloquently highlighted that developing the model's validations is crucial in WD predictions. Essential parameters, such as moisture content, drilling depth, condition of drill bits and factors related to the selection of the Resistograph-series model itself, locations, wood species and tree ages, should be considered. The previously published work on spruce wood (*Pinus brutia*), has shown a higher correlation between drilling resistance and density perpendicular than parallel to annual rings. The study also coincides with the findings discussed by Kappel and Mattheck (2003), provided drilling perpendicularly to the annual rings could achieve the best results. Drilling in linear path to annual rings is highly favoured as it will retain the drilling angle from being 90° distorted and obtain a clear succession of peaks and troughs in the profiles (Fig. 6). In addition, the density variability is most significant in the radial direction (Zeidler & Sedivka, 2015). High-resolution drilling profiles will deliver valuable density information and can be reliably interpreted as studied by Rinn et al. (1996). Therefore, drilling in radial direction from bark to pith could improve prediction on WD value, and researchers should study this approach thoroughly in future research.

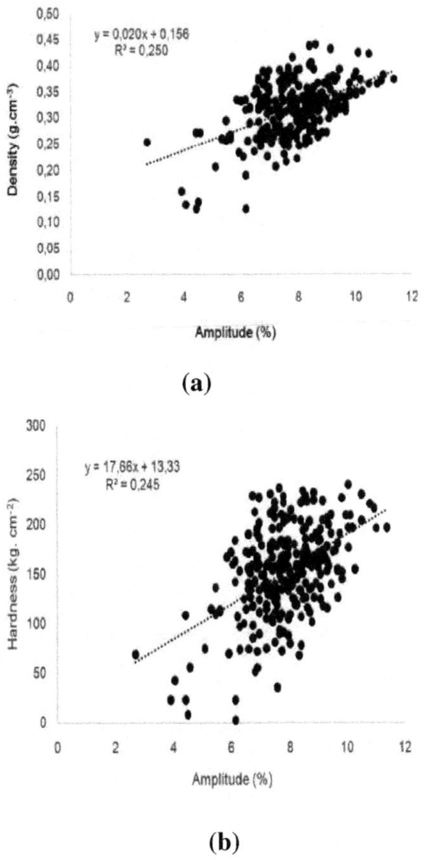

Fig. 5 **a** Linear relationship between density (g.cm^{-3}) and amplitude (%), **b** relationship between hardness (kg. cm^{-2}) and amplitude (%) (Karlinasari et al., 2017)

Aside from that, good processing on Resistogram would be a great advantage to provide reliably predicted WD value. Isik and Li (2003) documented the Resistogram can be processed using centred moving averages (CMOVAVE) and centred moving minimum averages (CMOVMIN). Their study corroborates the technique of deploying CMOVMIN helps improve the individual-tree level's problematic relationships. Nutto and Biechele (2015) contended that applying feed curve (feed force information) is proven helpful in processing the resistogram profiles for hardwood species. In addition, the instability of the drilling to be retained in radial direction will lead to poor interpretation on resistogram profile and may affect the accuracy of WD prediction. Resistance micro-drilling might require device configuration to maintain stability while drilling in a linear path. Emulating the invention of the smart increment borer by Kagawa and Fujiwara (2018), installing a reaction and expendable bar to absorb reaction force with the operator's leg on the ground could help address the problem.

Fig. 6 Drilling in a radial direction at breast height, 1.3 m from the ground (Joseph et al., 2022)

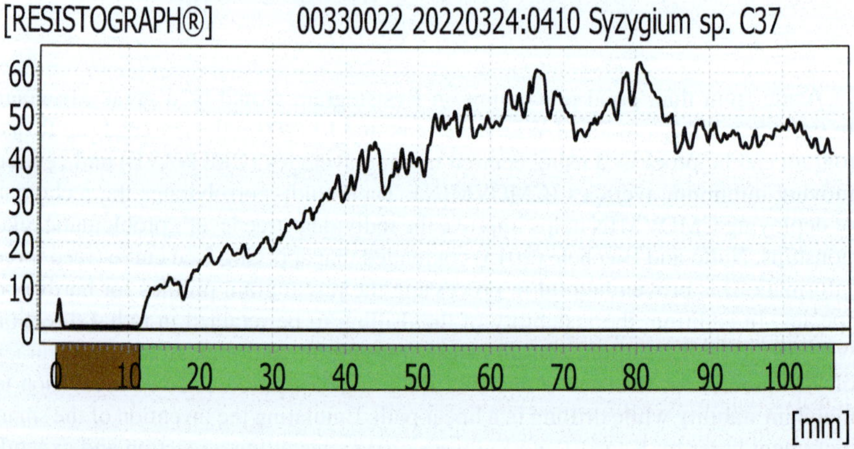

Fig. 7 Resistograph drilling resistance profile (brown coloured region; bark and green coloured region; intact wood) (Joseph et al., 2022)

Table 2 Factors affecting the reliability of Resistograph

Factors	Limitations
Knowledge of drilling profiles processing	Poor interpretation skills on drilling resistance due to lack of knowledge of wood anatomy, thus leading to difficulties in identifying incipient decay on drilling resistance profiles
Identification of Resistograph potential variables for WD predictive model development	Resistograph-generated variables alone are not sufficient to develop the WD predictive model (Joseph et al., 2022)
Condition of drill bits	Wear drill bits could affect the Resistograph's reading
Drilling depth and direction	The Resistograph drill bit could only reach 50 cm due to its standard lengths, making it unable to assess trees with a large diameter above 50 cm
Presence of abundant extractives in the tree (sap, resin)	The accumulation of wood extractives could increase shaft friction. Therefore, drill resistance will decrease with increasing mass fraction of extractives
Tree condition (hollow, truncated, decayed)	Severely decayed woods reduced drilling resistance (Karlinasari et al, 2017)
Misidentification of species	Introduce WD measurement error due to incorrect species information, which is based on common names and later identified by scientific names (Mora et al., 2014)

The general notion is that more potential methods could be explored and found regarding improving Resistogram profiles. It is imperative researchers should undertake rigorous WD assessment using micro-drilling in the future. Integrating potential variables such as geo-climatic factors to be fitted in linear regression analysis could generate various possible validation models, leading to the acquisition of the best predicted WD. Sound analyses of the drilling resistance profile demand in-depth theoretical knowledge in tree anatomy from experts, including the arborist and dendrologist. Well-interpretation of the drilling profiles will provide sufficient information on the WD of the tree itself as well as the tree health condition. Hence, this study asserts that the application of Resistograph requires both skills in handling the device in the field and, at the same time, must be knowledgeable on interpreting the resistance profiles.

Notably, the presence of extractives in wood is resultant from the formation of heartwood, which contributes to the density of the wood. Generally, high wood density contains high moisture content and extractives. While the drill bit penetrates beyond the pith, it will accumulate internal friction and cause erratic readings on the resistogram profiles as the drill bit travels further to the wood's dense part. Aside from that, the accumulated friction will depress the ability of micro resistance drilling to generate a high resolution of resistogram profile and eventually lead to excessive power consumption. Deterioration in power supply on resistance micro-drilling devices could be troublesome in the field, especially in the remote forest, and it could contribute to significant impediments in data collection. The operators should anticipate this scenario in high-density tropical hardwood trees with large diameters.

Even though Resistance micro-drilling can be used as a tool to predict the WD alone, prior research generally substantiates the belief that incremental borer, Pilodyn and SilviScan density can be used as an independent assessment to assess the WD of living trees (Fundova, Funda and Wu, 2018; Morales-Condea et al., 2014; Gao et al., 2017; Isik & Li, 2003; Rinn et al., 1996) (Fig. 7).

7 Potential Use of Resistance Micro Drilling on Predicting and Assessing WD for Forest AGB and Carbon Stock Estimation

Forest AGB estimation is the most practical approach adopted in Reducing Emissions from Deforestation, and Forest Degradation and Enhancing Carbon Stocks (REDD+) and is regularly quantified using tree allometric equations. Three variables frequently used in equations consist of tree diameter, height and wood specific gravity (WSG), also known as wood density. Amongst the two variables, WSG is perceived as an ideal predictor variable to estimate biomass. In light of the breakthrough of micro resistance drilling on various scientific studies relating to WD assessment on living trees, resistance micro-drilling is proven to exhibit a strong positive correlation with WD value.

A shaft diameter of 1.5 and a 3-mm-wide tip of resistance drill has enabled penetration into the wood without causing severe injury to the tree. The hole is barely found, making this tool ideal for non-destructive WD assessment in standing trees (Fig. 6.). The ability of the resistance micro-drilling to assess the accuracy of WD prediction without performing strenuous laboratory work has given an advantage in yielding rapid predicted WD values and developing AGB allometric models. The WD values could be calculated by converting drill resistance values into drill density values via a linear equation (Wang & Lin, 2001). The linear regression models will be evaluated to examine the relationship between WD and the Resistograph-generated variables (Figs. 8 and 9).

It would be intriguing to evaluate the potentiality and performance of Resistograph to assess *in-situ* WD on live trees with different species, ages, density, sampling location and diameter classes. Researchers have the opportunity to evaluate the tool's endurance in assessing the WD of large groups of standing trees to estimate the AGB and carbon stock accurately in tropical forests. Rapid assessment on WD using micro resistance drilling will provide critical information for the forest managers to facilitate decision-making on forest management activities. Screening and selecting ideal species regarding their density can be carried out seamlessly using the predicted WD

Fig. 8 3 mm hole on a tree after drilling resistance assessment (Joseph et al., 2022)

Fig. 9 The visible coring hole (Joseph et al., 2022)

value from the resistance micro drillings variables for forest restoration, harvesting and tree breeding programmes. In short, this is a plausible justification to assist the REDD+ mechanism in conserving the existing forest carbon stocks and sustainable forest management.

8 Conclusion

Resistance micro-drilling is envisaged as a profoundly promising tool to acquire infield WD value on standing trees in an enormous forest area with an abundance of tree stands. The underlying prospect of the non-destructive device is its operating mechanism which allows it to work well in protected areas and without interrupting the forest ecology. The researchers could explore the application of Resistograph in the context of green infrastructure and sustainable management in future research studies.

Acknowledgements This study is supported in part by the Ministry of Education Malaysia (Project No: FRGS/1/2019/WAB07/UMS/01/1).

References

Baker, T. R., Phillips, O. L., Malhi, Y., et al. (2004). Variation in wood density determines spatial patterns in Amazonian. *Global Change Biology, 10*(5), 545–562. https://doi.org/10.1111/j.1365-2486.2004.00751.x

Bouffier, L., Charlot, C., Raffin A, Rozenberg, P. et al. (2008). Can wood density be efficiently selected at early stage in Maritime pine (Pinus pinaster Ait.)?. *Annals of Forest Science, 65*(1), 106–113. https://doi.org/10.1051/forest:2007078

Chave, J. (2006). Measuring wood density for tropical forest trees: A field manual. *Sixth Framework Programme (2002–2006),* 1–6. PAN-AMAZONIA Project.

Couto, A. M., Trugilho, P. F., Neves, T. A., et al. (2013). Modeling of basic density of wood from Eucalyptus grandis and Eucalyptus urophylla using nondestructive methods. *Cerne, 19*(1), 27–34. https://doi.org/10.1590/S0104-77602013000100004

Downes, G. M., Harrington, J. J., Drew, D. M., et al. (2022). A comparison of radial wood property variation on Pinus radiata between an IML PD-400 'Resi' instrument and increment cores analysed by SilviScan. *Forests, 13*, 751. https://doi.org/10.3390/f13050751

Fay, N., & De Berker, N. (2018). *A review of the theory and practice of tree coring on live ancient and veteran trees*. Scottish Natural Heritage Research Report No. 789.

Fundova, I., Funda, T. & Wu, H. X. (2018). Non-destructive wood density assessment of Scots pine (Pinus sylvestris L.) using Resistograph and Pilodyn. *PLoS ONE, 13*(9). https://doi.org/10.1371/journal.pone.0204518

Gao, S., Wang, X., Wiemann, M., et al. (2017). A critical analysis of methods for rapid and nondestructive determination of wood density in standing trees. *Annals of Forest Science, 74*(2), 1–13. https://doi.org/10.1007/s13595-017-0623-4

Gwaze, D., & Stevenson, A. (2008). Genetic variation of wood density and its relationship with drill resistance in shortleaf pine. *Southern Journal of Applied Forestry, 32*(3), 130–133. https://doi.org/10.1093/sjaf/32.3.130

Hovde, T., Forsman, J., & Ross, R. et al. (2022). Sight versus sound: Do visual assessments of dead standing trees reflect acoustic nondestructive evaluations of wood quality?. *Forests, 13*(10). https://doi.org/10.3390/f13101680

Içel, B., & Güler, G. (2016). Nondestructive determination of spruce lumber wood density using drilling resistance (Resistograph) method. *Turkish Journal of Agriculture and Forestry, 40*(6), 900–907. https://doi.org/10.3906/tar-1606-76

Inagaki, T., Hartley, I. D., Tsuchikawa, S., et al. (2014). Prediction of oven-dry density of wood by time-domain terahertz spectroscopy. *Journal of Infrared, Milimeter, and Terahertz Waves, 35*, 949–961.

Isik, F., & Li, B. (2003). Rapid assessment of wood density of live trees using the Resistograph for selection in tree improvement programs. *Canadian Journal of Forest Research, 33*(12), 2427–2435. https://doi.org/10.1139/x03-176

Isik, F., Walker, T., & Mckeand, S. (2019). Genetic variation in acoustic time of flight and drill resistance of juvenile wood in a large loblolly pine breeding population. *Forestry Sciences, 65*(4), 469–482. https://doi.org/10.1093/forsci/fxz002

Joseph, J., Mui-How, P., & Debully Maripa, R. et al. (2022). Assessing intraspecific wood density variations of *Syzgium* sp. in tropical forest of Southwest Sabah. *IOP Conference Series: Earth and Environmental Science*. https://doi.org/10.1088/1755-1315/1053/1/012014

Kagawa, A., & Fujiwara, T. (2018). Smart increment borer: A portable device for automated sampling of tree-ring cores. *Journal of Wood Science, 64*(2), 1–7. https://doi.org/10.1007/s10086-017-1668-6

Karlinasari, L., Danu, D., & Nandika, D. (2017). Drilling resistance method to evaluate density and hardness properties of resinous wood of agarwood (Aquilaria malaccensis). *Wood Research, 62*(5), 683–690.

Kubus, M. (2009). The evaluation of using Resistograph when specifying the health condition of a monumental tree. *Notulae Botanicae Horti Agrobotanici Cluj-Napoca, 31*(1), 157–164. https://doi.org/10.15835/nbha3713112

Mattheck, C., & Kappel, R. (2003). Inspection of timber construction by measuring drilling resistance using Resistograph F300-S. *Transactions on the Built Environment, 66*. ISSN 1743–3509.

Morales-Condea, M. J., Rodríguez-Liñán, C., & Saporiti-Machado, J. (2014). Predicting the density of structural timber members in service; the combine use of wood cores and drill resistance data. *Materiales De Construccion, 64*(315), 1–11.

Nutto, L., & Biechele, T. (2015). Drilling resistance measurement and the effect of shaft friction–using feed force information for improving decay identification on hard tropical wood. In *Proceedings: 19th International Nondestructive Testing and Evaluation of Wood Symposium* (pp. 154–161). General Technical Report FPL-GTR-239. U.S. Department of Agriculture, Forest Service, Forest Products Laboratory.

Oh, J., Seo, J. W., & Kim, B. R. (2019). Review of the possibility of receipt investigation using Resistography. *Wood Engineering, 47*(1), 90–100.

Oliveira, J. T., Wang, X., & Vidaurre, G. B. (2017). Assessing specific gravity of young eucalyptus plantation trees using a resistance drilling technique. *Holzforschung, 71*(2), 137–145. https://doi.org/10.1515/hf-2016-0058

Rinn, F. (2012). Basics of micro-resistance drilling for timber inspection. *Holzforschung, 53*(3), 24–29.

Rinn, F., Schweingruber, F. H., & Schar, E. (1996). Resistograph and x-ray density charts of wood comparative evaluation of drill resistance profiles and x-ray density charts of different wood species. *Holzforschung, 50*(4), 303–311. https://doi.org/10.1515/hfsg.1996.50.4.303

Shijun, W., et al. (2010). Use of the pilodyn for assessing wood properties in standing trees of eucalyptus clones. *Journal of Forest Research (2010), 21*(1), 68–72. https://doi.org/10.1007/s11676-010-0011-5

Silva, C. E. S., Pace, J. H. C., & Latorraca, J. V. F. (2020). Comparison between Resistograph analysis with physical properties of the wood of Brazilian native tree species. *Wood Science and Technology, 27*(1), 1–7. https://doi.org/10.1590/2179-8087.005219

Tomczak, K., Tomczak, A., & Jelonek, T. (2022). Measuring radial variation in basic density of pendulate oak: Comparing increment core samples with the IML power drill. *Forests, 13*(4), 589. https://doi.org/10.3390/f13040589

Vlad, R., Pei, G., & Sidor, C. G. (2022). Assessing standing-tree wood density by microdrilling in tending forestry work carried out on Norway Spruce (*Picea abies* (L.) H. Karst) Stands. *Forests 2022*, *13*, 1450. https://doi.org/10.3390/f13091450

Wang, X., Carter, P., Ross, R. J., et al. (2007). Acoustic assessment of wood quality of raw forest materials—a path to increased profitability. *Forest Products Journal, 57*, 6–14.

Wang, S.Y., Lin, C.J. (2001). Application of the drill resistance method for density boundary evaluation of earlywood and latewood of Taiwania (Taiwania cryptomerioides Hay.) plantation wood. *Taiwan Journal of Science*, *16*(3), 196–199.

Wiemann, M.C., & Williamson, B.G. (2013). *Biomass determination using wood specific gravity from increment cores. General Technical Report FPL-GTR-225* (pp. 1–7). U.S. Department of Agriculture, Forest Service, Forest Products Laboratory.

Zeidler, A., & Sedivka, P. (2015). Influence of selected factors on wood density variability in grand fir (Abies grandis/Douglas/LindL.). *Wood Research*, *60*(1), 33–34.

Phytochemical Research for the Sustainability of *Moringa* Species Using Different Extraction Methods

Hannis Fadzillah Mohsin, Nurul Atika Bacho, Kathleen J. Jalani, and Ibtisam Abdul Wahab

Abstract Different extraction methods to screen the phytochemicals of *Moringa oleifera* (*M. oleifera*) leaves extracts have been conducted. This study aims to employ both maceration and ultrasonic-assisted techniques, followed by the phytochemical screening of the extracts by using thin-layer chromatography (TLC) and High-Performance Liquid Chromatography (HPLC). Compared to sonication, the maceration produced a larger yield of extracted material. The TLC analysis showed that a mobile phase consisting of toluene, butanol, and methanol (90:5:5) was the best solvent system for separating most of the constituents in the extracts. The TLC and HPLC profiles demonstrate that neither extraction procedure exhibits notable differences. The HPLC profile of gradient elution with the addition of 0.3% formic acid resulted in greater peak separation and more peak detection, as compared to the isocratic system. Quercetin was not identified as a significant constituent in this research sample, due to the relatively low absorbance. The findings can provide information on the pharmacological function of *M. oleifera* leaves, which are potentially valuable source in pharmaceutical and traditional medicines.

Keywords Chromatography, Moringa · Phytochemicals · Sustainability

1 Introduction

Medicinal plants remain the most accessible, dependable, and valuable source of active biomolecules for improving health, medicines, and nutritional supplements on a global state (Fahal et al., 2018). To date, around thirty-five to seventy thousand plant species have been evaluated for their medicinal potential (Veeresham, 2012). M*oringa oleifera* (*M. oleifera*) is one of the medicinal plants of the family Moringaceae. It is native to India or sub-Himalayan regions and can be found across the tropical and subtropical areas in dry, sandy soil (Lin et al., 2019; Xu et al.,

H. F. Mohsin · N. A. Bacho · K. J. Jalani · I. Abdul Wahab (✉)
Faculty of Pharmacy, Universiti Teknologi MARA (UiTM), 42300 Bandar Puncak Alam, Selangor Darul Ehsan, Malaysia
e-mail: ibtisam@uitm.edu.my

U. R. Hashim et al. (eds.), *Green Infrastructure*,
https://doi.org/10.1007/978-981-99-7003-2_21

2019). It has brittle branches, bipinnate or commonly tripinnate leaves, scented and yellowish-white flowers, tri-lobed capsules fruits referred to as pendulous pods, and whitish-grey bark (Anzano et al., 2021; Mishra et al., 2011; Mobarak et al., 2017; Parrota, 2009). The most utilized part of *M. oleifera* is the leaves, which contain a huge number of phytochemicals, including flavonoids, alkaloids, isothiocyanates, saponins, glucosinolate, phenolic acids, polyphenols, and tannins (Hassan et al., 2021; Rocchetti et al., 2020). These phytochemicals exhibit various biological activities such as anti-inflammatory, antioxidant, anti-diabetic, anti-cancer, hepatoprotective, hypocholesterolemia, and cardioprotective (Aju et al., 2019; Asrifah et al., 2017; Cheraghi et al., 2017; Hagoel et al., 2019; Oguntibeju et al., 2020). Additionally, *M. oleifera* leaves were effective in treating neurological disorders such as ischemic stroke, epilepsy, and Alzheimer's disease (González-Trujano et al., 2018; Kou et al., 2018; Mahaman et al., 2018; Zeng et al., 2019).

Traditionally, the leaf infusion of *M. oleifera* was used to treat conjunctivitis, snakebite, and wounds and as a poultice applied to the abdomen to expel intestinal worms (Padayachee & Baijnath, 2020). Furthermore, since fresh leaves of *M. oleifera* can boost milk production and treat anemia, they are suitable for pregnant and breast-feeding mothers (Alegbeleye, 2018). The leaves of *M. oleifera* were reported to contain essential nutritional components (Gopalakrishnan et al., 2016; Islam et al., 2021) like amino acids, carotenoids, potassium, 17-times higher calcium content than milk, 10-times higher vitamin A content than in carrots, and 7-times higher vitamin C content than in citrus, which help to treat malnutrition in children. Moreover, vitamin E, specifically α-tocopherol, which is involved in the modulation of gene expression, is also found in the leaves in amounts comparable to nuts. To analyze phytochemical constituents of plant extracts and prevent adulteration, standardization techniques such as phytochemical analysis and chromatographic techniques, for example, Thin Layer Chromatography (TLC) and High-Performance Liquid Chromatography (HPLC), have been developed in response to the growing interest in herbal medications (Pharma et al., 2016).

Furthermore, *M. oleifera* is a medicinal plant that is widely used in pharmaceutical, cosmeceutical, and traditional fields, due to its nutritional values and positive physiological effects contributed by its active compounds (Khan et al., 2017). However, comprehensive data on the phytochemical analysis of *M. oleifera* with different polarities of organic solvent was less, as the most recent study primarily employed ethanol and water as solvents, compared to other types of organic media (Dzuvor et al., 2021). Moreover, most studies of *M. oleifera* concentrated on its application and pharmacological activities (Liu et al., 2022).

Therefore, the purpose of this study was to perform different extraction methods on *M. oleifera* leaves. The procedure included laboratory techniques on maceration and ultrasonic-assisted extraction. The aim was to screen the phytochemical constituents of the extracts via TLC and HPLC, and to evaluate the medicinal potential of the plant. It is hoped that the outcome of this research would justify the traditional uses of *M. oleifera* (Brilhante et al., 2017; Meireles et al., 2020).

2 Materials and Methods

2.1 *Sample Preparation*

The *M. oleifera* leaf powder was purchased from an organic store. The fine leaf powder was weighed with an estimated weight of 30 g and then placed inside a 250 ml Erlenmeyer flask. This procedure was repeated six times.

2.2 *Extraction of* M. oleifera *Leaves*

Approximately 30 g of *M. oleifera* powdered leaves was transferred into three separated Erlenmeyer flask macerated with different polarity organic solvents: hexane, dichloromethane, and methanol, at room temperature. After a week, the extracts were filtered, and the filtrate was concentrated by a rotary evaporator. After that, the crude extracts were weighed and kept in a dark and dry place for further use.

The ultrasonic-assisted extraction method was adapted from the studies by Dadi et al. (2019) and Daghaghele et al. (2021). The leaf powder of *M. oleifera* was soaked in the organic solvents such as dichloromethane, methanol, and hexane in a ratio of 1:5 (g/ml) in the separate Erlenmeyer flask. Then, it was sonicated for about 30 min at room temperature in the 80 Hz. Next, the extracts were filtered, and the filtrate was evaporated with a rotary evaporator under a vacuum at 40 °C. Finally, the crude extracts were weighed and stored for further analysis.

2.3 *Crude Extracts Profiling via Thin Layer Chromatography (TLC)*

The qualitative analysis was adapted based on Mahdi et al. (2016) and Sierra-Campos et al. (2020). The TLC method was carried out on each extract of *M. oleifera* leaves, using TLC silica gel F_{254} as the coating substance. Toluene, butanol, and methanol (90:5:5 v/v) were utilized as mobile phase, optimizing the best solvent systems for *M. oleifera* extracts. All extracts were spotted on the same TLC plate and quercetin was used as the reference standard. Then, the plate was air-dried and was observed under ultraviolet (UV) light (254 nm). The Dragendorff's reagent was used to screen the alkaloids. In addition, the sulphuric anisaldehyde was sprayed on the TLC plate to detect the polyphenols and other type of phytochemicals, for example, the plant sterols and non-UV active compounds. The retention factor (Rf) value was calculated for each spot.

2.4 Phytochemical Screening via HPLC

The reverse-phase high-performance liquid chromatography with multiple wavelength detector (RP-HPLC-MWD) was used to analyze the chromatographic profile of *M. oleifera* methanol leaves extract. Quercetin, a reference compound, was used as an external standard in the analysis. The extract and standard were filtered through a 0.45 μm Millipore filter into the HPLC vials, and 10 μl of the sample was injected for the analysis (Sulastri et al., 2018). The HPLC analysis was performed at 36 °C and ran through a C-18 column (4.6 mm × 150 mm, 5 μm) as a stationary phase (Shervington et al., 2018). The gradient elution was conducted utilizing methanol and water as the mobile phase with a flow rate of 1 ml/min (Ghanjaoui et al., 2020). The chromatogram was recorded with a multiple wavelength detector (MWD) at 254 nm and the data analysis was performed using ChemStation software.

3 Results and Discussion

3.1 Extraction of M. oleifera Leaves

The percentage yield was calculated based on 30 g of the sample powder used for both extraction procedures. The maceration and sonication by methanol extracted most of the compounds, which were 4.47 g and 3.26 g, respectively, followed by dichloromethane and finally, hexane (Table 1). The maceration extracted higher components than sonication, which showed not much difference in yield. The maceration took a week, whereby, the sonication was conducted for about 30 min. This showed that sonication accelerated the solute dissolution and diffusion, by making the leaves' cell walls more permeable and causing cavitation. This makes ultrasound extraction just as effective as maceration (Bimakr et al., 2017; Zhang et al., 2018).

Table 1 Extraction yields of the crude extracts

Method of extraction	Solvents	Weight of crude extracts (g)	Percentage yield (%)
Maceration	Dichloromethane	1.26	4.20
	Methanol	4.47	14.90
	Hexane	0.71	2.37
Ultrasonic-assisted extraction (sonication)	Dichloromethane	0.93	3.10
	Methanol	3.26	10.87
	Hexane	0.69	2.30

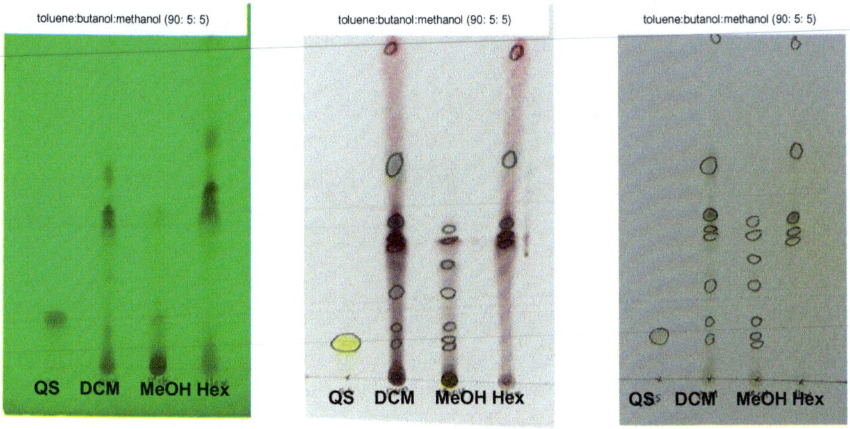

Fig. 1 The TLC plate of *M. oleifera* leaves extracts under UV light, 254 nm (left), TLC after sprayed with anisaldehyde reagent (middle), and TLC after sprayed with Dragendorff's reagent (right). The standard quercetin was placed in the first spot and labeled as QS

3.2 *Crude Extracts Profiling via TLC*

Five different solvent systems were used as developing systems; hexane: ethyl acetate (40:60), hexane: ethyl acetate (70:30), hexane: ethyl acetate: methanol (10:70:20), toluene: butanol: methanol (70:10:20), and toluene: butanol: methanol (90:5:5). Based on the optimization of the mobile phase, a system consisting of toluene, butanol, and methanol (90:5:5) gave the best separation of compounds under the UV light. Six, seven and nine spots were recorded from hexane (Hex), dichloromethane (DCM), and methanol (MeOH) extracts, respectively. The anisaldehyde spraying reagent was applied onto the TLC plate, and the pink-purplish spots appeared upon drying with hot air (Fig. 1). The spots possibly indicated the presence of non-UV active compounds such as terpenoids, steroids, and terpenes (Dougnon & Ito, 2019). On the other hand, the lower spot for both dichloromethane and methanol extracts indicated the same Rf value (0.11), as the standard quercetin (Fig. 1) suggesting the presence of this polyphenol in both extracts of *M. oleifera* leaves. Meanwhile, no changes were observed on the TLC plate after spraying with Dragendorff's reagent, which suggested the absence of alkaloids.

3.3 *Phytochemical Screening via HPLC*

The optimization was performed on methanolic extracts for both maceration and ultrasonic-assisted techniques. The isocratic elution with methanol and water (7:3) was unsuccessful in resolving the compounds in the extract from both extraction methods. The chromatogram showed multiple major and minor unseparated peaks

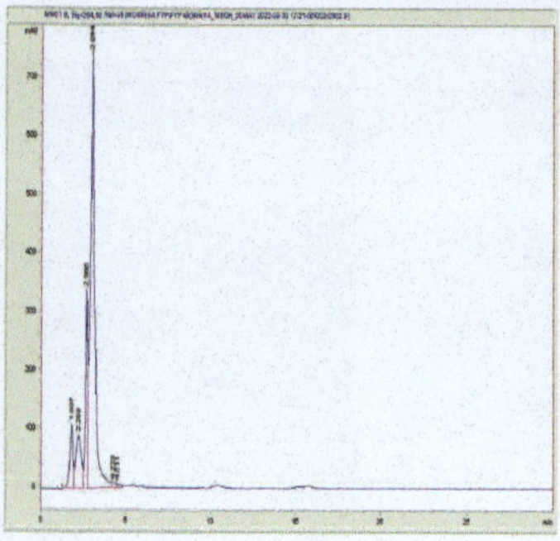

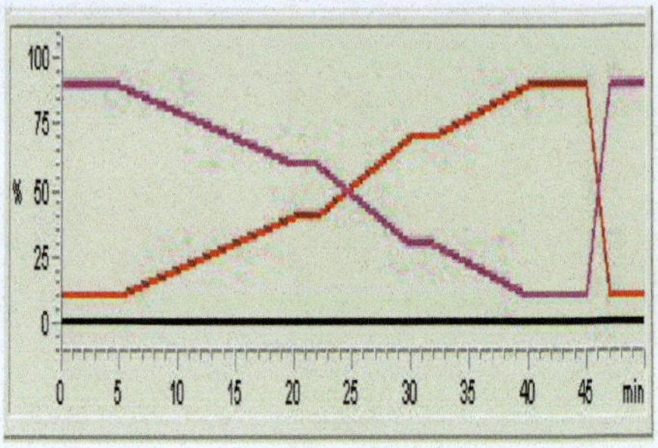

Fig. 2 The HPLC chromatogram of isocratic elution of *M. oleifera* methanolic extract (left). The gradient elution for HPLC analysis with methanol (red line): 0.3% formic acid in water (pink line) (right)

and unretained peaks eluted in less than 5 min of run time (Fig. 2). The unretained compounds on the C-18 column indicated that the characteristics of compounds in the extracts are relatively polar.

Therefore, a gradient elution consisting of methanol and water was improved. This method offered several advantages, including increased sensitivity, improved resolution, enhanced column efficiency, and the ability to separate complex samples (Moldoveanu & David, 2016). The experiments were performed with a neutral solvent and an acidic mobile phase, utilizing 0.3% of formic acid in water. Most of the vitamins, phenolic acids (gallic acid, chlorogenic acid, and caffeic acid), and flavonoids (quercetin) found in *M. oleifera* are organic acids (Anwar et al., 2007), containing hydroxyl and carboxyl groups (Sankhyan et al., 2013). Thus, a small amount of formic acid was added to reduce the pH and inhibit the ionization of acidic surface silanol species of the silica to prevent peak tailing and improve separation (Dolan, 2017; Wiczling et al., 2015). The method was set at stepwise gradient elution, whereas the methanol composition was gradually increased from 10 to 90% in 40 min of run time (Fig. 2).

In general, there was no significant difference between both extraction techniques; maceration and ultrasonic-assisted extraction, based on the chromatogram. The acidic mobile phase exhibited more peaks and good separation, compared to the neutral mobile phase (Fig. 3). The addition of 0.3% of formic acid in the mobile phase enhanced the peak separation of the methanolic extract of *M. oleifera*. Based on Fig. 3, 11 peaks were separated from 11.703 to 41.258 min in the acidic mobile phase, compared to only 6 considerably separated peaks, in the absence of acid.

In this analysis, the external standard, quercetin, with a retention time of 30.752 min, was compared. A small peak at 30.344 min, suggested the presence of quercetin, in the *M. oleifera* methanol extract (Fig. 4). The retention time was slightly shifted. This is because the quercetin might not exist, either in a free phenolic acid form, or it binds to glycosides, linked to one or more sugar molecules, via glycosidic bonds (Maiyo et al., 2016; Martono et al., 2019). Therefore, quercetin was not identified as a major compound in the extract since the peak has a very low absorbance.

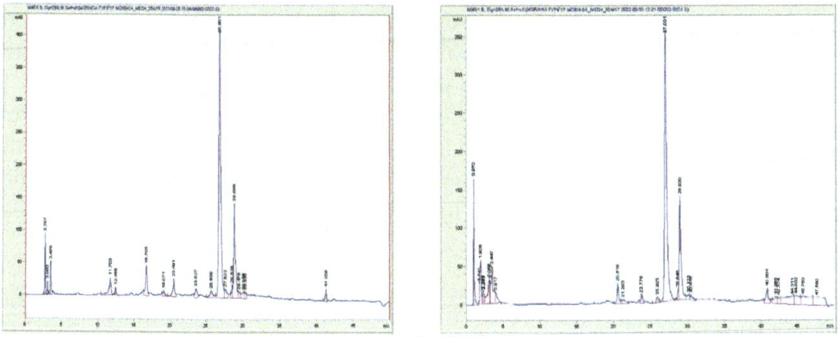

Fig. 3 The HPLC chromatogram with a gradient elution, containing methanol: 0.3% formic acid (left) and methanol: water (right)

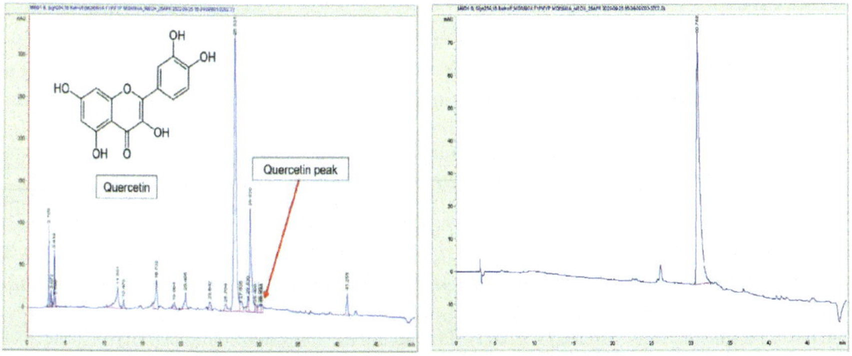

Fig. 4 The HPLC chromatograms for *M. oleifera* methanol extract (left) and the standard quercetin (right)

Meanwhile, two unknown major peaks were observed at retention times, equal to 26.801 and 28.806, respectively. In addition, the abundance of the quercetin in the extract was less than 50%, as compared to the standard quercetin.

4 Conclusion

The maceration extracted a higher yield, than the ultrasonic-assisted method, with not much difference. In the TLC study, a mobile phase made up of toluene, butanol, and methanol (90:5:5) was discovered as the best solvent system for separating most of the compounds in the extracts. The pink-purplish spots were noted after spraying the plate with anisaldehyde. This finding recommended the presence of non-UV active compounds such as terpenoids, steroids, and terpenes. Meanwhile, no changes were recorded on the TLC plate, after derivatization of Dragendorff's reagent. Therefore, this observation indicated the absence of alkaloids in the research samples. Quercetin was detected when the chromatographic data was referred to a standard quercetin. The TLC and HPLC profiles showed that there were not significantly different, in either of the extraction techniques. Compared to the isocratic system, the HPLC profile of gradient elution, with the addition of 0.3% formic acid, produced better peak separation and more peak detection. Nevertheless, quercetin was not identified as a major compound in the extract, following this HPLC analysis, since the peak has a relatively low absorbance.

Acknowledgements The authors would like to thank Faculty of Pharmacy, UiTM Puncak Alam, Selangor Darul Ehsan, Malaysia.

References

Aju, B. Y., Rajalakshmi, R., & Mini, S. (2019). Protective role of *Moringa oleifera* leaf extract on cardiac antioxidant status and lipid peroxidation in streptozotocin induced diabetic rats. *Heliyon, 5*(12), e02935.

Alegbeleye, O. O. (2018). How functional is *Moringa Oleifera*? A review of its nutritive, medicinal, and socioeconomic potential. *Food and Nutrition Bulletin, 39*(1), 149–170.

Anwar, F., Latif, S., Ashraf, M., & Gilani, A. H. (2007). Moringa oleifera: A food plant with multiple medicinal uses. *Phytotherapy Research, 21*(1), 17–25.

Anzano, A., Ammar, M., Papaianni, M., Grauso, L., Sabbah, M., Capparelli, R., & Lanzotti, V. (2021). *Moringa oleifera* Lam.: A phytochemical and pharmacological overview. *Horticulturae, 7*, 409.

Asrifah, I., Estiasih, T., & Sujuti, H. (2017). Potential hypocholesterolemic activity of flour from leaves of moringa (*Moringa oleifera* L.). *Malaysian Journal of Nutrition, 23*(3), 449–460.

Bimakr, M., Ganjloo, A., Zarringhalami, S., & Ansarian, E. (2017). Ultrasound-assisted extraction of bioactive compounds from Malva sylvestris leaves and its comparison with agitated bed extraction technique. *Food Science and Biotechnology, 26*(6), 1481–1490.

Brilhante, R., Sales, J., Pereira, V., Castelo-Branco, D., Cordeiro, R., Sampaio, C., Paiva, M., Santos, J., Sidrim, J., & Rocha, M. (2017). Research advances on the multiple uses of Moringa oleifera: A sustainable alternative for socially neglected population. *Asian Pacific Journal of Tropical Medicine, 10*(7), 621–630.

Cheraghi, M., Namdari, M., Daraee, H., & Negahdari, B. (2017). Cardioprotective effect of magnetic hydrogel nanocomposite loaded N, α-L-rhamnopyranosyl vincosamide isolated from *Moringa oleifera* leaves against doxorubicin-induced cardiac toxicity in rats: In vitro and in vivo studies. *Journal of Microencapsulation, 34*(4), 335–341.

Dadi, D. W., Emire, S. A., Hagos, A. D., & Eun, J. B. (2019). Effect of ultrasound-assisted extraction of *Moringa stenopetala* leaves on bioactive compounds and their antioxidant activity. *Food Technology and Biotechnology, 57*(1), 77.

Daghaghele, S., Kiasat, A. R., Safieddin Ardebili, S. M., & Mirzajani, R. (2021). Intensification of extraction of antioxidant compounds from *Moringa oleifera* leaves using ultrasound-assisted approach: BBD-RSM design. *International Journal of Fruit Science, 21*(1), 693–705.

Dolan, J. W. (2017). Back to basics: The role of pH in retention and selectivity. *LC-GC North America, 35*, 22–28.

Dougnon, G., & Ito, M. (2019). Medicinal uses, thin-layer chromatography and high-performance liquid chromatography profiles of plant species from Abomey-Calavi and Dantokpa Market in the Republic of Benin. *Journal of Natural Medicines*. https://doi.org/10.1007/s11418-019-01344-1

Dzuvor, C. K. O., Pan, S., Amanze, C., Amuzu, P., Asakiya, C., & Kubi, F. (2021). Bioactive components from *Moringa oleifera* seeds: Production, functionalities and applications—a critical review. *Critical Reviews in Biotechnology*. https://doi.org/10.1080/07388551.2021.1931804

Fahal, E. M., Saharan, N., Rani, B., Aklakur, M. D., & Chanu, T. I. (2018). Qualitative analysis of some important phytochemical constituents of *Moringa oleifera* (Lam) leaves and seeds. *International Journal of Chemical Studies, 6*(3), 648–650.

Ghanjaoui, M. E., Mandil, A., Mou, A. A. S., & Slimani, R. (2020). High performance liquid chromatography quality control. *International Journal of Advanced Chemistry, 8*(1), 1–11.

González-Trujano, M. E., Martínez-González, C. L., Flores-Carrillo, M., Luna-Nophal, S. I., Contreras-Murillo, G., & Magdaleno-Madrigal, V. M. (2018). Behavioral and electroencephalographic evaluation of the anticonvulsive activity of *Moringa oleifera* leaf non-polar extracts and one metabolite in PTZ-induced seizures. *Phytomedicine: International Journal Of Phytotherapy And Phytopharmacology 39*, 1–9.

Gopalakrishnan, L., Doriya, K., & Kumar, D. S. (2016). *Moringa oleifera*: A review on nutritive importance and its medicinal application. *Food Science and Human Wellness, 5*(2), 49–56.

Hagoel, L., Vexler, A., Kalich-Philosoph, L., Earon, G., Ron, I., Shtabsky, A., & Lev-Ari, S. (2019). Combined effect of Moringa oleifera and ionizing radiation on survival and metastatic activity of pancreatic cancer cells. *Integrative Cancer Therapies, 18*, 1534735419828829.

Hassan, M. A., Xu, T., Tian, Y., Zhong, Y., Ali, F. A. Z., Yang, X., & Lu, B. (2021). Health benefits and phenolic compounds of *Moringa oleifera* leaves: A comprehensive review. *Phytomedicine, 93*, 153771.

Islam, Z., Islam, S. M., Hossen, F., Mahtab-ul-Islam, K., Hasan, M., & Karim, R. (2021). *Moringa oleifera* is a prominent source of nutrients with potential health benefits. *International Journal of Food Science* 11.

Khan, W., Parveen, R., Chester, K., Parveen, S., & Ahmad, S. (2017). Hypoglycemic potential of aqueous extract of moringa oleifera leaf and in vivo GC-MS metabolomics. *Frontiers in Pharmacology, 8*, 577.

Kou, X., Li, B., Olayanju, J. B., Drake, J. M., & Chen, N. (2018). Nutraceutical or pharmacological potential of *Moringa oleifera* Lam. *Nutrients, 10*(3), 343.

Lin, H., Zhu, H., Tan, J., Wang, H., Wang, Z., Li, P., & Liu, J. (2019). Comparative analysis of chemical constituents of *Moringa oleifera* leaves from China and India by ultra-performance liquid chromatography coupled with quadrupole-time-of-flight mass spectrometry. *Molecules, 24*(5), 942.

Liu, R., Liu, J., Huang, Q., Liu, S., & Jiang, Y. (2022). Moringa oleifera: A systematic review of its botany, traditional uses, phytochemistry, pharmacology and toxicity. *Journal of Pharmacy and Pharmacology, 74*(3), 296–320.

Mahaman, Y. A. R., Huang, F., Wu, M., Wang, Y., Wei, Z., Bao, J., & Wang, X. (2018). *Moringa oleifera* alleviates homocysteine-induced Alzheimer's disease-like pathology and cognitive impairments. *Journal of Alzheimer's Disease, 63*(3), 1141–1159.

Mahdi, H., Yousif, E., Khan, N., Mahmud, R., Murugaiyah, V., & Asmawi, M. (2016). Optimizing extraction conditions of *Moringa oleifera* Lam leaf for percent yield, total phenolics content, total flavonoids content and total radical scavenging activity. *International Journal of Advanced Research, 4*(11), 682–695.

Maiyo, C., Moodley, R., & Singh, M. (2016). Cytotoxicity, antioxidant and apoptosis studies of quercetin-3-*O* glucoside and 4-(β-D-glucopyranosyl-1→ 4-α-L-rhamnopyranosyloxy)-benzyl isothiocyanate from *Moringa oleifera. Anti-Cancer Agents in Medicinal Chemistry (Formerly Current Medicinal Chemistry-Anti-Cancer Agents), 16*(5), 648–656.

Martono, Y., Yanuarsih, F. F., Aminu, N. R., & Muninggar, J. (2019). Fractionation and determination of phenolic and flavonoid compound from *Moringa oleifera* leaves. *Journal of Physics: Conference Series, 1307*(1), 012014.

Meireles, D., Gomes, J., Lopes, L., Hinzmann, M., & Machado, J. (2020). A review of properties, nutritional and pharmaceutical applications of *Moringa oleifera*: Integrative approach on conventional and traditional Asian medicine. *Adv Tradit Med (ADTM), 20*(4), 495–515.

Mishra, G., Singh, P., Verma, R., Kumar, S., Srivastav, S., Jha, K. K., & Khosa, R. L. (2011). Traditional uses, phytochemistry and pharmacological properties of *Moringa oleifera* plant: An overview. *Der Pharmacia Lettre, 3*(2), 141–164.

Mobarak, A. A., Shaltout, K., Ali, H. I., Baraka, D., & Aly, S. (2017). Morphological variability among *Moringa oleifera* (Lam) populations in Egypt. *Egyptian Journal of Botany, 57*(1), 241–257.

Moldoveanu, S. C., & David, V. (2016). *Selection of the HPLC method in chemical analysis*. Elsevier.

Oguntibeju, O. O., Aboua, G. Y., & Omodanisi, E. I. (2020). Effects of Moringa oleifera on oxidative stress, apoptotic and inflammatory biomarkers in streptozotocin-induced diabetic animal model. *South African Journal of Botany, 129*, 354–365.

Padayachee, B., & Baijnath, H. (2020). An updated comprehensive review of the medicinal, phytochemical and pharmacological properties of *Moringa oleifera*. *South African Journal of Botany, 129*, 304–316.

Parrota, J. A. (2009). *Moringa oleifera* Lam. Enzyklopädie der Holzgewächse: Handbuch und Atlas der Dendrologie 1–8.

Pharma, D., Bueno, P., Alvarez, M. R., Cruz, R., Macapulay, R., Vallesfin, F. J., & Heralde, F. (2016). Thin layer chromatography (TLC) and high-performance liquid chromatography (HPLC) profiling and phytochemical analysis of *Euphorbia hirta. Gliricidia Sepium and Moringa Oleifera Methanol Extracts Der Pharma Chemica, 8*, 456–461.

Rocchetti, G., Pagnossa, J. P., Blasi, F., Cossignani, L., Piccoli, R. H., Zengin, G., & Lucini, L. (2020). Phenolic profiling and in vitro bioactivity of *Moringa oleifera* leaves as affected by different extraction solvents. *Food Research International, 127*, 108712.

Sankhyan, N., Sharma, A., Seth, C. A., Chauhan, A., & Kulshrestha, S. (2013). Determination and comparison of vitamin c content from moringa oleifera by different methods. *International Journal of Agricultural Science and Research (IJASR), 3*(2), 67–70.

Shervington, L. A., Li, B. S., Shervington, A. A., Alpan, N., Patel, R., Muttakin, U., & Mulla, E. (2018). A comparative HPLC analysis of myricetin, quercetin and kaempferol flavonoids isolated from Gambian and Indian *Moringa oleifera* leaves. *International Journal of Chemistry, 10*(4), 28–37.

Sierra-Campos, E., Valdez-Solana, M., Avitia-Domínguez, C., Téllez-Valencia, A., Meza Velázquez, J., Aguilera-Ortíz, M., & Escobar-Ramírez, A. (2020). Standardization based on chemical markers of *Moringa oleifera* herbal products using bioautography assay, thin layer chromatography and high performance liquid chromatography-diode array detector. *Malaysian Journal of Analytical Sciences, 24*(3), 449–463.

Sulastri, E., Zubair, M. S., Anas, N. I., Abidin, S., Hardani, R., & Yulianti, R. (2018). Total phenolic, total flavonoid, quercetin content and antioxidant activity of standardized extract of *Moringa oleifera* leaf from regions with different elevation. *Pharmacognosy Journal, 10*(6s).

Veeresham, C. (2012). Natural products derived from plants as a source of drugs. *Journal of Advanced Pharmaceutical Technology & Research, 3*(4), 200.

Wiczling, P., Kubik, Ł., & Kaliszan, R. (2015). pH effects on chromatographic retention modes. *Analytical Separation Science* 263–278.

Xu, Y. B., Chen, G. L., & Guo, M. Q. (2019). Antioxidant and anti-inflammatory activities of the crude extracts of *Moringa oleifera* from Kenya and their correlations with flavonoids. *Antioxidants, 8*(8), 296.

Zeng, K., Li, Y., Yang, W., Ge, Y., Xu, L., Ren, T., & Yang, L. (2019). Moringa oleifera seed extract protects against brain damage in both the acute and delayed stages of ischemic stroke. *Experimental Gerontology, 122*, 99–108.

Zhang, Q. W., Lin, L. G., & Ye, W. C. (2018). Techniques for extraction and isolation of natural products: A comprehensive review. *Chinese Medicine, 13*(1), 1–26.

Performance of Kapok Fibres and Kapok Ash Wood as Oil Absorption Materials

Abdul Hadi Firuz Ahmad, Rohana Hassan, Nurbaiah Mohammad Noh, Nor Jihan Abd Malek, Anizahyati Alisibramulsi, and Ezahtul Shahreen Ab Wahab

Abstract Degradation of waste cooking oil in pipes may cause corrosion of metal and concrete elements. Research detected that the capacity of sorbent material to clean-up cooking oil from produced water towards different factors is associated with surface characteristics, oil type, oil film thickness, sorption time, and the temperature. Kapok (*Ceiba pentandra* (L.) *Gaertn*) fibre is a plant biomass used as sorbent material. The kapok fibre has a hollow structure with a large lumen that leads to its excellent oil absorbency and retention capacity. Due to its biodegradability, Kapok fibre can be ultimately disposed of for biomass energy recovery, thus leaving no secondary waste to pollute the environment. This study examined the absorption capacity of kapok fibre and kapok wood ash as well as a combination of both these materials into waste cooking oil. Economically, kapok is relatively cheap compared to polypropylene. The result shows that 10 g of kapok fibre, kapok ash wood, and a combination of both materials give a maximum absorption capacity of 25.70 g/g, 0.47 g/g, and 7.83 g/g respectively compared to the 20 and 30 g weight that gives lower value than 10 g weight of the material.

Keywords Filter · Absorption behaviour · Clean-up · Waste cooking oil · Kapok ash wood

A. H. F. Ahmad · R. Hassan (✉) · N. Mohammad Noh · A. Alisibramulsi
School of Civil Engineering, College of Engineering, Universiti Teknologi MARA, 40450 Shah Alam, Selangor, Malaysia
e-mail: rohan742@uitm.edu.my

R. Hassan · A. Alisibramulsi
Institute for Infrastructure Engineering and Sustainable Management (IIESM), Universiti Teknologi MARA, 40450 Shah Alam, Selangor, Malaysia

N. J. Abd Malek
Laman Seri Builder Group Sdn Bhd, 40000 Shah Alam, Selangor Darul Ehsan, Malaysia

E. S. Ab Wahab
Faculty of Civil Engineering Technology, Universiti Malaysia Pahang Al-Sultan Abdullah (UMPSA), Lebuhraya Tun Razak, 26300 Gambang, Kuantan, Pahang Darul Makmur, Malaysia

U. R. Hashim et al. (eds.), *Green Infrastructure*,
https://doi.org/10.1007/978-981-99-7003-2_22

1 Introduction

In modern society, oil is commonly used for the preparation of food. In today's world, the oil frying method is widely used due to the contribution of good taste, attractive colour, and better presentation of the food. As this method becomes increasingly popular, the accumulation of waste generated from cooking oil also increases. It was reported that waste cooking oil is widely produced all over the world (Kulkarni & Dalai, 2006). EU produces approximately 700,000–1,000,000 tonnes of waste cooking oil annually, inclusive of the oils from snack food and French fries. It was estimated that there were 40,000 tonnes per year of waste cooking oil produced in Asia countries such as China, Malaysia, Indonesia, Thailand, Hong Kong, India, etc. (Ahmad Razali, 2005). Cooking oil waste is improperly managed, resulting in the release of waste cooking oil into the environment, polluting the ecosystem, notably the land and water. An oil coating forms on the surface of the water, preventing oxygen from dissolving, leading to the loss of marine life. Furthermore, mixing oil and water raises the chemical oxygen demand (COD) of water, making it hazardous due to the presence of oil degradation by-products. Sea species absorb carcinogenic substances, which are subsequently transferred down the food chain to humans (Kulkarni & Dalai, 2006; Jafari, 2010; Marjadi & Dharaiya, 2010).

In recent years, the increasingly serious oil pollution from oil spills has been a cause of vast concern (ITOPF, 2013). Oil pollution from food production, catering industry, petroleum chemical and petroleum mining, etc. has drawn extensive attention of many researchers in recent years. Many methods, such as mechanical extraction, in situ combustion, and chemical degradation have been used to the clean-up oil from polluted areas. Owning to better economic and environmental benefits, the use of sorbent is considered as an effective method to concentrate, transfer, and absorb spilled oil (Adebajo, 2003). High-efficient oil sorbent is required to possess desirable characteristics, such as excellent hydrophobicity and oleophilicity, high uptake capacity, fast oil sorption rate, low cost, and high buoyancy. So far, various materials including natural materials (Choi & Cloud, 1992; Srinivasan & Viraraghavan, 2010), inorganic mineral materials (Moura & Lago, 2009; Teas et al., 2001), and organic synthetic fibres (Lin et al., 2012; Wei et al., 2003) have been used for oil sorption. Even so, inexpensive oil sorbents with advantageous oil-sorption properties are still in need. In this case, the utilization of renewable resources as agricultural products and waste is much more notable due to their great sorption capacity, high biodegradability, and low cost.

Currently, sorption materials can be divided into traditional absorbents, special wettable absorbents, and smart controllable absorbents (Xue et al., 2014). Traditional absorbents mainly include inorganic minerals (zeolites, silica aerogels, expanded vermiculites, exfoliated graphite, and activated carbon), synthetic polymers (nonwoven polypropylene and polyurethane foams), and natural materials (straw, sawdust, rice husks, bagasse, cotton, wool, kenaf, cattail, milkwood, silk-floss fibre, and populus seed fibre) (Annunciado et al., 2005; Ibrahim et al., 2010; Likon et al., 2013; Rengasamy et al., 2011; Said et al., 2009; Singh et al., 2013; Wahi et al.,

2013; Wei et al., 2003). These materials are commonly used; however, many of them suffer drawbacks from high water uptake, leading to low oil absorbencies and buoyancies. Special wettable materials (Arbatan et al., 2011; Su et al., 2012; Yang et al., 2014; Zhang et al., 2012) and smart controllable materials (Zhang et al., 2012) are recently developed fields. Compared with traditional absorbents, they generally possess superhydrophobicity, or controllable oil wettability, by having varying pH and temperature, or by employing a magnetic field and exhibiting excellent selectivity, high efficiency, and unique advantages in oil spill absorption; however, the complicated preparation process and high cost of these materials limit their practical applications.

Waste cooking oil causes some serious environmental problems including bad odour. It is said that one litre of oil poured into natural waters may pollute 500,000 L of water. Mainly it increases the organic load on water bodies and also forms a thin layer over water that reduces the required dissolved oxygen concentration for underwater living creatures (Carlos, 2011). Waste cooking oil disposed incorrectly into the kitchen sinks can solidify and hence block the sewer pipes. Further degradation of waste cooking oil in pipes may also cause corrosion of metal and concrete elements (Refaat, 2010). Thus, the removal of waste cooking oil from sewer streams at sewage treatment plants adds extra cost to it (Carlos, 2011).

Free fatty acid (FFA) content of waste cooking oil is very high as compared to normal cooking oil. If FFA in cooking oil is less than 15% it is said as yellow grace otherwise named as brown grace (15–60% FFA by weight) which has much greater viscosity. Proper waste management is thus necessary for waste cooking oil to overcome problems (Kulkarni & Dalai, 2006). Waste cooking oil is generated on a large scale by restaurants due to deep fryers. According to the Energy Information Administration in the United States capita, about 378 million L of waste cooking oil can be generated in the USA per day. Canada, European countries, and the UK are reported to produce approximately 135,000, 700,000–1,000,000, and 200,000 tonnes per year of waste cooking oil respectively (Arjun, 2008).

To date, several methods for fabricating effective oil clean-up materials have been used, including electrospinning of fibrous fibre (Wu et al., 2012; Zhu et al., 2011), esterification of cellulosic materials, and fabrication of superhydrophobic surfaces (Arbatan et al., 2011; Banerjee et al., 2006; Sathasivam & Haris, 2010; Zhu et al., 2011). However, most of these materials have low oil sorption capabilities, which must be enhanced in order to handle spilled oil effectively. Furthermore, because of their multistep preparation procedures and high cost, many existing techniques make it difficult to carry out final industrialized manufacturing. Kapok fibre is an agricultural product made from the fruits of the silk-cotton tree, and it has a high oil sorption capacity due to its hollowness and natural hydrophobic properties (Abdullah et al., 2010; Huang et al., 2006; Lim & Huang, 2007a, 2007b; Wang et al., 2012a). The hydrophobic–oleophilic feature of the fibre is critical for oil retention in the Kapok fibre arrangement. Nechchadi et al. (2020) have also documented oil removal from water using sorption on natural agrowaste materials.

According to the pollution study statistics in 2016 (Jauhar et al., 2022) every household uses a minimum of 2 kg of cooking oil per month while 45% of oil that cannot be used anymore will be discarded. In addition, problems arise when waste cooking oil is just poured into the sink or down the drain, which affects residents, communities, and municipalities. The disposal of cooking oil into the sink can cause the drainage system to clog lead to foul-smelling odours and promote the breeding of pests such as rats and cockroaches. This is indirectly a breeding ground for pests that will affect the health of the population. Waste cooking oil solidifies into thick layers inside drain pipes and sewers, constricting the water flow (Anonymous, 2022).

The aim of this study is to determine the absorption capacity and time required of kapok fibre, kapok ash wood, and the combination of both materials and to determine the oil absorption behaviour of the selected material. This study will gain more information on the chosen material which is kapok fibre, kapok ash wood, and the combination of Kapok fibre and Kapok ash wood on the performances to be used as natural organic oil sorbent material. Economically, Kapok is relatively cheap compared to polypropylene. Environmentally, Kapok fibre can be recovered from discarded bedding, upholstery, and life preservers for reuse as an oil-absorbing material.

Insoluble material where raw or a mixture of materials that recover liquids via the channels of absorption or adsorption or sometimes both (Rengasamy et al., 2011). Different sorts of sorbent materials have been tried and researched and additionally have been considered as best in oil removal application (Table 1). Organic natural materials, synthetic organic polymers, and inorganic mineral materials are the three types of sorbent materials. Straw, corn cob, sawdust, cotton fibre, wool fibre, coconut husk, and Kapok filaments are some of the natural sorbents used (Abdullah et al., 2010). Synthetic sorbents are widely regarded as the most effective means of absorbing oil. At times, a 40:1 weight ratio of oil to sorbent can be attained.

Table 1 Types of natural sorbent material and oil uptake (Abdullah et al., 2010)

Sorbent materials	Oil type	Oil uptake (g/g)	Form
Cotton	Diesel	30.6	Fibre
Kenaf bast	Diesel	7.2	Fibre
Kenaf core	Diesel	7.1	Fibre
Garlic peels	Crude	0.4	Sheets
Onion peels	Crude	0.5	Sheets
Wool	Motor	5.6	Fibre
Wool	Crude	31.8	Fibre
Milkweed	Crude	39.0	Fibre
Kapok	Crude	38.5	Fibre

1.1 Kapok Fibre

Malaysian Kapok is widely found in peninsular Malaysia, especially in the northern parts. It has leading up to now been perceived that the fibre of Kapok, which is of a silky character (being a types of silk-cotton tree naturally identified with the cotton plant, and which is become in extensive amounts in Java, Ceylon and different spots having comparable climatic conditions) are made of surprising delicate quality, warmth, and lightness, yet it has to this point been basically difficult to card and twist Kapok fibre in an immaculate condition, unmixed with different fibres. In these studies, Kapok fibre exhibited good water repellence, high oil adsorption capability, and well-reusable characteristics, demonstrating its potential as an alternative for application in oil pollution control (National Oceanic, 2010). Kapok or the scientific name *Ceiba Pentandra*, a natural and renewable material, has received increasing attention in recent years for its intrinsic properties (Bozaci, 2019). Conventionally, Kapok fibre is used as life-saving equipment, mattress/pillow stuffing, life preservers, bedding, and upholstery, and for insulation against sound and heat (School, 2019). Kapok fibre was detached and broke down minutely, and the physicochemical properties were determined by spectroscopic systems. Kapok fibre is a highly lignified organic seed fibre and mainly consists of cellulose, lignin, and xylan (Fengel & Wenzkowski, 1986; Gao et al., 2012). The chemical composition of Kapok fibre varies in different reports. One study found that Kapok fibre was chemically composed of 64% cellulose, 13% lignin, and 23% pentosan on a weight basis (Kobayashi et al., 1977), while another found that Kapok fibre was comprised of 35% cellulose, 21.5% lignin, and 22% xylan, with a high ratio of syringyl/guaiacyl units (4–6) and a high level of acetyl groups (13.0%) as compared with normal plant cell walls (about 2–4%) (Chung et al., 2008; Hori et al., 2000).

The differences might be related to the differences in Kapok sources and processing techniques. The crystallization degree of Kapok fibre has been determined to be 35.90%, the specific birefringence is 0.017, and the bulk density is 0.30 g/cm^3, considering its large lumen (Xiao et al., 2005). Kapok fibre is composed of two major layers with differing microfibrillar orientations. The outer layer is composed of cellulose microfibrils oriented transversely to the fibre axis, whereas the inner layer is composed of fibrils oriented nearly parallel to the fibre axis (Nilsson & Björdal, 2008). Optical microscopy and SEM observation found that Kapok fibre shows a cylindrical shape, a surface without texture and thickening groins at the ends (Fengel & Wenzkowski, 1986). The surface of Kapok fibre is smooth with a thick layer of wax, and the cross-section is oval to round with large lumen and thin wall (ca. 8–10 μm in diameter and ca. 0.8–1.0 μm in wall thickness) (Chung et al., 2008; Mwaikambo & Ansell, 1999).

Owing to its hollow lumen and waxy surface, Kapok fibre shows hydrophobic–oleophilic characteristics, and Kapok fibre assemblies possess good buoyancy and cannot be sunk in mixtures of oil and water (Rengasamy et al., 2011). This hollow structure differentiates the Kapok fibre from other natural fibres and endows it with a porosity of more than 80% (Xiang et al., 2013). As another unicellular fibre, cotton

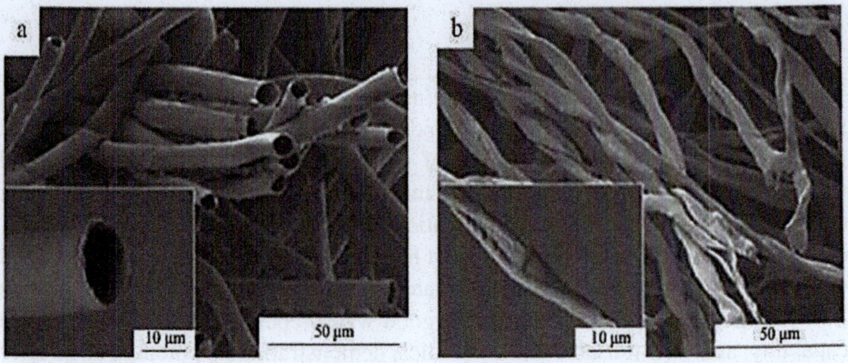

Fig. 1 SEM images of **a** Kapok fibres and **b** degreased cotton fibres (Xiang et al., 2013)

fibre shows a compressed ribbon-like morphology, rolling in a helicoidal manner around the axis (Chung et al., 2008), as shown in Fig. 1.

Pure Kapok fibre cannot be spun like cotton fibre because of its brittleness, low cohesivity, and strength, but it can be successfully combined with cotton fibre to make yarns. Increased Kapok content reduces yarn regularity and toughness while increasing yarn extensibility, resulting in lower total cost of production for the yarns (Dauda & Kolawole, 2003). However, the big lumen and waxy surface make it difficult for hydrophilic colouring agents or dyes to penetrate, resulting in a decreased dyeing efficiency for Kapok fibre (Lou, 2011). As it undergoes chemical reactions, it also contains particulate matter and an increased number of total polar solids, polymeric molecules (Kulkarni & Dalai, 2006).

For further improvement of oil sorption properties, the Kapok fibre can be coated to show a new lipophilic property. Using PBMA and PS as the modification agents, two kinds of oil absorbents with high sorption capacity were prepared by a facile solution-immersion process. Compared with raw Kapok fibre, the coated Kapok fibre showed higher oil-absorbing capacities for gasoline, diesel, soybean oil and paraffin oil, with an oil sorption capacity of up to 80 times their weight (Wang et al., 2013a, 2013b, 2013c, 2013d). Grafting copolymerization or esterification modification is also effective in increasing the hydrophobicity of Kapok fibre to enhance its oil sorption capacity (Wang et al., 2012a, 2013a). Wang et al. (2013b) has also reported on the preparation and properties of Kapok fibre enhanced oil sorption resins by suspended emulsion polymerization.

A typical diesel/water separation process indicates that the dynamic behaviour of oil/water separation in Kapok filters is classified into four stages: infiltration, separation, displacement, and equilibrium (Huang & Lim, 2006). In addition to excellent oil/water separation, the Kapok filters have excellent capability for reducing COD and turbidity during the process (Rahmah & Abdullah, 2010). The Kapok fibre can retain stability after fifteen cycles of reuse with only 30% of sorption capacity reduction (Abdullah et al., 2010). Kapok fibre can selectively absorb significant amounts of oil (40 g/g of fibre) from an oil suspension in freshwater and seawater (Hori et al.,

2000), with high oil-to-water sorption (O/W) ratios ranging from 19.35 to 201.53 (Ali et al., 2012).

For kapok fibre, the amounts of oil absorbed and retained within the Kapok assemblies depend on the oleophilicity of the Kapok fibre and the physical characteristics of the model oils including but not limited to density, viscosity, surface tension, and contact angle with the Kapok surface (Lim & Huang, 2007b). Compared with raw Kapok fibre, the coated Kapok fibres showed higher oil-absorbing capacities for gasoline, diesel, soybean oil, and paraffin oil, with an oil sorption capacity of up to 80 times their weight (Wang et al., 2013c). More importantly, the oil sorption capacity of Kapok fibre can be further enhanced by modifying its smooth surface into a rough surface with hierarchical micro- or nanostructures (Wang et al., 2013d). After the coating process, a new surface with hierarchical micro- and nanostructures is created. The nano-metre-sized roughness is derived from the combined silica nanoparticle-rough PBMA deposit sediment with low surface energy, which will offer more micro-areas that facilitate the retention of oil in the fibre assembly. Accordingly, more oils can be retained in the fibre assembly. To change the surface properties from hydrophobicity into superhydrophobicity, the surface of Kapok fibre can also be coated by the incorporation of silica nanoparticles and subsequent hydrophobic modification using hydrolysed DTMS (Wang et al., 2012b). This superhydrophobic fibre can quickly absorb diesel and soybean oil up to above 46.9 and 58.8 g/g.

2 Methodology

2.1 Specimen's Preparation

Kapok fibre has been bought from a local kapok retailer while kapok wood has been taken from Raub, Pahang, Malaysia. The material was brought to the laboratory to make ash wood from the kapok wood using a drilling machine. The other equipment prepared is a beaker, scale, and a net to retain the kapok fibre and ash wood from cluttering on their position. Waste cooking oil has been obtained from the nearest restaurant surrounding area in Shah Alam, Malaysia. The kapok fibre and kapok ash wood have been weighing using a scale with three different weights, 10, 20, and 30 g. The combination of both materials also has been weighed with the same proportion which is 50–50% of each material to get the total weight of 10, 20, and 30 g. Three materials were used with three constant weights and three constant times during this study. The proposed constant weight is 10, 20, and 30 g which is an interval of 10 g of each material. Each weight is tested with three proposed constant times that is 20, 40, and 60 min which is an interval time of every 20 min. For each material to get average value of absorption rate, at least three replications were carried out for each sample. The total experiments were 81 tests as shown in Table 2.

Table 2 The total testing and no of samples for each material

Material	Kapok fibre		Kapok Ash Wood		Combination of Materials	
Weight (g)	Time (min)	No. of samples	Time (min)	No. of samples	Time (min)	No. of samples
10	20	3	20	3	20	3
	40	3	40	3	40	3
	60	3	60	3	60	3
20	20	3	20	3	20	3
	40	3	40	3	40	3
	60	3	60	3	60	3
30	20	3	20	3	20	3
	40	3	40	3	40	3
	60	3	60	3	60	3
	Total	27	Total	27	Total	27

2.2 Absorption Test

Every material—Kapok fibre, Kapok ash wood, and combination of Kapok fibre and Kapok ash wood was weighed according to the required weight. The specimen is added into the beaker and left for a required contact time. The contact time used in this study ranged from 20 to 60 min, at 20 min interval. At the end of each contact time, the material in the beaker was passed through the sieve and allowed to drain for 5 min. The amount of sorb waste cooking oil recovered from the sorbent material was determined and recorded as Initial Weight (IW). Afterwards, subjecting it to be pressed using under a fixed load of 30 kg for 30 min. After pressing, the sorbent was re-weighed and recorded as Final Weight (FW); the weight of sorbate recovered was determined by weight difference. The amount of sorb waste cooking oil recovered per unit mass of each material was determined from Eq. 1:

$$q = \frac{\text{Initial Weight}(IW) - \text{Final Weight}(FW)}{\text{Mass of Sorbent}} \quad (1)$$

The experiment was performed in triplicate, the average and standard deviation were calculated and used. The amount of sorb oils retained per unit mass of sorbent material was obtained from Eq. 1: $q =$ Quantity of oil sorb per unit mass of sorbent material—Quantity of sorb oil recovered per unit mass of sorbent material. Figure 2 simply shows the flow of absorption test.

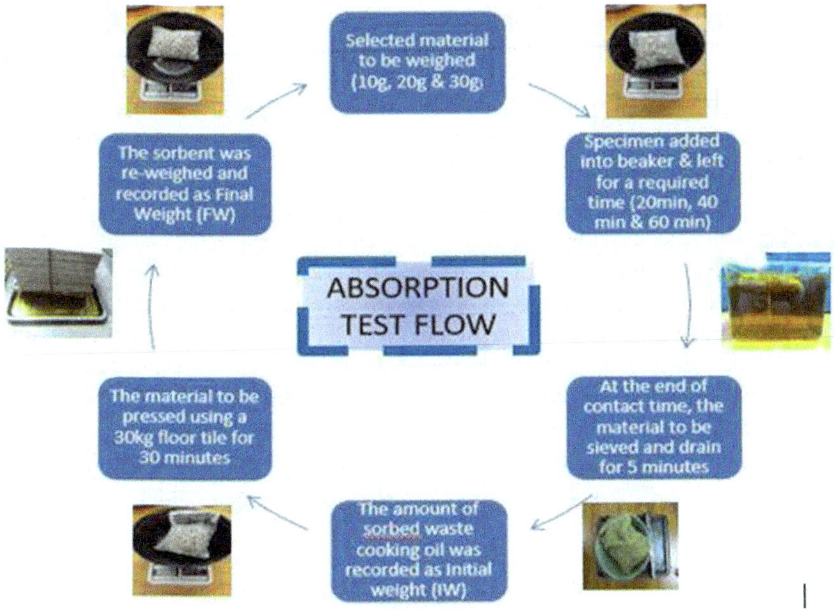

Fig. 2 Summary of absorption test flow

2.3 Oil Absorption Behaviour Study

This study also ascertained the mopping sorb behaviour of oil absorption between Kapok fibre, Kapok ash wood, and the combination of Kapok fibre and Kapok ash wood when the material was contacted with the waste cooking oil. Picture of these three materials has been taken so that the oil absorption behaviour can be visualized. Each material—Kapok fibre, Kapok ash wood, and a combination of Kapok fibre and ash wood was weighed according to the required weight which is 10, 20, and 30 g. The specimen is added into the beaker and left for a required contact time as stated above. The contact time used in this visualization study was every minute for first 10, 20, 40, and 60 min.

3 Results and Discussions

3.1 *Oil Absorption Test*

3.1.1 Kapok Fibre

The Kapok fibre absorption capacity, q, is presented in Table 3, the maximum absorption capacity, q, for the weight of 10, 20, and 30 g is 25.70 g/g at 40 min time, 19.73 g/g at 40 min time, and 16.07 g/g at 40 min time, respectively. Thus, the maximum absorption capacity of three different weight is 25.70 g/g with 10 g of Kapok fibre weight. The values of 10 g absorption capacity analysis are shown in Table 4. This result can be related to the previous study by Abdullah et al. (2010), referring to Table 1 of this study where the absorption result was 38.5 g/g and the previous study from Hori et al. (2000) justified that Kapok fibre can selectively absorb significant amounts of oil at 40 g/g. This study found about 1.5 g/g lesser than the absorption result from Hori et al. (2000) could be affected from the tested type of oil, the material weight, and the condition of the material.

The graph from Fig. 3, shows the relationship of oil absorption capacity, q (g/g), with time for the Kapok fibre material. The pattern shows that the absorption capacity reading was increased from 20 to 40 min times for each weight. The maximum oil

Table 3 The analysis of oil absorption capacity for Kapok fibre, Kapok ash, and its combinations

Time (min)	Kapok fibre (g/g)			Kapok ash wood (g/g)			Kapok fibre and Kapok ash wood (g/g)		
	10 g	20 g	30 g	10 g	20 g	30 g	10 g	20 g	30 g
20	20.23	17.72	12.03	0.37	18	0.17	6.1	6.43	6.53
40	25.7	19.73	16.07	0.47	0.22	0.18	7.83	6.85	7.46
60	23.27	19.32	14.9	0.4	0.2	0.16	7.5	5.77	6.89

Table 4 Absorption capacity, q, for Kapok fibre—weight of 10 g

Time (min)	Test no	IW (g)	FW (g)	IW–FW (g)	Mean	q (g/g)
20	1	361	153	208	202.33	20.23
	2	380	182	198		
	3	377	176	201		
40	1	461	196	265	257.00	25.70
	2	448	204	244		
	3	473	211	262		
60	1	451	208	243	232.67	23.27
	2	427	202	225		
	3	443	213	230		

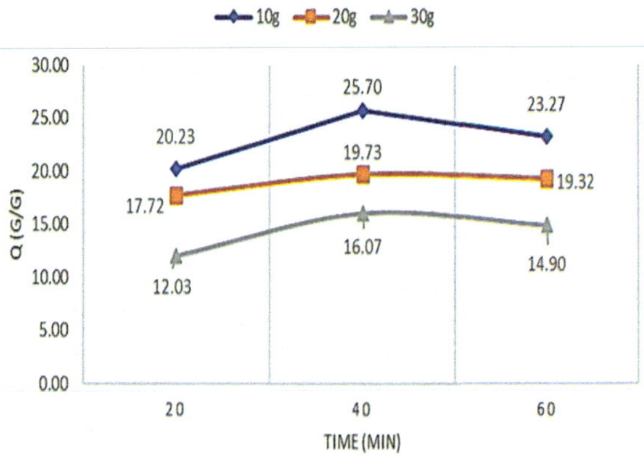

Fig. 3 Oil absorption capacity versus time for three different weights (Kapok Fibre)

absorption capacity for all materials was recorded at 40 min time. Beyond the 60 min, the reading of oil absorption capacity was slightly decreased. Therefore, from the observation, the Kapok fibre material can only absorb the oil at maximum capacity at 40 min time. On the other hand, it can be observed that the oil absorption capacity for 10 g weight is more effective than the other two weights. The oil absorption capacity for 30 g has the lowest reading while for 20 g weight was in the middle range of reading.

The bar chart from Fig. 4 shows the difference of oil absorption capacity, q (g/g), with time for the Kapok fibre material. The chart shows that the absorption capacity reading for 10 g, 20 g, and 30 g was increased to 21.27%, 10.22%, and 25.10% respectively from 20 to 40 min times. The highest increment in percentage was recorded for 30 g of Kapok fibre. When the specimens reach at maximum time, the absorption capacity reading for 10 g, 20 g, and 30 g was slightly decreased to 10.46%, 2.16%, and 7.83% respectively. The highest decrement in percentage was recorded for 10 g of kapok fibre.

It can be concluded that the Kapok fibre material can absorb the oil at a maximum rate of 40 min time. If the absorption time continues beyond 40 min, the oil absorption capacity rate will decrease. It can be related to a previous study by Lim and Huang (2007b), the amounts of oil absorbed and retained within the Kapok assemblies depend on the oleophilicity of the Kapok fibre and the physical characteristics of the model oils including but not limited to density, viscosity, surface tension, and contact angle with the Kapok surface.

3.1.2 Kapok Ash Wood

In view of single performance for Kapok ash wood, absorption capacity, q, is presented in Table 3; the maximum absorption capacity, q, for the weight of 10, 20, and 30 g is 0.47 g/g at 40, 0.22 g/g at 40, and 0.18 g/g at 40 min time respectively. Figures 5 and 6 show the relationship of oil absorption capacity, q (g/g), with time

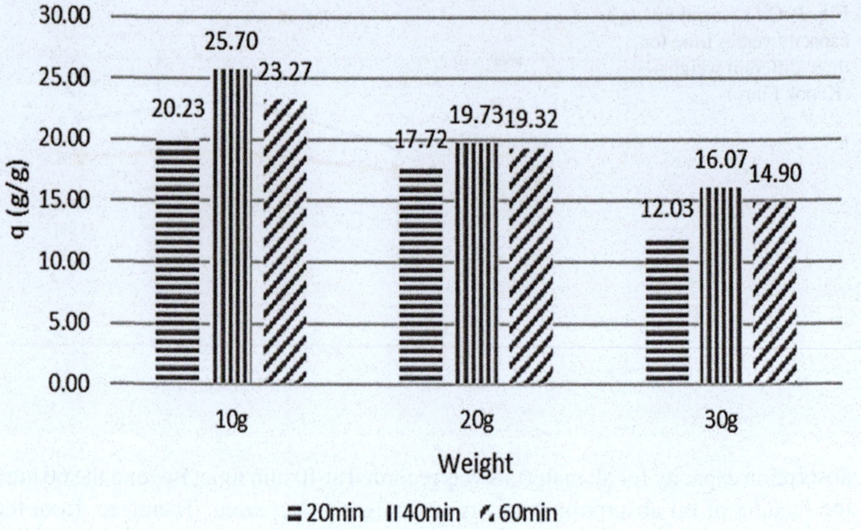

Fig. 4 The percentage of difference for Oil absorption capacity versus time for three different weights (Kapok Fibre)

for the Kapok ash wood material. The pattern shows that the absorption capacity reading was increased from 20 to 40 min times for each weight. The maximum oil absorption capacity for all materials was recorded at 40 min time. Beyond the 60 min, the reading of oil absorption capacity was slightly decreased. Therefore, from the observation the Kapok ash wood material can only absorb the oil at a maximum capacity of 40 min time. On the other hand, it can be observed that the oil absorption capacity for 10 g weight is more effective than the other two weights. The oil absorption capacity for 30 g has the lowest reading while for 20 g weight was in the middle range of reading.

Fig. 5 Oil absorption capacity versus time for three different weights (Kapok Ash Wood)

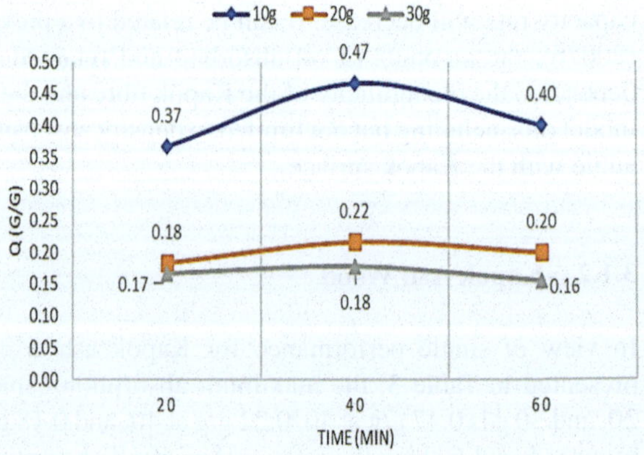

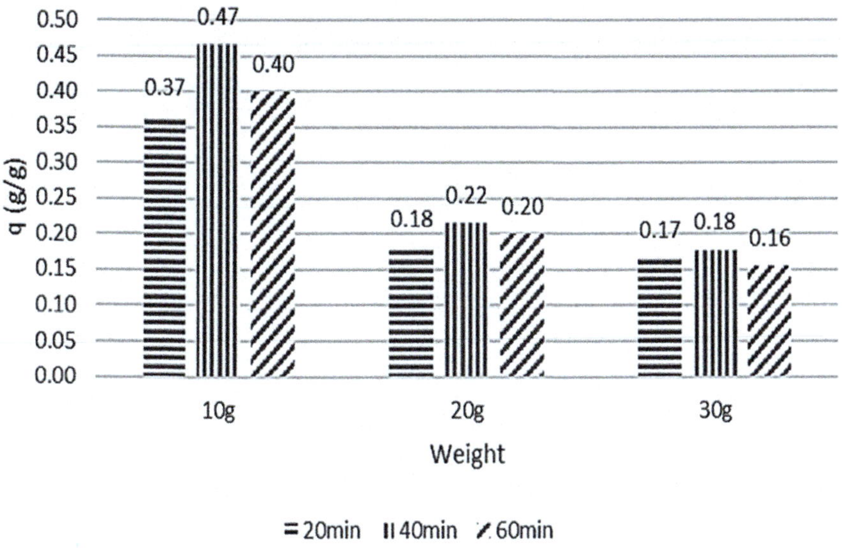

Fig. 6 The percentage of difference for oil absorption capacity versus time for three different weights (Kapok ash wood)

The highest decrement in percentage was recorded for 10 g of Kapok ash wood. It can be concluded that the Kapok ash wood material can absorb the oil at a maximum rate of 40 min time. If the absorption time continues beyond 40 min, the oil absorption capacity rate will decrease.

3.1.3 Combination of Kapok Fibre and Kapok Ash Wood

The combination performance of Kapok fibre and Kapok ash wood absorption capacity, q, is presented in Table 3, Fig. 5, and Fig. 6 respectively. Based on the recorded data, the maximum of absorption capacity, q, for weights of 10, 20, and 30 g is 7.83 g/g at 40 min, 6.85 g/g at 40 min time, and 7.46 g/g at 40 min time respectively.

Figure 7 shows the relationship of oil absorption capacity, q (g/g), with time for the combination of Kapok fibre and Kapok ash wood material. The pattern shows that the absorption capacity reading was increased from 20 to 40 min times for each weight. The maximum oil absorption capacity for all materials was recorded at 40 min time. Beyond the 60 min, the reading of oil absorption capacity was slightly decreased. Therefore, from the observation the Kapok ash wood material can only absorb the oil at maximum capacity at 40 min time. On the other hand, it can be observed that the oil absorption capacity for 10 g weight is more effective than the other two weights. The oil absorption capacity for 20 g has the lowest reading while for 30 g weight was in the middle range of reading.

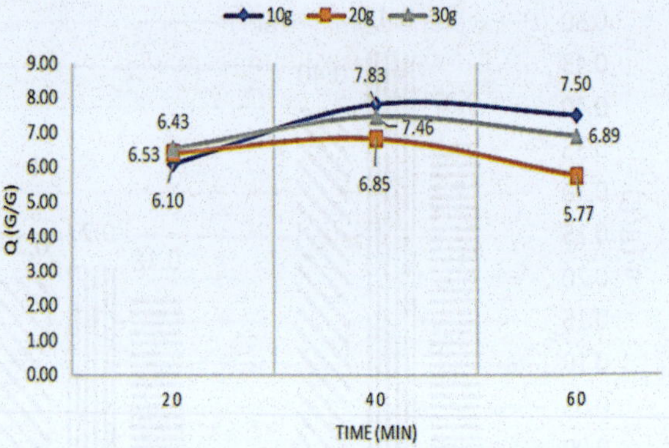

Fig. 7 Oil absorption capacity versus time for three different weights (combination)

Figure 8, shows the difference in oil absorption capacity, q (g/g), with time for the Kapok ash wood material. The chart shows that the absorption capacity reading for 10 g, 20 g, and 30 g was increased to 22.13%, 6.08%, and 12.37% respectively from 20 to 40 min times. The highest increment in percentage was recorded for 10 g of combination of Kapok fibre and Kapok ash wood. When the specimens reached at maximum time, the absorption capacity reading for 10 g, 20 g, and 30 g was slightly decreased to 4.44%, 18.79%, and 8.23% respectively. The highest decrement in percentage was recorded for 20 g of the combination of both materials. It can be concluded that the combination of Kapok fibre and Kapok ash wood material can absorb the oil at a maximum rate of 40 min time. If the absorption time continues beyond 40 min, the oil absorption capacity rate will decrease. The absorption rate is solely depending on the Kapok fibre material as the result and data shown under Kapok fibre is much higher compared to the Kapok ash wood material.

Above all, the highest decrement in percentage was recorded for Kapok ash wood materials. It can be concluded that the Kapok fibre material has the highest absorption

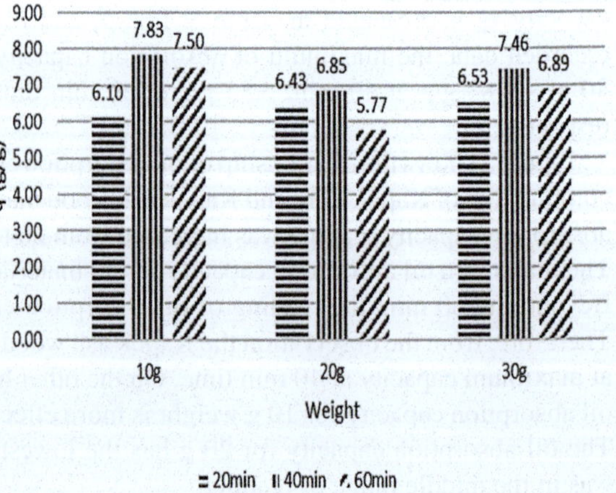

Fig. 8 The percentage of difference for Oil absorption capacity versus time for three different weights (combination)

rate for 10 g weight of material. It can be related to a previous study by Lim and Huang (2007b), where the hollow structure is beneficial due to its large effective pore volume, and the waxy surface can enhance its oil adherence ability. Between kapok ash and fibre, this study shown that the smaller surface of material will have lesser absorption capacity, as such it agrees with Wang et al. (2013d) that says the sorption capacity of Kapok fibre can be further enhanced by having rougher surface of material.

3.1.4 Behaviour of Kapok Fibre, Kapok Ash Wood, and It Combinations

The absorption behaviour of Kapok fibre in the waste cooking oil was seen as 25% of oil has been absorbed at 3 min time (Fig. 9). The absorption behaviour increased to 50% when it reached at 5 min time. After 10 min time, 75% of waste cooking oil has been absorbed and the recorded behaviour at 20 min time shows that the absorption behaviour has increased to 95%. Kapok fibre has fully sunk in the oil when the time reached 40 min and there were no changes in the behaviour for 60 min time. From the observation, the first 10 min absorption capacity shows rapid increment and when it reaches 20 min and 40 min time the absorption was getting slower. It can be compared with previous report from National Oceanic (2010), which says that Kapok fibre exhibited good water repellence, high oil adsorption capability, and well-reusable characteristics, demonstrating its potential. This study has proved that Kapok fibre has a significant absorption into the waste cooking oil.

The absorption behaviour of Kapok ash wood in the waste cooking oil visualized 90% of oil has been absorbed at the first minute (Fig. 10). The absorption behaviour of Kapok ash wood looks identical until 60-min time. From the observation, the first minute shows a rapid increment of 90% of absorption and there are no changes afterwards. The absorption behaviour of combination of Kapok fibre and Kapok ash wood in the waste cooking oil visualized 25% of oil has been absorbed at first minute's time (Fig. 11).

The absorption behaviour increased to 50% when it reached at 3 min time. After 5 min time, 75% of waste cooking oil has been absorbed and the recorded behaviour at 10 min time shows that the absorption behaviour has increased to 95%. The

Fig. 9 The visualization of the behaviour for Kapok Fibre

Fig. 10 The visualization of the behaviour of Kapok Ash wood

Fig. 11 The visualization of the behaviour for the combination of Kapok fibre and Kapok ash wood

combination of both materials has fully sink in the oil when the time reached 40 min and there were no changes in the behaviour for 60 min time. From the observation, the first 10 min absorption capacity shows rapid increment and when it reaches 20 and 40 min time the absorption was getting slower.

4 Conclusions

The absorption capacity has been tested with three different weight and three different time which is 10, 20, and 30 g and 20, 40, and 60 min respectively. The minimum absorption capacity of the selected materials is 12.03 g/g for Kapok fibre, while Kapok ash wood and a combination of Kapok fibre and Kapok ash wood have 0.16 g/g and 5.77 g/g respectively. The minimum reading was taken at the sample for 30 g weight. The maximum absorption capacity of the selected materials is 25.70 g/g for Kapok fibre, while Kapok ash wood and a combination of Kapok fibre and Kapok ash wood have 0.47 g/g and 7.83 g/g respectively. The maximum reading was taken at the sample for 10 g weight. It can be concluded that the sample of 10 g weight has significant absorption compared to the other weights. As for the results of oil absorption capacity and behaviour of the selected materials, it shows that Kapok fibre has significant absorption properties while the Kapok ash wood does not have any absorption properties as the material is directly sunk in the oil due to Kapok ash wood being light material. For the combination of Kapok fibre and Kapok ash wood, the results of absorption capacity are merely depending on the Kapok fibre, explained the combination results were at the middle range between Kapok fibre and Kapok ash wood result. It can be concluded that Kapok fibre is the best absorption material than the other.

Acknowledgements The author wishes to express the greatest appreciation to Universiti Teknologi Mara Malaysia and Ministry of Higher Education (MOHE) Malaysia for the financial support contributed through KPM; FRGS Grant: 600-RMI/FRGS 5/3 (181/2019).

References

Abdullah, M. A., Rahmah, A. U., & Man, Z. (2010). Physicochemical and sorption characteristics of Malaysian *Ceiba pentandra* (L.) Gaertn. as a natural oil sorbent. *Journal of Hazardous Materials, 177*, 683–691.

Adebajo, M. F. (2003). Porous materials for oil spill cleanup: A review of synthesis and absorbing properties. *Journal of Porous Materials, 10*, 159–170.

Ahmad Razali, K. (2005). Performance of special quality and standard palm olein in batch frying. *Palm Oil Developments, 43*, 10–15.

Ali, N., El-Harbawi, M., Jabal, A. A., & Yin, C. Y. (2012). Characteristics and oil sorption effectiveness of Kapok fibre, sugarcane bagasse and rice husks: Oil removal suitability matrix. *Environmental Technology, 33*(4–6), 481–486.

Anonymous. (2022). Used cooking oil recycling program. Solid waste authority of palm beach county. https://swa.org/173/Used-Cooking-Oil. Retrieved Jan 2022.

Annunciado, T., Sydenstricker, T., & Amico, S. (2005). Experimental investigation of various vegetable fibers as sorbent materials for oil spills. *Marine Pollution Bulletin, 50*, 1340–1346.

Arbatan, T., Fang, X., & Shen, W. (2011). Superhydrophobic and oleophilic calcium carbonate powder as a selective oil sorbent with potential use in oil spill cleanups. *Chemical Engineering Journal, 166*, 87–791.

Arjun, B. C. (2008). Waste cooking oil as an alternate feedstock for biodiesel production. *Energies, 1*, 3.

Banerjee, S. S., Joshi, M. V., & Jayaram, R. V. (2006). A treatment of oil spill by sorption technique using fatty acid grafted sawdust. *Chemosphere, 64*, 1026–1031.

Bozaci, E. (2019). Optimization of the alternative treatment methods for *Ceiba pentandra* (L.) Gaertn (Kapok) fiber using response surface methodology. *The Journal of The Textile Institute.*

Carlos, A. G. (2011). *Biodiesel: Feedstocks and process technology* (p. 23c). InTech.

Choi & Cloud. (1992). Natural sorbents in oil spill cleanup. *Environmental Science and Technology, 26*, 772–776.

Chung, B. Y., Cho, J. Y., Lee, M. H., Wi, S. G., Kim, J. H., & Kim, J. S. (2008). Adsorption of heavy metal ions onto chemically oxidized *Ceiba pentandra* (L.) Gaertn. (Kapok) fibers. *Journal of Applied Biological Chemistry, 51*(1), 28–35.

Dauda, B. M. D., & Kolawole, E. G. (2003). Processibility of Nigerian Kapok fibre. *Indian Journal of Fiber and Textile, 28*(2), 147–149.

Fengel, D., & Wenzkowski, M. (1986). Studies on Kapok-1. Electron microscopic observations. *Holzforschung, 40*(3), 137–141.

Gao, J., Zhao, T., & Chen, J. B. (2012). Composition, structure and property analysis of *Calotropis gigantea*, Kapok and cotton fibers. *Journal of Donghua University, 38*(2), 151–155.

Hori, K., Flavier, M. E., Kuga, S., Lam, T. B. T., & Iiyama, K. (2000). Excellent oil absorbent Kapok [*Ceiba pentandra* (L.) Gaertn.] fiber: Fiber structure, chemical characteristics, and application. *Journal of Wood Science*, *46*(5), 401–404.

Huang, X. F., & Lim, T. T. (2006). Perfomance and mechanism of hydrophobic–oleophilic Kapok filter for oil/water separation. *Desalination, 190*, 295–307.

Ibrahim, S., Wang, S., & Ang, H. M. (2010). Removal of emulsified oil from oily wastewater using agricultural waste barley straw. *Biochemical Engineering Journal, 49*, 78–83.

ITOPF. (2013). Oil tanker spill statistics 2013 from the International Tanker Owners Pollution Federation's (ITOPF) publications. Retrieved from http://www.itopf.com/information-services/publications/documents/OilSpillstats_2013.pdf

Jauhar, S., Mohd Zulkifli, N.A., Rosli, S. N. A., Mohd Mansor, F., Kamaruzaman, S. N., Rozali, N., & Miswadi, N. (2022). Recycling cooking oil as initiative for environment sustainability. Retrieved from https://www.usim.edu.my/news/research-news/recycling-cooking-oil-initiative-environment-sustainability/dated. 28 Jan 2022.

Jafari, N. (2010). Review of polution sources and controls in caspian sea region. *Journal of Ecology and the Natural Environment, 2*(2), 25–29.

Kobayashi, Y., Matsuo, R., & Nishiyama, M. (1977). Method for adsorption of oils. Japanese Patent, 52138081.

Kulkarni, M. G., & Dalai, A. K. (2006). Waste cooking oil-an economical source for biodiesel: A review. *Industrial and Engineering Chemistry Research, 45*(9), 2901–2913.

Likon, M., Remškar, M., Ducman, V., & Švegl, F. (2013). Populus seed fibers as a natural source for production of oil super absorbents. *Journal of Environmental Management, 114*, 158–167.

Lim, T. T., & Huang, X. F. (2007a). Evaluation of hydrophobicity/oleophilicity of Kapok and its performance in oily water filtration: Comparison of raw and solvent treated fibers. *Industrial Crops and Products, 26*, 125–134.

Lim, T. T., & Huang, X. F. (2007b). Evaluation of Kapok [*Ceiba pentandra* (L.) Gaertn.] as a natural hollow hydrophobic–oleophilic fibrous sorbent for oil spill cleanup. *Chemosphere, 66*, 955–963.

Lin, J. Y., Shang, Y. W., Ding, B., Yang, J. M., Yu, J. Y., & Al-Deyab, S. S. (2012). Nanoporous polystyrene fibers for oil spill cleanup. *Marine Pollution Bulletin, 64*, 347–352.

Lou, L. Q. (2011). Structure, properties of Kapok fiber and test on heat retention. *Shanghai Textile Science and Technology, 39*(5), 15–17.

Marjadi, D. S., & Dharaiya, N. A. (2010). Analysis of edible oil contaminated soil within North Gujarat Region. *Life Sciences Leaflets, 10*, 287–291.

Moura & Lago. (2009). Catalytic growth of carbon nanotubes and nanofibers on vermiculite to produce floatable hydrophobic "nanosponges" for oil spill remediation. *Applied Catalysis, B: Environmental, 90*, 436–440.

Mwaikambo, L. Y., & Ansell, M. P. (1999). The effect of chemical treatment on the properties of hemp, sisal, jute and Kapok for composite reinforcement. *Angewandte Makromolekulare Chemie, 272*(1), 108–116.

Nechchadi, B., Ghazzaf, H., Mazoir, N., Lhadi, E. K., El Krati, M., & Tahiri, S. (2020). Oil removal from water through the sorption on natural agrowaste materials. *Journal of Environmental and Engineering, 146*(10), 04020114. https://doi.org/10.1061/(ASCE)EE.1943-7870.0001802

National Oceanic A. (2010). Office of response and restoration, emergency response division, incident news: Arabian Gulf spills. Retrieved from office of response and restoration, emergency response division, incident news: Arabian Gulf spills.

Nilsson, T., & Björdal, C. (2008). The use of Kapok fibres for enrichment cultures of lignocellulose-degrading bacteri. *International Biodeterioration and Biodegradation, 61*(1), 11–16.

Rahmah, A. U., & Abdullah, M. A. (2010). Evaluation of Malaysian *Ceiba pentandra* (L.) Gaertn. for oily water filtration using factorial design. *Desalination*, 266(1–3), 51–55.

Refaat, A. A. (2010). Different techniques for the production of biodiesel from waste vegetable oil. *International Journal of Environmental Science and Technology, 7*(1), 183.

Rengasamy, R., Das, D., & Praba Karan, C. (2011). Study of oil sorption behavior of filled and structured fiber assemblies made from polypropylene, Kapok and milkweed fibers. *Journal of Hazardous Materials, 186*, 526–532.

Said, A.E.-A.A., Ludwick, A. G., & Aglan, H. A. (2009). Usefulness of raw bagasse for oil absorption: A comparison of raw and acylated bagasse and their components. *Bioresource Technology, 100*, 2219–2222.

Sathasivam, K., & Haris, M. R. H. M. (2010). Adsorption kinetics and capacity of fatty acid modified banana trunk fibers for oil in water. *Water, Air & Soil Pollution, 213*, 413–423.

School, T. (2019). Textile school. Retrieved from https://www.textileschool.com/183/Kapok-or-capok-fibres/

Singh, V., Kendall, R. J., Hake, K., & Ramkumar, S. (2013). Crude oil sorption by raw cotton. *Industrial and Engineering Chemistry Research, 52*, 6277–6281.

Srinivasan, & Viraraghavan. (2010). Adsorption kinetics and capacity of fatty acid modified banana trunk fibers for oil in water. *Water, Air & Soil Pollution, 213*, 413–423.

Su, C., Xu, Y., Zhang, W., Liu, Y., & Li, J. (2012). Porous ceramic membrane with superhydrophobic and superoleophilic surface for reclaiming oil from oily water. *Applied Surface Science, 258*, 2319–2323.

Teas, C., Kalligeros, S., Zanikos, F., Stournas, S., Lois, E., & Anastopoulos, G. (2001). Investigation of the effectiveness of absorbent materials in oil spills clean up. *Desalination, 140*, 259–264.

Wahi, R., Chuah, L. A., Choong, T. S. Y., Ngaini, Z., & Nourouzi, M. M. (2013). Oil removal from aqueous state by natural fibrous sorbent: An overview. *Separation and Purification Technology, 113*, 51–63.

Wang, J. T., Zheng, Y. A., & Wang, A. Q. (2012a). Effect of Kapok fiber treated with various solvents on oil absorbency. *Industrial Crops and Products, 40*, 178–184.

Wang, J. T., Zheng, Y. A., & Wang, A. Q. (2012b). Superhydrophobic Kapok fiber oil-absorbent: Preparation and high oil absorbency. *Chemical Engineering Journal, 213*, 1–7.

Wang, J. T., Zheng, Y. A., & Wang, A. Q. (2013a). Investigation of acetylated Kapok fibers on the sorption of oil in water. *Journal of Environmental Sciences, 25*(2), 246–253.

Wang, J. T., Zheng, Y. A., & Wang, A. Q. (2013b). Preparation and properties of Kapok fiber enhanced oil sorption resins by suspended emulsion polymerization. *Journal of Applied Polymer Science, 127*(3), 2184–2191.

Wang, J. T., Zheng, Y. A., & Wang, A. Q. (2013c). Coated Kapok fiber for removal of spilled oil. *Marine Pollution Bulletin, 69*(1–2), 91–96.

Wang, J. T., Zheng, Y. A., Kang, Y. R., & Wang, A. Q. (2013d). Investigation of oil sorption capability of PBMA/SiO_2 coated Kapok fiber. *Chemical Engineering Journal, 223*, 632–637.

Wei, Q. F., Mather, R. R., Fotheringham, A. F., & Yang, R. D. (2003). Evaluation of nonwoven polypropylene oil sorbents in marine oil-spill recovery. *Marine Pollution Bulletin, 46*, 780–783.

Wu, J., Wang, N., Wang, L., Dong, H., Zhao, Y., & Jiang, L. (2012). Electrospun porous structure fibrous film with high oil adsorption capacity. *ACS Applied Materials & Interfaces, 4*, 3207–3212.

Xiang, H. F., Wang, D., Liu, H. C., Zhao, N., & Xu, J. (2013). Investigation on sound absorption properties of Kapok fibers. *Chinese Journal of Polymer Science, 31*(3), 521–529.

Xiao, H., Yu, W. D., & Shi, M. W. (2005). Structures and performances of the Kapok fiber. *Journal of Textile and Research, 26*(4), 4–6.

Xue, Z., Cao, Y., Liu, N., Feng, L., & Jiang, L. (2014). Special wettable materials for oil/water separation. *Journal of Materials Chemistry A, 2*, 2445–2460.

Yang, J., Song, H., Yan, X., Tang, H., & Li, C. (2014). Superhydrophilic and superoleophobic chitosan-based nanocomposite coatings for oil/water seperations. *Cellulose, 21*, 1851–1857.

Zhang, L., Zhang, Z., & Wang, P. (2012). Smart surfaces with switchable superoleophilicity and superoleophobicity in aqueous media: Toward controllable oil/water separation. *NPG Asia Materials, 4*, e8.

Zhu, H. T., Qiu, S. S., Jiang, W., Wu, D. X., & Zhang, C. Y. (2011). Evaluation of electrospun polyvinyl chloride/polystyrene fibers as sorbent materials for oil spill cleanup. *Environmental Science and Technology, 45*, 4527–4531.

Physical and Chemical Characteristics of Podo Wood-Xylem Filtered

Rohana Hassan, Marfiah Ab Wahid, Jurina Jaafar, Nor Jihan Abd Malek, Ezahtul Shahreen Ab Wahab, and Adlin Sabrina Muhammad Roseley

Abstract Pine trees were found to successfully filter water through their wood-xylem. This study focuses on filtering water using wood-xylem of tropical timber, i.e. Podo species. Samples from Podo branches were cut for 20 mm diameter with 40 mm length. The water was filtered with normal drip without pressure. The filtered water was tested for physical and chemical characteristics of water quality. The physical characteristics tested were colour and turbidity whilst the chemical characteristics were tested on dissolved oxygen demand, biochemical oxygen demand, chemical oxygen demand and pH value. The observations on the physical properties of Podo xylem were visualized using Dino-Lite Digital Microscope before and after the filtering process. After filtration, the suspended material from the water samples appeared as a dark spot at the inlet of the tracheid. It was clearly spotted from the cross-section view. From the water quality test, it shows that the characteristics of both physical and chemical characteristics of the filtered water were not improved and failed to comply with the characteristics as stipulated in Drinking Water Quality Standard respectively.

Keywords Biological Oxygen Demand (BOD) · Chemical Oxygen Demand (COD) · Dissolved Oxygen (DO) · Water Filter

R. Hassan (✉)
Institute for Infrastructure Engineering and Sustainable Management (IIESM), Universiti Teknologi MARA, 40450 Shah Alam, Selangor Darul Ehsan, Malaysia
e-mail: rohan742@uitm.edu.my

R. Hassan · M. Ab Wahid · J. Jaafar
School of Civil Engineering, College of Engineering, Universiti Teknologi MARA, 40450 Shah Alam, Selangor, Malaysia

N. J. Abd Malek
Laman Seri Builder Group Sdn Bhd, 40000 Shah Alam, Selangor Darul Ehsan, Malaysia

E. S. Ab Wahab
Faculty of Civil Engineering Technology, Universiti Malaysia Pahang Al-Sultan Abdullah (UMPSA), Lebuhraya Tun Razak, 26300 Gambang Kuantan, Pahang Darul Makmur, Malaysia

A. S. Muhammad Roseley
Institute of Tropical Forestry and Forest Products (INTROP), Universiti Putra Malaysia (UPM), 43400 Serdang, Selangor, Malaysia

U. R. Hashim et al. (eds.), *Green Infrastructure*,
https://doi.org/10.1007/978-981-99-7003-2_23

1 Introduction

In order to obtain safe drinking water, it needs to be well filtered. Three main types of water filtering are rapid sand filter, slow sand filter and reverse osmosis (Grant & Cutler, 2005). The rapid sand filter is the most popular used in Malaysia. This type of filter has a filter rate of 40 times higher than slow sand filters, this is because of slow sand filter method requires a large area to construct. Usually, the slow sand filter method is applied in the rural area and the cost of this method is very low compared to others.

Wood in particular appears to be a promising biofilter matrix for several reasons. One of the main reasons is wood is an inexpensive and renewable resource. Current forest management practices produce large quantities of low-quality/low-value wood (US DA, 2003) that could be used for filter construction (Boving & Kevin Neary, 2006). Wood functions as a filter was studied by Boving and Zhang (2004) and Boving and Kevin Neary (2006). They used aspen wood cuttings to enhance the removal of dissolved contaminants in water bodies. It was utilized as wood filters to remove polycyclic aromatic hydrocarbons (PAH). Boving and Neary (2006) found that the PAH removal effectiveness was not affected by changes in water temperature or pH. The filter effectiveness increased with filter size and was highest in continuously submerged parts of the filter system (Boving & Kevin Neary, 2006).

Few studies reported on xylem; however, they discussed the plantation behaviour (Alhendawi et al., 2005; Jang & Choi, 2018; Rao et al., 2004) and not about its function as a filter material. A recent study reported on the effect of pore water but on the chemistry of the behaviour of kaolin-bentonite (Estabragh et al., 2018) which is different from the information shared in this research. Based on the previous study, only pine wood-xylem was reported to filter water and it was proven to be able to remove bacteria in the water (Boutilier et al., 2014). As mentioned by Boutilier et al. (2014), the process of wood drying and its influence on xylem conductivity needs further study. In particular, the flow through the xylem of different plants needs to be studied to identify locally available sources of xylem, which will truly enable the construction of filters from locally available materials. In the present study, Boutilier et al. (2014) report results using xylem derived from only Pine species and none are reported from tropical species.

The tropical wood species chosen for this study is Podo (Podocarpacea), a softwood species that has similar characteristics of filtering water as Pine. This study determined the physical and chemical characteristics of water filtered from wood-xylem encountered the colour, turbidity, dissolved oxygen, biochemical oxygen demand, chemical oxygen demand and pH value tests. Observation of the physical characteristics of Podo wood-xylem before and after the filtration process was captured using a Dino-Lite Digital Microscope.

As reported by Boutilier et al. (2014), the wood-xylem from Pine trees shows that approximately 3 cm^3 of sapwood can filter water with flow rates of about 4 L/d from 1 cm^2 filter areas at applied pressures of about 5 psi. This volume is sufficient to meet the clean drinking water needs of one person. The specimens for their study were from Pine branches and cut-off for a few selected lengths. Their findings demonstrate the potential of plant xylem to address the need for pathogen-free drinking water. Boutilier et al. (2014) also found that fluid-conducting xylem from the pine branch constitutes the majority of the cross-section. They concluded that in general the xylem tissue of coniferous trees, i.e. the sapwood is likely to be the most suitable xylem tissue for the construction of a water filtration device, at least for filtration of bacteria, protozoa and other pathogens on the micron or larger scale. From the observation of available research, existing study in this field is only limited to Pine trees. So far, no publication on the fundamental theory of tropical wood-xylem and their hydraulic conductivity efficiency in filtered water quality in comparison with standard water quality. Thus, this study is very important in analysing the anatomy of the selected wood and examines the filtered water quality.

2 Methodology

2.1 Materials

The branches of Podo a species of Podocarpaceae and Podocarpus spp wood were obtained from Selangor, Malaysia forest (Fig. 1). Podo is classified as tropical softwood other than Damar Minyak (*Agathis borneensis*) and Sempilor ((Dacrydium spp; Phyllocladius spp). Podo is selected as the filtered species based on its gymnosperms for softwood trees that have short tracheids to force water to flow through pit membranes even for small thicknesses (<1 cm) of xylem tissue (Boutilier et al., 2014).

The flow of sap in plants are primarily transpired through the cells of xylem, which encompass sapwood and heartwood portion of the stem. The Podocarpus spp is a gymnosperm (softwood) in which the xylem conduits are formed from single dead cells known as tracheids. Angiosperms (hardwood) conduit system, on the

Fig. 1 Branches of Podo

other hand, is known as a vessel element, which is cells of larger diameter arranged in a single file. These cells are arranged in parallel configuration and connected to each element via an opening known as pits. The pits are built with membranes with nanoscale pores with a specific function for filtration. Within the context of filtration, the gymnosperms are proven to be much better than angiosperms wood. This is due to angiosperms that have larger vessel elements which contributes to its efficiency at conducting sap, thus lowering the effect of filtration. The conduit system of gymnosperms, in contrast, is made entirely of tracheid cells which are smaller in diameter and shorter in length.

The built structure of tracheid will force water to flow through the cells via pit membranes. Higher resistance to sap flow is crucial for filtration, as the pit membrane may contribute to as much as 30–80% of resistance to sap flow. It can be concluded that the xylem tissue of softwood trees is suitable for the construction of a water filtration device, for the filtration of bacteria, protozoa and other pathogens on the micron or larger scale (Boutilier et al., 2014).

The branches samples were immediately immersed in water after cutting for the purposes of controlling their moisture content. The Podo's sapwood and heartwood colour is not too much different which is light yellow with light golden brown or greenish tinge. Its texture is very fine and even, without a significantly seen growth ring and separated with a thick layer of tracheids. The xylem of Podo was cut and used as the filtered material. The filtered water was observed on their physical and chemical parameters, which encountered colour, turbidity, pH, biochemical oxygen demand and chemical oxygen demand. The storm-water for the filtration test was collected from an open retention pond located at Shah Alam, Selangor, Malaysia.

2.2 *Wood-Xylem Filter Preparations*

The Podo branches consist of xylem and bark (Fig. 2) and were cut into 20 mm diameter with 40 mm length (Fig. 3a). The bark and cambium were peeled off (Fig. 3b) and each piece was inserted at the bottleneck and sealed (Fig. 3c). Collected water samples were filtered naturally and left to pass through the wood-xylem without extra energy and force. The water was filtered until the amount of water to be tested was enough for the physical and chemical tests.

2.3 *Imaging*

Xylem structure was visualized using Dino-Lite Digital Microscope. The digital Microscope able to inspect sample up to 500 × magnification with 5-megapixel resolutions. Two types of xylem samples were prepared. The first sample was visualized before the filtration process. The second sample was visualized after the filtration process. The samples were visualized at three different locations which are at the

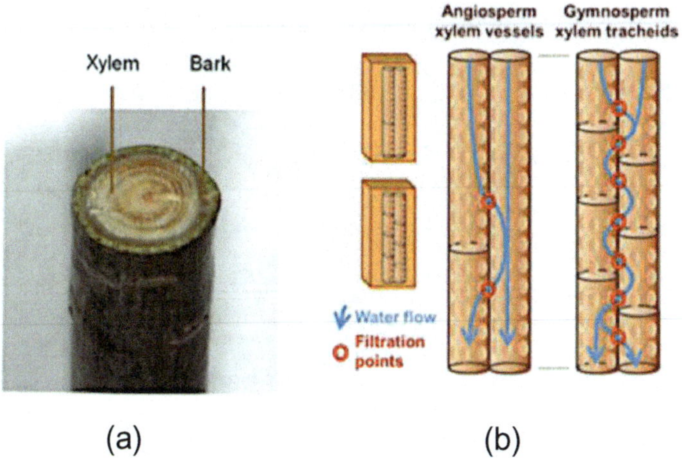

Fig. 2 **a** Xylem and bark and **b** structure of xylem vessels and tracheids in conifers (Boutilier et al., 2014)

outlet, inlet and radial surface. The sample was cut into two pieces for the radial surface.

2.4 *Oil Absorption Behaviour Study*

This study also ascertained the mopping sorb behaviour of oil absorption between Kapok fibre, Kapok ash wood and the combination of Kapok fibre and Kapok ash wood when the material was contacted with the waste cooking oil. Picture of these three materials has been taken so that the oil absorption behaviour can be visualized. Each material—Kapok fibre, Kapok ash wood and a combination of Kapok fibre and ash wood—was weighed according to the required weight which is 10, 20 and 30 g. The specimen is added into the beaker and left for a required contact time as stated above. The contact time used in this visualization study was for every minute and for first 10, 20, 40 and 60 min.

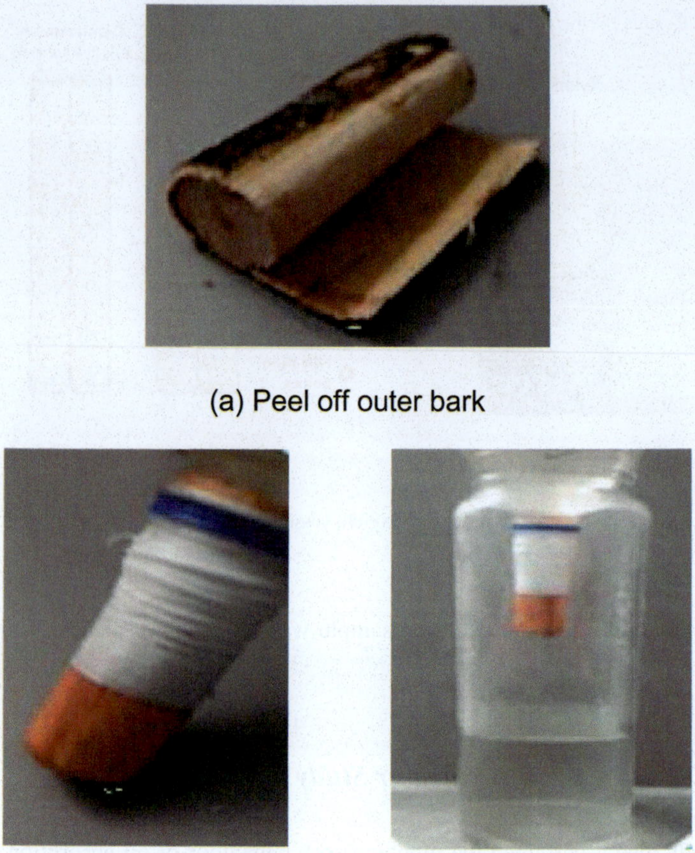

(a) Peel off outer bark

(b) Wood Sealed (c) Filtering in progress

Fig. 3 The process of filtering preparation

3 Results and Discussions

3.1 Water Quality

As the water was allowed to drip naturally and without any applied pressure, the sample water from the wood-xylem process was collected at a very slow dripping. Only about 10 ml of filtered water was collected within a day. The findings of the physical and chemical quality of the water sample before and after being filtered through Podo xylem of 20 mm diameter and 40 mm length is as shown in Table 1.

Table 1 Results for the physical and chemical quality of water sample before and after filtration

Type of test	This study				Water quality standard for drinking water (MoHM)
	Before filtration		Water filtration		
			40 mm length of xylem		
Physical test					
Turbidity (NTU)	14.2		13.2		5
Colour (ptCO)	201		376		15
Chemical test					
Chemical oxygen demand (COD)	51		168		0
Dissolved oxygen (DO)	DO_0	DO_5	DO_0	DO_5	0
	7.36	5.89	6.84	0.74	
pH Value	7.03		6.82		6.5–9.0
Biological oxygen demand (BOD)	8.80		36.60		0

3.2 *Physical Characteristics*

Table 1 and Fig. 4 show the water quality for physical characteristics observation contributes to only small improvement after the filtration test. The turbidity Nephelometric Turbidity Unit (NTU) has a 7% decrease from the original water samples, yet it is 154% higher than the allowable Drinking Water Quality Standard, as specified by Ministry of Health Malaysia (MoHM). The wood-xylem naturally contained deposit materials. However, the amounts of deposit materials are depending on the species of wood (Muhammed, 2006). There was a high possibility that the turbidity of water increases due to the presence of scattered particles during the filtration process (Purcell, 2003). Hence, double filtration is highly recommended. The purpose of the first filtration process is to flush out the natural existing deposit materials inside the wood-xylem. The second filtration process is to filter the water sample.

The result of colour after filtration has contributed to 87% higher ptCO. Consequently, it shows that the filtration process fails to provide better colour performance to the tested water. From the observations, the values obtained from colour tests are increasing with respect to the length of wood-xylem. There are two types of colours which are apparent colour and true colour. Apparent colour was formed because of the suspended solid and the true colour of the water was caused by the dissolved mineral from the plant (Purcell, 2003). Photosynthesis is a process of converting light to chemical energy. Chemical energy is produced in the form of sugar that contains oxygen, hydrogen and carbon. In addition, it also has an addition of mineral nutrients obtained from soil (Ridge, 2002). Hence, it is expected that the colour of the water samples was increased because of the presence of dissolved minerals in the wood-xylem.

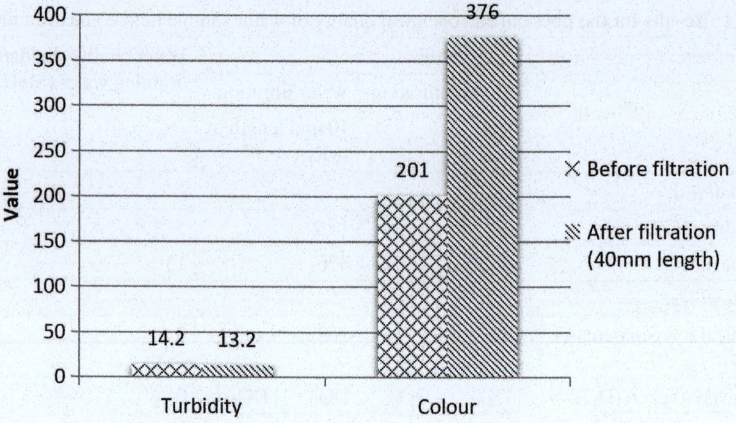

Fig. 4 Results of physical test (turbidity and colour)

3.3 Chemical Quality Tests

The chemical characteristics were based on Chemical Oxygen Demand (COD), Dissolved Oxygen (DO), pH Value and Biological Oxygen Demand (BOD) test. Result of chemical characteristic of COD is based on Fig. 5. It shows that the values of COD were triple increases (225%) after it was filtered with 40 mm length of xylem. The probability of the increased amount of COD after the filtration would be due to the presence of micro-organic increase during the filtration process since it was done in the laboratory under ambient temperature. It shows that the chemical oxygen test cannot achieve the prescribed Drinking Water Quality Standard requirement.

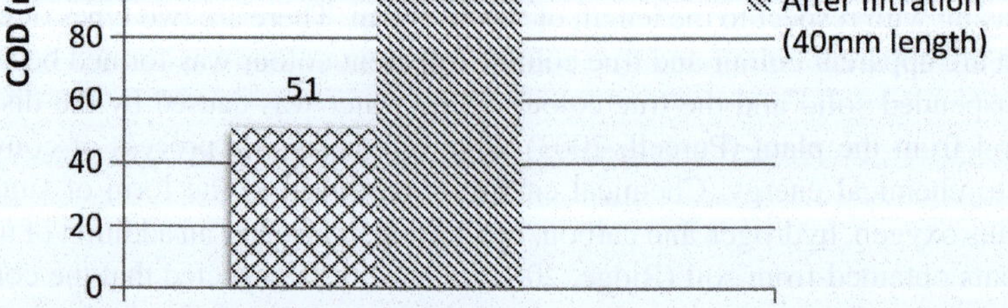

Fig. 5 Result for chemical oxygen demand test

Results of pH are presented in Fig. 6. It shows that the pH value decreased after the filtering process and turned from alkali to acidic. The pH value is classified as Class I for both after and before filtration process and it is in the range of pH as prescribed in Drinking Water Quality Standard requirement.

Figure 7 shows that the values of dissolved oxygen were reduced after filtration. The amounts of DO reduced represent the amounts of dissolved oxygen consumed by organisms. The initial value of DO for the unfiltered water was 7.36 mg/L reduced to 5.89 mg/L in five days. The organisms consumed 1.47 mg/L of dissolved oxygen.

As for the filtered water, the organisms consume 6.1 mg/L respectively. The values of dissolved oxygen that have been reduced for the filtered water were higher compared to unfiltered water. It proved that the value of organisms was increasing

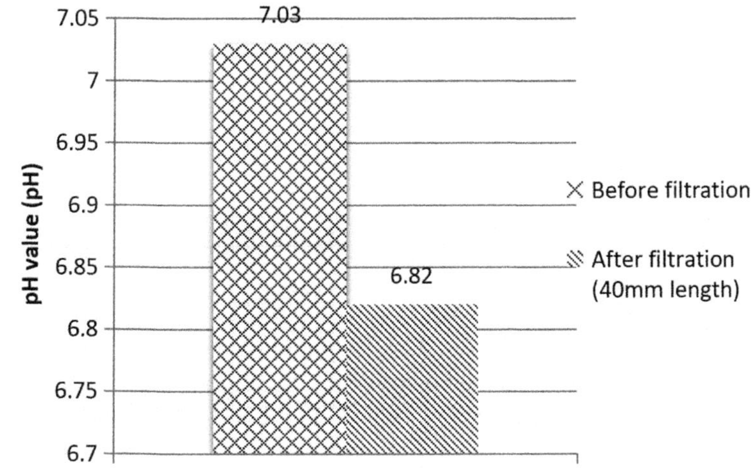

Fig. 6 Result of pH test

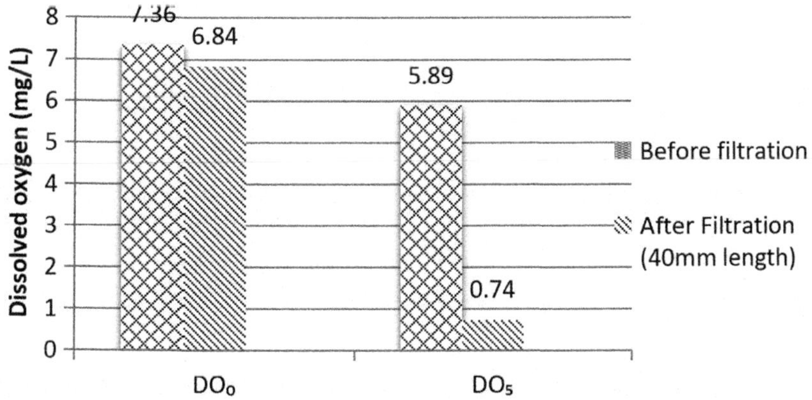

Fig. 7 Result of dissolved oxygen test

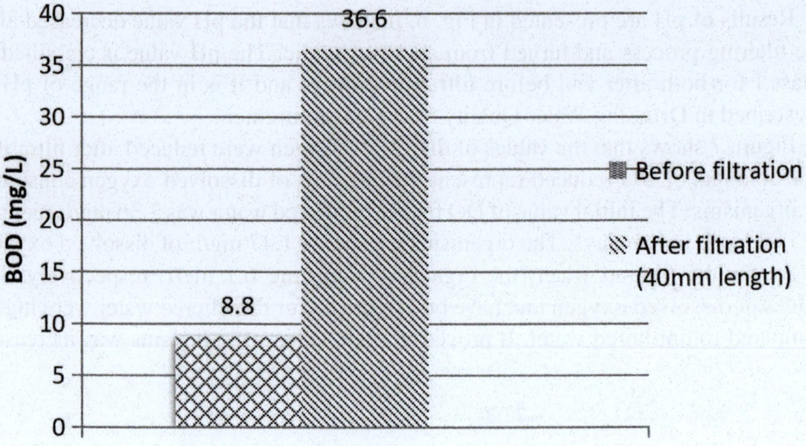

Fig. 8 Results of biochemical oxygen demand

during the filtration process. Overall, the dissolved oxygen test cannot achieve the Drinking Water Quality Standard (MoHM).

Figure 8 shows that the value of the BOD was increased from 8.8 mg/L to 36.6 mg/L when filtered. The biochemical demand value obtained represents the amount of oxygen needed by micro-organism in order to degrade any organic matter present in water biochemically. Based on the results, the amounts of oxygen needed for the degradation process were increasing after the filtration process. The higher the amount of organic matters, the higher the value of oxygen needed. The bacteria or fungi are normally attracted to the wet material. Since the filtration system is carried out at the room ambient temperature without any specific control environment, the growth of bacteria and fungi can occur in the filter material and in a sample of water that has been filtered. Hence, the biochemical oxygen demand test did not achieve the requirement of Drinking Water Quality Standard (MoHM).

3.4 *Observation and Imaging of Wood-Xylem*

The main purpose of imaging is to visualize the Podo wood-xylem cross-section. It is to observe the different performances of Podo xylem before and after filtration process. Figure 9a shows the radial surface of the xylem. The darker line shown in the figure is known as tracheid. Figure 9b shows the radial surface and cross-section surface. The dark brownish spot shown in the figure is known as a tracheid, zoomed up to 500× magnification with 5-megapixel resolutions.

Figure 10a–c shows the xylem of Podo before the filtering process for inlet cross-section, outlet cross-section and radial surface. The figure shows that the tracheid of the sample was clean from any dirt and dark spot.

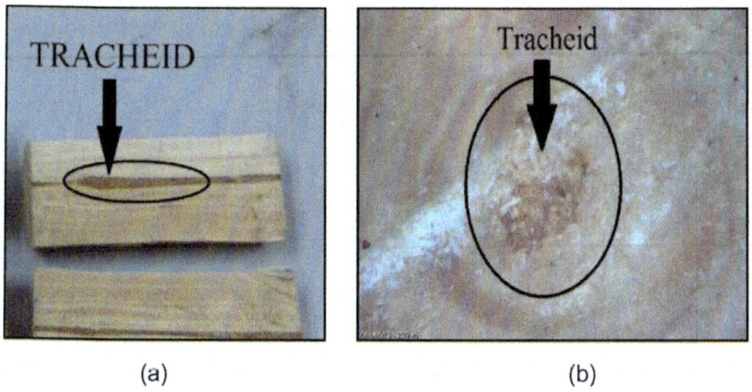

Fig. 9 Podo wood-xylem **a** radial surface **b** cross-section surface

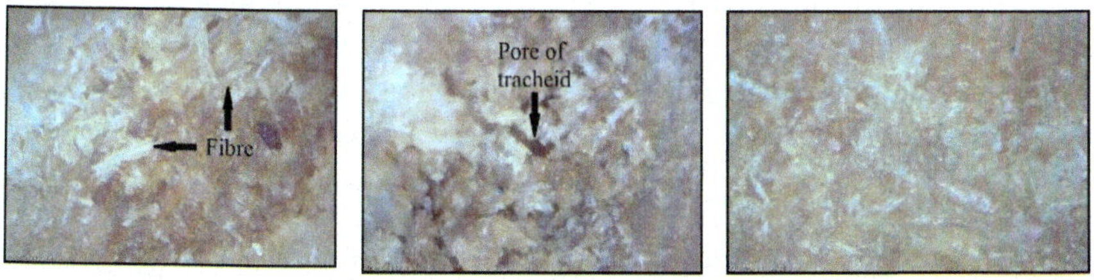

Fig. 10 Podo xylem before the filtering process. **a** Inlet cross-section surface. **b** Outlet cross-section surface. **c** Radial surface

After filtration, the dark spot at the inlet of the tracheid can be seen clearly from the cross-section view. Figure 11a–c shows the images of the xylem after the filtration process. The image of the cross-section for the inlet shows that the condition was clearly darker compared to the outlet condition (Fig. 11c). The dark spot formed at the xylem of Podo sample was the suspended material filtered from the water sample.

Throughout the test observation, the future flow through the xylem of different wood species needs to be studied to identify other locally available sources of xylem, which will truly enable the construction of filters. In this present study, only results using xylem derived from one species is reported. Further exploration of research

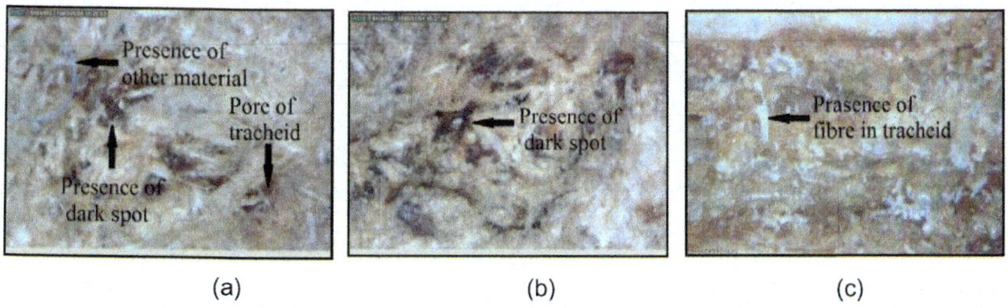

Fig. 11 Podo xylem after filtration process. **a** Inlet cross-section surface. **b** Outlet cross-section surface. **c** Radial surface

should be encountered to filter viruses using smaller nano-particle sizes of observation and its capability of filtering bacteria or whether conifer xylem can be impregnated with particles such as carbon black to improve rejection of viruses as suggested by Boutilier et al. (2014).

4 Conclusions

In conclusion, the result of physical and chemical characteristics of storm water from retention pond filtered using wood-xylem produced from Podo timber species are presented. The turbidity (NTU) has a 7% decrease from the original water samples, yet it is 154% higher than the allowable Drinking Water Quality Standard. The result of colour after filtration has contributed to 87% higher ptCO. Consequently, it shows that the filtration process fails to provide better physical performance of the tested water. Results of the chemical test show that the values of COD were triple increases (225%) after it was filtered through the wood-xylem, yet it reduced the pH of the filtered water from 7.03 to 6.82 accordingly. The values of DO were found reduced after filtration; however, the results of BOD were increased from 8.8 mg/L to 36.6 mg/L respectively. Consequently, imaging observation shows the presence of other materials and dark spots on the surface of the inlet and outlet of the wood-xylem cross-section after the test. It can be concluded that the filtered water from Podo wood-xylem did not match the characteristics of the Drinking Water Quality Standard (MoHM) requirement, accordingly.

Acknowledgements The author wishes to express the greatest appreciation to Universiti Teknologi Mara Malaysia and Ministry of Higher Education (MOHE) Malaysia for the financial support contributed through KPM; FRGS Grant: 600-RMI/FRGS 5/3 (181/2019).

References

Alhendawi, R. A., Kirkby, E. A., & Pilbeam, D. J. (2005). Evidence that sulfur deficiency enhances molybdenum transport in xylem sap of tomato plants. *Journal of Plant Nutrition, 28*(8), 1347–1353. https://doi.org/10.1081/PLN-200067449

Boutilier, M. S. H., Lee, J., Chambers, Venkatesh, V., & Karnik, R. (2014). Water filtration using plant xylem. department of mechanical engineering. *PLoS ONE*, *9*(2), e89934.

Boving, T. B., & Kevin Neary, K. (2006). Attenuation of polycyclic aromatic hydrocarbons from urban stormwater runoff by wood filters. *Journal of Contaminant Hydrology, 91*(2007), 43–57.

Boving, T. B., & Zhang, W. (2004). Removal of aqueous-phase polynuclear aromatic hydrocarbons using aspen wood fibers. *Chemosphere, 54*(2004), 831–839.

Drinking Water Quality Standard, Engineering Services Division, Ministry of Health Malaysia.

Estabragh, A. R., Soltani, A., & Javadi, A. A. (2018). Effect of pore water chemistry on the behaviour of a kaolin–bentonite mixture during drying and wetting cycles. *European Journal of Environmental and Civil Engineering*. https://doi.org/10.1080/19648189.2018.1428691

Grant, M., & Cutler, D. M. (2005). *Water, water everywhere*. Municipal Finance and Water Supply in America Cities.

Jang, G., & Choi, Y. D. (2018). Drought stress promotes xylem differentiation by modulating the interaction between cytokinin and jasmonic acid. *Plant Signaling & Behavior*, *13*, 3. https://doi.org/10.1080/15592324.2018.1451707

Muhammed, S. (2006). *Anatomi Kayu*. Pusat Penerbitan University (UPENA).

Purcell, P. (2003). *Design of water resources systems*. Thomas Telford Publishing.

Rao, K. S., Rajput, K. S., & Srinivas, T. (2004). Seasonal cambial anatomy and development of Xylem in Dalbergia sissoo growing under the influence of combined air pollutants. *Journal of Sustainable Forestry*, *18*(1), 73–88. https://doi.org/10.1300/J091v18n01_05

Ridge, I. (2002). *Plants*. Oxford University Press Inc.

US DA. (2003). National environmental policy act determination needed for fire management activities. *Federal Register*, *68*(108), 33814–33826.

World Health Organization. (1993–1998). *Guidelines for drinking water quality* (vols. 1–3). WHO.

Zainudin, Z. (2010). Benchmarking river water quality in Malaysia. Institut Jurutera Malaysia (IEM), Jurutera, 12–15. Retrieved: http://irep.iium.edu.my/2954/1/Feature-BenchmarkingRiverWater3pp.pdf

Effect of Tunnel Form Building Under 10 Past Earthquake Records Analyzed Using Ruaumoko 2D

A. Shamilah, A. S. Aweis, R. Che Amat, and N. Hamid

Abstract Countries that experience regular earthquakes with magnitudes greater than 7 on the Richter Scale are currently recognized to suffer significant destruction. These unpredicted calamities can have detrimental effects on both people and structures. However, by identifying the threat that might be damaging the current structures, the effect of seismic can be lessened. As a result, the effectiveness of tunnel form building (TFB) has been evaluated under seismic events of varying intensities. To establish the safety level of a tunnel form building, ten numbers from previous ground vibrations are chosen. Six previous earthquake records were chosen from East and West Malaysia, while four previous earthquake records were chosen from nations outside of Malaysia. The maximum lateral displacement and lateral loading capacity of the prototype building have been modeled and studied to determine the structural deformation of the TFB at peak ground acceleration for a few historical earthquake histories. It may be concluded that while this sort of structure survives local earthquakes, mass earthquake that originates in other countries than Malaysia causes severe damage or even cause the building to collapse.

Keywords Destruction · Earthquake · Deformation · Survive and collapse

1 Introduction

An industrialized construction method known as tunnel form building (TFB) uses steel forms with precise dimensions and smooth surfaces to cast the buildings differing from traditional reinforced concrete structures with no beams or columns employed. Both the Izmit earthquake in 1999 and the Bingol earthquake in 2003 recorded minimal non-structural failure of tunnel-shape structures (Ghorpade &

A. Shamilah (✉) · A. S. Aweis · R. Che Amat · N. Hamid
Faculty of Civil Engineering & Technology, Universiti Malaysia Perlis, Arau, Malaysia
e-mail: norha454@salam.uitm.edu.my; shamilah@unimap.edu.my

R. Che Amat
e-mail: roshazita@unimap.edu.my

U. R. Hashim et al. (eds.), *Green Infrastructure*,
https://doi.org/10.1007/978-981-99-7003-2_24

Swamy, 2018). While this was going on, comparable damage patterns to tunnel form buildings were seen in the 7.4 magnitude of Kocaeli Earthquake and the 7.1 magnitude of Duzce Earthquake in 1992. Additionally, similar behavior appears to have occurred in Romania due to earthquakes in 1977, 1986 and 1990 (Hamid et al., 2013).

Tunnel form building (TFB) has been used in multi-story residential buildings since 1960, including hotels, apartments and condominiums. Most of these buildings have been developed without seismic provision. Therefore, if an earthquake with a Richter magnitude larger than 5 were to strike Malaysia in the future, the country's current structures would no longer be protected from the resulting tragedy. Thus, the behavior of TFB needs to be examined. This study used historical earthquake records to analyze and forecast how TFB will behave in Malaysia during unpredictable earthquakes.

2 Methodology

Figure 1 shows the flowchart of modeling and analysis of tunnel form building using HYSTERES and Ruaumoko 2D program. The steps involved are selection of the existing tunnel form building in Malaysia using non-seismic code of practice, development of the hysteresis loops model using HYSTERES program, creation of data input, running the Ruaumoko 2D program under Mexico City, El Centro NS, Pacoima Dam, El Centro EW, Tawau, Sebuyau, Kunak, Batu Niah, Kundasang and Bukit Tinggi Earthquake past records. Then analyzed all the output obtained from Ruaumoko 2D using Dynaplot. Based on the output finding in the analysis, plot the lateral capacity level for TFB and develop the pseudo-spectral displacement and acceleration using Excel subjected to ten past earthquake records.

The hysteresis loops for each node in the prototype building were also created using the Dynaplot analysis. Plots of the pseudo-spectral displacement and pseudo-spectral acceleration for 10 chosen earthquake data were made for damping levels of 0, 2, 5, 10 and 20%. To evaluate the structural deformation of the TFB at peak ground acceleration under the chosen earthquake records, the maximum lateral load for pushing and pulling directions was obtained using the Dynaplot analysis. The maximum lateral load for each earthquake data was compared with the lateral capacity of tunnel form building to measure the safety level of TFB subjected to each selected earthquake data. PSA and PSD graphs were replotted using an excel spreadsheet to compare the acceleration and displacement for each earthquake data utilized in the analysis.

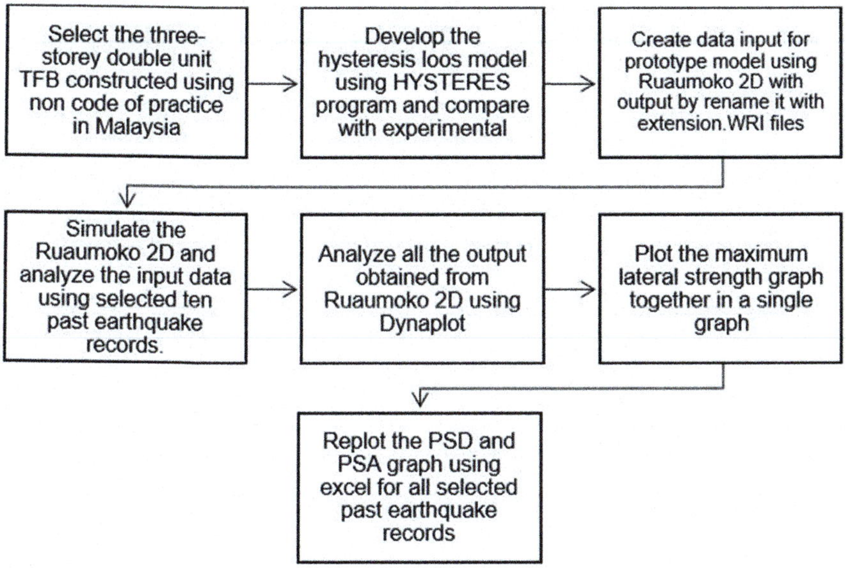

Fig. 1 Flowchart of modeling and analysis using Ruaumoko 2D program

3 Analysis of Result

3.1 Hysteresis Loops

Hysteres Based on the data gathered from the experimental study, the hysteresis loop was modeled using the hysteres program. Because it displayed behavior that was similar pattern to the experimental hysteresis loops, Pampanin Hysteresis Loops with reloading slip factor was chosen. The comparison of the experimental data with the hysteresis loops is shown in Fig. 2. The experimental outcome is shown by the blue color, whereas the modeling result is represented by the red color. Hysteresis loop modeling and experimentation appear to follow a similar trend. The plotted hysteresis loops showed that the experimental drift had a similar symmetrical structure, with a percentage difference of 1.2% reached at an ultimate drift of 1.75%. The difference in lateral displacement is measured at 0.165. According to testing data, the maximal lateral load and displacement are 6.09 and 52.33 mm, respectively. Therefore, the different stories of a TFB can be examined using this approach.

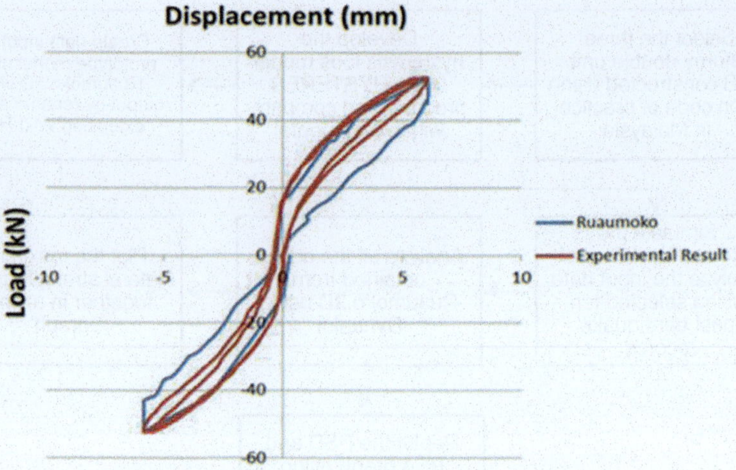

Fig. 2 Comparison of hysteresis loops between experimental and modeling

3.2 Lateral Capacity of Single Unit TFB Subjected to Multilevel of Seismic Events

According to Fig. 3, an investigation was conducted on TFB's resilience subjected to 10 previous earthquakes. When there is a local earthquake, tunnel form building can sustain its safety, but when there is a significant earthquake that originates from another country, it is severely damaged or even destroyed. Local seismic occurrences below the Richter scale of 5 are categorized as moderate, light, minor and very minor known as Tawau, Sebuyau, Kunak, Batu Niah, Kundasang and Bukit Tinggi earthquakes. Additionally, there were four more earthquakes with Richter magnitudes of greater than 5, known as Mexico, El Centro and Pacoima Dam earthquakes. The data supporting this discovery can be connected to the recent actual earthquake activities in Malaysia. The impact of several high-rise structures in Malaysia constructed using non-code of practice due to 6.2 magnitude of Bukit Tinggi, Sumatra earthquake on 25th February 2022 can be felt in Perak, Melaka, Negeri Sembilan, Johor and Selangor (Fatt, 2022).

3.3 Pseudo-Spectral Displacement

Pseudo-Spectral Displacement (PSD) is the maximum displacement due to earthquake to the moving structures. Figure 4 indicates the graphs of spectral displacement versus time graph for each of the ground motion records with 0, 2, 5, 10 and 20% damping ratio. The plotted graphs indicated the increment of percentage damping as the spectral displacement decreases. The maximum lateral displacement was 4.64 m

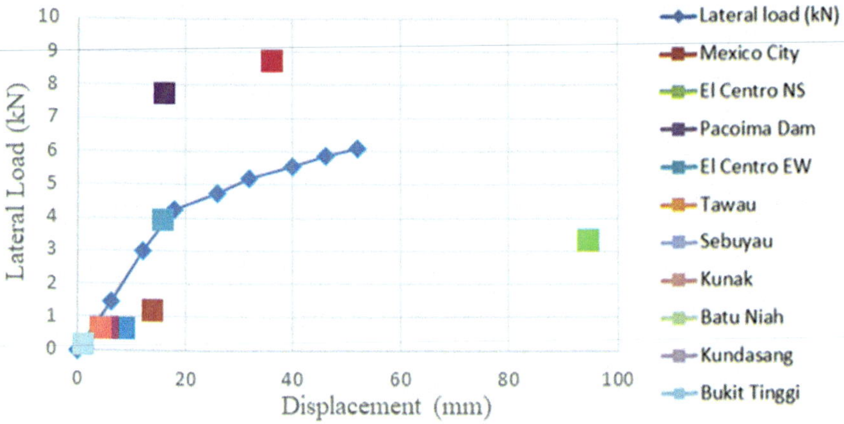

Fig. 3 The survivability level of TFB subjected to ten selected past earthquake records

obtained from Mexico Earthquake. The analysis data obtained from Ruaumoko reveal that the spectral displacement of the Mexico Earthquake, El Centro North South Earthquake, El Centro East West Earthquake and Pacoima Dam Earthquake have values of lateral displacement greater than 0.052 m and the prototype building is not stable under these earthquakes. Conversely, Tawau, Sebuyau, Kunak, Batu Niah, Kundasang and Bukit Tinggi earthquakes experienced lateral displacement below 0.015 m and seem to have survived under these earthquake records. Therefore, it can be concluded that the prototype building experience minor cracks and remains functional under local earthquake past records as compared to the earthquake records greater than 5 on Richter scale.

High displacement absorption of buildings during earthquake events may cause severe damage such happened in Ranau, Sabah. 2015 Ranau Earthquake caused severe damage to schools, quarters, hostels and rest houses due to the utilization of non-seismic code in the designing during construction phase of these buildings (Fatt, 2022).

Meanwhile, 60.4% (from 250 selected) of buildings in Ranau are classified as unsafe to be occupied due to active seismic activities in this region (Ghorpade & Swamy, 2018). Precisely, a survey study has been conducted in similar state (Sabah) indicated 10% roof damage, crack on wall (68%), floor (48%), column (23%) and 2% of houses were totally collapsed. These findings have increased the intention awareness of repair and strengthening approach among affected occupants in Sabah area (Hamid et al., 2013).

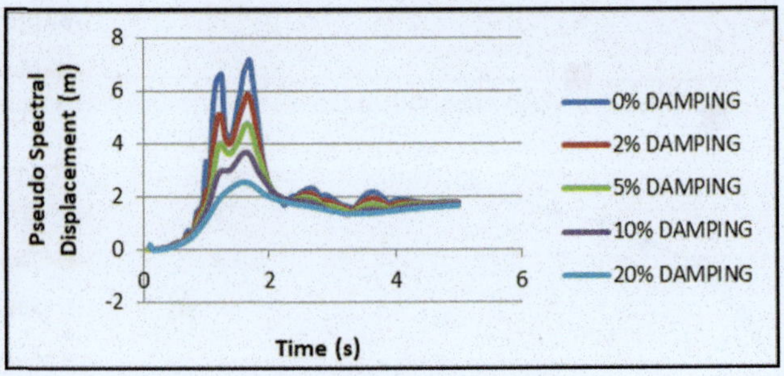

(a) MEXSCTIL.EQC

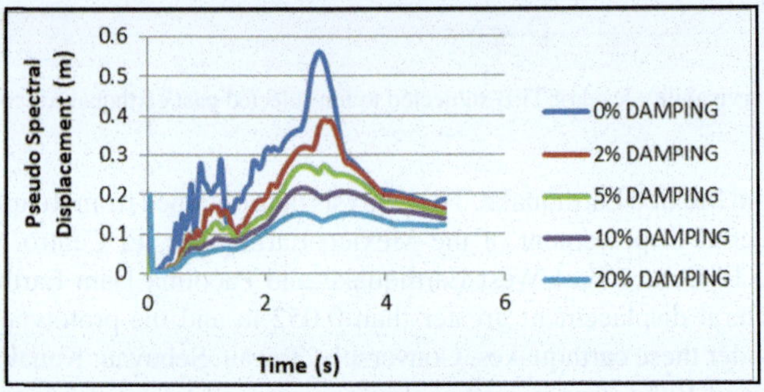

(b) EL40NSC.EQB

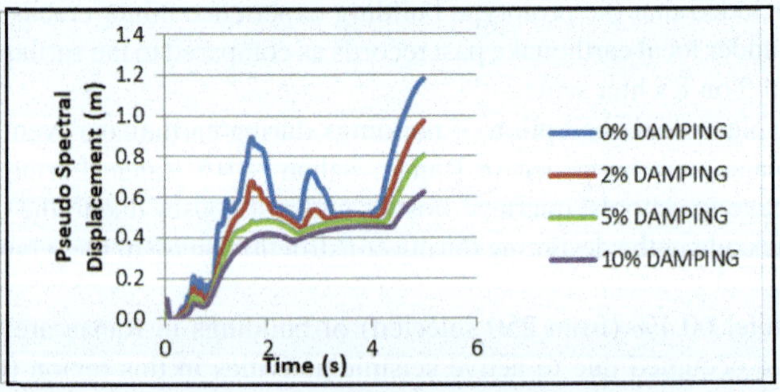

(c) PACMSW.EQB

Fig. 4 Profile of spectral displacement (PSD) for ten ground motion records under 5 different percentage damping coefficients

3.4 *Pseudo-Spectral Acceleration*

The pseudo-spectral acceleration of the prototype model is typically lower than the peak ground acceleration (PGA) of the earthquake. Building acceleration, velocity and lateral movement are all impacted by the damping coefficient ratio. In terms

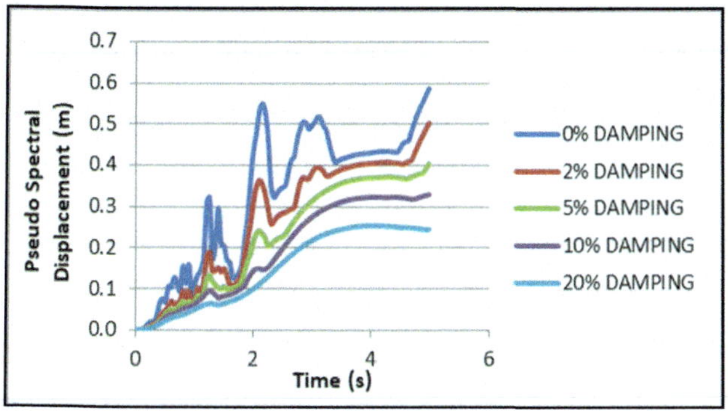

(d) EL40EWC.EQB

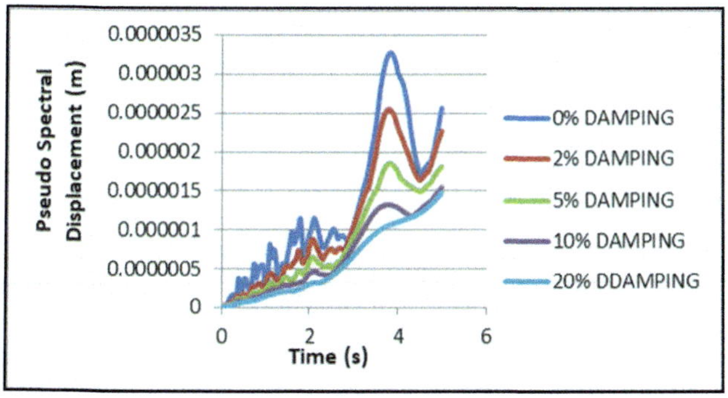

(e) TAWAU.EQF

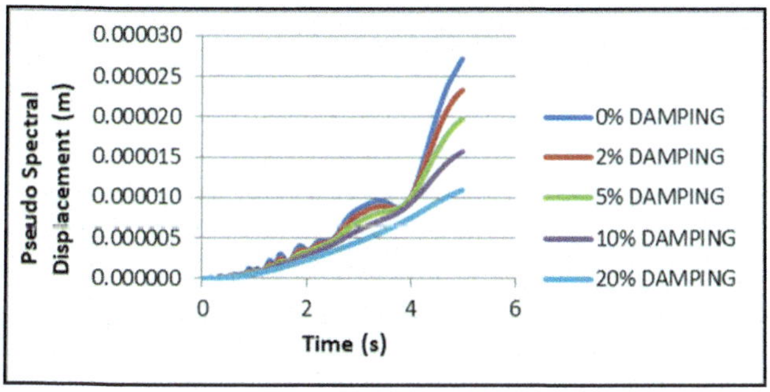

(f) SEBUYAU.EQF

Fig. 4 (continued)

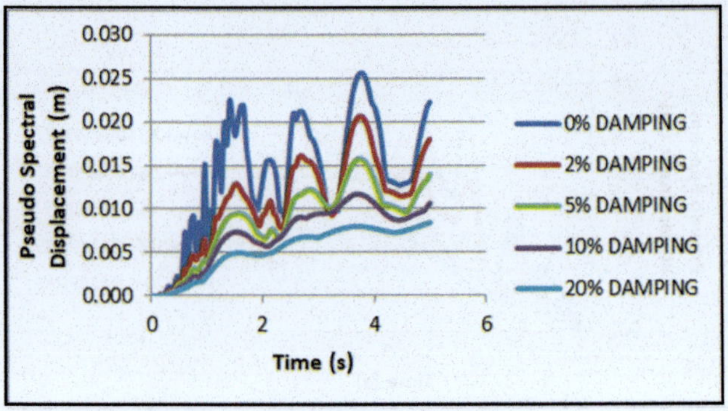

(g) KUNAK.EQF

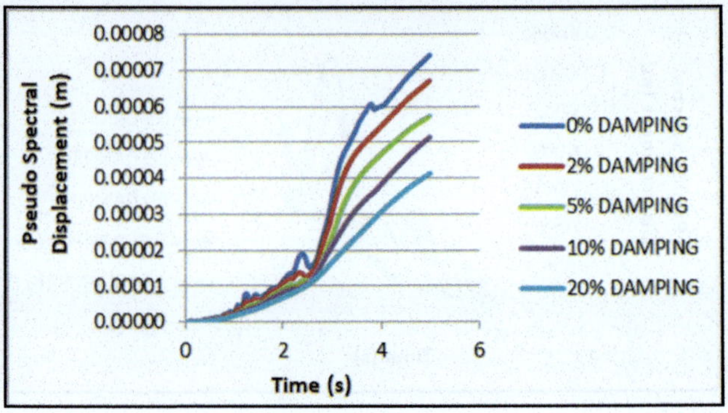

(h) BATUNIAH.EQF

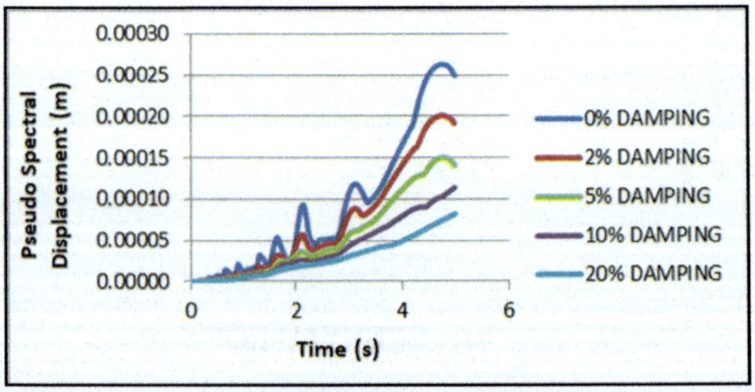

(i) KUNDASANG.EQF

Fig. 4 (continued)

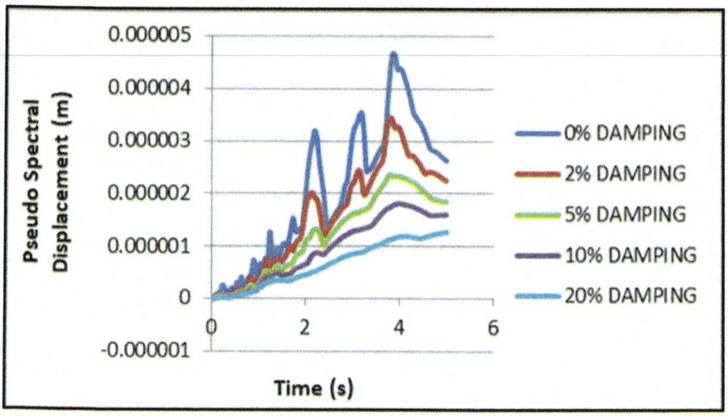

(j) BUKITTINGGI.EQF

Fig. 4 (continued)

of multiple degrees of freedom, this can be represented and related to Newton's Second Law of Motion (MDOF). The inertia force brought on by the structures' acceleration, the damping force and the stiffness force make up the total lateral force acting on the prototype model. The spectral acceleration and damping ratio are inversely correlated. Pseudo-Spectral acceleration (PSA) was plotted for 0, 2, 5, 10 and 20% damping ratio using output file from Dynaplot program. Figure 5 indicates the graph of Pseudo-Spectral Acceleration versus time for each selected earthquake.

The plotted graphs indicate the increment of percentage damping as the spectral displacement decreased. The analysis data obtained from Ruaumoko revealed that the spectral acceleration of Mexico, El Centro North South, El Centro East West and Pacoima Dam earthquakes gave the overall lateral displacement greater than 0.052 m causing the prototype building to not survive under these earthquakes. Conversely, Tawau, Sebuyau, Kunak, Batu Niah, Kundasang and Bukit Tinggi earthquakes experienced the lateral displacement of 0.00000142 to 0.03 m and seem to have survived under these earthquake records. This is because buildings with the greatest amount of damping energy can move less than those with the least amount of damping energy. Therefore, it is advised that the prototype building be fitted with energy dissipators to lessen acceleration while also minimizing structural damage. Larger energy dampers need to be installed in the building for major earthquakes such as Mexico Earthquake, El Centro North South Earthquake, El Centro East West Earthquake and Pacoima Dam Earthquake to reduce illusory acceleration.

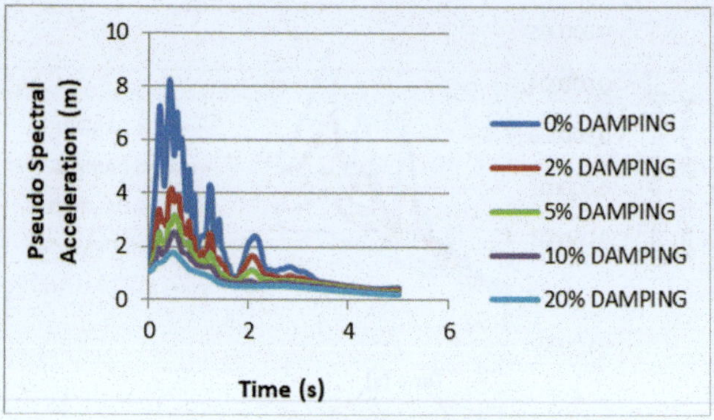

(a) MEXSCTIL.EQC

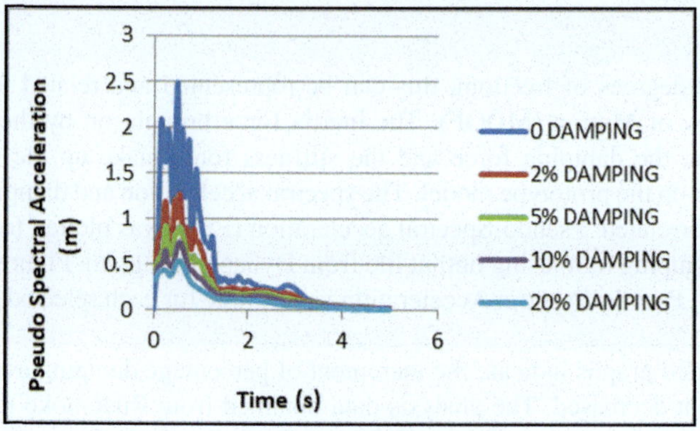

(b) EL40NSC.EQB

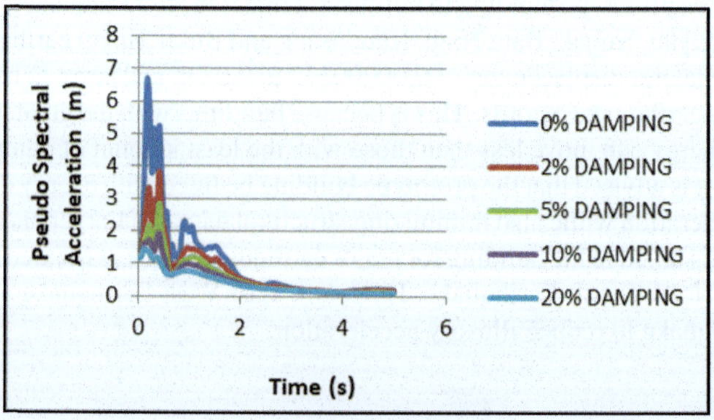

(c) PACMSW.EQB

Fig. 5 Profile of spectral acceleration (PSA) for ten ground motion records under 5 different percentage damping coefficients

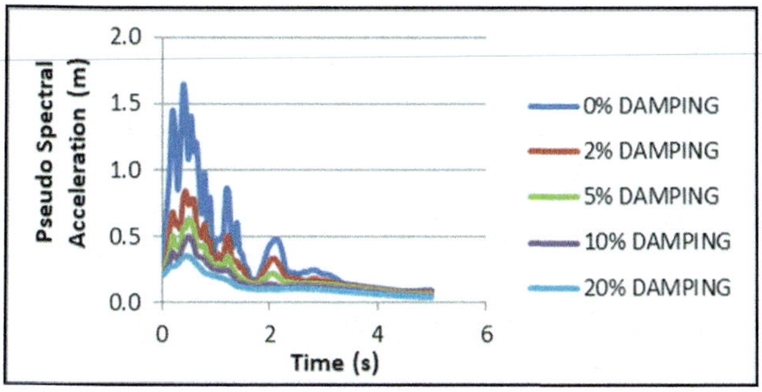

(d) EL40EWC.EQB

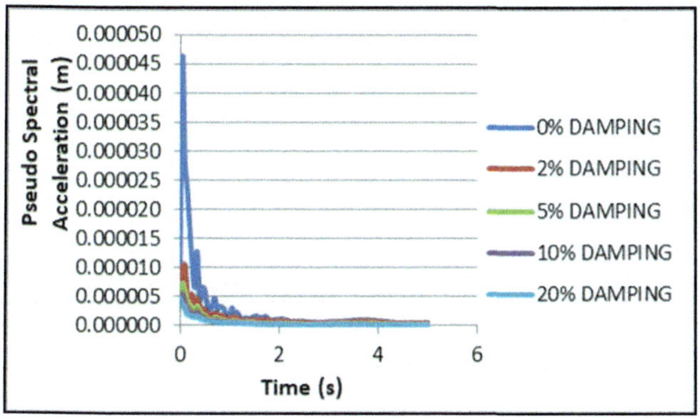

(e) TAWAU.EQF

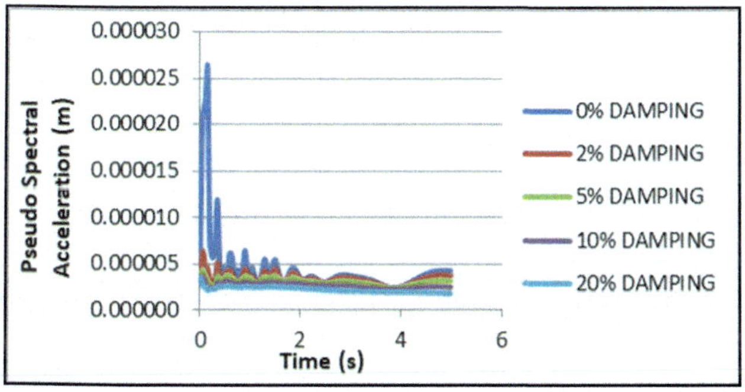

(f) SEBUYAU.EQF

Fig. 5 (continued)

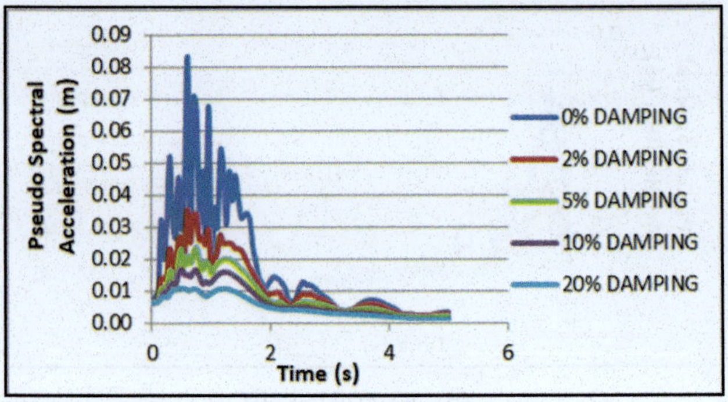

(g) KUNAK.EQF

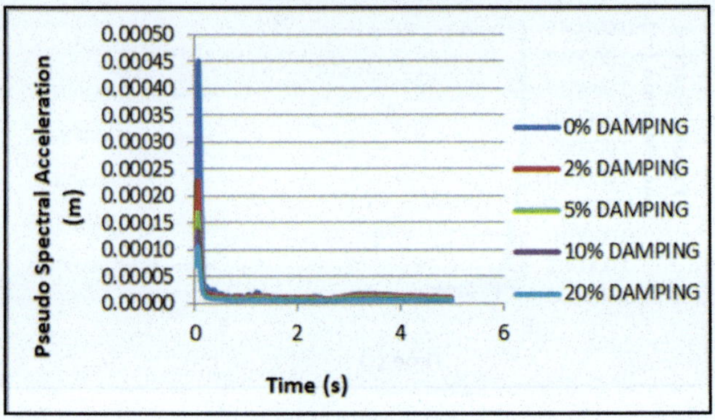

(h) BATUNIAH.EQF

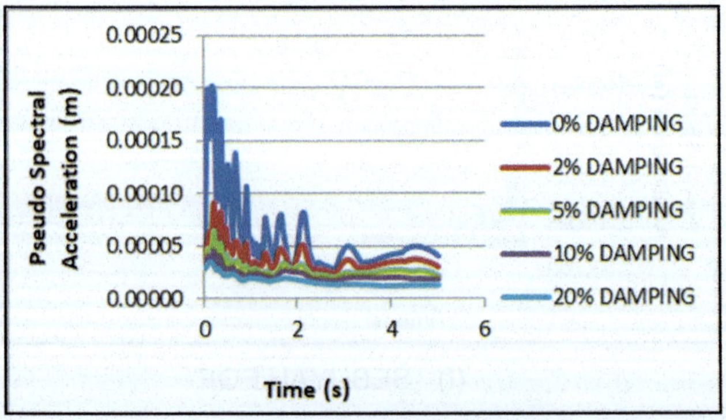

(i) KUNDASANG.EQF

Fig. 5 (continued)

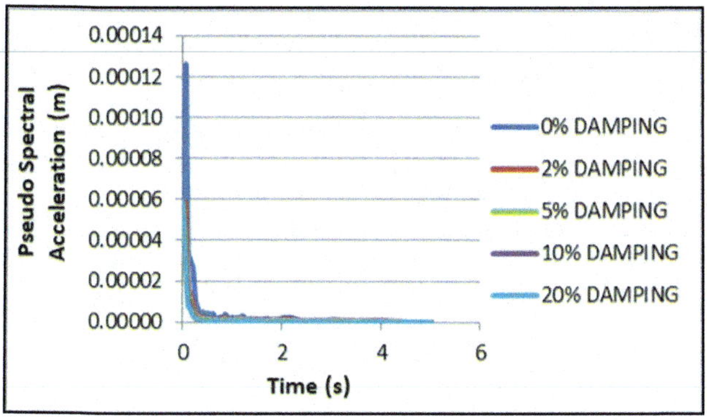

(j) BUKITTINGGI.EQF

Fig. 5 (continued)

4 Conclusion

By using the Ruaumoko 2D program and historical earthquake records, it is possible to determine the survivability of tunnel form buildings under multilayer seismic occurrences. To explore the dynamic behavior of tunnel form buildings (TFB) built in Malaysia without seismic provision, all the input data from the modeling analysis was carried out using Ruaumoko 2D under ten real earthquake records. The chosen ten historical ground motion data were used to represent the large, major, strong, moderate, light, small and very little earthquake events that took place globally as well as in Malaysia. The tunnel form building's actual seismic capability and dynamic behavior were determined by applying these genuine earthquakes to it.

The prototype specimen was subjected to ten historical earthquake recordings, and the yielding point of the specimen was found via dynamic plot analysis. Based on the dynamic study, it appears that this specimen may withstand small local earthquakes of less than Richter magnitude 5. However, this kind of structure was not secure against earthquakes that occurred outside of Malaysia and had Richter scale magnitudes ranging from 6 to 8. If Malaysia has an unexpected seismic event in the future, it is imperative to look into TFB's capacity for planning. The non-seismic code design in Malaysia will cause significant damage to the current building or, in the worst event, cause it to collapse entirely. Thus, the necessity for substantial construction, retrofits and maintenance results in higher operational expenses.

Acknowledgements The authors would like to acknowledge the support from the Ministry of Higher Education Malaysia for funding through the Fundamental Research Grant Scheme (FRGS) under grant number RACER/2019/TK01//UNIMAP//1.

References

Fatt, I. D. J. L. W. (2022). Buildings in Malaysia need seismic design to withstand earthquakes.

Ghorpade, A., & Swamy, B. (2018). Study on the performance of flat slab structure using pushover analysis with and without shear wall. *International Research Journal of Engineering and Technology (IRJET), 5*(7), 660–665.

Hamid, N. H., Hadi, N. D., & Ghani, K. D. (2013). Retrofitting of beam-column joint using CFRP and steel plate. *International Journal of Civil and Environmental Engineering, 7*(12), 941–946.

Experimental Analysis of Seismic Responses Interior Beam-Column Joint with and Without Fuse Bars Under In-Plane Lateral Cyclic Loading

Nor Hayati Abdul Hamid, Nurfarhana Diayana Hadi, Kay Dora Ghani, Patrick L. Y. Tiong, Rini Kusumawardani, and Norisham Ibrahim

Abstract Two full-scale super assemblage of interior beam-column joints with and without fuse bars were designed, constructed, analyzed, modeled and compared their performances under in-plane lateral cyclic loading. Based on experimental results, the interior joint with fuse bars performed better than without fuse bars in terms of structural damages, ultimate lateral strength capacity, elastic stiffness, secant stiffness, ductility and equivalent viscous damping. It was observed that the interior beam-column joint with fuse bars performed better than without fuse bars in terms of visual observations such as cracks propagation, width of cracks, spalling of concrete and mode of failure. Furthermore, experimental hysteresis loops were also validated with modeled hysteresis loops using Pampanin RC beam-joint (IHYST = 44) and HYSTERES Program. The percentage difference between these two hysteresis loops lies between 2 and 20%, which is acceptable because only one

N. H. Abdul Hamid (✉)
Institute For Infrastructure Engineering and Sustainable Management (IIESM), Universiti Teknologi MARA, 40450 Shah Alam, Selangor, Malaysia
e-mail: norha454@uitm.edu.my

N. D. Hadi · N. Ibrahim
School of Civil Engineering, College of Engineering, Universiti Teknologi MARA, 40450 Shah Alam, Selangor, Malaysia
e-mail: noris024@uitm.edu.my

K. D. Ghani
Universiti Teknologi MARA, Cawangan Pulau Pinang Kampus Permatang Pauh, 13500 Permatang Pauh, Pulau Pinang, Malaysia
e-mail: kaydora@uitm.edu.my

P. L. Y. Tiong
Doshin Rubber Products (M) Sdn Bhd, Lot PT 34252, Jalan Sekolah, Off Jalan Rantau Panjang, 42100 Klang, Selangor, Malaysia
e-mail: patrick@rubberdamping.com

R. Kusumawardani
Fakultas Teknik Sipil, Universitas Negeri Semarang, Sekaran, Gunung Pati, Semarang City, Central Java 50229, Indonesia
e-mail: rini.kusumawardani@mail.unnes.ac.id

U. R. Hashim et al. (eds.), *Green Infrastructure*,
https://doi.org/10.1007/978-981-99-7003-2_25

specimen was conducted and analyzed. Therefore, it is recommended to use fuse bars for construction of the RC buildings under moderate seismic regions especially in Sabah, Malaysia.

Keywords Ductility · Equivalent viscous damping · Fuse bars · Hysteresis loops · Stiffness

1 Introduction

Malaysia is located on a relatively stable Sundaland plate which forms the southern edge of the Eurasian plate where West Malaysia has low seismicity and East Malaysia has moderate seismicity (Mohamed, 2012). During the 2004 Banda Acheh Earthquake, residents at Subang and Damansara, Selangor, Malaysia, who stayed in high-rise buildings felt the tremors with distance from epicenter to these places being approximately between 400 to 650 km (Abas, 2001; Lay, 2015; Nabilah & Balendra, 2012). This devastating event also became a major concern among researchers in Singapore and Thailand where these countries feel threatened by far-field earthquake because it is located on soft and peat soil which can cause liquefaction (Balendra & Li, 2008; Joshilkar & Patil, 2018). Another earthquake known as the 2015 Ranau Earthquake, Sabah, Malaysia, occurred on 5th June 2015 with 5.9 scale Richter had caused structural damage to schools, teachers' quarters and hospitals. These damages include diagonal cracks, spalling of concrete and buckling of reinforcement bars at beam-column joints, columns and beams (Majid et al., 2015). All of these buildings were designed using BS8110 which is non-seismic code of practice and did not cater to seismic loads.

Due to structural damages that occurred to beam-column joints during past earthquakes, one of the innovative ways to reduce these damages is to use fuse bars at the beam-column joints. A fuse bar is a green infrastructure material which uses normal high-yield reinforcement bars and can be replaced if it is damaged or buckling after the earthquake. Furthermore, the fuse bar is safe, more economical, easy to make and can be assembled at the site. The concept of designing fuse bars to dissipate seismic energy while preserving the structural integrity had been initiated by many researchers (Connor et al., 1997; El-Bahey & Bruneau, 1302; Hamid xxxx; Sullivan et al., 2888). Various types of fuse bars can be applied to different types of structural components such as wall panels, concrete frames, steel frames, beam-column joints, columns and bridge piers. Fuse bars were installed inside precast hollow core wall panels which had been tested under in-plane and biaxial lateral cyclic loading (Hamid, 2006; Hamid & Ghani, 2013; Hamid & Mander, 2006, 2010). A similar concept of fuse bars together with prestressed unbonded tendons was also adopted to determine the seismic performance of cantilever walls (Rahman & Restrepo, 2007), confined-mansory rocking walls (Toranzo et al., 2009) and hybrid coupled wall systems (Ji et al., 2016). In timber structures, fuse bars and post-tensioned were inserted inside cantilever timber walls to behave well under in-plane lateral cyclic loading (Sarti et al., 2016) and LVL walls coupled with fuse bars (Iqbal et al., 2015). Thus, concrete wall panels and timber wall panels perform very well without any structural damage at the bottom corner of both sides of the wall.

However, fuse bars as green infrastructure material are also used as energy dissipators to resist seismic loads in interior, exterior and corner beam-column joints (Hadi et al., 2022; Hamid, 2010; Hamid et al., 2021). Fuse bars also can be placed either inside or outside of beam-column joint which had been tested under bidirectional cyclic loading (Luoman et al., 2008), fuse bars can be replaced in steel structure (Saravan et al., 2018) and replaceable steel coupling beam with end-plate connections (Alireza et al., 2016). In designing beam-column joint of timber frames, fuse bars and post-tensioned tendons were used to dissipate energy and self-centering mechanism (Aqbal et al., 2008; Beerschoten et al., 2011; Smith et al., 2012). Based on excellent performance of reinforced concrete and timber beam-column joints under in-plane lateral cyclic loading, the fuse bar is fully utilized in interior beam-column joints which designed using Eurocode 8 (current seismic code of practice).

The seismic response parameters such as hysteresis loops, lateral strength capacity, stiffness, ductility and equivalent viscous damping were compared with each other. Therefore, the intention of this paper is to make the comparison and experimentally analysis of the seismic performance of interior beam-column with and without fuse tested under quasi-static in-plane lateral cyclic loading. These two super-assemblage of interior beam-column joints were designed using Eurocode 8 with fuse bars and without fuse bars was designed using BS8110. They are designed, constructed, tested, analyzed and observed visual damages at Heavy Structural Laboratory, School of Civil Engineering, Universiti Teknologi Mara, Shah Alam, Selangor. The seismic parameters such as hysteresis loops, lateral strength capacity, stiffness, ductility, equivalent viscous damping and modeling will be compared for an interior beam-column joint with and without fuse bars in this study.

2 Materials and Methods

2.1 Design Fuse Bars

Fuse bars are made from characteristic high yield strength ($f_y = 460$ MPa) reinforcement bars machined down from 25 mm diameter to 20 mm diameter at middle length of the bars which created a "dog-bone" as shown in Fig. 1. The total length of fuse bars is 600 mm and the reduced diameter of 20 mm with a length of 200 mm at the middle section. The trimmed diameter of 20 mm was chosen because the designed ultimate lateral strength capacity of 300kN for interior joint with fuse bars. Figure 2 shows the actual size and dimension of fuse bars which had been trimmed down by a flexible trimming machine.

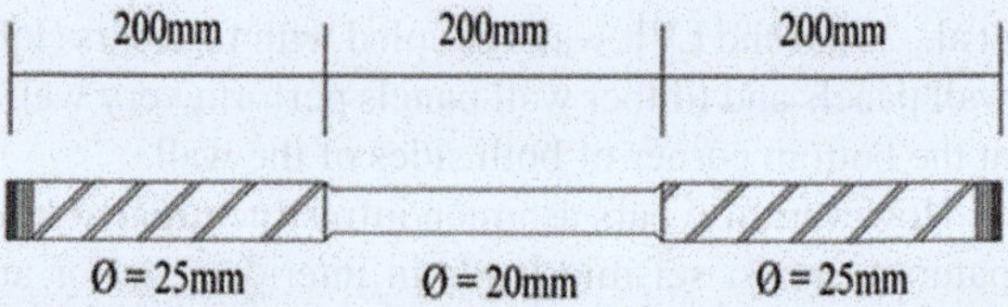

Fig. 1 Size and length of fuse bar

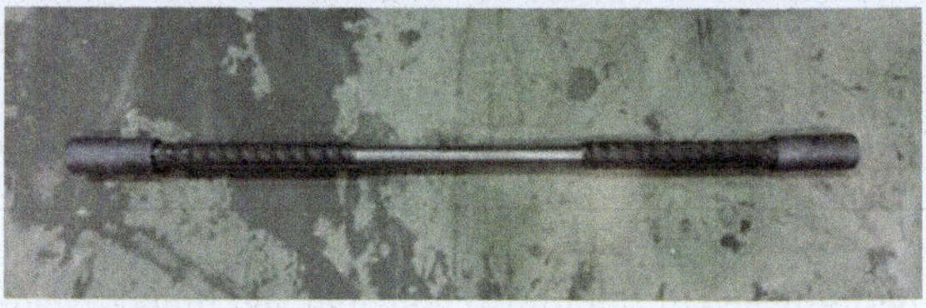

Fig. 2 The actual size of fuse bars used for the specimen

2.2 Construction of Two Specimens

Two specimens of interior beam-column joints consist of three beams, one column and one foundation beam using $f_{cu} = 30$ MPa and ready-mix concrete with medium slump. Figure 3 shows three-dimensions of the first specimen with fuse bars marked in red color labeled as EJ-3B. The sizes of the in-plane and out-of-plane beams are 400 × 400 × 3500 mm, dimension of the column is 400 × 400 × 3500 mm and size of foundation beam is 500 × 900 × 1800 mm. Meanwhile, Fig. 4 exhibits the second specimen with an interior beam-column joint with corbel and without fuse bars is labeled as BC2-3I. The dimensions of all columns are 400 × 400 × 3500 mm with corbels of 400 mm wide and 400 mm height. All the rectangular beams have a length of 3500 mm and two different cross-sections which are 500 × 750 mm and 400 × 750 mm, respectively.

The construction process of Specimen EJ-3B was started with the preparation of reinforcement caging for one foundation beam, three beams and one column. The construction process was continued with tying the main reinforcement bars for Beam 3, Beam 4 and Beam 5 since the main rebar for the column was already tied along with the transverse reinforcement bars earlier. Figure 5 shows four numbers of mechanical couplers were fitted to each top of reinforcement bars and another four couplers were fitted to the bottom of reinforcement bars with a diameter of 25 mm in Beam 3. After completing fitting the mechanical couplers into the bars, the next step was to position the eight numbers of fuse bars as shown in Fig. 2 at the middle of interior beam-column joint which categorized as a critical region. All eight numbers of fuse bars were fastened to the main reinforcement bars of Beam 3 which is parallel to Beam 5. Figure 6 shows the installation of fuse bars to the main reinforcement bars in the formwork for the erection process of Beam 4. Eleven numbers of LVDTs were placed along Beam 3, Beam 4, Beam 5, column and foundation beam to measure the in-plane and out-of-place displacement.

Figure 7 shows the systematic arrangement along the column labeled from LVDT 1 to LVDT 5; LVDT 6 and LVDT 7 were placed at the end of Beam 5 and Beam

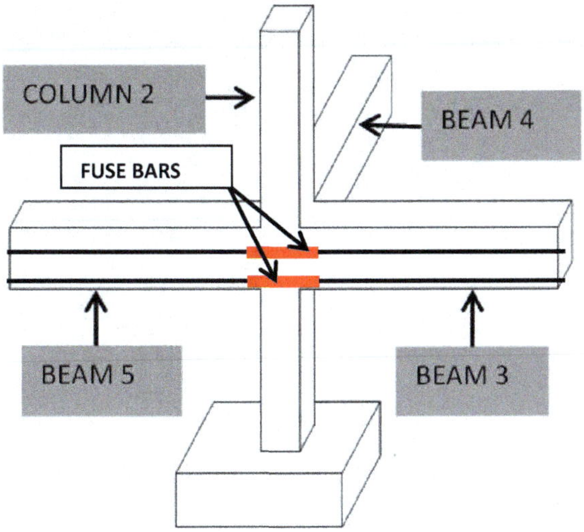

Fig. 3 Specimen EJ-3B with fuse bars

Fig. 4 Specimen BC2-3I without fuse bars

3, respectively. In the meantime, LVDT 8 and LVDT 9 were placed vertically and horizontally on the foundation beam and finally, LVDT 10 and LVDT 11 were positioned at the end of Beam 4 to measure out-of-place displacement. This experimental set-up was ready for testing on a strong floor.

The construction process of interior beam column with corbel labeled as Specimen BC2-3I without fuse bars was also conducted in the same laboratory (Kay Dora, 2015). This interior beam-column joint without fuse bars was designed using BS 8110 (non-seismic code of practice) and tested under in-plane lateral cyclic loading. Figure 8 shows the reinforcement caging for the construction of precast columns

Fig. 5 Mechanical couplers fitted to each rebar

Fig. 6 Installation of fuse bars at interior joint

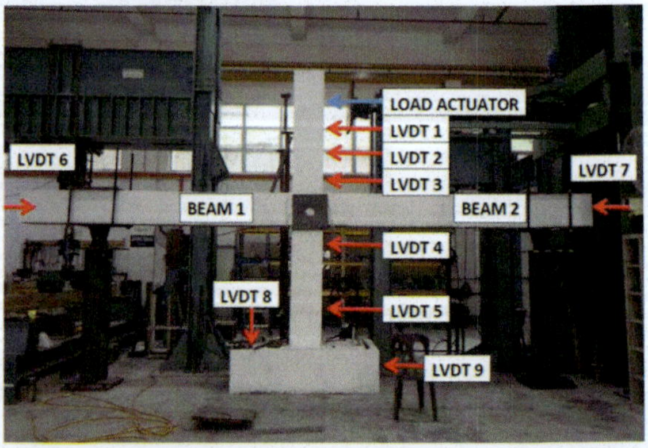

Fig. 7 Systematic arrangement of LVDTs and load actuator along the column and beams

Fig. 8 Reinforcement caging for beam-column joint with corbel using BS 8110

with three corbels for interior joint. High yield Grade 460 steel was used for the longitudinal reinforcement bars, while mild steel Grade 250 was bent to form the stirrup/shear reinforcement bars. Formwork made from plywood was fitted to the reinforcement caging of the three corbels. These corbels which were attached to the column were used to support the precast beams. The corbels were designed to cater for shear and vertical load from the beams.

Figure 9 shows the installation of precast Beam 3 and Beam 4 on top of two corbels. There was one extruder bar coming out of each of the corbel to fit with the beam located on top of it. Figure 10 shows schematic arrangements of LVDTs which were placed on the left-hand side along the column, foundation beam and at all ends of Beam 3, Beam 4 and Beam 5. A total number of 10 LVDTs were used to measure the in-plane lateral displacement with LVDTs labeled as T1, T2, T3, T4, T5, T9 and T10, out-of-plane lateral displacement was T8 and uplift of the foundation beam was T6 and T7.

Figure 11 shows the loading regime (graph of inter-story drift versus number of cycles) imposed for two super assemblages. Specimen EJ-3B with fuse bars and Specimen BC2-3I without fuse bars. The target drift is calculated by dividing the lateral displacement by the effective height of specimen and multiplied by 100. Each of the specimens was tested starting from the initial drift of ±0.01% and continuing with ±0.05% drift, ±0.1% drift, ±0.2% drift, ±0.5% drift, ±0.75% drift, ±1.00% drift, ±1.25% drift, ±1.5% drift, ±1.75% drift and ±2.00% drift. The test was repeated for two cycles for each drift. Both of the specimens were tested until there were strength degradation, spalling of concrete cover, buckling of reinforcement bars, severe cracks on the corbels, and finally failures at the joints.

Fig. 9 Installation of Beam 3 and Beam 4 on the corbel inside the laboratory

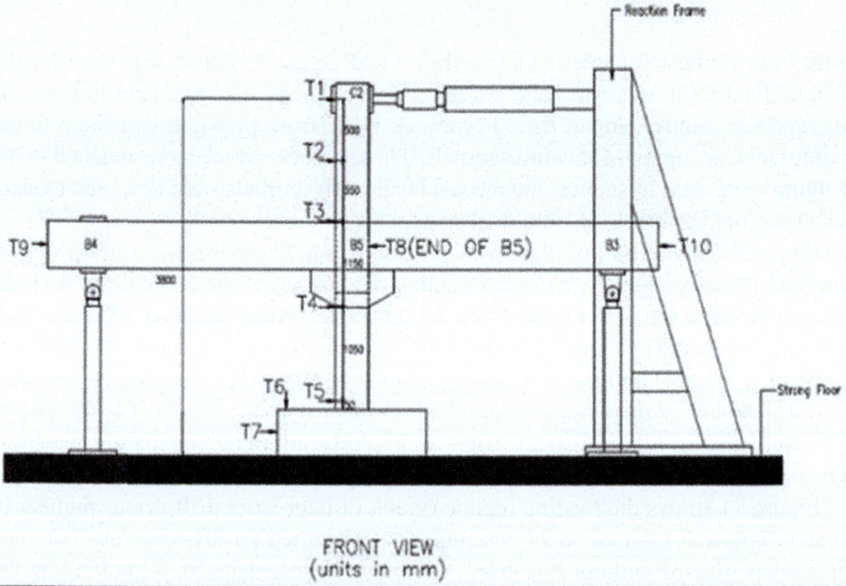

Fig. 10 Schematic arrangements of LVDTs a long column, beams and foundation beam

3 Results and Discussions

The experimental results for both Specimen EJ-3B and BC2-3I were compared, analyzed, interpreted and discussed their findings. The seismic performance parameters to be compared between them were hysteresis loops, lateral strength capacity and displacement, elastic and secant stiffness, ductility and equivalent viscous damping.

Fig. 11 Loading regime for testing procedure for Specimen EJ-3B and BC2-3I

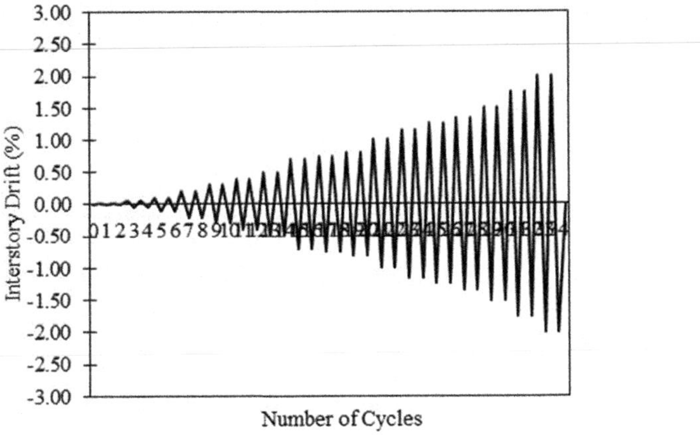

It is also important to compare and discuss the visual observation of damages of these two specimens with respect to the target drift imposed on them. The pictures of damages were recorded during testing and discussed in the following sub-section under this topic.

3.1 Comparison Visual Observation Damages of Specimen EJ-3B and BC2-3I

It is significant and vital to compare structural damage and crack propagation at different levels of drifts during testing for Specimen EJ-3B and Specimen BC2-3I. Figure 12 shows the visual observation damage for Specimen EJ-3B where hair-lines cracks were observed underneath Beam 5 at ±0.75% drift. Furthermore, there were no hairline cracks observed at interior joint for Specimen EJ-3B with fuse bars. However, there were three parallel diagonal cracks observed on the face of the interior joint Specimen BC2-3I at ±0.75% drift as shown in Fig. 13. Specimen BC2-3I without fuse bars was designed using BS 8110 where there is no specific reinforcement detailing at the interior joint with corbels which cause severe damage to occur under earthquake attacks.

Figure 14 shows visual observation of damages that occurred on the surface of in-plane Beam 5 closed to the joint with vertical parallel cracks at ±1.00% drift for Specimen EJ-3B. However, there were also no structural cracks observed on the interior joint with fuse bars. This interior joint was designed with closed stirrup/ shear reinforcement with 50 mm spacing apart and robust against shear and vertical force. Moreover, this joint was designed under Ductility Class Medium (DCM) in accordance with Eurocode 8 where its specified strong column and weak beam were designed. Contradictorily, the interior beam-column joint which labeled as BC2-3I with fuse bars suffered moderate structural damages with diagonal cracks to the

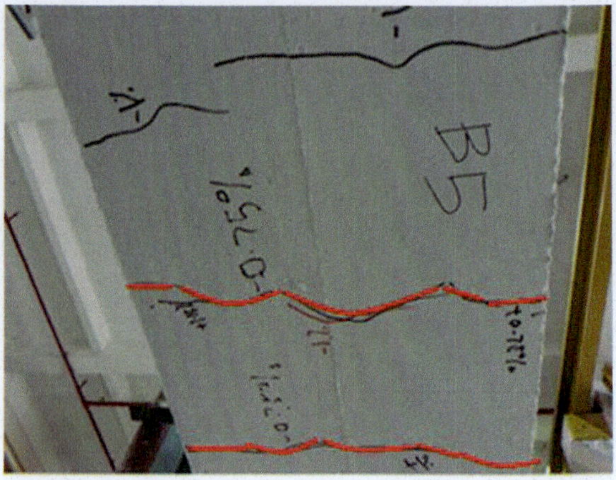

Fig. 12 Hairline crack observed below Beam 5 at ±0.75% drift for Specimen EJ-3B

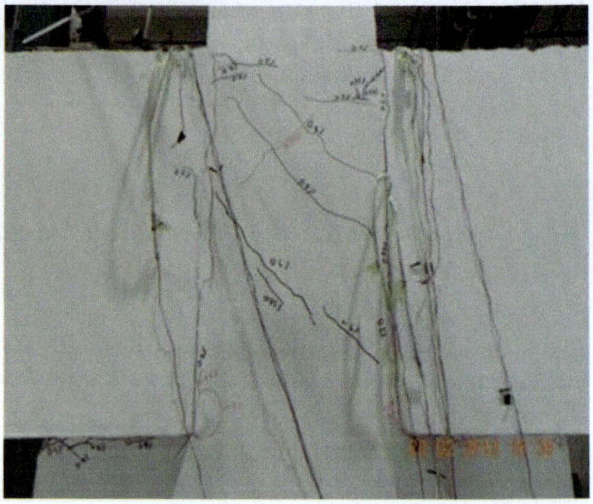

Fig. 13 Diagonal cracks were observed at the joint at ±0.75% drift for Specimen BC2-3I

corbels, horizontal crack at the bottom of the column and severe cracks and spalling of concrete cover occurred on top of the column above the corbels.

Figure 15 shows the horizontal cracks on the columns and corbels at ±1.00% drift. The main reason this type of joint suffers moderate damage is that there is no provision for seismic loads using BS8110. Furthermore, this joint is designed to cater only vertical load and shear subjected to the dead load of the structures and imposed load only.

Figure 16 exhibits the diagonal cracks were observed on the column below the joint at an ultimate drift of ±2.50% for Specimen EJ-3B. Several horizontal cracks and diagonal cracks occurred on the column due to lateral load which imposed on the top of the column. On the other hand, Fig. 17 shows the soft-story mechanism

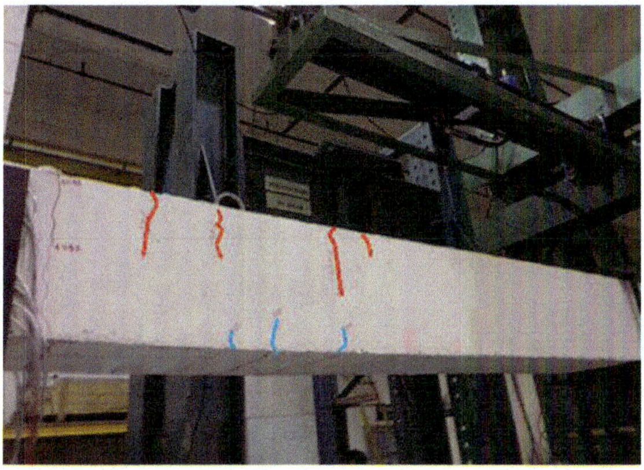

Fig. 14 Vertical parallel cracks on the surface Beam 5 at ±1.00% drift for Specimen EJ-3B

Fig. 15 Horizontal cracks on column and corbel at ±1.00% drift for Specimen BC2-3I

known as the captive column, which occurred at the top column above the interior joint at ±1.35% drift for Specimen BC2-3I. These common failures observed during experimental work are known as captive columns which normally occurs in a short column, open corridors, buildings on sloping grounds, short column created by stairway landing and others during an earthquake (Guevara & Garcia, 2005).

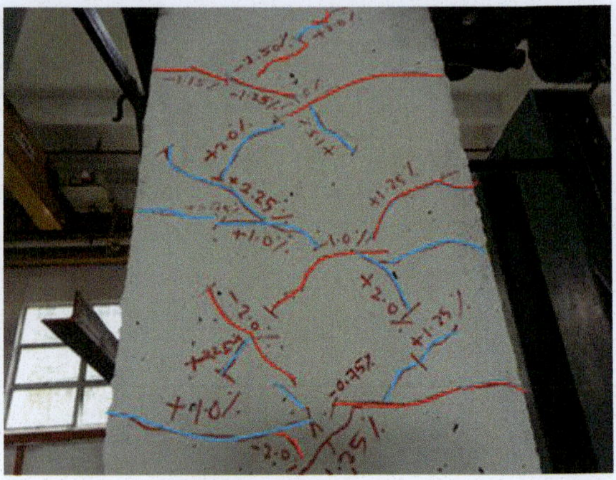

Fig. 16 Diagonal cracks observed on the column at ±2.50% drift for Specimen EJ-3B

Fig. 17 Captive column with severe damage at ±1.35% drift for Specimen BC2-3I

3.2 Comparison of Ultimate Lateral Strength Capacity

The comparison of ultimate lateral strength capacity and lateral displacement was made for Cycle 1 and Cycle 2 between Specimen EJ-3B and BC2-3I. Figure 18 shows the comparison of skeleton backbone curves for Cycle 1 between Specimen EJ-3B and BC2-3I. It can be noted that the ultimate strength capacity and lateral displacement for Specimen EJ-3B are higher than Specimen BC2-3I for Cycle 1. Meanwhile, Fig. 19 shows the comparison of the skeleton backbone curve for Cycle 2 between Specimen EJ-3B and BC2-3I. The development of empirical equations of these skeleton backbone curves is explained explicitly in a previous paper (Hadi

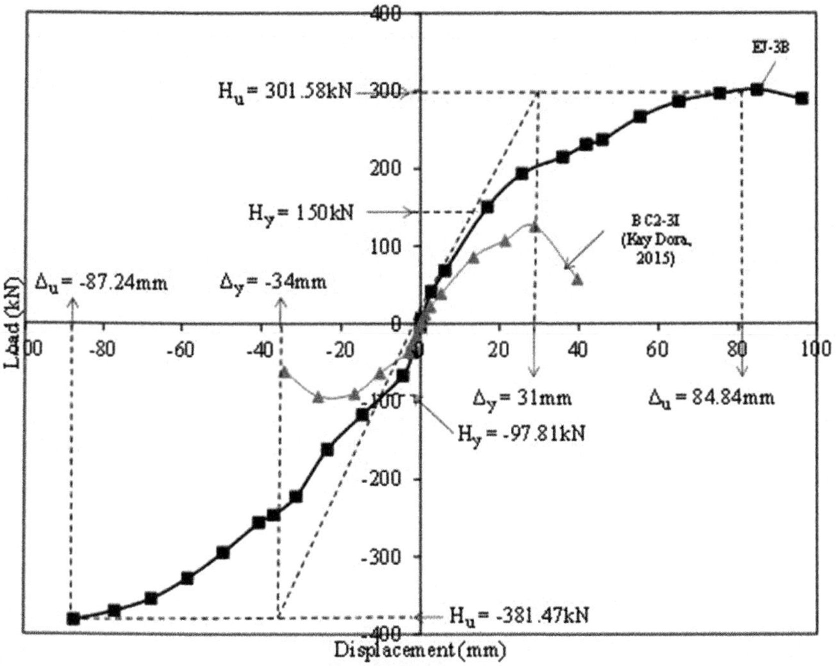

Fig. 18 Comparison of skeleton curve for Cycle 1 between Specimen EJ-3B and BC2-3I

et al., 2022). A similar pattern was observed in terms of ultimate strength capacity and lateral displacement between these two specimens. Previous research has also shown that replacing the steel reinforcement bar with fuse bars can improve the ultimate strength and stiffness of the structure under earthquake loading (Aghlara & Md Tahir, 2018). Table 1 shows comparison of lateral strength capacity between Specimen EJ-3B and BC2-3I for the first and second cycle in pushing and pulling directions.

3.3 *Comparison of Stiffness*

Table 2 shows the comparison values of elastic stiffness (Ke) and secant stiffness (Ksec) in terms of percentage difference for Cycle 1 and Cycle 2 between Specimen EJ-3B and Specimen BC2-3I. For EJ-3B Specimen in Cycle 1, the elastic stiffness values increased up to 43%, while for Cycle 2 the elastic stiffness went up to 33%. For Cycle 1 and Cycle 2, the stiffness increased by up to 64% for both cycles. The elastic stiffness of a column cross-section relies on the magnitude and distribution of stress applied to it and along with the extent of flexural and shear cracks (Le-Trung et al., 2010). Specimen EJ-3B can sustain a higher lateral load as compared to

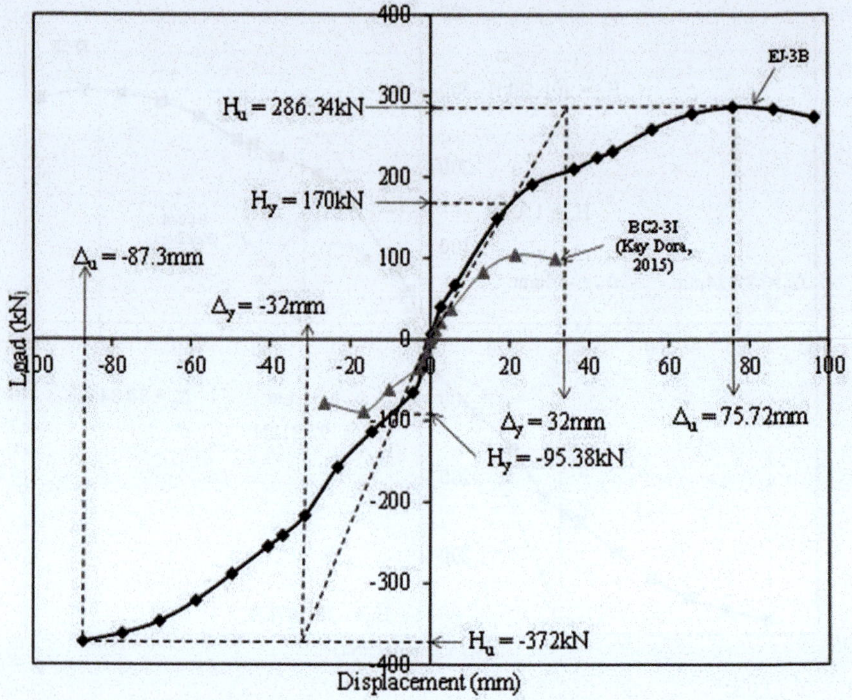

Fig. 19 Comparison of skeleton curve for Cycle 2 between Specimen Ej-3B and BC2-3I

Table 1 Comparison of lateral strength capacity between Specimen EJ-3B and BC2-3I

Cycles	Direction	H_u (kN)		Percentage difference (%)
		EJ-3B (with fuse bars)	BC2-3I (without fuse bars)	
Cycle 1	Pushing	301.58	125.54	58
	Pulling	−381.47	−93.78	75
Cycle 2	Pushing	286.34	104.27	64
	Pulling	−372.00	−89.02	76

BC2-3I under the 1% drift. It can be concluded that Specimen EJ-3B designed using Eurocode 8 has a higher stiffness of the beam-column joint increased significantly.

3.4 Comparison of Ductility

The displacement ductility of the specimen tested under in-plane lateral cyclic loading can be determined by dividing maximum ductility (Δu) by the yield ductility

Table 2 Comparison of K_e and K_{sec} between Specimen EJ-3B and Specimen BC2-3I

Cycle	Directions	Elastic stiffness (Ke)		Percentage difference	Secant stiffness (Ksec)		Percentage difference
		EJ-3B	BC2-3I	(%)	EJ-3B	BC2-3I	(%)
Cycle 1	Pushing	18.09	10.31	43	11.11	4.99	55
	Pulling	25.34	14.66	42	15.00	5.41	64
Cycle 2	Pushing	17.16	14.66	15	10.93	4.87	55
	Pulling	21.60	14.55	33	14.79	5.35	64

Table 3 Comparison of ductility between EJ-3B Specimen and BC2-3I Specimen

Cycle	Direction	Ductility, μ		Percentage difference (%)
		EJ-3B	BC2-3I	
Cycle 1	Pushing	3.08	1.99	35.54
	Pulling	2.57	2.14	16.70
Cycle 2	Pushing	3.01	1.91	36.61
	Pulling	2.73	1.72	37.12

(Δy) of each cycle of drift. Table 3 shows the displacement ductility for Specimen EJ-2B under Cycle 1 and Cycle 2 in pushing and pulling directions. The displacement ductility for Cycle 1 in pushing and pulling direction is 3.08 and 2.57, respectively. While for Cycle 2, the value of displacement ductility in pushing direction is equal to 3.01 and in pulling direction is 2.73. It can be noted that ductility in pushing and pulling direction for both cycles is higher than 2. Table 3 compares the displacement ductility for EJ-3B Specimen and BC2-3I Specimen. It was found that under both cycles, the ductility values are lower than 2 except for Cycle 1 pulling which is 2.14. However, all of the displacement ductility values for BC2-3I Specimen are generally lower as compared to displacement ductility values for EJ-3B Specimen. It indicates that the usage of fuse bars in EJ-3B Specimen managed to increase the ductility of the joint by up to 37.12%. Eurocode 8 has specified that for a structure to achieve Ductility Class Medium (DCM), the displacement ductility value must fall within the range of 2 to 6. Therefore, specimen EJ-3B Specimen is classified as a Ductility Class Medium structure, while specimen BC2-3I Specimen is classified as a structure with Ductility Class Low (DCL).

3.5 *Comparison of Equivalent Viscous Damping*

The equivalent viscous damping (ζ_{eq}) was calculated based on the derivation of Eq. 1 based on the derivation by Chopra (Chopra, 2012) as follows:

Table 4 Comparison of equivalent viscous damping for EJ-3B and BC2-3E under Cycle 1 and Cycle 2

Drift (%)	EJ-3B		Drift (%)	BC2-3I	
	Cycle 1	Cycle 2		Cycle 1	Cycle 2
	ζ_{eq} (%)	ζ_{eq} (%)		ζ_{eq} (%)	ζ_{eq} (%)
0.01	3.56	3.56	0.01	8.68	10.77
0.10	5.00	5.00	0.10	8.17	5.34
0.20	5.52	5.52	0.20	6.83	3.78
0.50	5.78	5.78	0.50	4.46	3.44
0.75	6.51	6.51	0.75	5.54	4.58
1.00	7.41	7.41	1.00	7.69	11.24
1.15	6.84	6.84			
1.25	6.60	6.60			
1.50	7.27	7.27			
1.75	7.31	7.31			
2.00	7.76	7.76			
2.25	8.54	8.54			
2.50	9.27	8.42			

$$\xi_{eq} = \frac{1}{4\pi}\frac{\text{Eso}}{\text{Ed}} \times 100\% \tag{1}$$

where Eso is the strain energy, calculated as the area of the triangle under lateral load versus displacement and Ed is energy dissipated calculated as the area under the hysteresis loops. Table 4 compares the hysteretic equivalent viscous damping between Cycle 1 and Cycle 2 for specimen EJ-3B from 0.05% drift to 2.5% drift. It was notified that the equivalent viscous damping for Cycle 1 is similar to the equivalent viscous damping for Cycle 2 at ±0.05%, ±0.10%, ±0.75%, ±1.0%, ±1.15% and ±1.25% drifts. Starting from ±0.2% drift and onwards, the equivalent viscous damping for Cycle 1 was slightly higher than Cycle 2. The maximum equivalent viscous damping is 9.27%, which occurred at 2.50% drift at Cycle 1 for Specimen EJ-3B.

3.6 *Comparison Between Experimental and Modeling of Hysteresis Loops*

Figure 20 shows the comparison of experimental hysteresis loops and modeled hysteresis loops for Specimen EJ-3B. There is a good agreement and fineness similarity between experimental hysteresis loops and modeled hysteresis loops using Pampanin RC Beam-column Joint Hysteresis Rule (IHYST = 44). Based on these

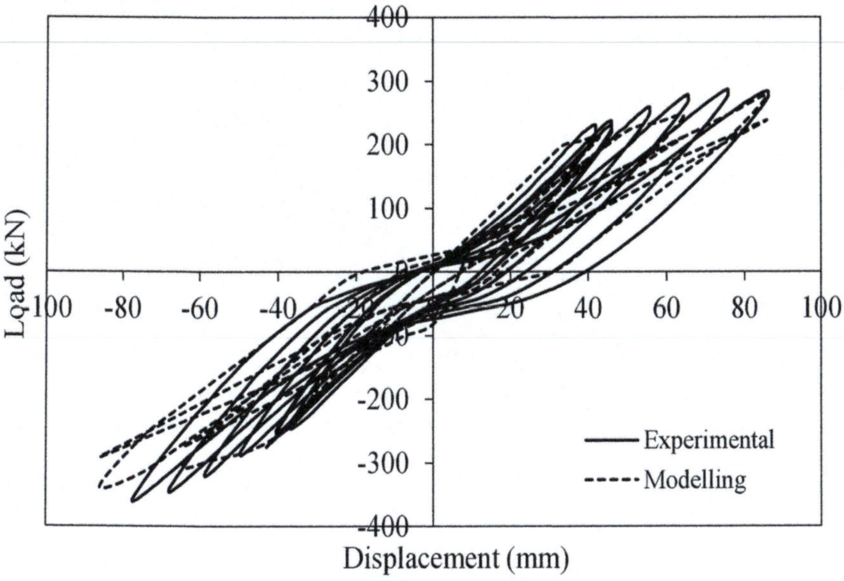

Fig. 20 Comparison between experimental and modeled hysteresis loops for Specimen EJ-3B

two hysteresis loops, the comparison of seismic parameters such as ultimate lateral strength capacity, stiffness, ductility and equivalent viscous damping at ±2.25% drifts in pushing direction for Specimen EJ-3B were further assessed. Figure 21 shows the comparison of experimental hysteresis loops and modeled hysteresis loops for Specimen BC2-3I for Cycle 1. A similar hysteresis rule was used known as Pampanin RC Beam-column Joint Hysteresis Rule (IHYST = 44) for Specimen BC2-3I. Based on these two hysteresis loops, there are similar seismic performance parameters that are compared with each other.

Table 5 shows the comparison of seismic parameters between experimental hysteresis loops and modeled hysteresis loops for Specimen EJ-3B and Specimen BC2-3I. For Specimen EJ-3B, the percentage difference of ultimate lateral strength capacity between experimental and modeled results is 8.68%, ductility is 4.38%, stiffness is 16.39% and, finally, the percentage difference for equivalent viscous damping is very good which is 1.51%. These percentage differences of seismic performance parameters are quite good agreement and acceptable correlation which lies ranging from 2 and 20% (Oberkampf & Barone, 2006). For Specimen BC2-3I, the percentage difference between experimental hysteresis loops and modeled hysteresis loops for ultimate strength capacity is 14.85%, ductility is 16.43%, stiffness is 4.39% and equivalent viscous damping is 4.29%. Based on the validation through experimental work by previous researcher (Oberkampf & Barone, 2006), the percentage differences of theses parameters are good agreement and acceptable correlation which lies between 2 and 20% because only one specimen was conducted.

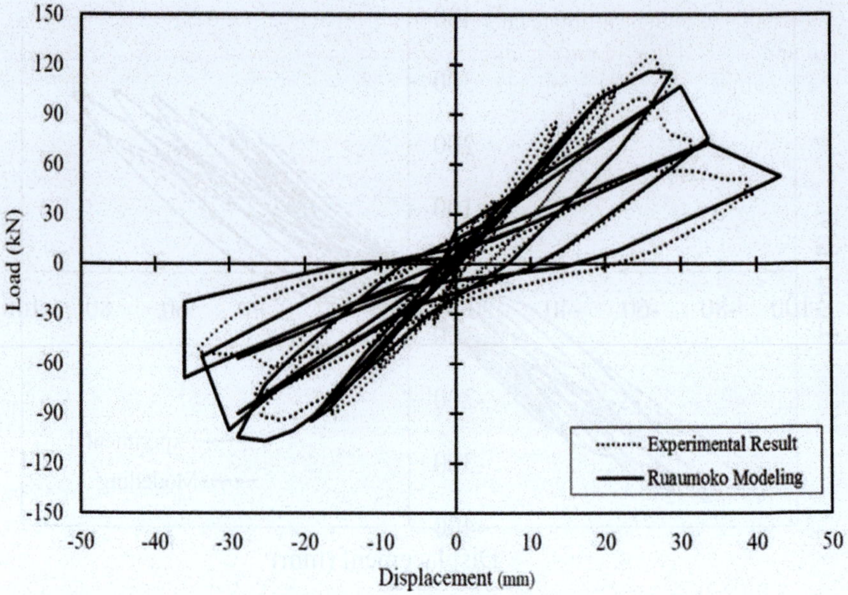

Fig. 21 Comparison between experimental and modeled hysteresis loops for Specimen BC2-3I

Table 5 Comparison seismic parameters between Specimen EJ-3B and Specimen BC2-3I

Parameters	SPECIMEN EJ-3B			SPECIMEN BC2-3I		
	Experimental results	Modeling results	Percentage difference (%)	Experimental results	Modeling results	Percentage difference (%)
Lateral strength (kN)	301.58	275.4	8.68	125.54	106.9	14.85
Ductility	2.74	2.86	4.38	1.40	1.63	16.43
stiffness (kN/mm)	3.05	3.55	16.39	4.33	4.14	4.39
Equivalent viscous damping (%)	9.27	9.41	1.51	7.69	8.02	4.29

4 Conclusions and Recommendations

This paper describes the design of fuse bars, experimental work, analysis of results, modeling and comparison of hysteresis loops between Specimen EJ-3B and Specimen BC2-3I. From the experimental result, it was discovered that EJ-3B Specimen can sustain the highest ultimate lateral load capacity up to 289.34kN in pushing direction and −381.47kN in pulling direction before experiencing strength degradation. These values can cause a reduction in the stiffness of interior beam-column joint in both cycles and directions. Nevertheless, the ductility has proven that the interior beam-column joint with fuse bars is a Ductility Class Medium (DCM) structure where the values of ductility range between 2 and 6. The ultimate load was reached at ±2.25% drift with major damage formed near the beam-column interface. The equivalent viscous damping, ζ_{eq} shows an increasing trend which represents the increasing energy dissipated by the beam-column joint with a damping ratio of 9.27%. This indicates that specimen EJ-3B can sustain much higher lateral loads with DCM with sufficient energy dissipation capacity compared to BC2-3I which was classified as a Ductility Class Low (DCL) structure and can only sustain lower lateral load with DCL. However, for numerical work, the selection of Pampanin RC Hysteresis Loops (IHYST = 44) is considered suitable for modeling the analytical hysteresis loops. This paper has identified that the use of fuse bars with 20 mm diameter resulted in a damping ratio of 9.27% for seismic-designed EJ-3B under DCM (PGA = 0.12 g). After comparing these experimental results and analysis of interior beam-column joints with other published literature reviews, the novelty of this research is that fuse bars made from common construction material namely a high-yield reinforcement bar be able to reduce structural damages and protect the reinforced concrete buildings against severe ground motions. Therefore, it is recommended that this type of fuse bars be used for moderate seismic regions such as Sabah, East Malaysia for the construction of low and medium-reinforced concrete buildings. It is proven that beam-column joint with fuse bars performed much better than beam-column joint without fuse bars.

Acknowledgements Special appreciation to Malaysian Higher Education who grant this research under Fundamental Research Grants (FRGS) with file No: 600-RMI/FRGS 5/3 (0091/2016) and the Research Management Institute for handling this research in the most competent ways. Thankfulness and acknowledgment to the laboratory staff members for their invaluable assistance during conducting this experimental work.

References

Abas, M. R. C. (2001). Earthquake monitoring in Malaysia. *Seismic Risk Seminar, Malaysia.*

Aghlara, R., Md Tahir, M. (2018). A passive metallic damper with replaceable steel bar components for earthquake protection of structures. *Engineering Structures, 159*, 185–197. https://doi.org/10.1016/j.engstruct.2017.12.049

Alireza, F., Farhad, F., & Pouya, P. (2016). Experimental study of a replaceable steel coupling beam with an end-plate connections. *Journal Construction Steel Research, 122*, 138–150. https://doi.org/10.1016/j.jcsr.2016.03.018

Aqbal, A., Pampanin, P., Palermo, A., & Buchanan, A. H. (2008). Experimental study of prestressed timber columns under bi-directional seismic loading. In *The 11th world conference on timber engineering, Italy* (vol. 10).

Balendra, T., & Li, Z. (2008). Seismic hazard of Singapore and Malaysia. In *EJSE special issue: Earthquake engineering in the low and moderate seismic regions of Southeast Asia and Australia* (pp. 57–63). https://doi.org/10.12989/eas.2018.15.1.067

Beerschoten, W. V., Palermo, A., Carrdaine, D., Sarti, F., & Buchanan, A. H. (2011). Experimental investigation on the stiffness of beam-column connections in post-tensioned timber frames. In *Structural engineers world congress, Italy.*

Chopra, A. K. (2012). *Dynamics of structures: Theory and applications to earthquake engineering* (4th edn). Pearson Education Publisher. ISBN-10: 0273774247

Connor, J. J., Wada, A., Iwata, M., & Huang, Y. H. (1997). Damage-controlled structures 1 : Preliminary design methodology for seismically active regions. *Journal of Structural Engineering, 123*, 423–431. https://doi.org/10.1061/asce0733-9445(1997)123:4(423)

El-Bahey, S., & Bruneau, M. (2010). Structural fuse concept for bridges. In *Proceedings of the 9th U.S. National and 10th Canadian Conference on Earthquake Engineering* (p. 1302).

Guevara, L. T., & Garcia, L. E. (2005). The captive-and short-column effects. *Earthquake Spectra, 21*(1), 141–160. https://doi.org/10.1193/1.1856533

Hadi, N. D., Hamid, N. H., Azmi, I. F., Hassan, R., Sutopo, Y., Chao, B., & Zainuddin, N. A. (2022). Comparison of seismic response corner beam-column joint with and without fuse bars under in-plane lateral cyclic loading. *Latin American Journal Solids and Structures, 19*(1). https://doi.org/10.1590/1679-78256581

Hamid, N. H. (2006a). Seismic damage avoidance design of warehouse buildings constructed using precast hollow core panels. PhD Thesis, Department of Civil Engineering, University of Canterbury, Christchurch, New Zealand

Hamid, N. H. A. (2006). In-plane and biaxial performance of seismic precast hollow core wall with different types of energy dissipaters. Jurnal Teknologi, *44*(B), 47–62. https://doi.org/10.11113/jt.v44.361

Hamid, N. H. A. (2010). Seismic performance of beam-column joints in reinforced concrete buildings subjected to reversible vertical cyclic loading. *Malaysia Journal of Civil Engineering, 22*(2), 264–290. https://doi.org/10.11113/mjce.v22.15802

Hamid, N. H., & Mander, J. B. (2006). Experimental study on bi-axial seismic performance of precast hollow core wall using shaking table. In *Proceedings of the 10ty east Asia-pacific conferences on structural engineering and construction (EASEC 2010)* (pp. 109–114).

Hamid, N. H. A., & Mander, J. B. (2010). Lateral seismic performance of multi-panel precast hollow core walls subjected to quasi-static cyclic loading. *Journal of Structural Engineering, 136*(7). https://doi.org/10.1061/(ASCE)ST.1943-541X.0000183

Hamid, N. H., Hadi, N. F. D., & Azmi, I. F. (2021). Earthquake fuse bar as smart material in improving seismic performance of beam-column joint during earthquake. In *Proceedings of the sustainable concrete materials and structures in construction 2020* (pp 67–82)

Hamid, N. H., & Ghani, K. D. (2013). Seismic behaviour of precast hollow core wall under biaxial lateral cyclic loading. *WIT Transactions on the Built Environment, 134*, 815–825. https://doi.org/10.2495/SAFE130721

Iqbal, A., Pampanin, S., Palermo, A., & Buchanan, A. H. (2015). Performance and design of LVL walls coupled with UFP dissipaters. *Journal of Earthquake Engineering, 19*(3), 383–409. https://doi.org/10.1080/13632469.2014.987406

Ji, X. D., Liu, D., & Sun, Y. (2016). Seismic performance assessment of a hybrid coupled wall system with replacement steel coupling beams versus traditional RC coupling beams. *Earthquake Engineering and Structural Dynamics*, 1–19. https://doi.org/10.1002/eqe.2801

Joshilkar, P. D., & Patil, C. S. (2018). Force displacement based design of steel moment resisting frame. *International Journal for Research in Applied Science and Engineering Technology, 6*(1), 1419–1424. https://doi.org/10.22214/ijraset.2018.1215

Kay Dora, A. G. (2015). Full-scale precast non-seismic beam-column joints with corbels under in-plane lateral cyclic loading. PhD Thesis, Universiti Teknologi Mara, Shah Alam, Selangor, Malaysia

Lay, T. (2015). The surge of great earthquakes from 2004 To 2014. *Earth and Planetary Science Letters, 409*, 133–146.

Le-Trung, K., Lee, K., Lee, J., Lee, D. H., & Woo, S. (2010). Experimental study of RC beam-column joints strengthened using CFRP composites. *Composites Part b: Engineering, 41*(1), 76–85. https://doi.org/10.1016/j.compositesb.2009.06.005

Luoman, L., Mander, J. B., & Dhakal, R. P. (2008). Bidirectional cyclic loading experiment on a 3D beam-column joint designed for damage avoidance. *Journal of Structural Engineering, ASCE, 134*(11). https://doi.org/10.1061/(ASCE)07339445(2008)134:11(1733)

Majid, T. A., Adnan, A., Adiyanto, M. I., Ramli, M. Z., & Ghuan, T. C. (2017). Preliminary damage assessment due to 2015 Ranau earthquake. *International Journal of Civil Engineering & Geo-Environmental* (Special Publication), 1–6.

Mohamed, A. (2012). Monitoring active faults in Ranau, Sabah using GPS. In *The 19th united nations regional cartographic conference for Asia and the Pacific Bangkok, Thailand*, (vol. 6, p. 8).

Nabilah, A. B., & Balendra, T. (2012). Seismic hazard analysis for Kuala Lumpur, Malaysia. *Journal of Earthquake Engineering, 16*, 1076–1094. https://doi.org/10.1080/13632469.2012.685208

Oberkampf, W. L., & Barone, M. F. (2006). Measures of agreement between computation and experiment: Validation metrics. *Journal of Computational Physics, 217*(2006), 5–36. https://doi.org/10.1016/j.jcp.2006.03.037

Rahman, A. M., & Restrepo, J. I. (2007). Seismic performance of self-centering structural walls incorporating energy dissipators. *Journal of Structural Engineering, 133*(11), 2007. https://doi.org/10.1061/(ASCE)07339445(2007)133:11(1560)

Saravan, M., Goswami, R., & Palani, G. S. (2018). Replaceable fuses in earthquake resistant steel structures: A review. *International Journal Steel Structure, 18*, 868–879. https://doi.org/10.1007/s13296-018-0035-9

Sarti, F., Palermo, A., & Pampanin, S. (2016). Development and testing of an alternative dissipative post-tensioned rocking timber wall with boundary columns. *Journal of Structural Engineering, ASCE, 142*(4), E4015011. https://doi.org/10.1061/(ASCE)ST.1943-541X.0001390

Smith, T., Ponzo, F.C., Cesare, A.D., Auetta, G., Pampanin, S., Carradine, D., Palermo, A., & Nigro, D. (2012). Seismic performance of a post-tensioned glue laminated beam to column joint experimental and numerical results. In *12th world conference on timber engineering, Auckland, New Zealand*.

Sullivan, T. J., Calvi, G. M., & Priestley, M. J. N. (2004). Initial stiffness versus secant stiffness in displacement based design. In *The 13th world conference on earthquake engineering* (p. 2888)

Toranzo, L. A., Restrepo, J. I., Mander, J. B., & Carr, A. J. (2009). Shake-table tests of confined-mansory rocking walls with supplementary hysterestic damping. *Journal of Earthquake Engineering, 13*(6), 882–898. https://doi.org/10.1080/13632460802715040

The manufacturer's authorised representative in the EU is Springer Nature Customer Service Centre GmbH, Europaplatz 3, 69115 Heidelberg, Germany. If you have any concerns regarding our products, please contact ProductSafety@springernature.com

Printed and bound by CPI Group (UK) Ltd, Croydon, CR0 4YY
01/10/2026
02225900-0017